Web API Development with ASP.NET Core 8

Learn techniques, patterns, and tools for building high-performance, robust, and scalable web APIs

Xiaodi Yan

‹packt›

Web API Development with ASP.NET Core 8

Copyright © 2024 Packt Publishing

Group Product Manager: Rohit Rajkumar
Senior Editor: Nathanya Dias
Technical Editor: K Bimala Singha
Copy Editor: Safis Editing
Project Coordinator: Sonam Pandey
Indexer: Hemangini Bari
Production Designer: Aparna Bhagat
Marketing Coordinators: Anamika Singh and Nivedita Pandey

First published: March 2024

Production reference: 1010324

Published by Packt Publishing Ltd.
Grosvenor House
11 St Paul's Square
Birmingham
B3 1RB, UK.

ISBN 978-1-80461-095-4

www.packt.com

When I signed the contract with the publisher, I never anticipated that authoring a book would be so challenging. I owe immense gratitude to my family for their unwavering support and understanding. As I dedicated most of my weekends and holidays to writing this book, I deeply appreciate their patience and encouragement despite the limited time we shared.

My heartfelt appreciation also extends to my friends and colleagues for their constant support and motivation. There were moments when I felt overwhelmed and on the verge of giving up, but their enduring patience and willingness to listen to my frustrations kept me going. Their encouragement was invaluable, and I am truly fortunate to have them in my life.

I also want to extend my heartfelt thanks to my editors, whose patience and understanding surpassed my expectations. My tendency to procrastinate undoubtedly presented challenges and I am sincerely grateful for their guidance and persistence throughout the editing process.

Additionally, I express my gratitude to the reviewers for their invaluable feedback. Their constructive suggestions have significantly enriched the content and contributed to its overall improvement.

As I reflect on this journey, I am immensely proud to have completed this book. It has been a labor of love, fueled by my passion for .NET and the incredible possibilities it offers. I hope you will find as much joy in exploring the world of .NET through these pages as I have experienced in writing about it.

Contributors

About the author

Xiaodi Yan is a seasoned software engineer with a proven track record of success in the IT industry. Since 2015, he has been awarded **Microsoft MVP**, showcasing his dedication to and expertise in .NET, AI, DevOps, and cloud computing. He is also a **Microsoft Certified Trainer** (MCT), Azure Solutions Architect Expert, and LinkedIn Learning instructor. Xiaodi often presents at conferences and user groups, leveraging his extensive experience to engage and inspire audiences. Based in Wellington, New Zealand, he spearheads the Wellington .NET User Group, fostering a vibrant community of like-minded professionals.

Connect with Xiaodi on LinkedIn at `https://www.linkedin.com/in/xiaodi-yan/` to stay updated on his latest insights.

To my family, whose unwavering support and understanding made this journey possible. Your constant support got me through those late nights and tough writing sessions.

And to you, the reader: it is my pleasure to share what I've learned with you. May this book inspire and empower you on your own journey in the world of ASP.NET Core.

About the reviewers

Rupenkumar Anjaria has more than 20 years of experience in architecting, designing, developing, and maintaining complex enterprise web applications. His primary technical experience includes working on .NET, C#, Angular, and SQL Server technologies, with several years of experience in ServiceNow and React, MongoDB, MVC, TFS, and Jenkins.

Aditya Oberai is a developer advocate at Appwrite and a Microsoft MVP. Having worked with various technologies, such as ASP.NET web APIs, .NET MAUI, Svelte, Node.js, and Microsoft Azure, he has spent the last five years empowering tech communities and hackathons across India and beyond. He can often be found interacting with and educating the community about APIs, serverless, cross-platform apps, community building, indie hacking, and open-source.

Ifeoluwa Osungade is a senior software engineer at Agoda, where he channels his deep expertise in C# and web development to craft software solutions that elevate the travel experience for countless customers worldwide. With an impressive career spanning over eight years in the software engineering domain, Ifeoluwa has made significant contributions across diverse industries such as technology, construction, and logistics. This breadth of experience underscores his ability to not only identify and define challenges but also create and implement innovative software solutions that streamline business operations, enhance efficiency, and boost profitability. His other areas of expertise are Scala, Python, SQL, and Azure.

Table of Contents

3

ASP.NET Core Fundamentals (Part 1)   91

4

ASP.NET Core Fundamentals (Part 2)   127

5

Data Access in ASP.NET Core (Part 1: Entity Framework Core Fundamentals) 163

6

Data Access in ASP.NET Core (Part 2 – Entity Relationships) 201

7

Data Access in ASP.NET Core (Part 3: Tips) 237

8

Security and Identity in ASP.NET Core   271

9

Testing in ASP.NET Core (Part 1 – Unit Testing)   333

10

Testing in ASP.NET Core (Part 2 – Integration Testing)   373

11

Getting Started with gRPC   407

12

Getting Started with GraphQL   453

13

Getting Started with SignalR   527

14

CI/CD for ASP.NET Core Using Azure Pipelines and GitHub Actions   575

15

ASP.NET Core Web API Common Practices   623

16

Error Handling, Monitoring, and Observability   663

17

Cloud-Native Patterns   717

Index   761

Other Books You May Enjoy   778

Preface

Welcome to the world of ASP.NET Core!

.NET Core has undergone significant evolution since its inception and has emerged as a robust solution for building a wide array of applications, including web, desktop, mobile, gaming, and AI applications. With the release of .NET 8 at the end of 2023, it solidified its position as one of the most powerful and versatile frameworks for modern applications.

ASP.NET Core, built on top of the .NET Core platform, inherits its strengths, offering cross-platform compatibility, exceptional performance, and a modular architecture. It has become a popular choice for building cloud-native applications that can run seamlessly on any operating system, including Windows, macOS, and Linux, and can be deployed on any cloud platform, such as Azure, AWS, or GCP.

As organizations increasingly adopt ASP.NET Core for web application development, the demand for ASP.NET Core developers is on the rise. Whether you're transitioning from the traditional .NET Framework or are new to web API development with ASP.NET Core, this book is tailored to meet your needs. It will guide you through building your first web API application with ASP.NET Core and provide you with the knowledge and skills to build robust, scalable, and maintainable web APIs.

While ASP.NET Core offers robust options for frontend development, such as Razor Pages, Blazor, and MVC, this book focuses on backend development. You'll explore a range of topics, including REST-based APIs, gRPC APIs, GraphQL APIs, and real-time APIs, gaining insights into essential concepts and best practices for building web APIs with ASP.NET Core.

Additionally, we'll delve into testing methodologies and tools such as unit testing and integration testing to ensure the quality and reliability of your web APIs. We'll also explore modern development practices such as CI/CD, containerization, monitoring, and cloud-native design patterns, essential for contemporary web API development.

While this book serves as a foundational resource, it merely scratches the surface of what ASP.NET Core has to offer. I encourage you to use it as a stepping stone to further explore the vast landscape of ASP.NET Core on your own. Experiment with the sample code provided and refer to the links in the book for more in-depth learning. Don't forget to explore the official documentation for the latest updates and features.

I hope you will find this book useful and it will inspire you to explore the world of ASP.NET Core. Happy reading!

Who this book is for

This book is for developers who want to learn how to build web APIs with ASP.NET Core 8 and create flexible, maintainable, scalable applications with the .NET platform. Prior basic knowledge of C#, .NET, and Git would be helpful.

What this book covers

Chapter 1, Fundamentals of Web APIs, provides an overview of web APIs, covering their historical background and various API styles, including REST-based APIs, gRPC APIs, GraphQL APIs, and real-time APIs. It will also discuss the process of designing web APIs.

Chapter 2, Getting Started with ASP.NET Core Web APIs, explores the fundamentals of ASP.NET Core, including the project setup, dependency injection, and minimal APIs. You will also learn how to create your first web API using ASP.NET Core and how to test it using a variety of tools.

Chapter 3, ASP.NET Core Fundamentals (Part 1), covers the fundamentals of ASP.NET Core, including routing, configuration, and environments.

Chapter 4, ASP.NET Core Fundamentals (Part 2), continues the discussion of ASP.NET Core fundamentals, covering logging and middleware.

Chapter 5, Data Access in ASP.NET Core (Part 1: Entity Framework Core Fundamentals), explores the utilization of **Entity Framework Core** (**EF Core**) for database interaction. You will gain insights into implementing CRUD operations using EF Core.

Chapter 6, Data Access in ASP.NET Core (Part 2: Entity Relationships), covers the configuration of EF Core to support various model relationships, including one-to-one, one-to-many, and many-to-many.

Chapter 7, Data Access in ASP.NET Core (Part 3: Tips), provides best practices for using EF Core in your web API, such as the `DbContext` pooling, raw SQL queries, bulk operations, and so on.

Chapter 8, Security and Identity in ASP.NET Core, covers the security considerations surrounding web APIs. You will delve into implementing authentication and authorization mechanisms using ASP.NET Core Identity to ensure the security of your web APIs.

Chapter 9, Testing in ASP.NET Core (Part 1 – Unit Testing), explores testing methodologies and tools, including xUnit and Moq. You will learn how to implement unit tests to ensure the quality of your web APIs.

Chapter 10, Testing in ASP.NET Core (Part 2 – Integration Testing), covers integration testing using xUnit and `WebApplicationFactory`. You will learn how to implement integration tests to test components of your web APIs.

Chapter 11, Getting Started with gRPC, explores gRPC, a modern high-performance RPC framework that can be used to build efficient APIs. You will learn how to create gRPC services and clients using ASP.NET Core.

Chapter 12, Getting Started with GraphQL, covers GraphQL, a powerful query language for APIs. You will learn how to create GraphQL APIs using ASP.NET Core.

Chapter 13, Getting Started with SignalR, explores SignalR, a real-time communication framework for ASP.NET Core. You will learn how to create real-time APIs and clients using ASP.NET Core.

Chapter 14, CI/CD for ASP.NET Core Using Azure Pipelines and GitHub Actions, covers the process of building, testing, and deploying your web API applications using Azure DevOps and GitHub Actions. It also introduces the use of Docker to containerize your web API applications.

Chapter 15, ASP.NET Core Web API Common Practices, provides best practices for building your ASP.NET Core web API applications. It covers topics such as asynchronous programming, caching, `HttpClientFactory`, and so on.

Chapter 16, Error Handling, Monitoring, and Observability, covers error handling, health checks, monitoring, and observability. You will learn how to handle errors in your web APIs and how to monitor and observe your web APIs using a variety of platforms and OpenTelemetry.

Chapter 17, Cloud-Native Patterns, explores advanced architecture and patterns essential for modern web API development. You will gain insights into cloud-native design patterns, **Domain-Driven-Design (DDD)**, **Command Query Responsibility Segregation (CQRS)**, Retry patterns, Circuit Breaker patterns, and so on.

Chapter 18, Leveraging Open-Source Frameworks, covers various open-source frameworks that can be used to streamline development and enhance productivity, including ABP Framework, Clean Architecture, Orchard Core, eShop, and .NET Aspire.

To get the most out of this book

You will need to have a basic understanding of programming using .NET and C# and be familiar with the concepts of **object-oriented programming (OOP)**. If you are new to C#, you can learn C# from Microsoft Learn and freeCodeCamp at `https://www.freecodecamp.org/learn/foundational-c-sharp-with-microsoft`.

Software/hardware covered in the book	Operating system requirements
.NET 8 SDK (`https://dotnet.microsoft.com/en-us/download/dotnet`)	Windows, macOS, or Linux
Visual Studio Code (`https://code.visualstudio.com/`)	Windows, macOS, or Linux

Software/hardware covered in the book	Operating system requirements
Visual Studio 2022 Community Edition (`https://visualstudio.microsoft.com/downloads/`)	Windows, macOS, or Linux
Seq (`https://datalust.co/download`)	Windows, Docker/Linux
Prometheus (`https://prometheus.io/download/`)	Windows, Docker/Linux
Grafana (`https://grafana.com/oss/grafana/`)	Windows, Docker/Linux
Jaeger (`https://www.jaegertracing.io/download/`)	Windows, Docker/Linux
Azure	
Azure DevOps	
GitHub	

In this book, we use LocalDB, which is a lightweight version of SQL Server. It is only available on Windows. If you are using a Mac or Linux, you can use a Docker container to run SQL Server. You can also use SQLite instead. To use SQLite, you need to update the connection string in the `appsettings.json` file and install the SQLite provider for EF Core, respectively. Please refer to *Chapter 5* for more details.

We advise you to type the code yourself or access the code via the GitHub repository (link available in the next section). Doing so will help you learn better and retain the knowledge longer.

Download the example code files

You can download the example code files for this book from GitHub at `https://github.com/PacktPublishing/Web-API-Development-with-ASP.NET-Core-8`. If there's an update to the code, it will be updated in the GitHub repository.

We also have other code bundles from our rich catalog of books and videos available at `https://github.com/PacktPublishing/`. Check them out!

Conventions used

There are a number of text conventions used throughout this book.

`Code in text`: Indicates code words in text, database table names, folder names, filenames, file extensions, pathnames, dummy URLs, user input, and Twitter handles. Here is an example: "Let us use the `IEnumerable` interface to query the database."

A block of code is set as follows:

```
using (var serviceScope = app.Services.CreateScope())
{
    var services = serviceScope.ServiceProvider;
    // Ensure the database is created.
    var dbContext = services.GetRequiredService<AppDbContext>();
    dbContext.Database.EnsureCreated();
}
```

Any command-line input or output is written as follows:

```
cd GrpcDemo.Client
dotnet add GrpcDemo.Client.csproj package Grpc.Net.Client
```

Bold: Indicates a new term, an important word, or words that you see onscreen. For instance, words in menus or dialog boxes appear in **bold**. Here is an example: "Click the **Continue** button."

> **Tips or important notes**
> Appear like this.

Get in touch

Feedback from our readers is always welcome.

General feedback: If you have questions about any aspect of this book, email us at customercare@ packtpub.com and mention the book title in the subject of your message.

Errata: Although we have taken every care to ensure the accuracy of our content, mistakes do happen. If you have found a mistake in this book, we would be grateful if you would report this to us. Please visit www.packtpub.com/support/errata and fill in the form.

Piracy: If you come across any illegal copies of our works in any form on the internet, we would be grateful if you would provide us with the location address or website name. Please contact us at copyright@packt.com with a link to the material.

If you are interested in becoming an author: If there is a topic that you have expertise in and you are interested in either writing or contributing to a book, please visit authors.packtpub.com.

Share Your Thoughts

Once you've read *Web API Development with ASP.NET Core 8*, we'd love to hear your thoughts! Scan the QR code below to go straight to the Amazon review page for this book and share your feedback.

https://packt.link/r/1-804-61095-X

Your review is important to us and the tech community and will help us make sure we're delivering excellent quality content.

Download a free PDF copy of this book

Thanks for purchasing this book!

Do you like to read on the go but are unable to carry your print books everywhere? Is your eBook purchase not compatible with the device of your choice?

Don't worry, now with every Packt book you get a DRM-free PDF version of that book at no cost.

Read anywhere, any place, on any device. Search, copy, and paste code from your favorite technical books directly into your application.

The perks don't stop there, you can get exclusive access to discounts, newsletters, and great free content in your inbox daily

Follow these simple steps to get the benefits:

1. Scan the QR code or visit the link below

https://packt.link/free-ebook/9781804610954

2. Submit your proof of purchase
3. That's it! We'll send your free PDF and other benefits to your email directly

1

Fundamentals of Web APIs

In today's world, **web APIs** are the backbone of the web. Millions of people use web APIs every day to purchase commodities, book a flight, get weather information, and more. In this chapter, we will learn about the fundamentals of web APIs. You might be wondering why we will start with the fundamental concepts. The answer is simple – we need to understand the basic concepts of web APIs before we build one.

This chapter introduces a couple of different web API styles, such as a REST-based API, a **remote procedure call** (**RPC**)-based API, a GraphQL API, and a real-time API. We will also learn about how to design them. If you would like to start developing a web API, feel free to jump to the next chapter.

In this chapter, we'll be covering the following topics:

- What is a web API?

- What is a REST API?

- Designing a REST-based API

- What are RPC and GraphQL APIs?

- What is a real-time API?

After reading this chapter, you will have a basic understanding of web APIs and be able to pick the right style for your project. Let's get started!

What is a web API?

API stands for **application programming interface**. A web API, as the name suggests, is a set of programming interfaces for the web. For example, when you book a flight on a website, the browser makes a request to the airline's server through a web API to access the airline's database. The airline's server then returns the information about the flight to the browser, allowing you to book your flight in it.

APIs have been delivered by organizations for decades. With the appearance of the World Wide Web, people needed a way to communicate between the server and the client.

We can build web APIs using different technologies, such as Java, Python, Ruby, PHP, .NET, and so on. Also, they have various styles. You might have heard of terms such as **SOAP**, **Web Service**, and **REST**. They are all based on the **HTTP** protocol but communicate in different ways.

In this book, we consider web APIs as a wider concept than REST. In the digital world, the way machines communicate with each other changes as either the demands or the infrastructure evolves. In the 1990s, people focused on how to improve the internal networks that used the same platforms. TCP/IP came to be the standard for this kind of communication. After a few years, people needed to find a way to optimize communication across multiple platforms. Web Services appeared, and they used the **Simple Object Access Protocol** (**SOAP**), which was defined for enterprises, and it ensured that programs built on different platforms could easily exchange data.

However, SOAP XML is quite heavy, which means it requires more bandwidth for its usage. In the early 2000s, **Windows Communication Foundation** (**WCF**) was released by Microsoft. This helped developers manage the complexities of working with SOAP. WCF is RPC-based but still uses SOAP as the underlying protocol. Over time, some old standards, such as SOAP, have been transitioned to REST APIs, which will be discussed in the next section. We will start with REST APIs and then move on to the other styles of web-based APIs, such as gRPC APIs, GraphQL APIs, and SignalR APIs.

What is a REST API?

REST, also known as **Representational State Transfer**, is an architectural style of web APIs that was created by Roy Fielding in his Ph.D. dissertation *Architectural Styles and the Design of Network-based Software Architectures* in 2000. Today, generally speaking, REST APIs are based on HTTP, but actually, Roy Fielding's paper just outlines the core concepts and constraints for understanding an architectural style, and it does not require any specific protocol for REST-based architecture, such as HTTP. However, since HTTP is the most widely used protocol for web APIs, we will use HTTP as the protocol for REST APIs.

Just keep in mind that REST is just a style, not a rule. When you build a web API, you do not have to follow the REST style. You can use any other style you like. You can build a web API that works well, but it might not be *REST enough*. REST is the recommended style because it helps us establish constraints, which contribute to the design of web APIs. It also helps developers easily integrate with other REST APIs if they follow the same style.

The core concept of REST is the term *representational state transfer*. Think about a web system, which is a collection of resources. For example, it might have a resource called *books*. The collection of books is a resource. A book is a resource too. When you request the list of the books, you select a link (for example, `http://www.example.com/books`), which will return a JSON string that contains all the books, resulting in the next resource's representation, such as the link of a specific book (for example, `http://www.example.com/books/1`). You can continue to request the book with this link. In this process, the representation state is transferred to the client and rendered for the user.

There are loads of resources that explain REST. If you would like to know more about REST, you can read the following article on Wikipedia: *REST: The Web Framework for Representational State Transfer* (`https://en.wikipedia.org/wiki/Representational_state_transfer`).

Let's take a look at the constraints of REST, following which we will show you a simple example of REST APIs.

The constraints of REST

Roy Fielding's paper defines the following six constraints for REST APIs:

- **Client-server**: This pattern enforces the principle of separation of concerns. The server and the client act independently. The client sends the request and the server responds, following which the client receives and interprets the response. The client does not need to know how the server works, and vice versa.

- **Statelessness**: The server does not maintain any state of the client. The client should provide the necessary information in the request. This stateless protocol is important to scale out the capacity of the server because it does not need to remember the session state of the clients.

- **Cacheability**: The response of the server must implicitly or explicitly contain information about whether the response is cacheable, allowing the client and intermediaries to cache the response. The cache can be performed on the client machine in memory, in browser cache storage, or in a **content delivery network** (**CDN**). It is also important to improve the scalability and performance of web APIs.

- **The layered system**: The client does not know how it is connected to the server. There may be multiple layers between the client and the server. For example, a security layer, a proxy, or a load balancer can be placed between the client and the server without impacting the client or server code.

- **Code on demand (optional)**: The client can request code from the server for client-side use. For example, the web browser can request JavaScript files to perform some tasks.

- **Uniform interface**: This one is essential for a RESTful system. It contains resource identification, resource manipulation through representations, self-descriptive messages, and hypermedia as the engine of the application state. It simplifies and decouples the architecture of the system, which enables each part to evolve independently.

If you feel these principles are a little bit distant or theoretical, let's look at an example.

A REST API example

The website `https://jsonplaceholder.typicode.com/` is a fake REST API that generates fake JSON data. Open the following link in your browser: `https://jsonplaceholder.typicode.com/posts`. You will see a JSON string returned:

```
[
  {
    "userId": 1,
    "id": 1,
    "title": "sunt aut facere repellat provident occaecati excepturi
optio reprehenderit",
    "body": "quia et suscipit\nsuscipit recusandae consequuntur
expedita et cum\nreprehenderit molestiae ut ut quas totam\nnostrum
rerum est autem sunt rem eveniet architecto"
  },
  {
    "userId": 1,
    "id": 2,
    "title": "qui est esse",
    "body": "est rerum tempore vitae\nsequi sint nihil reprehenderit
dolor beatae ea dolores neque\nfugiat blanditiis voluptate porro vel
nihil molestiae ut reiciendis\nqui aperiam non debitis possimus qui
neque nisi nulla"
  },
...
]
```

From the preceding request, we can get the resource for the collection of the posts.

Now, we can request a specific post by its ID. For example, we can request the post with ID 1 using the following URL: `https://jsonplaceholder.typicode.com/posts/1`. The response is as follows:

```
{
  "userId": 1,
  "id": 1,
  "title": "sunt aut facere repellat provident occaecati excepturi
optio reprehenderit",
  "body": "quia et suscipit\nsuscipit recusandae consequuntur expedita
et cum\nreprehenderit molestiae ut ut quas totam\nnostrum rerum est
autem sunt rem eveniet architecto"
}
```

And that's it! The URLs we used in the preceding examples are the identifiers of the resources. The responses (the JSON strings) are the representations of the resources. A resource is manipulated through hypertext representations that are transferred in messages between clients and servers.

> **Important note**
>
> Some documents use URI. A **Uniform Resource Identifier** (**URI**) is a unique sequence of characters that identifies a logical or physical resource used by web technologies. It might use a location, name, or both. A **Uniform Resource Locator** (**URL**) is a type of URI that points to a resource over a network. It defines its protocol, such as `http`, `https` or `ftp`. Nowadays, the term *URL* remains widely used, so, we will use that in this book. However, we should know they have different scopes. URI is the superset of URL.

To get a post resource, we send a `GET` request. There are some other methods for manipulating resources, such as `POST`, `PUT`, `PATCH`, and `DELETE`, as shown here:

HTTP method	URL	Operation	Description
`GET`	`/posts`	Read	Read the collection of the posts
`GET`	`/posts/1`	Read	Read a post by its ID
`GET`	`/posts/1/comments`	Read	Read the comments of the post
`POST`	`/posts`	Create	Create a new post
`PUT`	`/posts/1`	Update	Update a post by its ID
`PATCH`	`/posts/1`	Update (partial)	Update part of a post by its ID
`DELETE`	`/posts/1`	Delete	Delete a post by its ID

Table 1.1 – HTTP methods and URLs for manipulating resources

There are other methods that are less frequently used, such as `HEAD`, `OPTIONS`, and `TRACE`.

As we can see, the HTTP methods are mapped to the **create, update, read, and delete** (**CURD**) operations. But was it always this way?

Is my web API RESTful?

As already mentioned, REST is not a rule or a specification. There is no *official* standard for REST APIs. Contrary to popular opinion, it does not require JSON. Furthermore, it does not require the use of CRUD patterns. But REST implementation does make use of standards, such as HTTP, URL, JSON, XML, and so on. People apply HTTP methods and JSON to implement REST, but they may not intentionally apply the constraints as originally described in Fielding's paper. This leads people to disagree on whether their APIs are RESTful or not. Many developers describe their APIs as RESTful, even though these APIs do not satisfy all of the constraints described in Fielding's paper.

Frankly, it is not beneficial to argue whether a web API is *REST enough* or not. The goal is to make something work, rather than wasting time on a discussion of this kind of problem. Not everyone has read the original paper. Technology also evolves rapidly. There is a Chinese saying: *It doesn't matter whether it is a white cat or a black cat; as long as it catches mice, it is a good cat.*

However, it would be ideal if we follow conventions when we start a greenfield project. Generally, a REST-based API is defined with the following aspects:

- A base URL, which is the root of the API, such as `http://api.example.com`.

- Semantics of HTTP methods, such as `GET`, `POST`, `PUT`, `DELETE`, and so on.

- A media type, which defines state transition data elements, such as `application/json`, `application/xml`, and so on.

In this book, we will try to follow these conventions when we develop the REST APIs with ASP.NET Core.

Now that we have had an overview of a REST API, let's see how to design one following the conventions.

Designing a REST-based API

To build a REST-based API, there are many steps to take before we write code. The development team needs to communicate with stakeholders and analyze the requirements. Then, they need to write user stories (or job stories) to define the desired outcomes. This requires the insights of domain experts or subject matter experts. We will not cover this part in this book. Instead, next, we will focus on the API design, which is closer to what developers do.

In the past few years, the concept of **API-first** has gained more traction. The API-first approach means that the APIs are treated as first-class citizens for your project. This creates a contract for how the API is supposed to behave before any code is written. In this way, the development teams can work in parallel because the contract will be established first. Developers do not have to wait for the API to be released before integrating with frontend or mobile apps. They can mock and test the APIs based on the contract. Using tools such as **Swagger**, the process of building APIs can be automated, such as API documentation, mock APIs, SDKs, and so on. Automation can significantly speed up the development of APIs and applications, which helps to increase the speed to market.

Here are some steps we can follow to design a REST-based API:

1. Identify the resources.
2. Define the relationships between resources.
3. Identify operation events.
4. Design the URL paths for resources.
5. Map API operations to HTTP methods.
6. Assign response codes.
7. Document the API.

If you are familiar with the preceding steps, you can skip them. However, if you are not, read the following sections.

A popular API description format is **OpenAPI Specification (OAS)**. We can use it to describe API modeling and other details of an API. We do not need to include the implementation details at this stage, because we just want to make a contract. SwaggerHub (`https://app.swaggerhub.com/home`) is a tool we can use to design an API.

Identifying the resources

REST-based APIs center on resources. A resource is a collection of data, such as a collection of posts or a collection of users. A resource is identified by a URL. The client requests a resource using the URL, and the server responds with a representation of the resource. The representation of the resource is sent in the hypertext format, which can be interpreted by the widest possible range of clients.

It is important to identify the scope of the domain and the relationships between the resources. For example, if you are building a blog system, you may have a collection of posts, and each post has a collection of comments. The scope of an API may evolve as time goes by. More resources may be added to the current domain, or some resources will be removed. Also, relationships may change.

Let's start small. We can use the blog system as an example. After the requirement analysis, we can identify the following resources:

- Posts
- Categories
- Comments
- Users
- Tags

You may want to include some properties of each resource in this step. For example, a post has a title, a body, and a published datetime. A comment has a body, a publish datetime. A user has a name and an email. You may find more properties during development.

Defining the relationships between resources

Once the resources are identified, we can define the relationships between them. For example, a post has a collection of comments. A comment has a post. A user has a collection of posts.

The relationship is defined by how these resources relate to each other. Sometimes, these relationships exist in the database as well, but sometimes, they are specific to the REST resources only.

There are some terms we can use to describe the relationships:

- **Independent resource**: This resource can exist independently. It does not require another resource to exist. An independent resource can reference other independent or dependent resources. For example, a post is an independent resource, and it can reference its author. The authors resource is also an independent resource.

- **Dependent resource**: This resource requires another resource to exist. It can still reference other independent or dependent resources, but it cannot exist without the existence of the parent resource. For example, a comment requires a post as its parent resource; otherwise, it cannot exist. A comment can reference its author, which is an independent resource.

- **Principal key (or primary key)**: This resource has a unique identifier. It may be the primary key of the resource in the database. For example, a post has a `Id` property, which can uniquely identify itself.

- **Foreign key**: This dependent resource has a property to reference the principal resource, which stores the principal key of the principal resource. For example, a comment has a `PostId` property, which references a post.

There are three relationship types that these resources can have:

- **One-to-many**: This is when a resource has many related resources. For example, a user has many posts, but a post has only one author. This is also called the *parent-child (children)* relationship, which is the most common pattern for relationships we can see in the REST-based API world.

- **One-to-one**: This is when a resource has one related resource. For example, a post has a single author, and one house has only one address. The one-to-one relationship is a special case of the one-to-many relationship.

- **Many-to-many**: This is when a resource has many related resources and vice versa. For example, a blog has many tags, and a tag has many blogs. A movie can have many genres, and a genre can have many movies. In many systems, a user can have many roles, and a role can have many users.

Identifying operations

Next, we can think about what operations are needed for each resource. These operations may come from user stories that are defined beforehand. Generally, each resource has its CRUD operations. Note that the operations may include more beyond CRUD. For example, a post can be published, or it can be unpublished. A comment can be approved, or it can be rejected. During this process, we may need to create a new resource or property to reflect the operation.

It is important to consider the scope of the domain. CRUD operations are easy to understand, but for some complicated relationships, we may need help from domain experts.

When we work on these operations, we need to include important input and output details. We will use them in the next steps. However, it is not necessary to include all the details of each resource. We have enough time later to capture the complete design.

For the example of the blog system, we can identify these operations for the `Post` resource (including but not limited to the following):

Operation name	Resource(s)	Input	Output	Description
`createPost()`	`Post, category, user, tag`	Post detail		Create a new post
`listPosts()`	`Post`		A list of posts	List all posts
`listPostsByCategory()`	`Post and category`	Category ID	A list of posts	List posts by category
`listPostsByTag()`	`Post and tag`	Tag or Tag ID	A list of posts	List posts by tag
`searchPosts()`	`Post`	Search keyword	A list of posts	Search for Posts by title, author, and content
`viewPost()`	`Post, category, and user`	Post ID	Post detail	View a post detail
`deletePost()`	`Post`	Post ID		Delete a post
`updatePost()`	`Post and category`	Post detail		Update a post
`publishPost()`	`Post`	Post ID		Publish a post
`unpublishPost()`	`Post`	Post ID		Unpublish a post

Table 1.2 – Operations for the Post resource

For some operations, such as `createPost` and `deletePost`, the output is the results of the operation. This can be represented with the HTTP status code. We will discuss this later.

We can list more operations for other resources as well.

Designing the URL paths for resources

The next step is to design the URL paths for each resource. The clients use URLs to access the resources. Even though REST is not a standard, there are some guidelines or conventions for designing URL paths.

Using nouns instead of verbs

The operation events we identified in the previous step are some actions, such as `Create`, `List`, `View`, `Delete`, and so on. However, the URL paths are not usually presented by verbs. Because HTTP methods such as `GET`, `POST`, `PUT`, and `DELETE` are already verbs, it is not necessary to include verbs in the URL paths. Instead, we should use nouns to represent the resources – for example, `/posts`.

Using plural nouns to represent collections

If a resource is a collection, we should use plural nouns to represent the resource. For example, `/posts` is the URL path for the collection of posts. To get a single post by its ID, we can use `/posts/{postId}`.

Using logical nesting to represent relationships

For the resources that have a relationship, normally, the child resource (i.e., the dependent resource) should be nested under the parent resource, and the path should include the parent identifier. However, this does not reflect the database structure. For example, a post can have a collection of comments; the URL looks like `/posts/{postId}/comments`. It clearly shows that the comments are related to the post.

However, if the relationships are too deep or complicated, the nesting URL path can be too long. In this case, we can rethink how to better represent those resources. For example, if we want to retrieve an author's information from one comment, we could use `/posts/{postId}/comments/{commentId}/author`. But this goes too far. Instead, if we know the `UserId` of the author, we can use `/users/{userId}`. Avoid using deep nesting in URL paths because it makes an API more complicated and not readable.

Allowing filtering, sorting, and pagination

Returning all records simultaneously is not a good idea. We can use filtering, sorting, and pagination to return a subset of the records that a client needs. These operations can improve the performance of the APIs and provide a better user experience.

For example, if we want to search a list of posts for a specific keyword, we can use a query parameter, such as `/posts?search=keyword`. If we want to sort posts by the date, we can use `/posts?sort=date`. To get the second page of the posts, we can use `/posts?page=2`. These query parameters can be combined with each other.

What if I cannot find a proper verb in HTTP methods for an operation?

Generally, HTTP methods can represent CRUD operations. However, in the real world, there are many more complexities! For example, besides the basic CRUD operations, there are other operations, such as publishing or unpublishing a post. So, what HTTP methods should we use?

This is where things can get tricky. This subject is open to debate, but remember that we are not arguing whether an API is *RESTful* enough. We just want to make it work.

There are different approaches for these scenarios:

- One possible solution could be treating such operations like a sub-resource. So, you can use `/posts/{postId}/publish` to publish a post. GitHub uses the following URL to star a gist: `/gists/{gist_id}/star`. For more information, check out `https://docs.github.com/en/rest/gists/gists#star-a-gist`.

- The post should have an `IsPublished` field to indicate whether it is published. So, actually, the `publish` action is an update action, which updates the `IsPublished` field only. Then, you can treat it the same as the `updatePost()` operation.

Here are some resource URLs for the blog system:

Operation name	URL	Input	Output	Description
`createPost()`	`/posts`	Post detail		Create a new post
`listPosts()`	`/posts`		A list of posts	List all posts
`listPostsBy-Category()`	`/posts?categoryId={-categoryId}`	Category ID	A list of posts	List posts by category
`listPostsBy-Tag()`	`/posts?tag={tagId}`	Tag or Tag ID	A list of posts	List posts by tag
`searchPosts()`	`/posts?search={key-word}`	Search keyword	A list of posts	Search for posts by title, author, and content
`viewPost()`	`/posts/{postId}`	Post ID	Post detail	View a post detail
`deletePost()`	`/posts/{postId}`	Post ID		Delete a post
`updatePost()`	`/posts/{postId}`	Post detail		Update a post
`publishPost()`	`/posts/{postId}/publish`	Post ID		Publish a post
`unpublish-Post()`	`/posts/{postId}/unpub-lish`	Post ID		Unpublish a post

Table 1.3 – URLs for the Post resource

Some URLs are identical, such as `deletePost()` and `updatePost()`, because we will use HTTP methods to differentiate those operations.

Mapping API operations to HTTP methods

Next, we need to identify which HTTP method is appropriate for each operation. As we mentioned before, there are some common HTTP methods for CRUD operations. For example, when we request a resource, we should use the `GET` method. When we create a new resource, we should use the `POST` method.

When we map the API operations to HTTP methods, we also need to consider the safety of the operations and HTTP methods. There are three types of safety for HTTP operations:

- **Safe**: The operation is safe. It does not change any state of the resource. For example, if we send a `GET` request to `/posts/{postId}`, it will return the same result, no matter how many times the same request is sent. For some cases, the resource might be updated by a third party, and the next `GET` request will return the updated result. But this was not caused by the client, so it is important to understand whether the state change is caused by the client who sent the request.

- **Idempotent**: The operation is idempotent, which means it makes state changes to the target resource, but it will produce the same result if the same input is provided. Note that idempotency does not mean that the server must return the same response. For example, if we want to delete a post, we can send a `DELETE` request to `/posts/{postId}` to delete it. If the request is a success, we will get a `200 OK` or a `204 No Content` response. If we send the same request to `/posts/{postId}` again, it may return a `404 Not found` response because the resource was already deleted, but it will not cause any other side effects. If an operation is idempotent and the client knows whether the previous request failed, it is safe to reissue the request without any side effects.

- **Unsafe**: The operation is unsafe. It makes state changes to the target resource. If the request is sent again, it cannot guarantee the same result. For example, we can send a `POST` request to `/posts` to create a new post. If we send the same `POST` request again, it will create another new post with the same title and content, and so on.

All safe methods are also idempotent, but not all idempotent methods are safe. The following table lists the safety of each HTTP method:

HTTP method	Safe	Idempotent	Common operations
GET	Yes	Yes	Read, list, view, search, show, and retrieve
HEAD	Yes	Yes	HEAD is used to check the availability of a resource without actually downloading it.

HTTP method	Safe	Idempotent	Common operations
OPTIONS	Yes	Yes	OPTIONS is used to retrieve the available HTTP methods for a given resource.
TRACE	Yes	Yes	TRACE is used to get diagnostic information about a request/response cycle.
PUT	No	Yes	Update and replace
DELETE	No	Yes	Delete, remove, and clear
POST	No	No	Create, add, and update
PATCH	No	No	Update

Table 1.4 – Safety of HTTP methods

The following table shows how operations are mapped to HTTP methods:

Operation name	URL	HTTP method	Description
createPost()	/posts	POST	Create a new post
listPosts()	/posts	GET	List all posts
listPostsByCategory()	/posts?categoryId={categoryId}	GET	List posts by category
listPostsByTag()	/posts?tag={tagId}	GET	List posts by tag
searchPosts()	/posts?search={keyword}	GET	Search for posts by title, author, and content
viewPost()	/posts/{postId}	GET	View a post detail
deletePost()	/posts/{postId}	DELETE	Delete a post
updatePost()	/posts/{postId}	PUT	Update a post
publishPost()	/posts/{postId}/publish	PUT	Publish a post
unpublishPost()	/posts/{postId}/unpublish	PUT	Unpublish a post

Table 1.5 – Mapping HTTP methods for the Post resource

You may have seen some other cases, such as using POST to update a resource. That works, but it does not follow the HTTP standard. Generally speaking, we can state the following:

- GET is used to read resources.

- POST is used to create child resources with a server-defined URL, such as /posts.

- PUT is used to create or replace the resource with a client-defined URL, such as /posts/ {postId}.. In many cases, PUT can also be used to update a resource.

- PATCH is used to update parts of the resource with a client-defined URL, such as /posts/ {postId}.

Assigning response codes

It is time to assign the HTTP response codes for the operations. There are some main response code categories:

- 2xx **codes – success**: The action requested by the client was received, understood, and accepted.

- 3xx **codes – redirection**: The client must take additional action to complete the request. It is often used to indicate that the client should be redirected to a new location.

- 4xx **code – client errors**: The operation was not successful, but the client can try again.

- 5xx **codes – server errors**: The server has encountered an error or is incapable of performing the request. The client can retry in the future.

A common issue is that some developers invent their own response codes. For example, if we create a new post, we expect the server to return a 201 Created response code. Some developers may use 200 OK and include a status code in the response body. This is not a good idea. There are many layers between the server and the client. Using your own codes will probably cause problems for these middleware components. Make sure to use the right code for the right reason. Here are some common response codes:

HTTP response code	Description
200 OK	The standard response for successful HTTP requests.
201 Created	The request has been fulfilled, resulting in a new resource being created.
202 Accepted	The request has been accepted for processing, but processing has not been completed.
204 No Content	The server has successfully processed the request but does not return any content. This is common for delete operations.

HTTP response code	Description
400 Bad Request	The server cannot understand or process the request due to a client error, such as malformed syntax, a request size too large, or invalid input. The client should not repeat the request without modifications.
401 Unauthorized	The request requires user authentication.
403 Forbidden	The server understood the request but is refusing action. This may be due to the fact that the client does not have the necessary permissions or is attempting a prohibited action.
404 Not Found	The requested resource could not be found.
500 Internal Server Error	A generic error message. The server encountered an unexpected condition, so it cannot process the request, and no more specific messages are suitable at this time.
503 Service Unavailable	The server is currently unable to handle the request due to temporary overloading or maintenance of the server. The response should contain a Retry-After header if possible so that the client can retry after the estimated time,

Table 1.6 – Common HTTP response codes

Here is the table that shows the response codes for each operation:

Operation name	URL	HTTP method	Response	Description
createPost()	/posts	POST	Post, 201	Create a new post
listPosts()	/posts	GET	Post[], 200	List all posts
listPostsBy-Category()	/posts?category-Id={categoryId}	GET	Post[], 200	List posts by category
listPostsBy-Tag()	/posts?tag={tagId}	GET	Post[], 200	List posts by tag
searchPosts()	/posts?-search={keyword}	GET	Post[], 200	Search for posts by title, author, and content
viewPost()	/posts/{postId}	GET	Post, 200	View a post detail
deletePost()	/posts/{postId}	DELETE	204, 404	Delete a post
updatePost()	/posts/{postId}	PUT	200	Update a post

Operation name	URL	HTTP method	Response	Description
publishPost()	/posts/{postId}/publish	PUT	200	Publish a post
unpublish-Post()	/posts/{postId}/unpublish	PUT	200	Unpublish a post

Table 1.7 – Response codes for the Post resource

It is essential to utilize the correct response code in order to prevent any misunderstandings. This will ensure that all communication is clear and concise, thus avoiding any potential confusion.

What if I want to create my own status codes?

Technically, you can create your own status codes, but in practice, please stick as closely to the standards as possible. If you invent your own status codes, that would be risky. Your users might be in trouble consuming your APIs because they do not know your status codes. You should think about what the benefits are to have your own status codes. The convention is to respect the HTTP status codes defined in RFC. Before you create your own status codes, make sure you check the list of HTTP status codes first. Do not create your own status code unless you have strong reasons. You can find more information here: https://en.wikipedia.org/wiki/List_of_HTTP_status_codes.

However, there might be some special situations where you want to indicate a more specific status in the response. For example, you might have an API that can process a task, but it might fail for different reasons. You might want to indicate a more detailed message in the response to let your users know what happened, rather than returning a common 4xx code. You should think about the business logic carefully and differentiate between HTTP status codes and business status codes. If you cannot find a proper code in the HTTP status codes, and you do want to show a business-related status in the response, you can choose the HTTP status code to indicate the category of the response, and then attach a response body that contains your business status code. For example, you can return a response as shown here:

```
400 Bad Request
{ "error_code": 1, "message": "The post is locked and cannot be
updated."}
```

So, the HTTP status code represents the common status of the operation, and in the response body, you can include some information that is specific to your system. We will discuss how to handle errors using the Problem Details object in *Chapter 16*.

Documenting the API

OpenAPI is a popular REST API specification. It is a programming language-agnostic interface description for REST APIs, allowing both humans and computers to discover and understand the capabilities of a service without access to source code. Similar to an interface, it describes the inputs and outputs of an API, as well as how they should be transmitted. It is also known as the Swagger specification.

> **Swagger versus OpenAPI**
>
> Sometimes, *Swagger* and *OpenAPI* are used interchangeably. The Swagger project was developed in early 2010s to define a simple contract for an API that contains everything needed to produce or consume an API. It was donated to the OpenAPI initiative in 2015. So, OpenAPI refers to the API specification, and Swagger refers to the open-source and commercial projects from SmartBear, which work with the OpenAPI specification. In short, OpenAPI is a specification, and Swagger is tooling that uses the OpenAPI specification. Swagger UI is also one of the Swagger tools. At the time of writing, the latest version of OpenAPI was 3.1.0.

We can use SwaggerHub to design an API based on the previous steps. Here is an example, which defines a simple API for a blog system:

```
openapi: 3.0.0
servers:
  - description: SwaggerHub API Auto Mocking
    url: https://virtserver.swaggerhub.com/yanxiaodi/MyBlog/1.0.0
info:
  description: This is a simple API
  version: '1.0.0'
  title: Sample Blog System API
  contact:
    email: you@your-company.com
  license:
    name: Apache 2.0
    url: 'http://www.apache.org/licenses/LICENSE-2.0.html'
tags:
  - name: admins
    description: Secured Admin-only calls
  - name: developers
    description: Operations available to regular developers
paths:
  /posts:
    get:
      tags:
        - developers
      summary: searches posts
      operationId: searchPost
      description: |
        By passing in the appropriate options, you can search for
        available blogs in the system
      parameters:
```

```
            - in: query
              name: searchString
              description: pass an optional search string for looking up
  post
              required: false
              schema:
                type: string
            - in: query
              name: skip
              description: number of records to skip for pagination
              schema:
                type: integer
                format: int32
                minimum: 0
            - in: query
              name: limit
              description: maximum number of records to return
              schema:
                type: integer
                format: int32
                minimum: 0
                maximum: 50
        responses:
          '200':
            description: search results matching criteria
            content:
              application/json:
                schema:
                  type: array
                  items:
                    $ref: '#/components/schemas/Post'
          '400':
            description: bad input parameter
```

The other file of this file has been omitted.

The preceding API documentation is a YAML file, which defines two models (resources) – Post and Category – and two operations – GET for searching posts and POST for creating a new post. For each operation, there are details about the input and output, including the expected response codes.

After the API design is done, we can share the API documentation with other developers for integrations, as shown here:

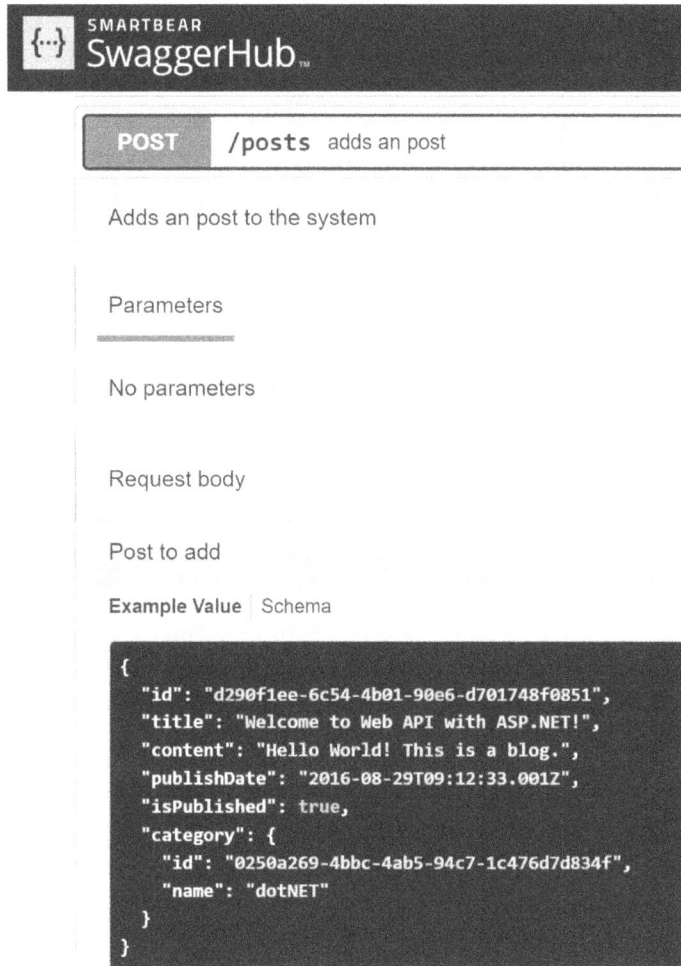

Figure 1.1 – The SwaggerHub UI

Note that you might need to add more properties, based on your user stories and domains, before you share the API documentation with other teams. The API contract should be quite stable; otherwise, it will impact the consumers.

We have explained how to design a REST API. If you would like to learn how to start developing with ASP.NET Core, you can move on to *Chapter 2*.

REST API is one of the most popular API styles. In the next section, we will introduce other API styles, such as RPC APIs, GraphQL APIs, and real-time APIs.

RPC and GraphQL APIs

While REST-based APIs are widely used in many scenarios today, it is not the only style of web API. For some scenarios, RPC-based APIs or GraphQL APIs may be better suited. It is important to understand the advantages and disadvantages of each style of API so that you can choose the right styles for your scenarios.

What is an RPC-based API?

RPC has existed for many years. It is the earliest, simplest form of web interaction. It is like a local call in some other languages, but it executes over a network. The client is given a list of available methods. Each method takes pre-defined, typed, and ordered parameters, returning a structured response result. So, the client can run on a different machine or in a different process but still work with the server, such as in the same application.

In this way, the client is tightly coupled to the server. If the server changes these methods or any parameters, the client will be affected. Developers must update the code of the client to match the new server methods. This can be a disadvantage of RPC-based APIs, but it can offer better performance.

The remote procedures are defined with the **interface definition language** (IDL). The IDL defines the methods and parameters of the remote procedure. Normally, some code generators can generate the client and server stubs based on the IDL. The code is strongly typed, which provides better type safety and error handling.

To implement an RPC-based API, there are some specifications for different languages. For example, WCF was a popular framework for RPC a couple of years ago. Some other popular frameworks include XML-RPC, SOAP PRC, JSON-RPC, and gRPC.

Because RPC is like a local method call, you often see verbs in the method names. Unlike REST, RPC supports various operations beyond CRUD. Here is an example of a JSON-RPC request and response:

Request:

```
POST https://rpc.example.com/calculator-service HTTP/1.1
Content-Type: application/json
Content-Length: ...Accept: application/json
{"jsonrpc": "2.0", "method": "subtract", "params": [42, 23], "id": 1}
```

Response:

```
{ "jsonrpc": "2.0", "result": 19, "id": 3 }
```

One of the most popular RPC frameworks is gRPC, which we will discuss in the next section.

What is gRPC?

One of the most popular RPC frameworks is gRPC. It is a high-performance, open-source modern RPC framework for building network services and distributed applications. gRPC was initially created by Google, which used a RPC framework called Stubby. In March 2015, Google decided to make it open-source, resulting in gRPC, which is now used in many organizations outside of Google.

gRPC has some awesome features, such as the following:

- **Interoperability**: gRPC uses a **Protocol Buffer** (**protobuf**) file to declare services and messages, which enables gRPC to be completely language- and platform-agnostic. You can find gRPC tools and libraries for all major programming languages and platforms.

- **Performance**: `protobuf` is a binary format, which has a smaller size and faster performance than JSON. It is not readable by humans, but it is readable by computers. HTTP/2 also supports multiplexing requests over a single connection. It needs fewer resources, even in slower networks.

- **Streaming**: gRPC is based on the HTTP/2 protocol, which makes it support bidirectional streaming.

- **Productivity**: Developers author protobuf files (`.proto`) to describe services with input and output. Then, they can use the `.proto` files to generate stubs for different languages or platforms. It is similar to the OpenAPI specification. Teams can focus on business logic and work on the same service in parallel.

- **Security**: gRPC is designed to be secure. HTTP/2 is built on top of **Transport Layer Security** (**TLS**) end-to-end encrypted connection. It also supports client certificate authentication.

With these benefits, gRPC is a good choice for microservices-style architecture. It can efficiently connect services in and across data centers, even across load balancers. It is applicable in the last mile of distributed systems because service-to-service communication needs low latency and high performance. Also, the polyglot systems may have multiple languages or platforms, and gRPC can support different languages and platforms.

Here is an example of a gRPC `.proto` file:

```
syntax = "proto3";

option csharp_namespace = "GrpcGreeter";

package greet;

service Greeter {
  // Sends a greeting message.
  rpc SayHello (HelloRequest) returns (HelloReply);
}
```

```
// The request message containing the user's name.
message HelloRequest {
  string name = 1;
}

// The response message containing the greeting.
message HelloReply {
  string message = 1;
}
```

ASP.NET Core provides great support for gRPC. With the `.proto` file, the .NET types for services, clients, and messages can be automatically generated in a .NET project. We will learn how to develop the gRPC service in *Chapter 11*.

However, gRPC is not a silver bullet. There are several factors we need to consider before we choose gRPC:

- **Tight coupling due to protocol changes**: The client and server are tightly coupled because of the protocol. Once the protocol changes, the client and server must be updated, even just changing the order of the parameters.

- **Non-human readable format**: `protobuf` is a non-human readable format, so debugging is not convenient. Developers need additional tools to analyze the payloads.

- **Limited browser support**: gRPC heavily relies on HTTP/2, but browsers cannot support gRPC natively. That is why it is more suitable for service-to-service communication. There are some projects, such as `grpcweb`, that can provide a library to perform conversions between gRPC and HTTP/1.1.

- **No edge caching**: gRPC uses the `POST` method, which is not cacheable for the clients.

- **A steeper learning curve**: Unlike REST, which is human-readable, many teams find gRPC challenging to learn. They need to learn protobuf and HTTP/2 and look for proper tools to deal with the message content.

So, should we choose gRPC or REST?

Should I use gRPC instead of a REST API?

It is challenging to choose gRPC over REST for most server-client scenarios. A REST-based API is well supported by all browsers, so it has a wider adoption nowadays. If you need to support browser clients, REST is a better choice compared with gRPC. However, gRPC has some features that are useful in certain circumstances, such as high-performance communication, real-time communication, low bandwidth, and multiple language environments. So, it is a good choice for microservices-style architecture.

In a microservices architecture, the services are loosely coupled, and each one does a specific task or processes specific data. They need to be able to communicate with each other with simplicity and efficiency without considering browser compatibility. gRPC is suitable for this scenario because it is based on HTTP/2, which is a high-performance protocol and provides bidirectional streaming, binary messaging, and multiplexing. **Dapr**, which is a portable, event-driven runtime for microservices, implements gRPC APIs so that apps can communicate with each other via gRPC. We will not discuss Dapr in this book, but if you are interested, you can find more information here: https://dapr.io/.

In conclusion, using gRPC or REST depends on your use-case demands.

gRPC API design

The gRPC API design process is very similar to the REST API design process. In fact, the first three steps are similar to the REST API design process. We need to identify the resources, define the relationships between resources, and identify operation events as well.

Next, use the information from the first three steps to design and document the gRPC API. When we convert the operation events to gRPC operations, there are some differences. REST API uses HTTP methods to represent the operations. In gRPC, these operations are like the methods of the service, which means we can use verbs in the method names. For example, the method to get a post can be represented as GetPost().

gRPC uses protobuf as the IDL. When we design a gRPC API, we actually need to author the .proto files. These .proto files consist of two parts:

- The definition of the gRPC service

- The messages that are used in the service

This is similar to the REST OpenAPI definition but has a different syntax. Each request needs a type-defined message that includes the sorted input parameters. Each response returns a message, an array of messages, or an error status response. We can have a .proto file for the blog system, as shown here:

```
option csharp_namespace = "GrpcMyBlogs";
syntax = "proto3";
package myBlogs;
service Greeter {
  rpc GetPost(GetPostRequest) returns (Post);
  rpc CreatePost(CreatePostRequest) returns (Post);
  rpc UpdatePost(UpdatePostRequest) returns (Post);
  rpc SearchPosts(SearchPostsRequest) returns (SearchPostsResponse);
  // More methods...
  ...
}
message GetPostRequest {
  string post_id = 1;
}
message CreatePostRequest {
  string category_id = 1;
  string title = 2;
```

```
      string content = 3;
      // More properties below
      ...
  }
  message UpdatePostRequest {
    string post_id = 1;
    string category_id = 2;
    string title = 3;
    string content = 4;
    // More properties below
    ...
  }
  message SearchPostsRequest {
    string keyword = 1;
  }
  message Post {
    string post_id = 1;
    Category category = 2;
    string title = 3;
    string content = 4;
    // More properties below
    ...
  }
  message Category {
    string category_id = 1;
    string name = 2;
    // More properties below
    ...
  }
  message SearchPostsResponse {
    int32 page_number = 1;
    int32 page_size = 2 [default = 10];
    repeated Post posts = 3;
  }
```

That's it! Now, the `.proto` file has a basic gRPC service definition, including the message definitions. Next, we can use various tools to generate the code for the gRPC service and the client. In the development stage, we may need to frequently change the gRPC protocol definition by updating the `.proto` files. The changes will be reflected in the generated code. So, please consider carefully before you publish the service for consumption. We will discuss more about gRPC in *Chapter 11*. If you would like to start gRPC development with .NET 8 now, please jump to that chapter.

Next, let's look at the GraphQL API.

What is a GraphQL API?

Think about the scenarios that use REST APIs. We may find some issues:

- **Over-fetching**: This is when a client downloads more information than is actually required. Let's say we have a `/posts` endpoint, which returns a list of the posts. When we show a post list page, we only need some properties such as `Title`, `PublishDate`, and `Category`. But the posts returned from the endpoint may contain more information about the posts, such as `IsPublished`, which is useless for clients.

- **Under-fetching and the n+1 problem**: Under-fetching is when the endpoint does not provide sufficient required information. The client will have to make more requests to fetch the missing information. For example, the API may provide a `/posts/{postId}` endpoint that shows the post information and a `/posts/{postId}/related` endpoint that shows related posts. If we want to show a post detail, the client will need to call the `/posts/{postId}` endpoint, but the related posts are not provided in the response. So, the client will have to make another request to `/posts/{postId}/related` to get the related posts. The N+1 problem often refers to the parent-child relationship. The endpoint that returns the collection resource does not provide enough information about the child resources for the clients. For example, the `/posts` endpoint returns a list of posts, but the response does not have the content summary for each post. To show the content summary on the post list page, the client will have to call the `/posts/{postId}` endpoint for each post to get the content summary. So, the total number of requests will be $n + 1$, where n is the number of posts.

The over- and under-fetching problems are some of the most common problems with REST-based APIs. Because REST-based APIs center on resources, for each endpoint, the response structure is fixed and encoded in the URL, so it is not flexible for client requirements.

These problems can be overcome with a GraphQL API. A GraphQL API is another API style, which offers powerful query capabilities. It supports fetching data with a flexible structure based on the client's needs. It can fetch data by a resource identifier, paginated listing, filtering, and sorting. It also supports mutating data as well, like CRUD in REST.

Introduction to GraphQL

GraphQL is a powerful query language for executing queries with a flexible data structure. It was developed internally by Facebook in 2012, following which it was released publicly in 2015. Now, it is open-source and maintained by a large community of companies and individuals from all around the world.

GraphQL solves the over- and under-fetching problem by providing more flexibility and efficiency. It is not tied to any database or storage engine, nor to any specific languages. There are many libraries to implement GraphQL services and clients. A GraphQL service defines types and fields on your resource types, and then provides functions for each field on each type.

Unlike REST, which uses *resources* as its core concept and defines URLs that return a fixed data structure for each resource, the conceptual model of GraphQL is an entity graph. So, all GraphQL operations are performed through a single HTTP POST- or GET-based endpoint, which is usually / graphql. It is completely flexible and allows the client to decide what data structure it needs. The GraphQL service receives GraphQL queries to validate whether a query refers to the correct types and fields defined, and then it executes the functions to return the correct data structure. The format of requests and responses is JSON.

Besides solving the over- and under-fetching problem, GraphQL has some other advantages:

- GraphQL reduces the complexity of maintaining API versions. There is only one endpoint and one version of the graph. It allows the API to evolve without breaking the existing clients.

- GraphQL uses a strong type system to define the types and fields in a schema using SDL. The schema behaves as the contract, which reduces the miscommunication between the client and the server. Developers can develop frontend applications by mocking the required data structures. Once the server is ready, they can switch to the actual API.

- GraphQL does not define the specific application architecture, which means it can work on top of an existing REST API to reuse some code.

- The payload is smaller because clients get what they exactly request without over-fetching.

However, there are also some disadvantages of using GraphQL:

- GraphQL presents a high learning curve for REST API developers.

- The implementation of the server is more complicated. The query could be complex.

- GraphQL uses a single endpoint, which means it cannot leverage the full capabilities of HTTP. It does not support HTTP content negotiation for multiple media types beyond JSON.

- It is challenging to enforce the authorization because, normally, the API gateway enforces access control based on URLs. Rate-limiting is also often associated with the path and HTTP methods. So, you need more consideration to adopt the new style.

- Caching is complicated to implement because the service does not know what data clients need.

- File uploads are not allowed, so a separate API for file handling is needed.

GraphQL API design

A GraphQL query is flexible, so clients can send various queries based on their requirements. To design the GraphQL API, we need to define the GraphQL schema first, which is the core of every GraphQL API.

GraphQL uses the GraphQL SDL to define GraphQL schemas. SDL has a type system that allows us to define the data structure, like the other strongly typed languages, such as C#, Java, TypeScript, Go, and so on.

We can define the following types and fields in the GraphQL schema:

```
type Query {
  posts: [Post]
  post(id: ID!): Post
}

type Post {
  id: ID!
  title: String
  content: String
  category: Category
  publishDate: String
  isPublished: Boolean
  relatedPosts: [Post]
}

type Category {
  id: ID!
  name: String
}
```

A GraphQL request uses a query language to describe the desired fields and structure the client needs. The following is a simple query:

```
{
  posts {
    id
    title
    content
    category {
      id
      name
    }
    publishDate
    // If we do not need the `isPublished` field, we can omit it.
    // isPublished
    relatedPosts {
      id
      title
      category {
        id
        name
      }
```

```
        publishDate
      }
    }
  }
}
```

In the preceding query, we can omit the `isPublished` field and include related posts in the response, so the client does not need to send more requests.

To modify data or perform a calculation logic, GraphQL establishes a convention called *mutation*. We can think of mutation as a way to update the data. The following request is a mutation:

```
mutation {
  createPost(
    categoryId: ID!
    title: String!
    content: String!
  ) {
    post {
      id
      title
      content
      category {
        id
        name
      }
      publishDate
      isPublished
    }
  }
}
```

There are some tools to generate GraphQL documents and test the service, such as GraphiQL,GraphQL Playground, and so on. We will not discuss GraphQL much further now. In *Chapter 12*, we will learn how to use ASP.NET Core 8 to develop GraphQL APIs.

Next, we will discuss another API style, which is the real-time API.

Real-time APIs

We have introduced some web API styles, such as REST-based APIs, gRPC APIs, and GraphQL APIs. They all follow the request/response pattern – the client sends a request to the server, and the server returns a response. This pattern is easy to understand. However, this pattern may not be suitable for some scenarios.

Let's say we have an application that contains two parts – the server, which is a fire station, and the clients, which are the fire engines. How can we notify the fire engines when an incident occurs?

If we use the request/response pattern, the clients need to send requests to the server to get the latest notification regarding the incident. But what is the best frequency of sending requests? 1 minute, or 10 seconds? Think about how urgent the incident is. If the fire engine gets a notification after a 10-second delay, this might be a problem, as the fire might be more serious and urgent! So, what about sending requests every 1 second? Then, the server would be very busy, and most of the time, it just returns a `No incident` response. With the request/response pattern, the server cannot push the notifications to the clients, so it is not suitable for this case. This leads us to the problem of API polling.

The problem with API polling

The request/response pattern has limitations. The server cannot notify the clients about what changes are happening on the server side. If the client needs to get the latest data, it has to frequently send requests to the server before receiving any updates.

For example, if the client would like to know when a new post is published, it needs to call `/posts/latest` to get the latest post. The client may set up an interval to send the request periodically. This pattern is called API polling, which is a common solution for clients that need to be updated for resource changes.

API polling does not have many differences from the common REST APIs. It can be implemented based on the request/response pattern. However, it is not the ideal solution for this kind of scenario. Normally, the frequency of resource changes is not predictable, so it is hard to decide the frequency of the requests. If the interval is too short, the client may send too many unnecessary requests and the server will handle too many queries. However, if the interval is too long, the clients cannot get the latest changes in time. In particular, if the application needs to notify clients in real-time, then the system would be very busy.

There are more challenges when we use API polling:

- The logic to check the resource changes is complex. It may implement the logic in the server, so the server needs to check the timestamp in the request and then query the data, based on the timestamp. Alternatively, the client queries all the data and compares the collection with the data from the previous request. It brings a lot of complexity.
- It is hard to check whether a specific event has occurred – for example, creating resources and updating resources.
- Rate-limiting may block the client from sending too many requests at the desired intervals.

The ideal way to solve the problem of API polling is to allow the server to send events to the clients in real-time, rather than constantly polling and implementing the logic to check for changes. This is a different pattern from the request/response pattern. It supports real-time communication between servers and clients, which enables new possibilities for the application. This is what a real-time API can do.

What is a real-time API?

A real-time API goes beyond the traditional REST APIs. It provides some benefits:

- The application can respond to internal events in real-time. For example, if a new post is published, the client can get the notification immediately.

- It can improve API efficiency by reducing the number of requests. The clients do not need API polling to check the resource changes. Instead, the server sends messages to clients when some events occur. It reduces the resources required during communication.

Some technologies can implement real-time APIs, such as long polling, **Server-Sent Events** (**SSE**), WebSocket, SignalR, and gRPC streaming.

Let's take a quick look at these.

Long polling

The API polling problem we described previously is called short polling or regular polling, which is easy to implement but less efficient. The client cannot receive an update from the server in real-time. To overcome this issue, long polling is another choice.

Long polling is a variation of short polling, but it is based on *Comet*, which is a web application model in which a long-held HTTPS request allows a web server to push data to a browser, without it having to request it explicitly. Comet contains multiple techniques to implement long polling. It also has many names, such as Ajax push, HTTP streaming, and HTTP server push.

To use long polling, the client sends the request to the server but with the expectation that the server may not respond immediately. When the server receives the request, if there is no new data for the client, the server will keep the connection alive. If there is something available, the server will send the response to the client and complete the open request. The client receives the response and usually makes a new request right away or after a pre-defined interval to establish the connection again. The operation is repeated. In this way, it can effectively emulate the server push feature.

There are some considerations when using long polling. The server needs to manage multiple connections and preserve the session state. If the architecture becomes more complex (for example, when multiple servers or load balancers are in use), then it leads to the session stickiness issue, which means the subsequent client requests with the same session must be routed to the same server to which the original request was handled. It is hard to scale the application. Also, it is hard to manage the message order. If the browser has two tabs open and sends multiple requests simultaneously to write data, the server will not know which request is the latest.

Long polling is supported by many web browsers. In recent years, SSE and WebSocket have been widely adopted, so long polling is not the first choice anymore. Now, it is usually accompanied by other technologies or as a fallback. For example, SignalR uses long polling as a fallback when WebSocket and SSE are not available.

SSE

SSE is a server push technology that allows the server to send events to the web browser. SSE was first proposed in 2004 as part of the *WHATWG Web Applications 1.0*. It is based on the EventSource API, which is a standard API of HTML5. The Opera web browser implemented this feature in 2006. Now, all modern browsers support SSE.

In SSE, the client behaves as a subscriber, initializing the connection by creating a new JavaScript `EventSource` object, passing the URL of the endpoint to the server over a regular HTTP `GET` request with the media type of `text/event-stream`. Once connected, the server keeps the connection open and pushes new events separated by a newline character to the client, until it has no more events to send, or until the client explicitly closes the connection by calling the `EventSource.close()` method.

If the client lost the connection for any reason, it could reconnect to receive new events. To recover from the failure, the client can provide a `Last-Event-ID` header to the server to specify the last event ID that the client received. Then, the server can use this information to determine whether the client missed any events.

SSE is suitable for scenarios where real-time notifications have to be sent to the client when the data changes from the server, in order to keep a user interface in sync with the latest data state. Examples include Twitter updates, stock price updates, news feeds, alerts, and so on.

The limitation of SSE is that it is unidirectional, so it cannot be used to send data from the client to the server. Once the client connects to the server, it can receive responses only, but it cannot send new requests on the same connection. If you need bidirectional communication, WebSocket may be a better option.

WebSocket

WebSocket is a protocol that provides full-duplex communication between a client and a server within a single TCP connection. It allows the client to send requests to the server, while the server can push events and responses back to the client in real-time. WebSocket was first referenced as a TCP-based socket API in the HTML5 specification. In 2008, the WebSocket protocol was standardized by W3C. Google Chrome was the first browser to support WebSocket in 2009. Now, WebSocket is supported in most modern browsers, including Google Chrome, Microsoft Edge, Firefox, Safari, and Opera.

Unlike the HTTP protocol, WebSocket enables a two-way ongoing conversation between the client and server. The communication is usually done over a TCP port `443` connection (or `80` if there is an unsecure connection), so it can be easily configured in a firewall.

From a WebSocket perspective, the message content is opaque. A subprotocol is required to specify an agreement between the client and server. WebSocket can support both text and binary format subprotocols. As part of the initial handshake process, the client can specify which subprotocols it supports. Then, the server must pick one of the protocols that the client supports. This is called subprotocol negotiation. You can find many subprotocols officially registered here: `https://www.iana.org/assignments/websocket/websocket.xml`.

The WebSocket protocol defines `ws` and `wss` as the URI schemas that are used for unencrypted and encrypted connections, respectively. It is always recommended to use `wss` to make sure the transport security layer encrypts data. For example, we can use the following code to create a WebSocket connection in JavaScript:

```
const socket = new WebSocket('wss://websocket.example.com/ws/
updates');
```

WebSocket does not define how to manage the events for the connections, such as reconnection, authentication, and so on. The client and server would need to manage those events. There are various libraries to implement WebSocket for different languages. For example, *Socket.IO* (`https://socket.io`) is a popular library that implements WebSocket servers and clients in JavaScript, Java, Python, and so on.

WebSocket is a great choice for real-time communications, such as online games, sales updates, sports updates, online chat, real-time dashboards, and so on.

Uni-directional versus bidirectional

Uni-directional communication is like a radio. SSE is uni-directional because the server broadcasts data to clients, but clients cannot send data to the server. Bidirectional communication supports two-way communication. There are two types of bidirectional communication – half-duplex and full-duplex.

Half-duplex communication is like a walkie-talkie. Both the server and client can send messages to each other, but only one party may send messages at a time.

Full-duplex communication is like a telephone. The message can be sent from either side at the same time. WebSocket is full-duplex.

---

gRPC streaming

We introduced gRPC in the previous section. As we mentioned, gRPC is based on the HTTP/2 protocol, which provides a foundation for long-lived, real-time communication. Unlike HTTP/1.1, which requires a new TCP socket connection for each request, one HTTP/2 connection can be used for one or more simultaneous requests, so this avoids the overhead of creating new connections for each request. Also, HTTP/2 supports pushing data to clients without them having to request it. It is a huge improvement over the request/response pattern of HTTP/1.1.

gRPC takes advantage of the HTTP/2 protocol to support bidirectional communication. A gRPC service supports these streaming combinations:

- Unary (No streaming)

- Server-to-client streaming

- Client-to-server streaming

- Bidirectional streaming

Both gRPC and WebSocket support full-duplex communication, but, unlike WebSocket, gRPC uses `protobuf` by default, so it does not need to select a subprotocol. However, browsers have no built-in support for gRPC, so gRPC streaming is often used for service-to-service communication.

Which real-time communication technology is best for your application?

There are a couple of choices for your real-time application. So, how do we choose? It is important to note that it depends on the circumstances and constraints of your application. For example, do you need a push-only application or bidirectional communication? Do you want to support most browsers or just server-to-service communication? Do you need to push data to multiple clients or just one client?

Fortunately, Microsoft provides **SignalR** in ASP.NET Core to implement real-time communication. SignalR is an open-source library that enables real-time communication between clients and servers. It can automatically manage the connections and allow servers to send messages to all connected clients or a specific group of clients. Note that SignalR encapsulates multiple technologies, including WebSocket, SSE, and long polling. It hides the details and the complex implementations of these protocols. As a result, we do not need to worry about which technology is used for real-time communication. SignalR automatically chooses the best transport method for your application. WebSocket is the default protocol. If WebSocket is unavailable, SignalR will fall back to SSE, and then long polling.

SignalR is a good choice for these scenarios:

- When clients require high-frequency updates or alerts/notifications from the server – for example, games, social networks, voting, auctions, maps, and so on

- Dashboard and monitoring applications – for example, system dashboard applications, instant diagram applications, sales data monitoring applications, and so on

- Collaborative applications – for example, chat applications,whiteboard applications, and so on

ASP.NET Core also provides good support to gRPC. So, the next question is, how do you choose between gRPC and SignalR?

Here are some thoughts you may want to consider:

- If your need to build a real-time application that supports multiple clients (browsers), you may use SignalR because it is well-supported by browsers, and gPRC is not.

- If you need to build a distributed application or a microservices architecture application where you want to communicate between multiple servers, you may use gRPC because it is more suitable for server-to-server communication, and is more efficient than SignalR in this scenario.

Summary

In this chapter, we introduced some different API styles, including REST-based APIs, gRPC APIs, and GraphQL APIs, and explored how to design them. We also introduced a couple of different ways to implement real-time communication, including WebSocket, gRPC streaming, and SignalR. So far, we have not touched on much code, but we have reviewed the basic concepts of web APIs.

In the next chapter, we will start to learn how to use ASP.NET Core to implement them.

2

Getting Started with ASP.NET Core Web APIs

ASP.NET Core is a cross-platform, open-source web application framework for building modern, cloud-enabled web applications and APIs. It is primarily used with the C# programming language. ASP.NET Core provides features to help you build web apps in various ways – for example, through ASP.NET MVC, web APIs, Razor Pages, Blazor, and so on. This book will mainly cover web APIs. In this chapter, we will learn how to build a simple REST web API with ASP.NET Core.

In this chapter, we'll be covering the following topics:

- Setting up the development environment
- Creating a simple REST web API project
- Building and running the project
- Understanding the MVC pattern
- **Dependency injection (DI)**
- Introduction to minimal APIs

This chapter will provide you with the necessary information to create a basic REST web API project with ASP.NET Core. By the end of this chapter, you should have a better understanding of the steps required to create your first ASP.NET Core web API project.

Technical requirements

You are expected to know the basic concepts of **.NET Framework** or **.NET Core**, and **object-oriented programming (OOP)**. You should also have a basic understanding of the **C#** programming language. If you are not familiar with these concepts, you can refer to the following resources:

- **.NET fundamentals**: `https://learn.microsoft.com/en-us/dotnet/fundamentals/`

- **C#:** `https://learn.microsoft.com/en-us/dotnet/csharp/`
- **OOP (C#):** `https://learn.microsoft.com/en-us/dotnet/csharp/fundamentals/tutorials/oop`

The code examples in this chapter can be found at `https://github.com/PacktPublishing/Web-API-Development-with-ASP.NET-Core-8/tree/main/samples/chapter2`.

Setting up the development environment

.NET Core is fully cross-platform and can run on Windows, Linux, and macOS, so you can use any of these platforms to develop ASP.NET Core applications. The code samples in this book are written on Windows 11. However, you can run the same code on Linux and macOS.

There are also several IDEs available for ASP.NET Core, such as Visual Studio, **Visual Studio Code** (**VS Code**), Visual Studio for Mac, and Rider. In this book, we will mainly use VS Code.

> **Why not Visual Studio?**
>
> Visual Studio is a powerful IDE for the .NET platform. It provides a bunch of tools and features to elevate and enhance every stage of software development. However, VS Code is more lightweight and is open-source and cross-platform. We will use VS Code to understand the concepts of ASP.NET Core, then migrate to Visual Studio to use its rich features. If you are familiar with Visual Studio or any other IDE, feel free to use it.

Here is a list of software, SDKs, and tools you need to install:

- **VS Code:** `https://code.visualstudio.com/download`
- **.NET 8 SDK:** `https://dotnet.microsoft.com/en-us/download`

Both VS Code and the .NET 8 SDK are cross-platform, so please choose the correct one for your OS. When you install VS Code, please make sure you check the **Add to PATH** option.

If you use Windows, you may want to install **Windows Terminal** to run the command line. Windows Terminal is available for Windows 10 and above, and it provides a better user experience. But it is optional because you can also use the command line directly.

Configuring VS Code

Strictly speaking, VS Code is a code editor. It cannot recognize all the coding languages. Therefore, you'll need to install some extensions to support your development workflow. You can browse and install extensions by clicking on the **Extensions** icon in the **Activity** bar on the left-hand side of the VS Code interface. Then, you will see a list of the most popular extensions on the VS Code Marketplace:

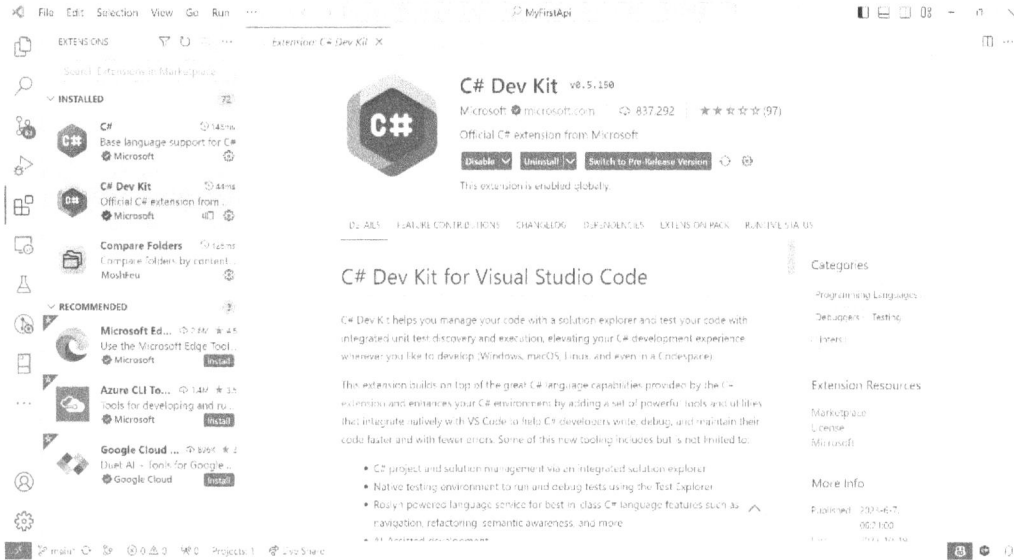

Figure 2.1 – Overview of the C# Dev Kit extension for VS Code

You need to install this extension to support .NET development:

- **C# Dev Kit**: This is the official C# extension for VS Code provided by Microsoft. When you install C# Dev Kit, the following extensions will automatically be installed:

  - **C# extension**: This extension provides C# language support powered by OmniSharp

  - **IntelliCode for C# Dev Kit**: This extension provides AI-assisted IntelliSense for C#

  - **.NET Runtime Install Tool**: This extension provides a unified way to install local, private versions of the .NET runtime

The C# Dev Kit extension provides a lot of features to help you develop .NET applications. Press *Ctrl + Shift + P* (on Windows) or *Command + Shift + P* (on macOS) to open the Command Palette, then type .net to see the commands provided by the C# Dev Kit extension. You can use these commands to create new projects, generate assets for build and debug, run tests, and more.

You can also install the following extensions to improve your productivity:

- **EditorConfig for VS Code**: This extension provides EditorConfig support for VS Code. EditorConfig helps teams of multiple developers maintain consistent coding styles when working on the same project across various editors and IDEs.

- **GitHub Copilot**: GitHub Copilot is your AI pair programmer. You can get code suggestions in real-time based on your context and comments in VS Code. This extension is not free, but you can try it for free for 30 days. If you are a student, a teacher, or a maintainer of a popular open-source project, you can get it for free.

To configure EditorConfig, you can create a file named `.editorconfig` in the root folder of the project. You can find a sample EditorConfig file at `https://learn.microsoft.com/en-us/dotnet/fundamentals/code-analysis/code-style-rule-options`.

Checking the .NET SDK

Once you install the .NET SDK, you can check the version by running the following command:

```
dotnet --version
```

You should be able to see the version number as follows:

```
8.0.101-rc.2.23502.2
```

Microsoft releases new versions of .NET SDKs frequently. If you encounter a different version number, that is acceptable.

You can list all available SDKs by running the following command:

```
dotnet --list-sdks
```

The preceding command will list all the available SDKs on your machine. For example, it may show the following output if have multiple .NET SDKs installed:

```
6.0.415 [C:\Program Files\dotnet\sdk]
7.0.402 [C:\Program Files\dotnet\sdk]
8.0.100 [C:\Program Files\dotnet\sdk]
8.0.101 [C:\Program Files\dotnet\sdk]
```

Multiple versions of .NET SDKs can be installed at the same time. We can specify the version of the .NET SDKs in the project file.

> **Which version of the SDKs should I use?**
>
> Every Microsoft product has a lifecycle. .NET and .NET Core provides **Long-term support (LTS)** releases that get 3 years of patches and free support. When this book was written, .NET 7 is still supported, until May 2024. Based on Microsoft's policy, even numbered releases are LTS releases. So .NET 8 is the latest LTS release. The code samples in this book are written with .NET 8.0.
>
> To learn more about .NET support policies, please visit `https://dotnet.microsoft.com/en-us/platform/support/policy`.

We are now prepared to start developing ASP.NET Core applications. Let's get to work!

Creating a simple REST web API project

In this section, we will use the **.NET command-line interface (.NET CLI)** to create a basic web API project and see how it works.

The .NET CLI is a command-line tool that helps you to create, develop, build, run, and publish .NET applications. It is included in the .NET SDK.

You have multiple ways to run .NET CLI commands. The most common way is to run the command in the terminal window or command prompt. Also, you can run the command in VS Code directly. VS Code provides an integrated terminal that starts at the root of your workspace. To open the terminal in VS Code, you can do any one of the following:

- Press *Ctrl* + ` (on Windows) or *Command* + ` (on macOS) to open the terminal
- Use the **View** | **Terminal** menu item to open the terminal
- From the Command Palette, use the **View: Toggle Terminal** command to open the terminal

In the terminal, navigate to a folder where you want to create the project, then create a web API project by running the following command:

```
dotnet new webapi -n MyFirstApi -controllers
cd MyFirstApi
code .
```

The preceding commands create a new web API project and open it in VS Code. `dotnet new` provides many options to create various types of projects, such as web APIs, console apps, class libraries, and so on.

There are some options we can use to specify the project:

- `-n|--name <OUTPUT_NAME>`: The name for the created output. If not specified, the name of the current directory is used.

- `-o|--output <OUTPUT_PATH>`: The output path for the created project. If not specified, the current directory is used.

- `-controllers|--use-controllers`: Indicates whether to use controllers for actions. If not specified, the default value is `false`.

- `-minimal|--use-minimal-apis`: Indicates whether to use minimal APIs. The default value is `false`, but the `-controllers` option will override the `-minimal` option. If neither `-controllers` nor `-minimal` is specified, the default value of the `-controllers` option, which is `false`, will be used, so a minimal API will be created.

> **Important note**
>
> Since .NET 6.0, ASP.NET Core 6.0 provides a new way to create web API projects, which is called **minimal APIs**. It is a simplified approach for building APIs without controllers. We will introduce minimal APIs later. For now, we will use the traditional way to create a web API project with controllers. So, we need to specify the `--use-controllers` option.

To learn more about the `dotnet new` command, check this page: `https://docs.microsoft.com/en-us/dotnet/core/tools/dotnet-new`. We will introduce more details on the `dotnet` command in the following sections.

When you use VS Code to open the project, the C# Dev Kit extension can create a solution file for you. This feature makes VS Code more friendly to C# developers. You can see the following structure in the Explorer view:

The reason is that VS 2022 will create a sln file for the project, but .NET CLI does not. When using VS Code to open the project, the C# DevKit will create the sln file. I think it's worth mentioning it here.

The C# Dev Kit extension provides a new feature, the solution explorer, which is located at the bottom. This feature is especially useful when working with multiple projects in one solution. You can drag and drop the **SOLUTION EXPLORER** to the top to make it more visible.

When you use VS Code to open the project, the C# Dev Kit extension can create a solution file for you. This feature makes VS Code more friendly to C# developers. You can see the following structure in the Explorer view:

EXPLORER   ...

∨ OPEN EDITORS

∨ MYFIRSTAPI

    > bin

    > Controllers

    > Models

    > obj

    ∨ Properties

       {} launchSettings.json

    > Services

    {} appsettings.Development.json

    {} appsettings.json

    ≋ MyFirstApi.csproj   M

    MyFirstApi.http   U

> OUTLINE

> TIMELINE

∨ SOLUTION EXPLORER

  ∨ ▤ MyFirstApi   U

    ∨ ▥ MyFirstApi   ⬚ ⬚ M

      > ⬡ Dependencies

      > Properties

      > Controllers

      > Models

      > Services

      > {} appsettings.json

      ≋ MyFirstApi.http   U

      C# Program.cs

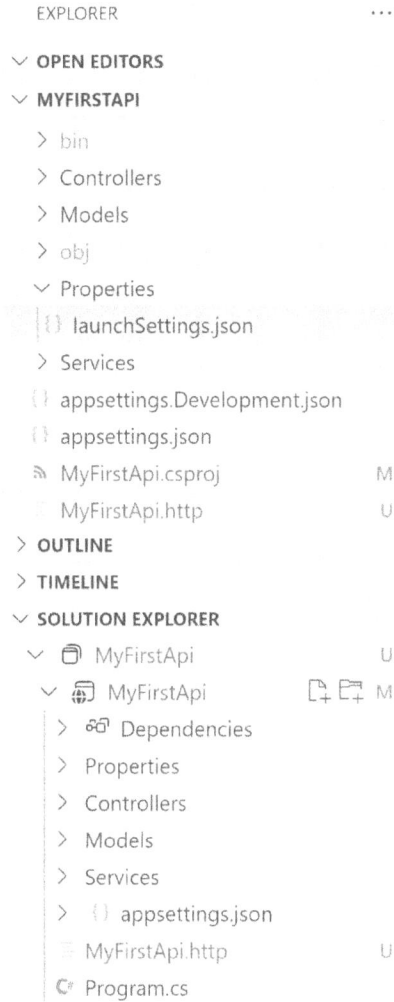

Figure 2.2 – The solution explorer and the folder structure

Next, we can start to build and run the project.

Building and running the project

In this section, we will learn how to build and run the project and introduce some useful tools to help you test the APIs. To make it compatible with all platforms, we will use .NET CLI commands to build and run the project. We will also learn how to debug the project in VS Code.

Building the project

The easiest way to build and run the project is to use the `dotnet` command. You can run the following command to build the project:

```
dotnet build
```

The preceding command will build the project and its dependencies and generate a set of binaries. You can find these binaries in the `bin` folder. The `bin` folder is the default output folder for the `dotnet build` command. You can use the `--output` option to specify the output folder. However, it is recommended to use the default `bin` folder. The binaries are some **Intermediate Language** (**IL**) files with a `.dll` extension.

You might see the following popups when you use VS Code to open the project:

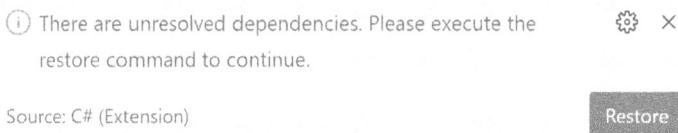

Figure 2.3 – VS Code prompts to restore dependencies

This is because VS Code inspects that the project is a .NET project, and it is trying to restore the dependencies. You can click the **Restore** button to restore the dependencies. Similarly, if you see other prompts from VS Code to add assets to debug the project, please select **Yes** in the dialog:

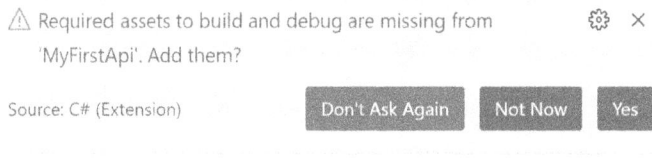

Figure 2.4 – VS Code prompts to add required assets to build and debug

Some commands, such as `dotnet build`, `dotnet run`, `dotnet test`, and `dotnet publish`, will implicitly restore dependencies. So don't worry if you missed out on these prompts.

If no errors or warnings are shown, that means the build is successful.

Running the project

You can run the following command to run the project:

```
dotnet run
```

The `dotnet run` command is a convenient way of running the project from the source code. Keep in mind that it is useful in development, but not for production. The reason is that if the dependencies are outside of the shared runtime, the `dotnet run` command will resolve the dependencies from the NuGet cache. To run the application in production, you need to create a deployment package with the `dotnet publish` command and deploy it. We will explore the deployment process in future chapters.

You should be able to see the following output:

```
Building...
info: Microsoft.Hosting.Lifetime[14]
      Now listening on: https://localhost:7291
info: Microsoft.Hosting.Lifetime[14]
      Now listening on: http://localhost:5247
info: Microsoft.Hosting.Lifetime[0]
      Application started. Press Ctrl+C to shut down.
info: Microsoft.Hosting.Lifetime[0]
      Hosting environment: Development
info: Microsoft.Hosting.Lifetime[0]
      Content root path: C:\example_code\chapter2\MyFirstApi\
MyFirstApi
info: Microsoft.Hosting.Lifetime[0]
      Application is shutting down...
```

There is a link in the output, such as `http://localhost:5247`. The port number was randomly generated when we created the project. In a browser, navigate to `http://localhost:<your_port>/swagger`. You will see the web API documentation with **Swagger UI**, which offers a web-based UI to provide information and tools to interact with the API. You can use Swagger UI to test APIs:

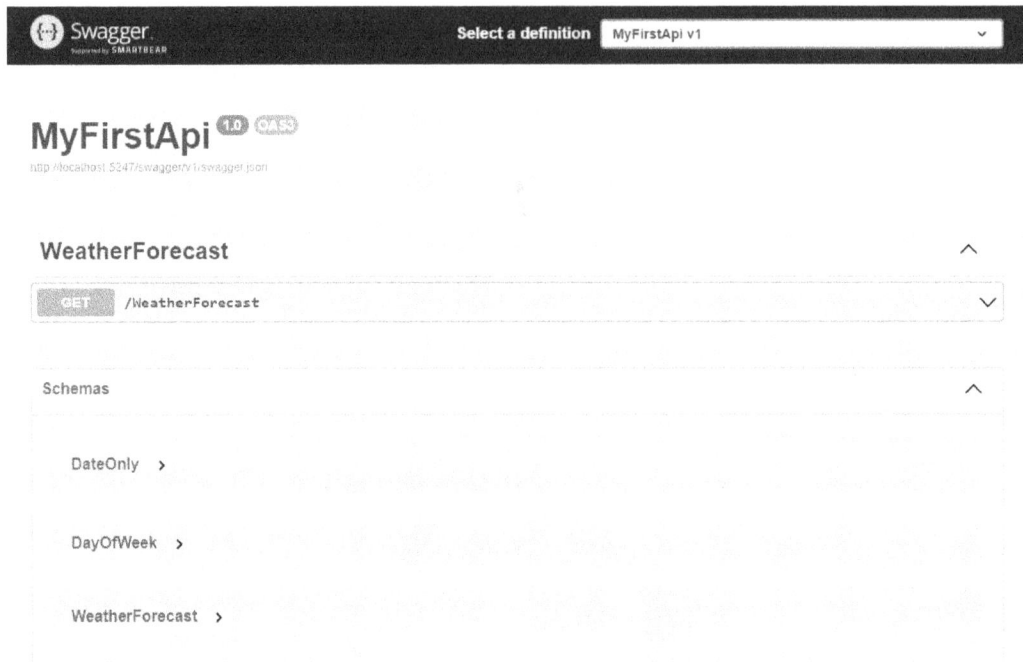

Figure 2.5 – Swagger UI

The API project is now running! You can see the web API template provides a /WeatherForecast endpoint. If you navigate to the http://localhost:5247/WeatherForecast link in the browser, you will see the API response.

To support HTTPS, you may need to trust the HTTPS development certificate by running the following command:

```
dotnet dev-certs https --trust
```

You will see a dialog if the certificate was not previously trusted. Select **Yes** to trust the development certificate:

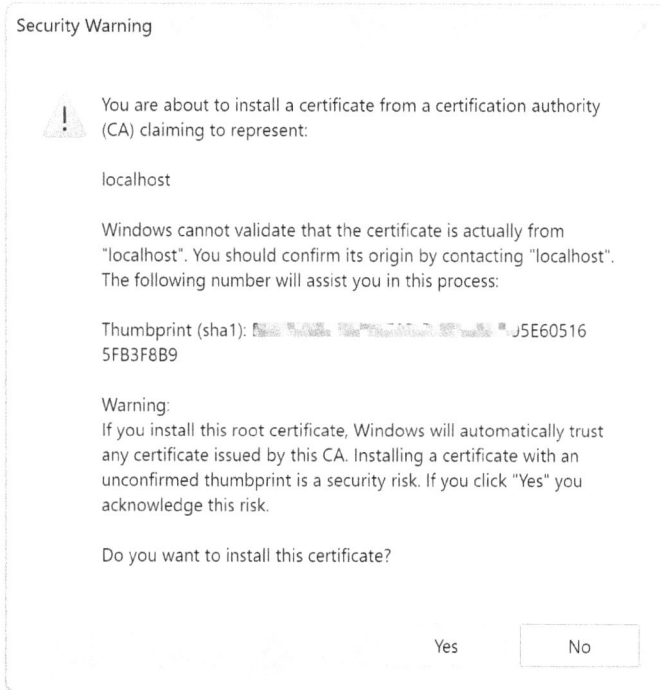

Figure 2.6 – Installing the certificate for local development

Please note that the preceding command does not work on Linux. See your Linux distribution documentation for more details.

Changing the port number

The port number is defined in the `launchSettings.json` file in the `Properties` folder. You can change the port number by editing the file. Based on the convention, when the web API project was created, a port from `5000` to `5300` will be selected for HTTP, and from `7000` to `7300` for HTTPS. Here is an example of the `launchSettings.json` file:

```
{
  "$schema": "https://json.schemastore.org/launchsettings.json",
  ...
  "profiles": {
    "https": {
      "commandName": "Project",
      "dotnetRunMessages": true,
      "launchBrowser": true,
      "launchUrl": "swagger",
```

```
      "applicationUrl": "https://localhost:7291;http://
localhost:5247",
      "environmentVariables": {
        "ASPNETCORE_ENVIRONMENT": "Development"
      }
    },
    ...
  }
}
```

You can update the port number here. Just keep in mind that the port number should be unique on your machine to avoid conflicts.

Hot Reload

When you use `dotnet run` to run the project, if you change the code, you need to stop the project and start it again. If your project is complicated, it takes time to stop and restart. To speed up development, you can use the `dotnet watch` command to enable the Hot Reload feature.

.NET Hot Reload is a feature that allows you to apply code changes to a running app without restarting the app. It was first provided with .NET 6. Instead of using `dotnet run`, you can use `dotnet watch` to activate Hot Reload in development. Once you update the code, the web browser will automatically refresh the page. However, Hot Reload does not support all code changes. In some cases, `dotnet watch` will ask you if you want to restart the application. There are some options: `Yes`, `No`, `Always`, and `Never`. Choose the appropriate option for the code change you want to apply, as shown next:

```
dotnet watch   File changed: .\Services\IService.cs.
dotnet watch   Unable to apply hot reload because of a rude edit.
   Do you want to restart your app - Yes (y) / No (n) / Always (a) /
Never (v)?
```

The API project is now running, and we can start to test the API.

Testing the API endpoint

The browser can send a `GET` request easily, but it is not as simple for `POST` endpoints. There are various ways to call the API for testing purposes, such as Swagger UI, Postman, and other tools. In this section, we will introduce some tools you can use in the development stage.

Swagger UI

We introduced how to use SwaggerHub to design APIs in *Chapter 1*. From version 5.0, ASP.NET Core enables OpenAPI support by default. It uses the `Swashbuckle.AspNetCore` NuGet package, which provides the Swagger UI to document and test the APIs.

We can use Swagger UI to test the API directly. Expand the first `/WeatherForecast` API in Swagger UI and click the **Try it out** button. You will see an **Execute** button. Click the button, and you will see the following response:

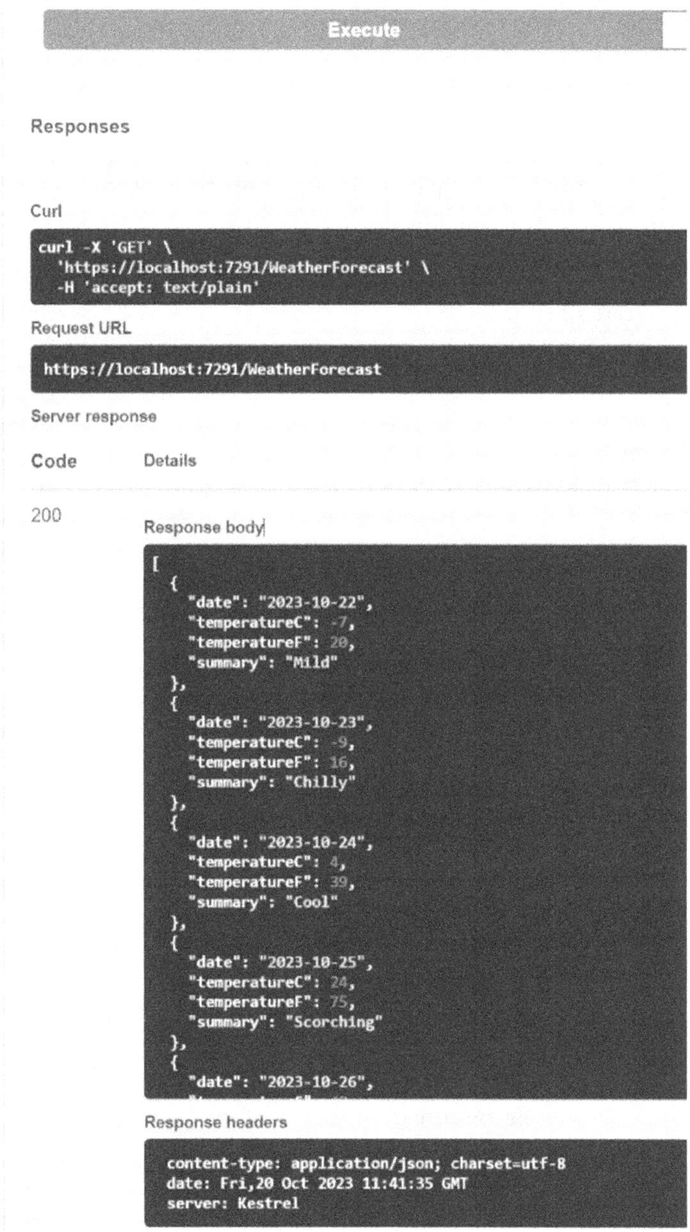

Figure 2.7 – Testing an endpoint in Swagger UI

Figure 2.7 demonstrates that the API is functioning correctly and is providing the expected response. To learn more about Swagger and OpenAPI, you can check the following links:

- **Swagger**: `https://swagger.io/`
- **OpenAPI**: `https://www.openapis.org/`
- **SmartBear**: `https://www.smartbear.com/`

Postman

Postman is a powerful API platform for building and using APIs. It is widely used by many individual developers and organizations. You can download it here: `https://www.postman.com/downloads/`.

Click the + button to create a new tab. Use `http://localhost:5247/WeatherForecast` as the URL. Then, click the **Send** button. You will see the response next:

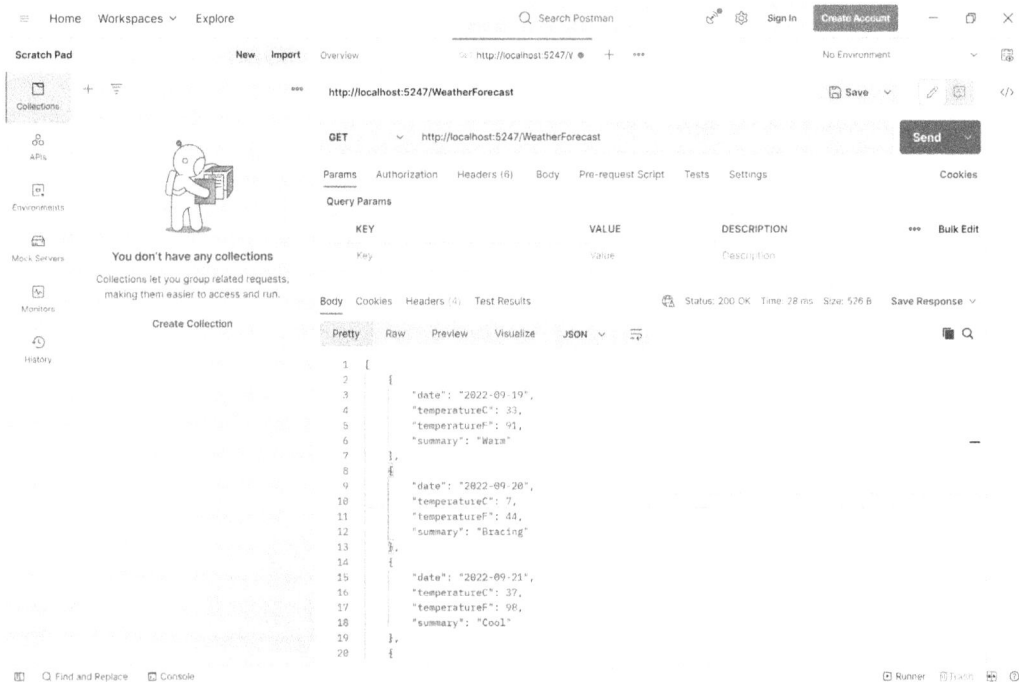

Figure 2.8 – Using Postman to call the API

Postman provides a rich set of features to test APIs. To learn more about Postman, check the official documentation: `https://learning.postman.com/docs/getting-started/introduction/`.

HttpRepl

HTTP Read-Eval-Print Loop (**HttpRepl**) is a command-line tool that allows you to test APIs. It is lightweight and cross-platform, so it can run anywhere. It supports the GET, POST, PUT, DELETE, HEAD, OPTIONS, and PATCH HTTP verbs.

To install HttpRepl, you can use the following command:

```
dotnet tool install -g Microsoft.dotnet-httprepl
```

After the installation, you can use the following command to connect to our API:

```
httprepl <ROOT URL>/
```

<ROOT URL> is the base URL of the web API, such as the following:

```
httprepl http://localhost:5247/
```

After the connection is built, you can use the ls or dir command to list the endpoints, such as the following:

```
http://localhost:5247/> ls
.                    []
WeatherForecast    [GET]
```

The preceding command shows the WeatherForecast endpoint supports a GET operation. Then, we can use the cd command to navigate to the endpoint, such as the following:

```
http://localhost:5247/> cd WeatherForecast
/WeatherForecast     [GET]
```

Then, we can use the get command to test the endpoint, such as the following:

```
http://localhost:5247/WeatherForecast> get
```

The output looks like this:

```
http://localhost:5247/> ls
.                     []
WeatherForecast    [GET]

http://localhost:5247/> cd WeatherForecast
/WeatherForecast      [GET]

http://localhost:5247/WeatherForecast> get
HTTP/1.1
Content-Type: application/json; charset=utf-8
Date: Fri, 20 Oct 2023 20:04:14 GMT
Server: Kestrel
Transfer-Encoding: chunked

[
  {
    "date"  "2023-10-22"
    "temperatureC"  41
    "temperatureF"  105
    "summary"  "Balmy"
  }
  {
    "date"  "2023-10-23"
    "temperatureC"  7
    "temperatureF"  44
    "summary"  "Scorching"
  }
  {
    "date"  "2023-10-24"
    "temperatureC"  20
    "temperatureF"  67
    "summary"  "Sweltering"
  }
```

Figure 2.9 – Output of HttpRepl

To disconnect, press *Ctrl + C* to exit.

You can find more information about HttpRepl at `https://docs.microsoft.com/en-us/aspnet/core/web-api/http-repl/`.

Thunder Client

If you prefer to do everything in VS Code, **Thunder Client** is a great tool for testing APIs. Thunder Client is a lightweight REST API client extension for VS Code, allowing users to test their APIs without having to leave VS Code. This makes it an ideal choice for developers who want to streamline their workflow:

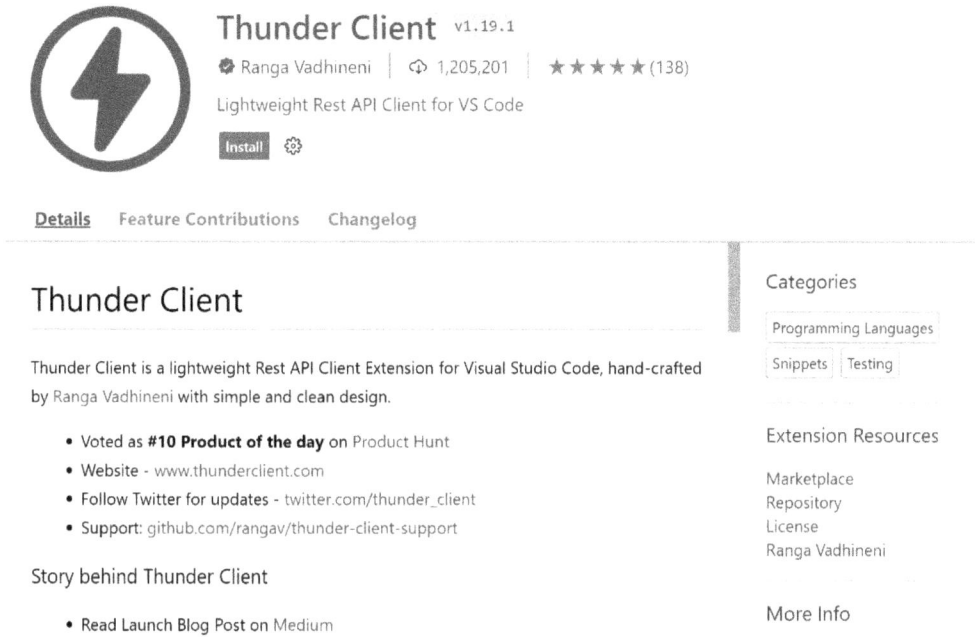

Figure 2.10 – The Thunder Client extension for VS Code

After the installation, click the **Thunder Client** icon on the **Action** bar. From the sidebar, click the **New Request** button. The following UI will be shown:

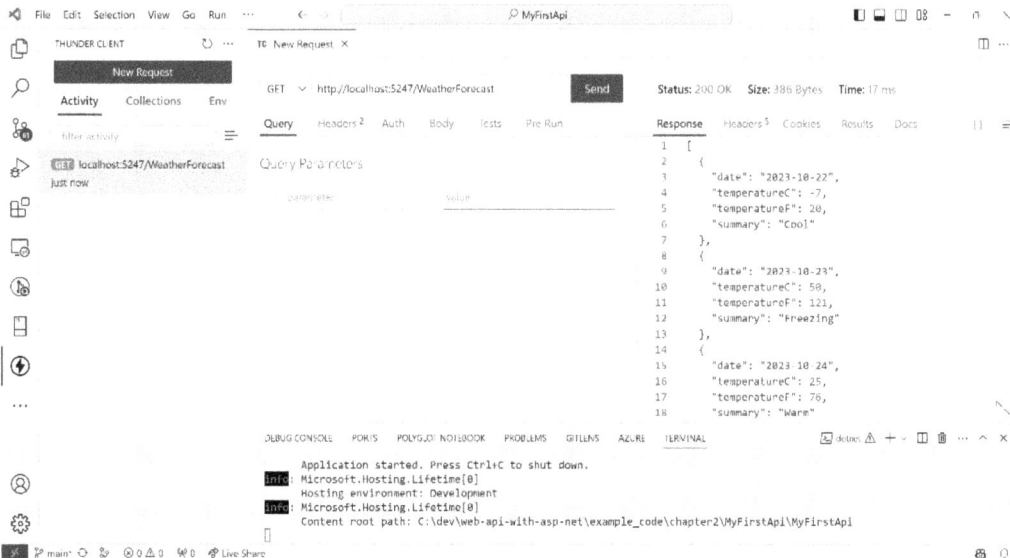

Figure 2.11 – Testing the API with Thunder Client

To learn more about Thunder Client, visit their GitHub page: `https://github.com/rangav/thunder-client-support`.

Using .http files in VS 2022

If you use Visual Studio 2022, you can use the `.http` file to test the API. The `.http` file is a text file that contains definitions of HTTP requests. The latest ASP.NET Core 8 template project provides a default `.http` file. You can find it in the `MyFirstApi` folder. The content of the file is as follows:

```
@MyFirstApi_HostAddress = http://localhost:5247


GET {{MyFirstApi_HostAddress}}/weatherforecast/
Accept: application/json


###
```

The first line defines a variable named `MyFirstApi_HostAddress` with the value of the root URL of the API. The second line defines a `GET` request to the `/weatherforecast` endpoint. The third line defines an `Accept` header. In this case, it accepts the `application/json` content type. Open this file in Visual Studio 2022, and you will see the **Send Request** button on the left side of the request. Click the button, and you will see the response as follows:

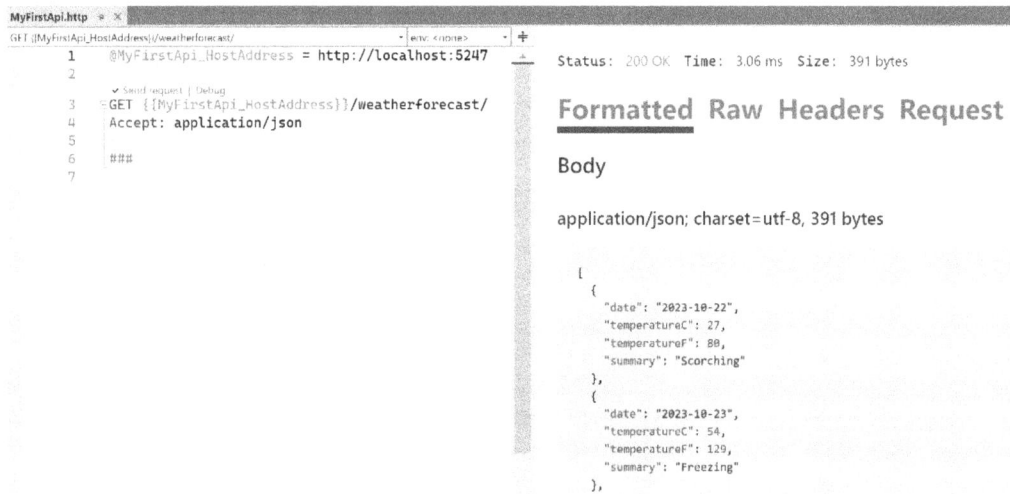

Figure 2.12 – Using the .http file to test the API in Visual Studio 2022

However, when this book was written, the `.http` files lacked some features, such as environment variables. Also, this feature is only available in Visual Studio 2022, so we will not use it in this book. But if you are interested in exploring this feature further, please refer to the Microsoft docs at `https://learn.microsoft.com/en-us/aspnet/core/test/http-files` for more information.

We have introduced some tools to test APIs. Let's now learn how to debug APIs.

Debugging

VS Code has a built-in debugging feature that allows you to debug code. Unlike Visual Studio, it needs a `launch.json` configuration for debugging. When you open an ASP.NET Core project in VS Code, it will prompt you to add some assets. If you choose **Yes**, VS Code can generate a `launch.json` file in the `.vscode` folder.

If you missed it, you can add it manually from the Debug view:

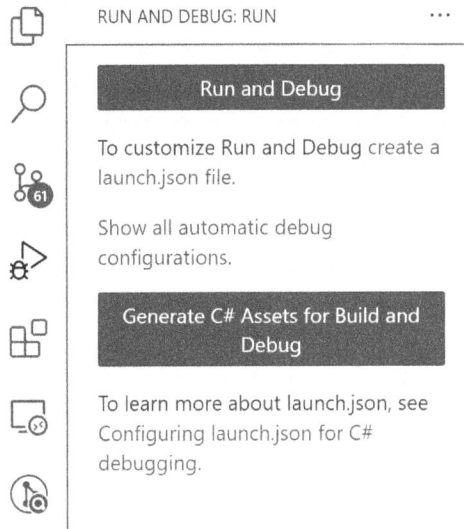

Figure 2.13 – Creating a launch.json file from the Debug view

If you cannot see the buttons in *Figure 2.13*, you can open the Command Palette by pressing *Ctrl + Shift + P* (on Windows) or *Command + Shift + P* (on macOS), then type `.net` and choose **.NET: Generate Assets for Build and Debug**. It will then generate a `launch.json` file in the `.vscode` folder:

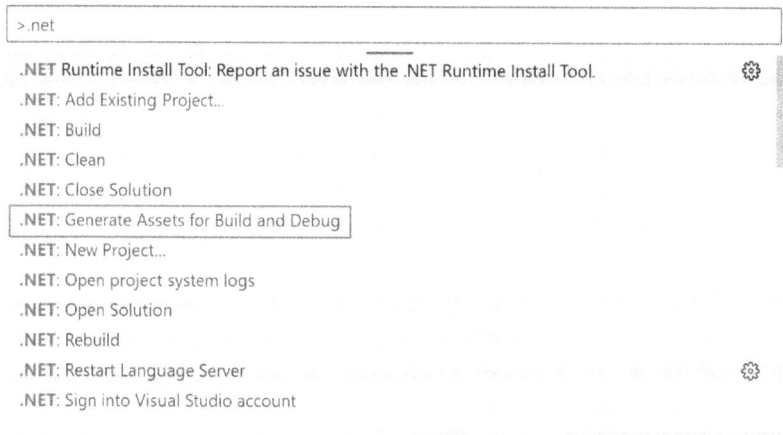

Figure 2.14 – Generating a launch.json file from the Command Palette

The content of the default `launch.json` configuration is shown next:

```
{
    "version": "0.2.0",
    "configurations": [
        {
            // Use IntelliSense to find out which attributes
exist for C# debugging
            // Use hover for the description of the existing
attributes
            // For further information visit https://github.
com/dotnet/vscode-csharp/blob/main/debugger-launchjson.md.
            "name": ".NET Core Launch (web)",
            "type": "coreclr",
            "request": "launch",
            "preLaunchTask": "build",
            // If you have changed target frameworks, make sure
to update the program path.
            "program": "${workspaceFolder}/bin/Debug/net8.0/
MyFirstApi.dll",
            "args": [],
            "cwd": "${workspaceFolder}",
            "stopAtEntry": false,
            // Enable launching a web browser when ASP.NET
Core starts. For more information: https://aka.ms/VSCode-CS-
LaunchJson-WebBrowser
            "serverReadyAction": {
                "action": "openExternally",
```

```
                    "pattern": "\\bNow listening on:\\
s+(https?://\\S+)"
                },
                "env": {
                    "ASPNETCORE_ENVIRONMENT": "Development"
                },
                "sourceFileMap": {
                    "/Views": "${workspaceFolder}/Views"
                }
            },
            {

                "name": ".NET Core Attach",
                "type": "coreclr",
                "request": "attach"
            }
        ]
    }
```

This file specifies the configuration for debugging. Some important attributes are described next:

- The program attribute specifies the path to the executable file

- The args attribute specifies the arguments to pass to the executable file

- The cwd attribute specifies the working directory

- The env attribute specifies the environment variables

We do not need to change anything in this file now.

Set up a breakpoint in the app. For example, we can set a breakpoint in the Get() method in the WeatherForecastController.cs file by clicking in the left margin of the code window. Once the breakpoint is set, you will see a red dot before the line number in the left margin:

```
21      [HttpGet(Name = "GetWeatherForecast")]
        0 references
22  ∨   public IEnumerable<WeatherForecast> Get()
23      {
● 24  ∨     return Enumerable.Range(1, 5).Select(index => new WeatherForecast
25          {
26              Date = DateOnly.FromDateTime(DateTime.Now.AddDays(index)),
27              TemperatureC = Random.Shared.Next(-20, 55),
28              Summary = Summaries[Random.Shared.Next(Summaries.Length)]
29          })
30          .ToArray();
31      }
32  }
```

Figure 2.15 – Setting up a breakpoint in VS Code

To debug the app, open the **Debug** view by selecting the **Debugging** icon on the left-side menu. Make sure you choose the correct debugging configuration from the drop-down menu. For this case, please select **.NET Core Launch (web)**. Then, select the green arrow at the top of the pane:

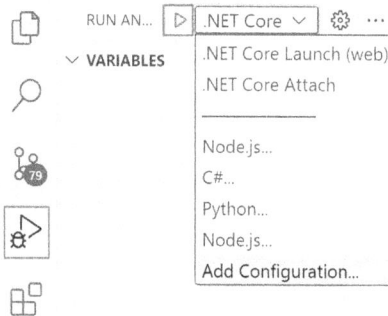

Figure 2.16 – Debugging the API in VS Code

Send a request from any of the tools in the previous section, and you will see the program execution stops when it reaches the breakpoint, as shown next:

Figure 2.17 – Hitting the breakpoint in VS Code

The **Locals** section of the **VARIABLES** window will display the values of variables that are defined in the current context.

You can also type a variable in the **DEBUG CONSOLE** window to check the value directly. To execute the next step, you can use the control toolbar at the top of the VS Code window. You can run the code line by line to monitor its execution. It is helpful if we need to know how the program works.

Now we have learned how to build, run, and test APIs, it is time to look at the code of APIs.

Understanding the MVC pattern

ASP.NET Core MVC is a rich framework for building web applications with the **Model-View-Controller** (**MVC**) design pattern. The MVC pattern enables web applications to separate the presentation from the business logic. An ASP.NET Core web API project follows the basic MVC pattern, but it does not have views, so it only has a Model layer and a Controller layer. Let's look at this in a bit more detail:

- **Models**: Models are classes that represent the data that is used in the application. Normally, the data is stored in a database.
- **Controllers**: Controllers are classes that handle the business logic of the application. Based on the convention of ASP.NET Core, controllers are stored in the `Controllers` folder. *Figure 2.18* shows an example of the MVC pattern in an web API project. However, the view layer is not included in the web API project. The request from the client will be mapped to the controller, and the controller will execute the business logic and return the response to the client.

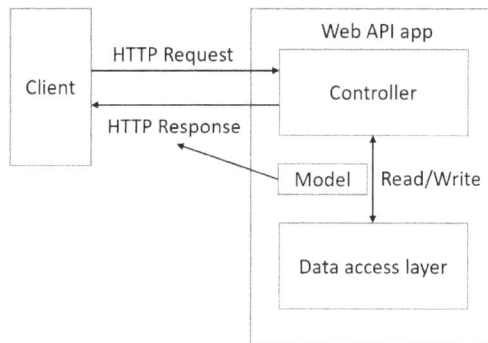

Figure 2.18 – The MVC pattern

Next, we will look at the code of the model and the controller in an ASP.NET Core web API project.

The model and the controller

In the ASP.NET Core template project, you can find a file named `WeatherForecast.cs`. This file is a model. It is a pure C# class that represents a data model.

The controller is the `WeatherForecastController.cs` file located in the `Controllers` folder. It contains the business logic.

It looks like this:

```
[ApiController]
[Route("[controller]")]
public class WeatherForecastController : ControllerBase
{
    // Some code is ignored
    private readonly ILogger<WeatherForecastController> _logger;

    public
WeatherForecastController(ILogger<WeatherForecastController> logger)
    {
        _logger = logger;
    }

    [HttpGet(Name = "GetWeatherForecast")]
    public IEnumerable<WeatherForecast> Get()
    {
        return Enumerable.Range(1, 5).Select(index => new
WeatherForecast
        {
            Date = DateOnly.FromDateTime(DateTime.Now.AddDays(index)),
            TemperatureC = Random.Shared.Next(-20, 55),
            Summary = Summaries[Random.Shared.Next(Summaries.Length)]
        })
        .ToArray();
    }
}
```

The constructor of the controller class has a parameter named `ILogger<WeatherForecastController> logger`. This parameter is used to log messages. It is injected with DI by the ASP.NET Core framework. We will talk about DI in the next section.

This class has an `[ApiController]` attribute that indicates that it is a web API controller. It also has a `[Route("[controller]")]` attribute that indicates the URL of the controller.

The `Get()` method has a `[HttpGet(Name = "GetWeatherForecast")]` attribute that indicates the name of the endpoint, and the `Get()` method is a GET operation. This method returns a list of weather forecasts as the response.

Note that the `[Route("[controller]")]` attribute is marked on the controller class. It means the path of the controller is `/WeatherForecast`. Currently, there is no `[Route]` attribute on the `Get()` method. We will learn more about routing in future sections.

We should now have a basic understanding of how ASP.NET Core web API works. The client sends the request to the web API, and the request will be mapped to the controller and the method. The controller will execute the business logic and return the response. We can use some methods to get, save, update, and delete data from the database in the controllers.

Next, let us create a new API endpoint by adding a new model and controller.

Creating a new model and controller

In *Chapter 1*, we showed an example REST API on `https://jsonplaceholder.typicode.com/posts`. It returns a list of posts, as shown next:

```
[
  {
    "userId": 1,
    "id": 1,
    "title": "sunt aut facere repellat provident occaecati excepturi
optio reprehenderit",
    "body": "quia et suscipit\nsuscipit recusandae consequuntur
expedita et cum\nreprehenderit molestiae ut ut quas totam\nnostrum
rerum est autem sunt rem eveniet architecto"
  },
  {
    "userId": 1,
    "id": 2,
    "title": "qui est esse",
    "body": "est rerum tempore vitae\nsequi sint nihil reprehenderit
dolor beatae ea dolores neque\nfugiat blanditiis voluptate porro vel
nihil molestiae ut reiciendis\nqui aperiam non debitis possimus qui
neque nisi nulla"
  },
  ...
]
```

Let us implement a similar API. First, we need to create a new model. Create a new folder named `Models` in the project. Then, create a new file named `Post.cs` in the `Models` folder:

```
namespace MyFirstApi.Models;

public class Post
{
    public int UserId { get; set; }
    public int Id { get; set; }
    public string Title { get; set; } = string.Empty;
    public string Body { get; set; } = string.Empty;
}
```

> **File-scoped namespace declaration**
>
> From C# 10, you can use a new form of namespace declaration, as shown in the previous code snippet, which is called a file-scoped namespace declaration. All the members in this file are in the same namespace. It saves space and reduces indentation.

> **Nullable reference types**
>
> You may be wondering why we assign an empty string to the `Title` and `Body` properties. This is because the properties are of type `string`. If we do not initialize the property, the compiler will complain:
>
> ```
> Non-nullable property 'Title' must contain a non-null value
> when exiting constructor. Consider declaring the property as
> nullable.
> ```
>
> By default, the ASP.NET Core web API project template enabled the **nullable reference types** annotation in the project properties. If you check the project file, you will find `<Nullable>enable</Nullable>` in the `<PropertyGroup>` section.
>
> Nullable reference types were introduced in C# 8.0. They can minimize the likelihood of errors that cause the runtime to throw a `System.NullReferenceException` error. For example, if we forget to initialize the `Title` property, we may get a `System.NullReferenceException` error when we try to access a property of it, such as `Title.Length`.
>
> With this feature enabled, any variable of a reference type is considered to be non-nullable. If you want to allow a variable to be nullable, you must append the type name with the ? operator to declare the variable as a nullable reference type; for example, `public string Title? { get; set; }`, which explicitly marks the property as nullable.
>
> To learn more about this feature, see `https://docs.microsoft.com/en-us/dotnet/csharp/nullable-references`.

Next, create a new file named `PostController.cs` in the `Controllers` folder. You can manually add it, or install the `dotnet-aspnet-codegenerator` tool to create it. To install the tool, run the following commands from the project folder:

```
dotnet add package Microsoft.VisualStudio.Web.CodeGeneration.Design
dotnet tool install -g dotnet-aspnet-codegenerator
```

The preceding commands install a NuGet package required for scaffolding. The `dotnet-aspnet-codegenerator` tool is a scaffolding engine that is used to generate code.

Then, run the following command to generate a controller:

```
dotnet-aspnet-codegenerator controller -name PostsController -api
-outDir Controllers
```

The preceding command generates an empty controller. The `-name` option specifies the name of the controller. The `-api` option indicates that the controller is an API controller. The `-outDir` option specifies the output directory. Update the content of the controller as follows:

```
using Microsoft.AspNetCore.Mvc;
using MyFirstApi.Models;

namespace MyFirstApi.Controllers;

[Route("api/[controller]")]
[ApiController]
public class PostsController : ControllerBase
{
    [HttpGet]
    public ActionResult<List<Post>> GetPosts()
    {
        return new List<Post>
        {
            new() { Id = 1, UserId = 1, Title = "Post1", Body = "The
first post." },
            new() { Id = 2, UserId = 1, Title = "Post2", Body = "The
second post." },
            new() { Id = 3, UserId = 1, Title = "Post3", Body = "The
third post." }
        };
    }
}
```

Target-typed new expressions

When we create a new `List` instance of a specific type, we will normally use code like this:

```
var list = new List<Post>
{

 new Post() { Id = 1, UserId = 1, Title = "Post1", Body = "The
first post." },

};
```

When the list is declared as `List<Post>`, the type is known, so it is not necessary to use `new Post()` here when adding new elements. The type specification can be omitted for constructors, such as `new()`. This feature was introduced in C# 9.0.

The controller is named `PostsController`. The convention is the resource name with the `Controller` suffix. It is marked with the `ApiController` attribute, which indicates that the controller is a web API controller. It also has a `[Route("api/[controller]")]` attribute that indicates the URL of the controller. `[controller]` is like a placeholder, which will be replaced with the name of the controller in the routing. So, the route of this controller is `/api/posts`.

In this controller, we have a method named `GetPosts()`. This method returns a list of posts as the response. The method is marked with the `[HttpGet]` attribute, which indicates that this method is a GET operation. It does not have any route template, because it will match `/api/posts`. For other methods, we can use the `[Route("[action]")]` attribute to specify the route template.

The return type of the `GetPosts()` method is `ActionResult<IEnumerable<Post>>`. ASP.NET Core can automatically convert the object to JSON and return it to the client in the response message. Also, it can return other HTTP status codes, such as `NotFound`, `BadRequest`, `InternalServerError`, and so on. We will see more examples later.

If you run `dotnet run` or `dotnet watch`, then navigate to Swagger UI, such as `https://localhost:7291/swagger/index.html`, you will see the new API listed. The API is accessible at `/api/posts`.

Currently, the `/api/posts` endpoint returns a hardcoded list of posts. Let us update the controller to return a list of posts from a service.

Creating a service

Create a `Services` folder in the project. Then, create a new file named `PostService.cs` in the `Services` folder, as shown next:

```
using MyFirstApi.Models;

namespace MyFirstApi.Services;

public class PostsService
{
    private static readonly List<Post> AllPosts = new();

    public Task CreatePost(Post item)
    {
        AllPosts.Add(item);
        return Task.CompletedTask;
    }

    public Task<Post?> UpdatePost(int id, Post item)
```

```
    {
        var post = AllPosts.FirstOrDefault(x => x.Id == id);
        if (post != null)
        {
            post.Title = item.Title;
            post.Body = item.Body;
            post.UserId = item.UserId;
        }

        return Task.FromResult(post);
    }

    public Task<Post?> GetPost(int id)
    {
        return Task.FromResult(AllPosts.FirstOrDefault(x => x.Id ==
id));
    }

    public Task<List<Post>> GetAllPosts()
    {
        return Task.FromResult(AllPosts);
    }

    public Task DeletePost(int id)
    {
        var post = AllPosts.FirstOrDefault(x => x.Id == id);
        if (post != null)
        {
            AllPosts.Remove(post);
        }

        return Task.CompletedTask;
    }
}
```

The `PostsService` class is a simple demo service that manages the list of posts. It has methods to create, update, and delete posts. To simplify the implementation, it uses a static field to store the list of posts. This is just for demonstration purposes; please do not use this in production.

Next, we will follow the API design to implement CRUD operations. You can review the *REST-based API design* section of the previous chapter.

Implementing a GET operation

The design for the viewPost() operation is as follows:

Operation name	URL	HTTP method	Input	Response	Description
viewPost()	/posts/{postId}	GET	PostId	Post, 200	View a post detail

Table 2.1 – The design for the viewPost() operation

Update the PostController class as follows:

```
using Microsoft.AspNetCore.Mvc;
using MyFirstApi.Models;
using MyFirstApi.Services;

namespace MyFirstApi.Controllers;

[Route("api/[controller]")]
[ApiController]
public class PostsController : ControllerBase
{
    private readonly PostsService _postsService;

    public PostsController()
    {
        _postsService = new PostsService();
    }

    [HttpGet("{id}")]
    public async Task<ActionResult<Post>> GetPost(int id)
    {
        var post = await _postsService.GetPost(id);
        if (post == null)
        {
            return NotFound();
        }

        return Ok(post);
    }
    // Omitted for brevity
}
```

In the constructor method of the controller, we initialize the `_postsService` field. Note that we use the `new()` constructor to create an instance of the service. That means the controller is coupled with the `PostsService` class. We will see how to decouple the controller and the service in the next chapter.

Then, create a `GetPost()` method that returns a post with the specified ID. It has a `[HttpGet("{id}")]` attribute to indicate the URL of the operation. The URL will be mapped to `/api/posts/{id}`. `id` is a placeholder, which will be replaced with the ID of the post. Then, `id` will be passed to the `GetPost()` method as a parameter.

If the post is not found, the method will return a `NotFound` response. ASP.NET Core provides a set of built-in response messages, such as `NotFound`, `BadRequest`, `InternalServerError`, and so on.

If you call the API now, it will return `NotFound` because we have not created a post.

Implementing a CREATE operation

The design for the `createPost()` operation is as follows:

Operation name	URL	HTTP method	Input	Response	Description
`createPost()`	`/posts`	POST	Post	Post, 201	Create a new post

Table 2.2 – The design for the `createPost()` operation

Create a new method named `CreatePost()` in the controller. As the controller has been mapped to `api/posts`, we do not need to specify the route of this method. The content of the method is as follows:

```
[HttpPost]
public async Task<ActionResult<Post>> CreatePost(Post post)
{
    await _postsService.CreatePost(post);

    return CreatedAtAction(nameof(GetPost), new { id = post.Id },
post);
}
```

When we call this endpoint, the `post` object will be serialized in the JSON format that is attached to the POST request body. In this method, we can get the post from the request and then call the `CreatePost()` method in the service to create a new post. Then, we will return the built-in `CreatedAtAction`, which returns a response message with the specified action name, route values, and post. For this case, it will call the `GetPost()` action to return the newly created post.

Now, we can test the API. For example, we can send a POST request in Thunder Client.

Change the method to POST. Use the following JSON data as the body:

```
{
  "userId": 1,
  "id": 1,
  "title": "Hello ASP.NET Core",
  "body": "ASP.NET Core is a cross-platform, high-performance,
open-source framework for building modern, cloud-enabled, Internet-
connected apps."
}
```

Click the **Send** button. Note that the status of the response is 201 Created:

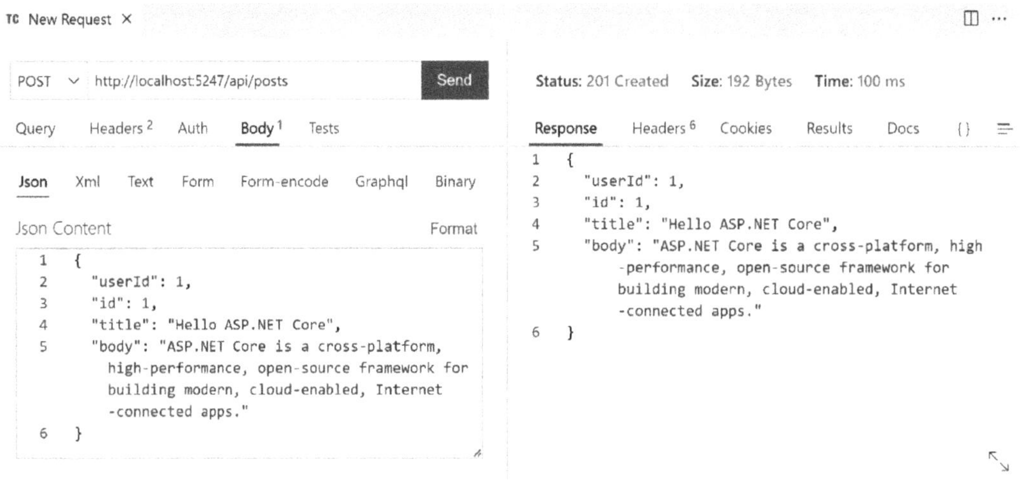

Figure 2.19 – Sending a POST request

Then, send a GET request to the api/posts/1 endpoint. We can get a response like this:

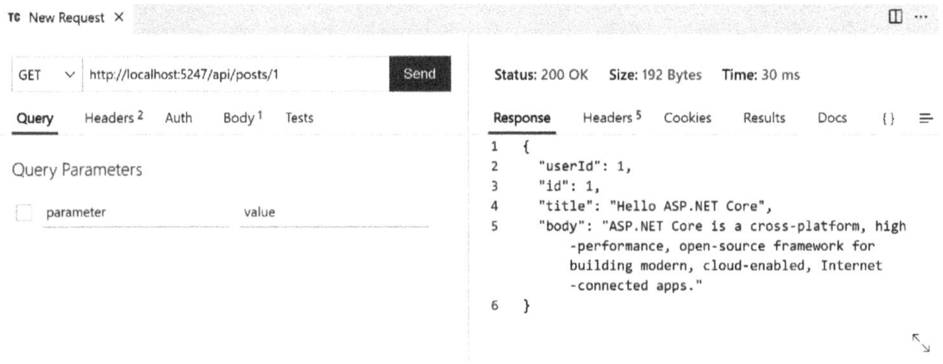

Figure 2.20 – Sending a GET request

Please note that the post we created is stored in the memory of the service. Because we have not provided a database to store the data, if we restart the application, the post will be lost.

Next, let us see how to implement an update operation.

Implementing an UPDATE operation

The design for the updatePost() operation is as follows:

Operation name	URL	HTTP method	Input	Response	Description
updatePost()	/posts/{postId}	PUT	Post	Post, 200	Update a new post

Table 2.3 – The design for the updatePost() operation

Create a new UpdatePost() method in the controller, as shown next:

```
[HttpPut("{id}")]
public async Task<ActionResult> UpdatePost(int id, Post post)
{
    if (id != post.Id)
    {
        return BadRequest();
    }

    var updatedPost = await _postsService.UpdatePost(id, post);
    if (updatedPost == null)
    {
        return NotFound();
    }

    return Ok(post);
}
```

This method has a [HttpPut("{id}")] attribute to indicate that it is a PUT operation. Similarly, id is a placeholder, which will be replaced with the ID of the post. In the PUT request, we should attach the serialized content of the post to the request body.

This time, let us test the API with HttpRepl. Run the following command to connect to the server:

```
httprepl https://localhost:7291/api/posts
connect https://localhost:7291/api/posts/1
put -h Content-Type=application/json -c "{"userId": 1,"id": 1,"title":
"Hello ASP.NET Core 8","body": "ASP.NET Core is a cross-platform,
high-performance, open-source framework for building modern, cloud-
enabled, Internet-connected apps."}"
```

You will see this output:

```
HTTP/1.1 200 OK
Content-Type: application/json; charset=utf-8
Date: Thu, 18 Aug 2022 11:25:26 GMT
Server: Kestrel
Transfer-Encoding: chunked

{
  "userId": 1,
  "id": 1,
  "title": "Hello ASP.NET Core 8",
  "body": "ASP.NET Core is a cross-platform, high-performance,
open-source framework for building modern, cloud-enabled, Internet-
connected apps."
}
```

Then, we can update the GetPosts() method as follows:

```
[HttpGet]
public async Task<ActionResult<List<Post>>> GetPosts()
{
    var posts = await _postService.GetAllPosts();
    return Ok(posts);
}
```

We have implemented GET, POST, and PUT operations. Next, you can try to implement the DeletePost() method using the DELETE operation by yourself.

Dependency injection

In the preceding example of the controller, there is a _postsService field that is initialized in the constructor method of the controller by using the new() constructor:

```
private readonly PostsService _postsService;
public PostsController()
{
    _postsService = new PostsService();
}
```

That says the PostsController class depends on the PostsService class, and the PostsService class is a dependency of the PostsController class. If we want to replace PostsService with a different implementation to save the data, we have to update the code of PostsController. If the PostsService class has its own dependencies, they must also be initialized by the PostsController class. When the project grows larger, the dependencies will become more complex. Also, this kind of implementation is not easy to test and maintain.

Dependency injection (**DI**) is one of the most well-known design patterns in the software development world. It helps decouple classes that depend on each other. You may find the following terms being used interchangeably: **Dependency Inversion Principle** (**DIP**), **Inversion of Control** (**IoC**), and DI. These terms are commonly confused even though they are related. You can find multiple articles and blog posts that explain them. Some say they are the same thing, but some say not. What are they?

Understanding DI

The Dependency Inversion Principle is one of the *SOLID* principles in **object-oriented** (**OO**) design. It was defined by Robert C. Martin in his book *Agile Software Development: Principles, Patterns, and Practices, Pearson*, in 2002. The principle states, *"high-level modules should not depend on low-level modules; both should depend on abstractions. Abstractions should not depend on details. Details should depend upon abstractions."*

In the preceding controller, we said `PostsController` depends on `PostsService`. The controller is the high-level module, and the service is the low-level module. When the service is changed, the controller must be changed as well. Keep in mind that the term *inversion* does not mean that the low-level module will depend on the high level. Instead, both of them should depend on abstractions that expose the behavior needed by high-level modules. If we invert this dependency relationship by creating an interface for the service, both the controller and the service will depend on the interface. The implementation of the service can change as long as it respects the interface.

IoC is a programming principle that inverts the flow of control in an application. In traditional programming, custom code is responsible for instantiating objects and controlling the execution of the main function. IoC inverts the flow of control as compared to traditional control flow. With IoC, the framework does the instantiation, calling custom or task-specific code.

It can be used to differentiate a framework from a class library. Normally, the framework calls the application code, and the application code calls the library. This kind of IoC is sometimes referred to as the Hollywood principle: *"Don't call us, we'll call you."*

IoC is related to DIP, but it is not the same. DIP concerns decoupling dependencies between high-level modules and low-level modules through shared abstractions (interfaces). IoC is used to increase the modularity of the program and make it extensible. There are several technologies to implement IoC, such as **Service Locator**, DI, the template method design pattern, the strategy design pattern, and so on.

DI is a form of IoC. This term was coined by Martin Fowler in 2004. It separates the concerns of constructing objects and using them. When an object or a function (the client) needs a dependency, it does not know how to construct it. Instead, the client only needs to declare the interfaces of the dependency, and the dependency is injected into the client by external code (an injector). It makes it easier to change the implementation of the dependency. It is often similar to the strategy design pattern. The difference is that the strategy pattern can use different strategies to construct the dependency, while DI typically only uses a single instance of the dependency.

There are three main types of DI:

- **Constructor injection**: The dependencies are provided as parameters of the client's constructor
- **Setter injection**: The client exposes a setter method to accept the dependency
- **Interface injection**: The dependency's interface provides an injector method that will inject the dependency into any client passed to it

As you can see, these three terms are related, but there are some differences. Simply put, DI is a technique for achieving IoC between classes and their dependencies. ASP.NET Core supports DI as a first-class citizen.

DI in ASP.NET Core

ASP.NET Core uses constructor injection to request dependencies. To use it, we need to do the following:

1. Define interfaces and their implementations.
2. Register the interfaces and the implementations to the service container.
3. Add services as the constructor parameters to inject the dependencies.

You can download the example project named `DependencyInjectionDemo` from the folder `samples/chapter2/ DependencyInjectionDemo/DependencyInjectionDemo` in the chapter's GitHub repository.

Follow the steps below to use DI in ASP.NET Core:

1. First, we will create an interface and its implementation. Copy the `Post.cs` file and the `PostService.cs` file from the previous `MyFirstApi` project to the `DependencyInjectionDemo` project. Create a new interface named `IPostService` in the `Service` folder, as shown next:

   ```
   public interface IPostService
   {
       Task CreatePost(Post item);
       Task<Post?> UpdatePost(int id, Post item);
       Task<Post?> GetPost(int id);
       Task<List<Post>> GetAllPosts();
       Task DeletePost(int id);
   }
   ```

   Then, update the `PostService` class to implement the `IPostService` interface:

   ```
   public class PostsService : IPostService
   ```

   You may also need to update the namespace of the `Post` class and the `PostService` class.

2. Next, we can register the IPostService interface and the PostService implementation to the service container. Open the Program.cs file, and you will find that an instance of WebApplicationBuilder named builder is created by calling the WebApplication. CreateBuilder() method. The CreateBuilder() method is the entry point of the application. We can configure the application by using the builder instance, and then call the builder.Build() method to build the WebApplication. Add the following code:

    ```
    builder.Services.AddScoped<IPostService, PostsService>();
    ```

    The preceding code utilizes the AddScoped() method, which indicates that the service is created once per client request and disposed of upon completion of the request.

3. Copy the PostsController.cs file from the previous MyFirstApi project to the DependencyInjectionDemo project. Update the namespace and the using statements. Then, update the constructor method of the controller as follows:

    ```
    private readonly IPostService _postsService;
    public PostsController(IPostService postService)
    {
        _postsService = postService;
    }
    ```

    The preceding code uses the IPostService interface as the constructor parameter. The service container will inject the correct implementation into the controller.

DI has four roles: services, clients, interfaces, and injectors. In this example, IPostService is the interface, PostService is the service, PostsController is the client, and builder. Services is the injector, which is a collection of services for the application to compose. It is sometimes referred to as a DI container.

The PostsController class requests the instance of IPostService from its constructor. The controller, which is the client, does not know where the service is, nor how it is constructed. The controller only knows the interface. The service has been registered in the service container, which can inject the correct implementation into the controller. We do not need to use the new keyword to create an instance of the service. That says the client and the service are decoupled.

This DI feature is provided in a NuGet package called Microsoft.Extensions. DependencyInjection. When an ASP.NET Core project is created, this package is added automatically. If you create a console project, you may need to install it manually by using the following command:

```
dotnet add package Microsoft.Extensions.DependencyInjection
```

If we want to replace the IPostService with another implementation, we can do so by registering the new implementation to the service container. The code of the controller does not need to be changed. That is one of the benefits of DI.

Next, let us discuss the lifetime of services.

DI lifetimes

In the previous example, the service is registered using the `AddScoped()` method. In ASP.NET Core, there are three lifetimes when the service is registered:

- **Transient**: A transient service is created each time it is requested and disposed of at the end of the request.

- **Scoped**: In web applications, a scope means a request (connection). A scoped service is created once per client request and disposed of at the end of the request.

- **Singleton**: A singleton service is created the first time it is requested or when providing the implementation instance to the service container. All subsequent requests will use the same instance.

To demonstrate the difference between these lifetimes, we will use a simple demo service:

Create a new interface named `IDemoService` and its implementation named `DemoService` in the `Services` folder, as shown next:

IDemoService.cs:

```
namespace DependencyInjectionDemo.Services;

public interface IDemoService
{
    SayHello();
}
```

DemoService.cs:

```
namespace DependencyInjectionDemo.Services;

public class DemoService : IDemoService
{
    private readonly Guid _serviceId;
    private readonly DateTime _createdAt;

    public DemoService()
    {
        _serviceId = Guid.NewGuid();
        _createdAt = DateTime.Now;
    }

    public string SayHello()
```

```
    {
        return $"Hello! My Id is {_serviceId}. I was
created at {_createdAt:yyyy-MM-dd HH:mm:ss}.
";
    }
}
```

The implementation will generate an ID and a time when it was created, and output it when the SayHello() method is called.

1. Then, we can register the interface and the implementation to the service container. Open the Program.cs file and add the code as follows:

    ```
    builder.Services.AddScoped<IDemoService, DemoService>();
    ```

2. Create a controller named DemoController.cs. Now, we can add the service as constructor parameters to inject the dependency:

    ```
    [ApiController]
    [Route("[controller]")]
    public class DemoController : ControllerBase
    {
        private readonly IDemoService _demoService;

        public DemoController(IDemoService demoService)
        {
            _demoService = demoService;
        }

        [HttpGet]
        public ActionResult Get()
        {
            return Content(_demoService.SayHello());
        }
    }
    ```

For this example, if you test the /demo endpoint, you will see the GUID value and the creation time in the output change every time:

```
http://localhost:5147/> get demo
HTTP/1.1 200 OK
Content-Length: 91
Content-Type: text/plain; charset=utf-8
Date: Fri, 20 Oct 2023 22:06:46 GMT
Server: Kestrel
```

```
Hello! My Id is 6ca84d82-90cb-4dd6-9a34-5ea7573508ac. I was
created at 2023-10-21 11:06:46.
```

```
http://localhost:5147/> get demo
HTTP/1.1 200 OK
Content-Length: 91
Content-Type: text/plain; charset=utf-8
Date: Fri, 20 Oct 2023 22:07:02 GMT
Server: Kestrel
```

```
Hello! My Id is 9bc5cf49-661d-45bb-b9ed-e0b3fe937827. I was
created at 2023-10-21 11:07:02.
```

We can change the lifetime to AddSingleton(), as follows:

```
builder.Services.AddSingleton<IDemoService, DemoService>();
```

The GUID values and the creation time values will be the same for all requests:

```
http://localhost:5147/> get demo
HTTP/1.1 200 OK
Content-Length: 91
Content-Type: text/plain; charset=utf-8
Date: Fri, 20 Oct 2023 22:08:57 GMT
Server: Kestrel
```

```
Hello! My Id is a1497ead-bff6-4020-b337-28f1d3af7b05. I was created at
2023-10-21 11:08:02.
```

```
http://localhost:5147/> get demo
HTTP/1.1 200 OK
Content-Length: 91
Content-Type: text/plain; charset=utf-8
Date: Fri, 20 Oct 2023 22:09:12 GMT
Server: Kestrel
```

```
Hello! My Id is a1497ead-bff6-4020-b337-28f1d3af7b05. I was created at
2023-10-21 11:08:02.
```

As the DemoController class only requests the IDemoService interface once for each request, we cannot differentiate the behavior between scoped and transient services. Let us look at a more complex example.

1. You can find the example code in the `DependencyInjectionDemo` project. There are three interfaces along with their implementations:

```
public interface IService
{
    string Name { get; }
    string SayHello();
}

public interface ITransientService : IService
{
}

public class TransientService : ITransientService
{
    private readonly Guid _serviceId;
    private readonly DateTime _createdAt;

    public TransientService()
    {
        _serviceId = Guid.NewGuid();
        _createdAt = DateTime.Now;
    }

    public string Name => nameof(TransientService);
    public string SayHello()
    {
        return $"Hello! I am {Name}. My Id is {_serviceId}. I
was created at {_createdAt:yyyy-MM-dd HH:mm:ss}.";
    }
}

public interface ISingletonService : IService
{
}

public class SingletonService : ISingletonService
{
    private readonly Guid _serviceId;
    private readonly DateTime _createdAt;

    public SingletonService()
    {
        _serviceId = Guid.NewGuid();
```

```csharp
        _createdAt = DateTime.Now;
    }

    public string Name => nameof(SingletonService);

    public string SayHello()
    {
        return $"Hello! I am {Name}. My Id is {_serviceId}. I
was created at {_createdAt:yyyy-MM-dd HH:mm:ss}.";
    }
}

public interface IScopedService : IService
{
}

public class ScopedService : IScopedService
{
    private readonly Guid _serviceId;
    private readonly DateTime _createdAt;
    private readonly ITransientService _transientService;
    private readonly ISingletonService _singletonService;

    public ScopedService(ITransientService transientService,
ISingletonService singletonService)
    {
        _transientService = transientService;
        _singletonService = singletonService;
        _serviceId = Guid.NewGuid();
        _createdAt = DateTime.Now;
    }

    public string Name => nameof(ScopedService);

    public string SayHello()
    {
        var scopedServiceMessage = $"Hello! I am {Name}. My Id
is {_serviceId}. I was created at {_createdAt:yyyy-MM-dd
HH:mm:ss}.";
        var transientServiceMessage = $"{_transientService.
SayHello()} I am from {Name}.";
        var singletonServiceMessage = $"{_singletonService.
SayHello()} I am from {Name}.";
        return
```

```
                    $"{scopedServiceMessage}{Environment.
        NewLine}{transientServiceMessage}{Environment.NewLine}
        {singletonServiceMessage}";
            }
        }
```

2. In the `Program.cs` file, we can register them to the service container as follows:

```
builder.Services.AddScoped<IScopedService, ScopedService>();
builder.Services.AddTransient<ITransientService,
TransientService>();
builder.Services.AddSingleton<ISingletonService,
SingletonService>();
```

3. Then, create a controller named `LifetimeController.cs`. The code is shown next:

```
[ApiController]
[Route("[controller]")]
public class LifetimeController : ControllerBase
{
    private readonly IScopedService _scopedService;
    private readonly ITransientService _transientService;
    private readonly ISingletonService _singletonService;

    public LifetimeController(IScopedService scopedService,
ITransientService transientService,
        ISingletonService singletonService)
    {
        _scopedService = scopedService;
        _transientService = transientService;
        _singletonService = singletonService;
    }

    [HttpGet]
    public ActionResult Get()
    {
        var scopedServiceMessage = _scopedService.SayHello();
        var transientServiceMessage = _transientService.
SayHello();
        var singletonServiceMessage = _singletonService.
SayHello();
        return Content(
            $"{scopedServiceMessage}{Environment.
NewLine}{transientServiceMessage}{Environment.NewLine}
{singletonServiceMessage}");
    }
}
```

In this example, `ScopedService` has two dependencies: `ITransientService` and `ISingletonService`. So, when `ScopedService` is created, it will ask for the instances of these dependencies from the service container. On the other hand, the controller also has dependencies: `IScopedService`, `ITransientService`, and `ISingletonService`. When the controller is created, it will ask for these three dependencies. That means `ITransientService` and `ISingletonService` will be needed twice for each request. But let us check the output of the following requests:

```
http://localhost:5147/> get lifetime
HTTP/1.1 200 OK
Content-Length: 625
Content-Type: text/plain; charset=utf-8
Date: Fri, 20 Oct 2023 22:20:44 GMT
Server: Kestrel


Hello! I am ScopedService. My Id is df87d966-0e86-4f08-874f-
ba6ce71de560. I was created at 2023-10-21 11:20:44.
Hello! I am TransientService. My Id is 77e29268-ad48-423c-94e5-
de1d09bd3ba5. I was created at 2023-10-21 11:20:44. I am from
ScopedService.
Hello! I am SingletonService. My Id is 95a44c5b-8678-48c6-a2f0-
cc6b90423773. I was created at 2023-10-21 11:20:44. I am from
ScopedService.
Hello! I am TransientService. My Id is e77564d1-e146-4d29-b74b-
a07f8f6640c1. I was created at 2023-10-21 11:20:44.
Hello! I am SingletonService. My Id is 95a44c5b-8678-48c6-a2f0-
cc6b90423773. I was created at 2023-10-21 11:20:44.


http://localhost:5147/> get lifetime
HTTP/1.1 200 OK
Content-Length: 625
Content-Type: text/plain; charset=utf-8
Date: Fri, 20 Oct 2023 22:20:57 GMT
Server: Kestrel


Hello! I am ScopedService. My Id is e5f802ed-5e4c-4abd-9213-
8f13f97c1008. I was created at 2023-10-21 11:20:57.
Hello! I am TransientService. My Id is daccb91b-438f-4561-9c86-
13b02ad8e358. I was created at 2023-10-21 11:20:57. I am from
ScopedService.
Hello! I am SingletonService. My Id is 95a44c5b-8678-48c6-a2f0-
cc6b90423773. I was created at 2023-10-21 11:20:44. I am from
ScopedService.
Hello! I am TransientService. My Id is 94e9e6c1-729a-4033-8a27-
550ea10ba5d0. I was created at 2023-10-21 11:20:57.
Hello! I am SingletonService. My Id is 95a44c5b-8678-48c6-a2f0-
cc6b90423773. I was created at 2023-10-21 11:20:44.
```

We can see that in each request, `ScopedService` was created once, while `ITransientService` was created twice. In both requests, `SingletonService` was created only once.

Group registration

As the project grows, we may have more and more services. If we register all services in `Program.cs`, this file will be very large. For this case, we can use group registration to register multiple services at once. For example, we can create a service group named `LifetimeServicesCollectionExtensions.cs`:

```
public static class LifetimeServicesCollectionExtensions
{
    public static IServiceCollection AddLifetimeServices(this
IServiceCollection services)
    {
        services.AddScoped<IScopedService, ScopedService>();
        services.AddTransient<ITransientService, TransientService>();
        services.AddSingleton<ISingletonService, SingletonService>();

        return services;
    }
}
```

This is an extension method for the `IServiceCollection` interface. It is used to register all services at once in the `Program.cs` file:

```
// Group registration
builder.Services.AddLifetimeServices();
```

In this way, the `Program.cs` file will be smaller and easier to read.

Action injection

Sometimes, one controller may need many services but may not need all of them for all actions. If we inject all the dependencies from the constructor, the constructor method will be large. For this case, we can use action injection to inject dependencies only when needed. See the following example:

```
[HttpGet]
public ActionResult Get([FromServices] ITransientService
transientService)
{
    ...
}
```

The `[FromServices]` attribute enables the service container to inject dependencies when needed without using constructor injection. However, if you find that a service needs a lot of dependencies, it may indicate that the class has too many responsibilities. Based on the **Single Responsibility Principle** (**SRP**), consider refactoring the class to split the responsibilities into smaller classes.

Keep in mind that this kind of action injection only works for actions in the controller. It does not support normal classes. Additionally, since ASP.NET Core 7.0, the [FromServices] attribute can be omitted as the framework will automatically attempt to resolve any complex type parameters registered in the DI container.

Keyed services

ASP.NET Core 8.0 introduces a new feature known as keyed services, or named services. This feature allows developers to register services with a key, allowing them to access the service with that key. This makes it easier to manage multiple services that implement the same interface within an application, as the key can be used to identify and access the service.

For example, we have a service interface named IDataService:

```
public interface IDataService
{
    string GetData();
}
```

This IDataService interface has two implementations: SqlDatabaseService and CosmosDatabaseService:

```
public class SqlDatabaseService : IDataService
{
    public string GetData()
    {
        return "Data from SQL Database";
    }
}
```

```
public class CosmosDatabaseService : IDataService
{
    public string GetData()
    {
        return "Data from Cosmos Database";
    }
}
```

We can register them to the service container using different keys:

```
builder.Services.AddKeyedScoped<IDataService,
SqlDatabaseService>("sqlDatabaseService");
builder.Services.AddKeyedScoped<IDataService,
CosmosDatabaseService>("cosmosDatabaseService");
```

Then, we can inject the service by using the `FromKeyedServices` attribute:

```
[ApiController]
[Route("[controller]")]
public class KeyedServicesController : ControllerBase
{
    [HttpGet("sql")]
    public ActionResult
GetSqlData([FromKeyedServices("sqlDatabaseService")]
IDataService dataService) =>
        Content(dataService.GetData());


    [HttpGet("cosmos")]
    public ActionResult
GetCosmosData([FromKeyedServices("cosmosDatabaseService")]
IDataService dataService) =>
        Content(dataService.GetData());
}
```

The `FromKeyedServices` attribute is used to inject the service by using the specified key. Test the API with HttpRepl, and you will see the output as follows:

```
http://localhost:5147/> get keyedServices/sql
HTTP/1.1 200 OK
Content-Length: 22
Content-Type: text/plain; charset=utf-8
Date: Fri, 20 Oct 2023 22:48:49 GMT
Server: Kestrel


Data from SQL Database


http://localhost:5147/> get keyedServices/cosmos
HTTP/1.1 200 OK
Content-Length: 25
Content-Type: text/plain; charset=utf-8
Date: Fri, 20 Oct 2023 22:48:54 GMT
Server: Kestrel


Data from Cosmos Database
```

The keyed services can be used to register singleton or transient services as well. Just use the `AddKeyedSingleton()` or `AddKeyedTransient()` method respectively; for example:

```csharp
builder.Services.AddKeyedSingleton<IDataService,
SqlDatabaseService>("sqlDatabaseService");
builder.Services.AddKeyedTransient<IDataService,
CosmosDatabaseService>("cosmosDatabaseService");
```

It is important to note that if an empty string is passed as the key, a default implementation for the service must be registered with a key of an empty string, otherwise the service container will throw an exception.

Microsoft releases new versions of .NET SDKs frequently. If you encounter a different version number, that is acceptable.

The preceding command will list all the available SDKs on your machine. For example, it may show the following output if have multiple .NET SDKs installed.

> **Important note**
>
> Every Microsoft product has a lifecycle. .NET and .NET Core provides **Long-term support (LTS)** releases that get 3 years of patches and free support. When this book was written, .NET 7 is still supported, until May 2024. Based on Microsoft's policy, even numbered releases are LTS releases. So .NET 8 is the latest LTS release. The code samples in this book are written with .NET 8.0.

When you use VS Code to open the project, the C# Dev Kit extension can create a solution file for you. This feature makes VS Code more friendly to C# developers. You can see the following structure in the Explorer view:

It uses the `Swashbuckle.AspNetCore` NuGet package, which provides the Swagger UI to document and test the APIs.

Follow the steps below to use DI in ASP.NET Core:

We can see that in each request, `ScopedService` was created once, while `ITransientService` was created twice. In both requests, `SingletonService` was created only once.

Using primary constructors to inject dependencies

Beginning with .NET 8 and C# 12, we can use the primary constructor to inject dependencies. A primary constructor allows us to declare the constructor parameters directly in the class declaration, instead of using a separate constructor method. For example, we can update the `PostsController` class as follows:

```csharp
public class PostsController(IPostService postService) :
```

```
ControllerBase
{
    // No need to define a private field to store the service
    // No need to define a constructor method
}
```

You can find a sample named `PrimaryConstructorController.cs` in the `Controller` folder of the `DependencyInjectionDemo` project.

When using the primary constructor in a class, note that the parameters passed to the class declaration cannot be used as properties or members. For example, if a class declares a parameter named `postService` in the class declaration, it cannot be accessed as a class member using `this.postService` or from external code. To learn more about the primary constructor, please refer to the documentation at `https://learn.microsoft.com/en-us/dotnet/csharp/programming-guide/classes-and-structs/instance-constructors#primary-constructors`.

Primary constructors can save us from writing fields and constructor methods. So, we'll use them in the following examples.

Do not use `new` to create service B, otherwise, service A will be tightly coupled with service B.

Resolving a service when the app starts

If we need a service in the `Program.cs` file, we cannot use constructor injection. For this situation, we can resolve a scoped service for a limited duration at app startup, as follows:

```
var app = builder.Build();
using (var serviceScope = app.Services.CreateScope())
{
    var services = serviceScope.ServiceProvider;

    var demoService = services.GetRequiredService<IDemoService>();
    var message = demoService.SayHello();
    Console.WriteLine(message);
}
```

The preceding code creates a scope and resolves the `IDemoService` service from the service container. Then, it can use the service to do something. After the scope is disposed of, the service will be disposed of as well.

DI tips

ASP.NET Core uses DI heavily. The following are some tips to help you use DI:

- When designing your services, make the services as stateless as possible. Do not use static classes and members unless you have to do so. If you need to use a global state, consider using a singleton service instead.

- Carefully design dependency relationships between services. Do not create a cyclic dependency.

- Do not use new to create a service instance in another service. For example, if service A depends on service B, the instance of service B should be injected into service A with DI. Do not use new to create service B, otherwise, service A will be tightly coupled with service B.

- Use a DI container to manage the lifetime of services. If a service implements the `IDisposable` interface, the DI container will dispose of the service when the scope is disposed of. Do not manually dispose of it.

- When registering a service, do not use new to create an instance of the service. For example, `services.AddSingleton(new ExampleService());` registers a service instance that is not managed by the service container. So, the DI framework will not be able to dispose of the service automatically.

- Avoid using the service locator pattern. If DI can be used, do not use the `GetService()` method to obtain a service instance.

You can learn more about the DI guidelines at `https://docs.microsoft.com/zh-cn/dotnet/core/extensions/dependency-injection-guidelines`.

Why there is no configuration method for the logger in the template project?

ASP.NET Core provides a built-in DI implementation for the logger. When the project was created, logging was registered by the ASP.NET Core framework. Therefore, there is no configuration method for the logger in the template project. Actually, there are more than 250 services that are automatically registered by the ASP.NET Core framework.

Can I use third-party DI containers?

It is highly recommended that you use the built-in DI implementation in ASP.NET Core. But if you need any specific features that it does not support, such as property injection, `Func<T>` support for lazy initialization, and so on, you can use third-party DI containers, such as *Autofac* (`https://autofac.org/`).

Introduction to minimal APIs

In the previous section, *Creating a simple web API project*, we created a simple web API project using the `dotnet new webapi -n MyFirstApi -controllers` command. The `-controllers` option (or `--use-controllers`) indicates that the project will use controller-based routing. Alternatively, the `-minimal` or `--use-minimal-apis` option can be used to create a project that uses minimal APIs. In this section, we will introduce minimal APIs.

Minimal APIs is a new feature introduced in ASP.NET Core 6.0. It is a new way to create APIs without using controllers. Minimal APIs are designed to be simple and lightweight with minimal dependencies. They are a good choice for small projects or prototypes, and also for projects that do not need the full features of controllers.

To create a minimal API project, we can use the following command:

```
dotnet new webapi -n MinimalApiDemo -minimal
```

There is no `Controllers` folder in the project. Instead, you can find the following code in the `Program.cs` file:

```
app.MapGet("/weatherforecast", () =>
{
    var forecast =  Enumerable.Range(1, 5).Select(index =>
        new WeatherForecast
        (
            DateOnly.FromDateTime(DateTime.Now.AddDays(index)),
            Random.Shared.Next(-20, 55),
            summaries[Random.Shared.Next(summaries.Length)]
        ))
        .ToArray();
    return forecast;
})
.WithName("GetWeatherForecast")
.WithOpenApi();
```

The preceding code uses the `MapGet()` method to map the GET request to the `/weatherforecast` endpoint. The `MapGet()` method is an extension method of the `IEndpointRouteBuilder` interface. This interface is used to configure the endpoints in the application. Its extension method, `MapGet()`, returns an `IEndpointConventionBuilder` interface that allows us to use fluent APIs to configure the endpoint by using other extension methods, such as `WithName()` and `WithOpenApi()`. The `WithName()` method is used to set the name of the endpoint. The `WithOpenApi()` method is used to generate an OpenAPI document for the endpoint.

Creating a simple endpoint

Let us create a new /posts endpoint that supports CRUD operations. First, add the following code to the end of the Program.cs file to define a Post class:

```
public class Post
{
    public int Id { get; set; }
    public string Title { get; set; } = string.Empty;
    public string Content { get; set; } = string.Empty;
}
```

Add the following code to the Program.cs file:

```
var list = new List<Post>()
{
    new() { Id = 1, Title = "First Post", Content = "Hello World" },
    new() { Id = 2, Title = "Second Post", Content = "Hello Again" },
    new() { Id = 3, Title = "Third Post", Content = "Goodbye World" },
};


app.MapGet("/posts",
    () => list).WithName("GetPosts").WithOpenApi().WithTags("Posts");
app.MapPost("/posts",
    (Post post) =>
    {
        list.Add(post);
        return Results.Created($"/posts/{post.Id}", post);
    }).WithName("CreatePost").WithOpenApi().WithTags("Posts");
app.MapGet("/posts/{id}", (int id) =>
{
    var post = list.FirstOrDefault(p => p.Id == id);
    return post == null ? Results.NotFound() : Results.Ok(post);
}).WithName("GetPost").WithOpenApi().WithTags("Posts");
app.MapPut("/posts/{id}", (int id, Post post) =>
{
    var index = list.FindIndex(p => p.Id == id);
    if (index == -1)
    {
        return Results.NotFound();
    }
    list[index] = post;
    return Results.Ok(post);
```

```
}).WithName("UpdatePost").WithOpenApi().WithTags("Posts");
app.MapDelete("/posts/{id}", (int id) =>
{
    var post = list.FirstOrDefault(p => p.Id == id);
    if (post == null)
    {
        return Results.NotFound();
    }
    list.Remove(post);
    return Results.Ok();
}).WithName("DeletePost").WithOpenApi().WithTags("Posts");
```

The preceding code defines five endpoints:

- `GET /posts`: Get all posts
- `POST /posts`: Create a new post
- `GET /posts/{id}`: Get a post by ID
- `PUT /posts/{id}`: Update a post by ID
- `DELETE /posts/{id}`: Delete a post by ID

We use the `WithTags` extension method to group these endpoints into a tag named `Posts`. In this example, a list is used to store the posts. In a real-world application, we should use a database to store the data.

Using DI in minimal APIs

Minimal APIs support DI as well. You can find the `IPostService` interface and its `PostService` implementation in the `Services` folder. Here is an example of using DI in minimal APIs:

```
app.MapGet("/posts", async (IPostService postService) =>
{
    var posts = await postService.GetPostsAsync();
    return posts;
}).WithName("GetPosts").WithOpenApi().WithTags("Posts");


app.MapGet("/posts/{id}", async (IPostService postService, int id) =>
{
    var post = await postService.GetPostAsync(id);
    return post == null ? Results.NotFound() : Results.Ok(post);
```

```csharp
}).WithName("GetPost").WithOpenApi().WithTags("Posts");


app.MapPost("/posts", async (IPostService postService, Post post) =>
{
    var createdPost = await postService.CreatePostAsync(post);
    return Results.Created($"/posts/{createdPost.Id}", createdPost);
}).WithName("CreatePost").WithOpenApi().WithTags("Posts");


app.MapPut("/posts/{id}", async (IPostService postService, int id,
Post post) =>
{
    try
    {
        var updatedPost = await postService.UpdatePostAsync(id, post);
        return Results.Ok(updatedPost);
    }
    catch (KeyNotFoundException)
    {
        return Results.NotFound();
    }
}).WithName("UpdatePost").WithOpenApi().WithTags("Posts");


app.MapDelete("/posts/{id}", async (IPostService postService, int id)
=>
{
    try
    {
        await postService.DeletePostAsync(id);
        return Results.NoContent();
    }
    catch (KeyNotFoundException)
    {
        return Results.NotFound();
    }
}).WithName("DeletePost").WithOpenApi().WithTags("Posts");
```

In the preceding code, the `IPostService` interface is used as a parameter of the action method. The DI container will inject the correct implementation into the action method. You can run the project and test the endpoints. It should have the same behavior as the controller-based project.

What is the difference between minimal APIs and controller-based APIs?

Minimal APIs are simpler than controller-based APIs, allowing us to map endpoints to methods directly. This makes minimal APIs a good choice for quickly creating simple APIs or demo projects. However, minimal APIs do not support the full range of features that controllers provide, such as model binding, model validation, and so on. These features may be added in the future. Therefore, we will mainly use controller-based APIs and not discuss minimal APIs in detail in this book. If you want to learn more about minimal APIs, please refer to the official documentation at `https://learn.microsoft.com/en-us/aspnet/core/fundamentals/minimal-apis`.

Summary

In this chapter, we created a simple web API project and introduced how to run the project locally and call the APIs with different clients. We implemented basic CRUD operations using an in-memory list. Also, we explained how to use DI in ASP.NET Core. We explored the lifetime of services and learned some tips. In addition, we introduced minimal APIs. In the next chapter, we will delve further into the built-in components of ASP.NET Core.

3

ASP.NET Core Fundamentals (Part 1)

In the previous chapter, we learned how to create a basic REST API using ASP.NET Core. ASP.NET Core provides a lot of features that make it easy to build web APIs.

In this chapter, we will be covering the following topics:

- Routing
- Configuration
- Environments

Routing is used to map incoming requests to the corresponding controller actions. We will discuss how to use attribute routing to configure the routing for ASP.NET Core web APIs. **Configuration** is used to provide the initial settings for an application on its startup, such as database connection strings, API keys, and other settings. Configuration is often used with **environments**, such as development, staging, and production. At the conclusion of this chapter, you will have the skills to create RESTful routes for your ASP.NET Core web APIs and utilize the ASP.NET Core configuration framework to manage configurations for different environments.

Technical requirements

The code examples in this chapter can be found at `https://github.com/PacktPublishing/Web-API-Development-with-ASP.NET-Core-8`.

You can use Visual Studio 2022 or **VS Code** to open the solutions.

Routing

In *Chapter 2,* we introduced how to create a simple ASP.NET Core web API project using the default controller-based template. The project uses some attributes, such as [Route("api/controller")], [HttpGet], and so on, to map incoming requests to the corresponding controller actions. These attributes are used to configure the routing for the ASP.NET Core web API project.

Routing is a mechanism that monitors incoming requests and determines which action method is to be invoked for those requests. ASP.NET Core provides two types of routing: conventional routing and attribute routing. Conventional routing is typically used for ASP.NET Core MVC applications, while ASP.NET Core web APIs use attribute routing. In this section, we will discuss attribute routing in more detail.

You can download the RoutingDemo sample project from /samples/chapter3/RoutingDemo/ in the chapter's GitHub repository.

What is attribute routing?

Open the Program.cs file in the RoutingDemo project. You will find the following code:

```
app.MapControllers();
```

This line of code adds endpoints for controller actions to the IEndpointRouteBuilder instance without specifying any routes. To specify the routes, we need to use the [Route] attribute on the controller class and the action methods. The following code shows how to use the [Route] attribute on the WeatherForecastController class:

```
[ApiController]
[Route("[controller]")]
public class WeatherForecastController : ControllerBase
{
  // Omitted for brevity
}
```

In the preceding code, the [controller] token is a placeholder for the controller name. In this case, the controller name is WeatherForecast, so the [controller] route template is replaced with WeatherForecast. That means the route for the WeatherForecastController class is /WeatherForecast.

ASP.NET Core has some built-in route tokens, such as [controller], [action], [area], [page], and so on. These tokens are enclosed in square brackets ([]) and will be replaced with the corresponding values. Note that these tokens are reserved route parameter names and should not be used as a route parameter like {controller}.

In ASP.NET Core REST web APIs, we usually use a [Route("api/[controller]")] template to represent API endpoints. You can find the PostsController class in the Controllers folder. The following code shows the routing attribute of the PostsController class:

```
[ApiController]
[Route("api/[controller]")]
public class PostsController : ControllerBase
{
  // Omitted for brevity
}
```

The route for the PostsController class is /api/Posts. This is an indication that the endpoint is a REST API endpoint. Whether you use /api as the route prefix or not is up to you. There is no standard for this.

Some developers prefer to use lowercase for route templates, such as /api/posts. To achieve this, the route value can be explicitly specified; for example, [Route("api/posts")]. However, it seems a bit tedious to specify the route value for each controller class. Fortunately, ASP.NET Core provides a way to configure the route value globally. Add the following code to the Program.cs file:

```
builder.Services.AddRouting(options => options.LowercaseUrls = true);
```

The preceding code converts all route templates to lowercase. Actually, the text matching in ASP.NET Core routing is case-insensitive. So, this change only affects the generated paths' URLs, such as the URLs in Swagger UI and the **OpenAPI Specification**. You can use either /api/Posts or /api/posts to hit the same controller route.

Multiple routes can be applied to one controller class. The following code shows how to apply multiple routes to the PostsController class:

```
[ApiController]
[Route("api/[controller]")]
[Route("api/some-posts-whatever")]
public class PostsController : ControllerBase
{
  // Omitted for brevity
}
```

In this case, the PostsController class has two routes: /api/posts and /api/some-posts-whatever. It is not recommended to have multiple routes for the same controller class as this can lead to confusion. If you require multiple routes for the same controller class, please ensure that you have strong reasons for doing so.

In ASP.NET Core REST APIs, we usually do not use the [action] token because the action name is not included in the route template. Similarly, do not use the [Route] attribute for action methods. Instead, we use the HTTP method to distinguish action methods. We will discuss this in the following section.

Mapping HTTP methods to action methods

REST APIs are centered on resources. When we design a REST API, we need to map the CRUD operations to the HTTP methods. In ASP.NET Core, we can use the following HTTP verb attributes to map HTTP methods to action methods:

- [HttpGet] maps an HTTP GET method to an action method
- [HttpPost] maps an HTTP POST method to an action method
- [HttpPut] maps an HTTP PUT method to an action method
- [HttpDelete] maps an HTTP DELETE method to an action method
- [HttpPatch] maps an HTTP PATCH method to an action method
- [HttpHead] maps an HTTP HEAD method to an action method

The following code shows how to use the [HttpGet] attribute to map the HTTP GET method to the GetPosts() action method:

```
[HttpGet]
public async Task<ActionResult<List<Post>>> GetPosts()
{
  // Omitted for brevity
}
```

In ASP.NET Core REST APIs, each action must have an HTTP verb attribute. If you do not specify an HTTP verb attribute, the framework cannot determine which method should be invoked for the incoming request. In the preceding code, a GET request to the /api/posts endpoint is mapped to the GetPosts() action method.

The following code shows how to use the [HttpGet] attribute to map the HTTP GET method to the GetPost() action method with a route template:

```
[HttpGet("{id}")]
public async Task<ActionResult<Post>> GetPost(int id)
{
  // Omitted for brevity
}
```

The preceding HttpGet attribute has an {id} route template, which is a route parameter. The route parameter is enclosed in curly braces ({ }). The route parameter is used to capture the value from the incoming request. For example, a GET request to the /api/posts/1 endpoint is mapped to the GetPost(int id) action method, and the value 1 is captured by the {id} route parameter.

The following code shows how to use the [HttpPut] attribute to publish a post:

```
[HttpPut("{id}/publish")]
public async Task<ActionResult> PublishPost(int id)
{
  // Omitted for brevity
}
```

The preceding HttpPut attribute has an {id}/publish route template. The publish literal is used to match the publish literal in the incoming request. So, a PUT request to the /api/posts/1/publish endpoint is mapped to the PublishPost(int id) action method, and the value 1 is captured by the {id} route parameter.

When defining a route template, please make sure there are no conflicts. For example, we want to add a new action method to get posts by a user ID. If we use the following code, it will not work:

```
[HttpGet("{userId}")] // api/posts/user/1
public async Task<ActionResult<List<Post>>> GetPostsByUserId(int
userId)
```

This is because we already have a GetPost() action method that uses [HttpGet("{id}")]. When sending a GET request to the /api/posts/1 endpoint, the request matches multiple actions, so you will see a 500 error as follows:

```
Microsoft.AspNetCore.Routing.Matching.AmbiguousMatchException: The
request matched multiple endpoints. Matches:

RoutingDemo.Controllers.PostsController.GetPost (RoutingDemo)
 RoutingDemo.Controllers.PostsController.GetPostsByUserId
(RoutingDemo)
```

To fix it, we need to specify a different template, such as [HttpGet("user/{userId}")].

Route constraints

In the previous section, we introduced how to use a route parameter to capture the value from an incoming request. A [HttpGet("{id}")] attribute can match a GET request to the /api/posts/1 endpoint. But what if the request is a GET request to the /api/posts/abc endpoint?

As the id parameter is of type int, the framework will try to convert the captured value to an int value. If the conversion fails, the framework will return a 400 Bad Request response. So, a GET request to the /api/posts/abc endpoint will fail and return a 400 Bad Request response.

We can add route constraints to route parameters to restrict the values of the route parameters. For example, we can add a route constraint to the id parameter to ensure that the id parameter is an integer. The following code shows how to add a route constraint to the id parameter:

```
[HttpGet("{id:int}")]
public async Task<ActionResult<Post>> GetPost(int id)
{
    // Omitted for brevity
}
```

Now, the id parameter must be an integer. A GET request to the /api/posts/abc endpoint will return a 404 Not Found response because the route does not match.

ASP.NET Core provides a set of built-in route constraints, such as the following:

- int: The parameter must be an integer value.
- bool: The parameter must be a Boolean value.
- datetime: The parameter must be a DateTime value.
- decimal: The parameter must be a decimal value. Similarly, there are double, float, long, and so on.
- guid: The parameter must be a GUID value.
- minlength(value): The parameter must be a string with a minimum length; for example, {name:minlength(6)}, which means the name parameter must be a string and the length of the string must be at least 6 characters. Similarly, there are maxlength(value), length(value), length(min, max), and so on.
- min(value): The parameter must be an integer with a minimum value; for example, {id:min(1)}, which means the id parameter must be an integer and the value must be greater than or equal to 1. Similarly, there are max(value), range(min, max), and so on.
- alpha: The parameter must be a string with one or more letters.
- regex(expression): The parameter must be a string that matches the regular expression.
- required: The parameter must be provided in the route; for example, {id:required}, which means the id parameter must be provided in the route.

If the value of the route parameter does not match the route constraint, the action method will not accept the request, and a 404 Not Found response will be returned.

Multiple route constraints can be applied together. The following code shows how to apply multiple route constraints to the `id` parameter, which means the `id` parameter must be an integer and the value must be greater than or equal to 1 and less than or equal to 100:

```
[HttpGet("{id:int:range(1, 100)}")]
 public async Task<ActionResult<Post>> GetPost(int id)
 {
   // Omitted for brevity
 }
```

Route constraints can be used to make a route more specific. However, they should not be used to validate the input. If the input is invalid, the API should return a `400 Bad Request` response rather than a `404 Not Found` response.

Binding source attributes

We can define parameters in the action. See the following action method:

```
[HttpGet("{id}")]
 public async Task<ActionResult<Post>> GetPost(int id)
```

The `GetPost()` method has a parameter named `id`, which matches the parameter in the `{id}` route template. So, the value of `id` will come from the route, such as 1 in the `/api/posts/1` URL. This is called parameter inference.

ASP.NET Core offers the following binding source attributes:

- `[FromBody]`: The parameter is from the request body
- `[FromForm]`: The parameter is from the form data in the request body
- `[FromHeader]`: The parameter is from the request header
- `[FromQuery]`: The parameter is from the query strings in the request
- `[FromRoute]`: The parameter is from the route path
- `[FromServices]`: The parameter is from the **DI** container

For example, we can define a pagination action method as follows:

```
[HttpGet("paged")]
 public async Task<ActionResult<List<Post>>> GetPosts([FromQuery] int
pageIndex, [FromQuery] int pageSize)
 {
     // Omitted for brevity
 }
```

The preceding code means the pageIndex parameter and the pageSize parameter should be from query strings in the URL, such as /api/posts/paged?pageIndex=1&pageSize=10.

When an [ApiController] attribute is applied to a controller class, a set of default inference rules will be applied, so we do not need to explicitly add these binding source attributes. For example, the following code shows a POST action method:

```
[HttpPost]
public async Task<ActionResult<Post>> CreatePost(Post post)
{
    // Omitted for brevity
}
```
The post parameter is a complex type, so [FromBody] inferred that the post should be from the request body. But [FromBody] is not inferred for simple data types, such as int, string, and so on. We will define an action method as follows:
```
[HttpPost("search")]
public async Task<ActionResult<Post>> SearchPosts(string keyword)
```

The keyword parameter is a simple type, so [FromQuery] inferred that the keyword parameter should be from the query strings in the URL, such as /api/posts/search?keyword=xyz. If we want to force the keyword parameter to be from the request body, we can use the [FromBody] attribute as follows:

```
[HttpPost("search")]
public async Task<ActionResult<Post>> SearchPosts([FromBody] string
keyword)
```

Then, the keyword parameter must be from the request body. Note that this is a bad example because we usually do not use the request body to pass a simple type parameter.

The default inference rules of those binding source attributes are listed next:

- For complex type parameters, if the type is registered in the DI container, [FromServices] is inferred.

- For complex type parameters that are not registered in the DI container, [FromBody] is inferred. It does not support multiple [FromBody] parameters.

- For types such as IFormFile and IFormFileCollection, [FromForm] is inferred.

- For any parameters that appear in the route, [FromRoute] is inferred.

- For any parameters of simple types, such as int, string, and so on, [FromQuery] is inferred.

If a parameter can be inferred based on these rules, the binding source attribute can be omitted. Otherwise, we need to explicitly specify the binding source attribute.

Routing is a very important concept in REST APIs. Ensure that routes are well designed, intuitive, and easy to understand. This will help the consumers of your REST APIs use them easily.

Next, we will check the configuration in ASP.NET Core.

Configuration

ASP.NET Core provides a comprehensive configuration framework that makes it easy to work with configuration settings. A configuration is considered a key-value pair. These configuration settings are stored in a variety of sources, such as JSON files, environment variables, and command-line arguments, or in the cloud, such as Azure Key Vault. In ASP.NET Core, these sources are referred to as **configuration providers**. Each configuration provider is responsible for loading configuration settings from a specific source.

ASP.NET Core supports a set of configuration providers, such as the following:

- The file configuration provider, such as, `appsettings.json`
- The User secrets
- The environment variables configuration provider
- The command-line configuration provider
- The Azure App Configuration provider
- The Azure Key Vault configuration provider

The configuration of ASP.NET Core is provided by the `Microsoft.Extension.Configuration` NuGet package. You do not need to install this package explicitly as it is already installed with the default ASP.NET Core template, which provides several built-in configuration providers, such as `appsettings.json`. These configuration providers are configured in priority order. We will discuss this in more detail in the *Understanding the priorities of configuration and environment variables* section. First, let us look at how to use `appsettings.json`.

Run the following command to create a new ASP.NET Core web API project:

```
dotnet new webapi -n ConfigurationDemo -controllers
```

You can download the example project named `ConfigurationDemo` from the `/samples/chapter3/ConfigurationDemo` folder in the chapter's GitHub repository.

Using appsettings.json

By default, ASP.NET Core apps are configured to read configuration settings from `appsettings.json` using `JsonConfigurationProvider`. The `appsettings.json` file is located in the project's root directory, which is a JSON file that contains key-value pairs. The following code shows the default content of the `appsettings.json` file:

```
{
  "Logging": {
    "LogLevel": {
      "Default": "Information",
      "Microsoft.AspNetCore": "Warning"
    }
  },
  "AllowedHosts": "*"
}
```

You will find another `appsettings.Development.json` file, which will be used for the development environment. We will introduce the environment in the following section.

Let us add a `"MyKey": "MyValue"` key-value pair to the `appsettings.json` file. This key-value pair is an example configuration that we will read in the code using `JsonConfigurationProvider`:

```
{
  "Logging": {
    "LogLevel": {
      "Default": "Information",
      "Microsoft.AspNetCore": "Warning"
    }
  },
  "AllowedHosts": "*",
  "MyKey": "MyValue"
}
```

Create a new controller named `ConfigurationController` in the `Controllers` folder. In this controller, we will read the configuration value from the `appsettings.json` file and return it as a string. The following code shows the `ConfigurationController` class:

```
using Microsoft.AspNetCore.Mvc;

namespace ConfigurationDemo.Controllers;

[ApiController]
[Route("[controller]")]
```

```
public class ConfigurationController(IConfiguration configuration) :
ControllerBase
{
    [HttpGet]
    [Route("my-key")]
    public ActionResult GetMyKey()
    {
        var myKey = configuration["MyKey"];
        return Ok(myKey);
    }
}
```

To access the configuration settings, we need to inject the IConfiguration interface into the constructor of the controller. The IConfiguration interface represents a set of key/value application configuration properties. The following code shows how to access the configuration settings:

```
var myKey = configuration["MyKey"];
```

Run the application and send a request to the /Configuration/my-key endpoint. You can use any HTTP client, such as Postman, Thunder Client in VS Code, or HttpRepl. The following code shows how to use HttpRepl:

```
httprepl http://localhost:5116
cd Configuration
get my-key
```

You will see the following response:

```
HTTP/1.1 200 OK
Content-Type: text/plain; charset=utf-8
Date: Fri, 23 Sep 2022 11:22:40 GMT
Server: Kestrel
Transfer-Encoding: chunked

MyValue
```

The configuration supports hierarchical settings. For example, consider the following configuration settings:

```
{
  "Database": {
    "Type": "SQL Server",
    "ConnectionString": "This is the database connection string"
  }
}
```

To access the `Type` and `ConnectionString` properties, we can use the following code:

```
[HttpGet]
[Route("database-configuration")]
public ActionResult GetDatabaseConfiguration()
{
    var type = configuration["Database:Type"];
    var connectionString = configuration["Database:ConnectionString"];
    return Ok(new { Type = type, ConnectionString = connectionString
});
}
```

Note that we use a colon (`:`) to separate the hierarchical settings.

Run the application and send a request to the `/Configuration/database-configuration` endpoint. If you use HttpRepl, you can use the following command:

```
httprepl http://localhost:5116
cd Configuration
get database-configuration
```

The following code shows a response from HttpRepl:

```
HTTP/1.1 200 OK
Content-Type: application/json; charset=utf-8
Date: Fri, 23 Sep 2022 11:35:55 GMT
Server: Kestrel
Transfer-Encoding: chunked

{
  "type": "SQL Server",
  "connectionString": "This is the database connection string"
}
```

Using the `IConfiguration` interface, we can access the configuration settings with the `configuration[key]` format. However, hardcoding the keys is not a good practice. To avoid hardcoding, ASP.NET Core supports the options pattern, which can provide a strongly typed way to access hierarchical settings.

Using the options pattern

To use the options pattern, we need to create a class that represents the configuration settings. The following code shows how to create a class named `DatabaseOption`:

```
namespace ConfigurationDemo;

public class DatabaseOption
```

```
{
    public const string SectionName = "Database";
    public string Type { get; set; } = string.Empty;
    public string ConnectionString { get; set; } = string.Empty;
}
```

The `SectionName` field is used to specify the section name in the `appsettings.json` file. This field is not mandatory. But if we do not define it here, we will need to pass a hardcoded string for the section name when we bind the configuration section. To better leverage the strong typing, we can define a `SectionName` field. The `Type` and `ConnectionString` properties are used to represent the `Type` and `ConnectionString` fields in the `appsettings.json` file.

Note that an option class must be non-abstract with a public parameterless constructor.

There are multiple ways to use the options pattern. Let's continue.

Using the ConfigurationBinder.Bind() method

First, Let's use the `ConfigurationBinder.Bind()` method, which attempts to bind a given object instance to configuration values by recursively matching property names against configuration keys.

In the `ConfigurationController` class, add the following code:

```
[HttpGet]
[Route("database-configuration-with-bind")]
public ActionResult GetDatabaseConfigurationWithBind()
{
    var databaseOption = new DatabaseOption();
    // The `SectionName` is defined in the `DatabaseOption` class,
which shows the section name in the `appsettings.json` file.
    configuration.GetSection(DatabaseOption.SectionName).
Bind(databaseOption);
    // You can also use the code below to achieve the same result
    // configuration.Bind(DatabaseOption.SectionName, databaseOption);
    return Ok(new { databaseOption.Type, databaseOption.
ConnectionString });
}
```

Run the application and send a request to the `/Configuration/database-configuration-with-bind` endpoint. You will see the same response as in the previous section, *Using appsettings. json*. In this way, we can use the strongly typed option class to access the configuration settings, such as `databaseOption.Type`.

Using the ConfigurationBinder.Get<TOption>() method

We can also use the `ConfigurationBinder.Get<TOption>()` method, which attempts to bind the configuration instance to a new instance of type T. If this configuration section has a value, then that value will be used; otherwise, it attempts to bind the configuration instance by matching property names against configuration keys recursively. The code is shown next:

```
[HttpGet]
[Route("database-configuration-with-generic-type")]
public ActionResult GetDatabaseConfigurationWithGenericType()
{
    var databaseOption = configuration.GetSection(DatabaseOption.
SectionName).Get<DatabaseOption>();
    return Ok(new { databaseOption.Type, databaseOption.
ConnectionString });
}
```

Run the application and send a request to the `/Configuration/database-configuration-with-generic-type` endpoint. You will see the same response as in the *Using appsettings.json* section.

Using the IOptions<TOption> interface

ASP.NET Core provides built-in DI support for the options pattern. To use DI, we need to register the `DatabaseOption` class in the `Services.Configure()` method of the `Program.cs` file. The following code shows how to register the `DatabaseOption` class:

```
// Register the DatabaseOption class as a configuration object.
// This line must be added before the `builder.Build()` method.
builder.Services.Configure<DatabaseOption>(builder.Configuration.
GetSection(DatabaseOption.SectionName));

var app = builder.Build();
```

Next, we can use DI to inject the `IOptions<DatabaseOption>` interface into the `ConfigurationController` class. The following code shows how to inject the `IOptions<DatabaseOption>` interface:

```
[HttpGet]
[Route("database-configuration-with-ioptions")]
public ActionResult
GetDatabaseConfigurationWithIOptions([FromServices]
IOptions<DatabaseOption> options)
{
    var databaseOption = options.Value;
```

```
        return Ok(new { databaseOption.Type, databaseOption.
ConnectionString });
    }
```

Run the application and send a request to the /Configuration/database-configuration-with-ioptions endpoint. You will see the same response as in the *Using appsettings.json* section.

Using other options interfaces

We have introduced several ways to use the options pattern. What differences do they have?

Run the application and test the preceding endpoints. You will see all responses are the same, which contains a Type property with the value SQL Server.

Keep the application running. Let's make a change to the appsettings.json file. Change the Type property from SQL Server to MySQL. Save the file and send the requests to these endpoints again. You will find the following results:

- database-configuration returns the *new* value MySQL

- database-configuration-with-bind returns the *new* value MySQL

- database-configuration-with-generic-type returns the *new* value MySQL

- database-configuration-with-ioptions returns the *old* value SQL Server

Let's try to use the IOptionsSnapshot<T> interface to replace the IOptions<TOption> interface. The IOptionsSnapshot<TOption> interface provides a snapshot of options for the current request. The following code shows how to use the IOptionsSnapshot<TOption> interface:

```
[HttpGet]
[Route("database-configuration-with-ioptions-snapshot")]
public ActionResult
GetDatabaseConfigurationWithIOptionsSnapshot([FromServices]
IOptionsSnapshot<DatabaseOption> options)
{
    var databaseOption = options.Value;
    return Ok(new { databaseOption.Type, databaseOption.
ConnectionString });
}
```

Run the application again. Change the Type property in the appsettings.json file. Send requests to the /Configuration/database-configuration-with-ioptions-snapshot endpoint. You will find the response is the *new* value.

Okay, we now know the difference between the IOptions<TOption> interface and the IOptionsSnapshot<TOption> interface:

- The IOptions<TOption> interface provides a way to access options, but it cannot get the latest value if the setting value has been changed when the application is running.
- The IOptionsSnapshot<TOption> interface provides a snapshot of options for the current request. The IOptionsSnapshot<TOption> interface is useful when we want to get the latest options for the current request.

But why?

The ASP.NET Core framework provides built-in support for appsetting.json using JsonConfigurationProvider, which reads the configuration values from the appsettings.json file. When the framework registers JsonConfigurationProvider, the code looks like this:

```
config.AddJsonFile("appsettings.json", optional: true, reloadOnChange:
true)
    .AddJsonFile($"appsettings.{env.EnvironmentName}.json", optional:
true, reloadOnChange: true);
```

The reloadOnChange parameter is set to true, which means the configuration values will be reloaded if the appsettings.json file has been changed. So, the ConfigurationBinder.Bind() method and the ConfigurationBinder.Get<TOption>() method can get the latest value.

However, when the IOptions<TOption> interface is registered by the ASP.NET Core framework, it is registered as a *singleton* service, which means an instance of IOption<TOption> will be created only once. You can inject it into any service lifetime, but it cannot read the latest value if the setting value has been changed.

In contrast, the IOptionsSnapshot<TOption> interface is registered as a *scoped* service, so it cannot be injected into a singleton service. It is useful in scenarios if you want to get the latest options for each request.

It looks like IOptionsSnapshot<TOption> is better than IOptions<TOption>. Not really. IOptionsSnapshot<TOption> can only cache options for the current request. It may cause performance issues because it is recomputed per request. So, you need to choose the interface to use wisely. If the options are not changed, you can use the IOptions<TOption> interface. If the options are changed frequently and you want to ensure the app gets the latest value per request, you can use the IOptionsSnapshot<TOption> interface.

There is another options interface called IOptionsMonitor<TOption>. It is a combination of the IOptions<TOption> and the IOptionsSnapshot<TOption> interfaces. It provides the following features:

- It is a singleton service, which can be injected into any service lifetime
- It supports reloadable configuration

Here is an example of using the `IOptionsMonitor<TOption>` interface:

```
[HttpGet]
[Route("database-configuration-with-ioptions-monitor")]
public ActionResult
GetDatabaseConfigurationWithIOptionsMonitor([FromServices]
IOptionsMonitor<DatabaseOption> options)
{
    var databaseOption = options.CurrentValue;
    return Ok(new { databaseOption.Type, databaseOption.
ConnectionString });
}
```

The `IOptionsMonitor<TOption>` interface provides the `CurrentValue` property to get the latest value. It also provides the `OnChange(Action<TOption, string> listener)` method to register a listener that will be called whenever the options are reloaded. Normally, you do not need to use the `OnChange()` method unless you want to do something when the options are reloaded.

Using named options

Sometimes, we need to use multiple database instances in our application. Consider the following scenario:

```
{
  "Databases": {
    "System": {
      "Type": "SQL Server",
      "ConnectionString": "This is the database connection string for
the system database."
    },
    "Business": {
      "Type": "MySQL",
      "ConnectionString": "This is the database connection string for
the business database."
    }
  }
}
```

Rather than creating two classes to represent the two database options, we can use the named options feature. The following code shows how to use the named options feature for each section:

```
public class DatabaseOptions
{
    public const string SystemDatabaseSectionName = "System";
    public const string BusinessDatabaseSectionName = "Business";
```

```
        public string Type { get; set; } = string.Empty;
        public string ConnectionString { get; set; } = string.Empty;
}
```

Then, register the named options feature in the `Program.cs` file:

```
builder.Services.Configure<DatabaseOptions>(DatabaseOptions.
SystemDatabaseSectionName, builder.Configuration.
GetSection($"{DatabaseOptions.SectionName}:{DatabaseOptions.
SystemDatabaseSectionName}"));
builder.Services.Configure<DatabaseOptions>(DatabaseOptions.
BusinessDatabaseSectionName, builder.Configuration.
GetSection($"{DatabaseOptions.SectionName}:{DatabaseOptions.
BusinessDatabaseSectionName}"));
```

The following code shows how to access the named options:

```
[HttpGet]
[Route("database-configuration-with-named-options")]
public ActionResult
GetDatabaseConfigurationWithNamedOptions([FromServices]
IOptionsSnapshot<DatabaseOptions> options)
{
    var systemDatabaseOption = options.Get(DatabaseOptions.
SystemDatabaseSectionName);
    var businessDatabaseOption = options.Get(DatabaseOptions.
BusinessDatabaseSectionName);
    return Ok(new { SystemDatabaseOption = systemDatabaseOption,
BusinessDatabaseOption = businessDatabaseOption });
}
```

Run the application and send a request to the `/Configuration/database-configuration-with-named-options` endpoint. You will find the response contains the two database options, as shown next:

```
{
  "systemDatabaseOption": {
    "type": "SQL Server",
    "connectionString": "This is the database connection string for
the system database."
  },
  "businessDatabaseOption": {
    "type": "MySQL",
    "connectionString": "This is the database connection string for
the business database."
  }
}
```

Now, Let's summarize the options feature in ASP.NET Core:

	Server lifetime	Reloadable configuration	Named options
`IOptions<TOption>`	Singleton	No	No
`IOptionsSnapshot<TOption>`	Scope	Yes	Yes
`IOptionsMonitor<TOption>`	Singleton	Yes	Yes

Table 3.1 – Summary of the options feature in `ASP.NET` Core

Next, we will discuss how to register a group of options to make the `Program.cs` file cleaner.

Group options registration

In *Chapter 2*, we introduced how to use group registration to register multiple services in an extension method. The group registration feature is also available for the options feature. The following code shows how to use the group registration feature to register multiple options:

```
using ConfigurationDemo;
namespace DependencyInjectionDemo;

public static class OptionsCollectionExtensions
{
    public static IServiceCollection AddConfig(this IServiceCollection
services, IConfiguration configuration)
    {
        services.Configure<DatabaseOption>(configuration.
GetSection(DatabaseOption.SectionName));
        services.Configure<DatabaseOptions>(DatabaseOptions.
SystemDatabaseSectionName, configuration.
GetSection($"{DatabaseOptions.SectionName}:{DatabaseOptions.
SystemDatabaseSectionName}"));
        services.Configure<DatabaseOptions>(DatabaseOptions.
BusinessDatabaseSectionName, configuration.
GetSection($"{DatabaseOptions.SectionName}:{DatabaseOptions.
BusinessDatabaseSectionName}"));
        return services;
    }
}
```

Then, register the options in the `Program.cs` file:

```
builder.Services.AddConfig(builder.Configuration);
```

Now, the `Program.cs` file is much cleaner.

Other configuration providers

We mentioned that ASP.NET Core supports multiple configuration providers. The configuration provider for reading the `appsettings.json` file is `JsonConfigurationProvider`, which is derived from the `FileConfigurationProvider` base class. There are some other implementations of the `FileConfigurationProvider` base class, such as `IniConfigurationProvider`, `XmlConfigurationProvider`, and so on.

Besides `JsonConfigurationProvider`, the ASP.NET Core framework automatically registers the following configuration providers:

- A **user secrets configuration provider** is used to read secrets from the local secrets file when the app runs in the `Development` environment

- A **non-prefixed environment variables configuration provider** is used to read environment variables that do not have a prefix

- A **command-line configuration provider** is used to read command-line arguments

Let us see more details about these configuration providers.

User secrets configuration provider

It is not a good practice to store sensitive information in the `appsettings.json` file. For example, if the database connection string is stored in the `appsettings.json` file, developers may accidentally commit the database connection string (or other sensitive information, secrets, and so on) to the source control system, which will cause security issues.

Instead, we can use the user secrets feature to store sensitive information in the local secrets file. The user secrets feature is only available in the `Development` environment. By default, the ASP.NET Core framework registers the user secrets configuration provider after the JSON configuration provider. Therefore, the user secrets configuration provider has higher priority than the JSON configuration provider, so it will override the JSON configuration provider if the same configuration key exists in both providers.

To use user secrets, we need to use the Secret Manager tool to store the secrets in a local secrets file. Run the following command in the project folder to initialize a local secrets file:

```
dotnet user-secrets init
```

The preceding command creates a `UserSecretsId` property in the `.csproj` file. By default, the value of the `UserSecretsId` property is a GUID, such as the following:

```
<PropertyGroup>
  <TargetFramework>net8.0</TargetFramework>
  <Nullable>enable</Nullable>
  <ImplicitUsings>enable</ImplicitUsings>
```

```
  <UserSecretsId>f3351c6a-2508-4243-8d80-89c27758164d</UserSecretsId>
</PropertyGroup>
```

Then, we can use the Secret Manager tool to store secrets in the local secrets file. Run the following command from the project folder to store secrets:

```
dotnet user-secrets set "Database:Type" "PostgreSQL"
dotnet user-secrets set "Database:ConnectionString" "This is the
database connection string from user secrets"
```

After running the preceding commands, a `secrets.json` file is created in the `%APPDATA%\Microsoft\UserSecrets\<UserSecretsId>` folder. The `secrets.json` file contains the following content:

```
{
  "Database:Type": "PostgreSQL",
  "Database:ConnectionString": "This is the database connection string
from user secrets"
}
```

Note that the JSON structure is flattened.

Location of the secrets.json file

If you use Linux or macOS, the `secrets.json` file is created in the `~/.microsoft/usersecrets/<UserSecretsId>` folder.

Run `dotnet run` to run the application and send a request to the `/Configuration/database-configuration` endpoint. You will find the response contains the database options from the user secrets, which overrides the database options from the `appsettings.json` file and contains a PostgreSQL database type.

The local secrets file is out of the project folder and not committed to the source control system. Keep in mind that the Secret Manager tool is only for development purposes. Developers should have the responsibility to protect the local secrets file.

There are some commands to operate the local secrets file. You need to run the following commands from the project folder:

```
# List all the secrets
dotnet user-secrets list
# Remove a secret
dotnet user-secrets remove "Database:Type"
# Clear all the secrets
dotnet user-secrets clear
```

> **Important note**
>
> If you download the code example for this section, the secrets file is not included in the repository. You need to run the `dotnet user-secrets init` command to initialize the secrets file on your local machine.

Environment variables configuration provider

.NET and ASP.NET Core define some environment variables that can be used to configure the application. These specific variables have a prefix of `DOTNET_`, `DOTNETCORE_`, or `ASPNETCORE_`. Variables that have the `DOTNET_` or `DOTNETCORE_` prefix are used to configure the .NET runtime. Variables that have the `ASPNETCORE_` prefix are used to configure ASP.NET Core. For example, the `ASPNETCORE_ENVIRONMENT` variable is used to set the environment name. We will discuss the environment in the *Environments* section.

For those environment variables that do not have the `ASPNETCORE_` prefix, ASP.NET Core can also use the environment variables configuration provider to read them. Environment variables have a higher priority than the `appsettings.json` file. For example, we have the following configuration in the `appsettings.json` file:

```
{
  "Database": {
    "Type": "SQL Server",
    "ConnectionString": "This is the database connection string."
  }
}
```

If we set the `Database__Type` environment variable to `MySQL`, the `Database__Type` value in the `appsettings.json` file will be overridden by the environment variable value. The following code shows how to access an environment variable in PowerShell:

```
$Env:<variable-name>
```

To represent the hierarchical keys of environment variables, it is recommended to use `__` (double underscore) as a separator because it is supported by all platforms. Please do not use `:` because it is not supported by Bash.

You can use the following command to set an environment variable in PowerShell:

```
$Env:Database__Type="SQLite"
```

To check if the environment variable is set correctly, run the following command:

```
$Env:Database__Type
```

> **Important note**
>
> If you use Bash, you need to use the following command to set the environment variable:
>
> `export Database__Type="SQLite"`
>
> For more information, please refer to `https://linuxize.com/post/how-to-set-and-list-environment-variables-in-linux/`.
>
> Also, please note that, unlike in Windows, environment variable names are case-sensitive on macOS and Linux.

You will see the output is `SQLite`. Now, in the same PowerShell session, you can use `dotnet run` to run the application and send a request to the `/Configuration/database-configuration` endpoint. You will find the response contains the `SQLite` value, even though the `appsettings.json` file contains the `SQL Server` value. That means the environment variable value overrides the `appsettings.json` file value.

Command-line configuration provider

Command-line arguments have a higher priority than environment variables. By default, the configuration settings set on the command line override values set with other configuration providers.

Following the example in the previous section, the `Database__Type` value is set to `SQL Server` in the `appsettings.json` file. We also set the `Database__Type` environment variable to `SQLite`. Now, let us change the value to `MySQL` in the command-line argument using the following command:

```
dotnet run Database:Type=MySQL
```

Send a request to the `/Configuration/database-configuration` endpoint. You will find the response contains the `MySQL` value, which means the command-line argument value overrides the environment variable value.

Command-line arguments can also be set in the following ways:

```
dotnet run --Database:Type MySQL
dotnet run /Database:Type MySQL
```

If `--` or `/` is used for the key, the value can follow a space. Otherwise, the value must follow the `=` sign. Please do not mix the two ways in the same command.

Azure Key Vault configuration provider

Azure Key Vault is a cloud-based service that provides secure storage for secrets, such as passwords, certificates, and keys. The Azure Key Vault configuration provider is used to read the secrets from an Azure Key Vault. It is a good choice for running the application in production.

To use it, you need to install the following NuGet packages:

- `Azure.Extensions.AspNetCore.Configuration.Secrets`
- `Azure.Identity`

Azure App Configuration is a cloud-based service that provides a centralized configuration store for managing application settings and feature flags. App Configuration complements Azure Key Vault. It aims to simplify tasks of working with complex application settings. You need to install the following NuGet packages to use the Azure App Configuration provider:

- `Microsoft.Azure.AppConfiguration.AspNetCore`

We will not cover details of the Azure Key Vault and Azure App Configuration providers in this book. For more information, please refer to the following links:

- `https://learn.microsoft.com/en-us/azure/key-vault/general/`
- `https://docs.microsoft.com/en-us/azure/azure-app-configuration/`

Environments

In the previous section, we introduced how to read the configuration settings from various resources, including the `appsettings.json` file, user secrets, environment variables, and command-line arguments. In this section, we will discuss environments in more detail.

Run the following command to create a new ASP.NET Core web API project:

```
dotnet new webapi -n EnvironmentDemo -controllers
```

You can download the example project named `EnvironmentDemo` from the `/samples/chapter3/EnvironmentsDemo` folder in the chapter's GitHub repository.

We have mentioned the default ASP.NET Core web API template contains an `appsettings.json` file and an `appsettings.Development.json` file. When we run the application using `dotnet run`, the application runs in the `Development` environment. So the configuration settings in the `appsettings.Development.json` file override the configuration settings in the `appsettings.json` file.

Add the following section to the `appsettings.Development.json` file:

```
{
  // Omitted for brevity
  "Database": {
    "Type": "SQL Server",
```

```
      "ConnectionString": "This is the database connection string from
base appsettings.json."
    }
}
```

Then, add the following section to the `appsettings.Development.json` file:

```
{
  // Omitted for brevity
  "Database": {
    "Type": "LocalDB",
    "ConnectionString": "This is the database connection string from
appsettings.Development.json"
    }
}
```

> **Important note**
>
> Note that user secrets have higher priority than the `appsettings.Development.json`
> file. So, if you configured the local user secrets in the previous section, please clear the secrets.

Create a new controller named `ConfigurationController.cs` and add the following code:

```csharp
using Microsoft.AspNetCore.Mvc;

namespace EnvironmentsDemo.Controllers;

[ApiController]
[Route("[controller]")]
public class ConfigurationController(IConfiguration configuration) :
ControllerBase
{
    private readonly IConfiguration _configuration;

    public ConfigurationController(IConfiguration configuration)
    {
        _configuration = configuration;
    }

    [HttpGet]
    [Route("database-configuration")]
    public ActionResult GetDatabaseConfiguration()
    {
        var type = configuration["database:Type"];
        var connectionString =
configuration["Database:ConnectionString"];
```

```
        return Ok(new { Type = type, ConnectionString =
connectionString });
    }
}
```

Run the application using dotnet run. Send the request to the /Configuration/database-configuration endpoint. You will find the response contains a LocalDB value, which means the appsettings.Development.json file overrides the appsettings.json file.

So, where is the environment name Development set?

Understanding the launchSettings.json file

Open the launchSettings.json file in the Properties folder. You will find the ASPNETCORE_ENVIRONMENT environment variable is set to Development:

```
    "profiles": {
      "http": {
        "commandName": "Project",
        "dotnetRunMessages": true,
        "launchBrowser": true,
        "launchUrl": "swagger",
        "applicationUrl": "http://localhost:5161",
        "environmentVariables": {
          "ASPNETCORE_ENVIRONMENT": "Development"
        }
      },
      "https": {
        "commandName": "Project",
        "dotnetRunMessages": true,
        "launchBrowser": true,
        "launchUrl": "swagger",
        "applicationUrl": "https://localhost:7096;http://
localhost:5161",
        "environmentVariables": {
          "ASPNETCORE_ENVIRONMENT": "Development"
        }
      },
      "IIS Express": {
        "commandName": "IISExpress",
        "launchBrowser": true,
        "launchUrl": "swagger",
        "environmentVariables": {
          "ASPNETCORE_ENVIRONMENT": "Development"
```

```
      }
    }
  }
```

The `launchSettings.json` file is used to configure the local development environment. It is not deployable. The default `launchSettings.json` file contains three profiles: `http`, `https`, and `IIS Express`:

- The `http` profile is used to run the application with the HTTP protocol
- The `https` profile is used to run the application with the HTTPS protocol
- The `IIS Express` profile is used to run the application in IIS Express

The `commandName` field in the `http` and `https` profiles is `Project`, which means the Kestrel server is launched to run the application. Similarly, the `IISExpress` value in the `IIS Express` profile means the application expects IIS Express to be the web server.

> **What is the Kestrel server?**
>
> Kestrel is a cross-platform web server for ASP.NET Core. Kestrel is included and enabled by default in ASP.NET Core project templates. ASP.NET Core can also be hosted in IIS (or IIS Express), but IIS is not cross-platform. So, Kestrel is the preferred web server for ASP.NET Core applications.

When running `dotnet run`, the first profile with the `commandName` value, `Project`, is used. For the demo project, the `http` profile is used. The `ASPNETCORE_ENVIRONMENT` environment variable is set to `Development` in the `http` profile. So, the application runs in the `Development` environment. You can see the output in the console:

```
Building...
info: Microsoft.Hosting.Lifetime[14]
      Now listening on: http://localhost:5161
info: Microsoft.Hosting.Lifetime[0]
      Application started. Press Ctrl+C to shut down.
info: Microsoft.Hosting.Lifetime[0]
      Hosting environment: Development
info: Microsoft.Hosting.Lifetime[0]
      Content root path: C:\dev\web-api-with-asp-net\example_code\
chapter3\EnvironmentsDemo
```

We can specify the profile to use when running the application using the `--launch-profile` argument:

```
dotnet run --launch-profile https
```

Note that this approach is only available for Kestrel profiles. You cannot use this argument to run the application in IIS Express.

If you use VS 2022 to open the project, you can choose which profile to use, like this:

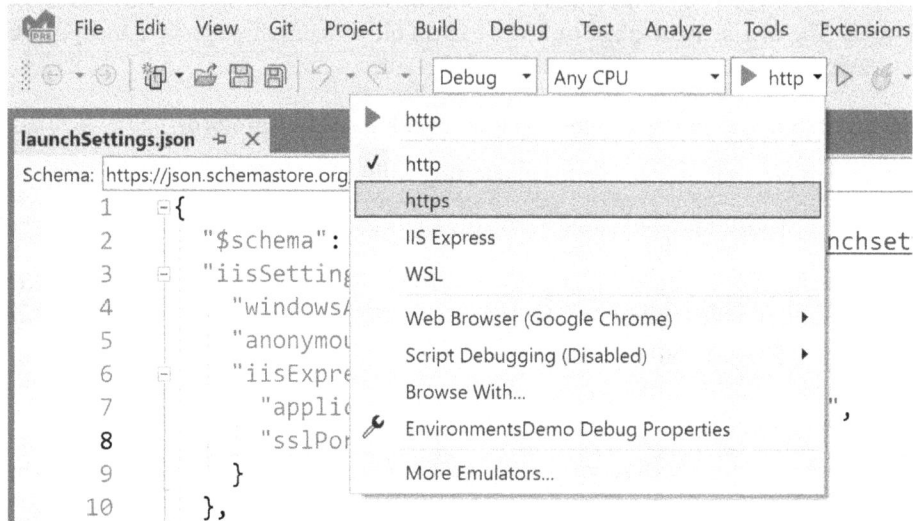

Figure 3.1 – Choosing a profile to use in Visual Studio 2022

Next, let's explore how to configure the application to run in the Production environment.

Setting the environment

There are several ways to change the environment. Let's create a new file named `appsettings. Production.json`. This configuration file will be used for the Production environment. Add the following section to the file:

```
{
  // Omitted for brevity
  "Database": {
    "Type": "PostgreSQL",
    "ConnectionString": "This is the database connection string from
  appsettings.Production.json"
  }
}
```

Next, we will specify the environment as Production to apply this configuration.

Using the launchSettings.json file

For development purposes, we can create a new profile in the launchSettings.json file, which specifies the ASPNETCORE_ENVIRONMENT variable as Production. Add the following section to the launchSettings.json file:

```
// Omitted for brevity
  "profiles": {
    // Omitted for brevity
    "production": {
      "commandName": "Project",
      "dotnetRunMessages": true,
      "launchBrowser": true,
      "launchUrl": "swagger",
      "applicationUrl": "https://localhost:7096;http://
localhost:5161",
      "environmentVariables": {
        "ASPNETCORE_ENVIRONMENT": "Production"
      }
    }
    // Omitted for brevity
  }
```

Use the following command to run the application in the Production environment:

```
dotnet run --launch-profile production
```

You will see in the console that the application runs in the Production environment:

```
PS C:\dev\web-api-with-asp-net\example_code\chapter3\EnvironmentsDemo>
dotnet run --launch-profile production
Building...
warn: Microsoft.AspNetCore.Server.Kestrel.Core.KestrelServer[8]
      The ASP.NET Core developer certificate is not trusted. For
information about trusting the ASP.NET Core developer certificate, see
https://aka.ms/aspnet/https-trust-dev-cert.
info: Microsoft.Hosting.Lifetime[14]
      Now listening on: https://localhost:7096
info: Microsoft.Hosting.Lifetime[14]
      Now listening on: http://localhost:5161
info: Microsoft.Hosting.Lifetime[0]
      Application started. Press Ctrl+C to shut down.
info: Microsoft.Hosting.Lifetime[0]
      Hosting environment: Production
info: Microsoft.Hosting.Lifetime[0]
      Content root path: C:\dev\web-api-with-asp-net\example_code\
chapter3\EnvironmentsDemo
```

It looks good. Let's try to access the /Configuration/database-configuration endpoint. You will see the response is from the appsettings.Production.json file.

Using the ASPNETCORE_ENVIRONMENT environment variable

You can also set the ASPNETCORE_ENVIRONMENT environment variable to Production in the current session, as shown next:

```
$Env:ASPNETCORE_ENVIRONMENT = "Production"
dotnet run --no-launch-profile
```

Furthermore, you can set the environment variable globally in your system. Use the following command to set the environment variable globally:

```
[Environment]::SetEnvironmentVariable("ASPNETCORE_ENVIRONMENT",
"Production", "Machine")
```

The Machine argument sets the environment variable globally. You can also set the environment variable for the current user by using the User argument.

Using the --environment argument

Another way is to set the environment with the --environment argument:

```
dotnet run --environment Production
```

Using the launch.json file in VS Code

If you use VS Code to open the project, you can set the environment in the launch.json file in the .vscode folder. When you open an ASP.NET Core project, VS Code will prompt you to add required assets to debug the project. A launch.json file and a tasks.json file will be added to the .vscode folder. If you do not see the prompt, you can open the Command Palette and run the **.NET: Generate Assets for Build and Debug** command.

Open the launch.json file and you will find the following content:

```
{
  // Omitted for brevity
  "configurations": [
    {
      "name": ".NET Core Launch (web)",
      "type": "coreclr",
      // Omitted for brevity
      "env": {
        "ASPNETCORE_ENVIRONMENT": "Development"
      },
```

```
        // Omitted for brevity
    }
  ]
}
```

You can add a new configuration following the existing one. Change the `ASPNETCORE_ENVIRONMENT` field to `Production` and use `.NET Core Launch (Production)` as the name. Save the file. You can now run the application in the `Production` environment by clicking the green arrow in the debug panel:

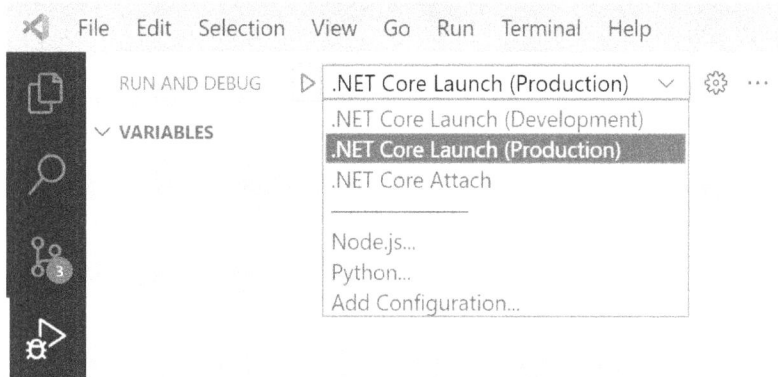

Figure 3.2 – Running the application in a specific environment in VS Code

---

Important note

The `launch.json` file is only used in VS Code **RUN AND DEBUG**. It is not used when you run the application using the `dotnet run` command.

Using the launchSettings.json file in Visual Studio 2022

Visual Studio 2022 provides a **Launch Profiles** dialog to set up the environment variables. You have multiple ways to open the **Launch Profiles** dialog:

- Open the **Debug** menu | <*Your project name*> **Debug Properties**.

- Click the arrow next to the green arrow in the debug panel and select <*Your project name*> **Debug Properties**.

- Right-click the project in the **Solution Explorer** window and select **Properties**. In the **Debug / General** tab, click the **Open debug launch profiles UI** link.

Then, you can see the **Launch Profiles** dialog:

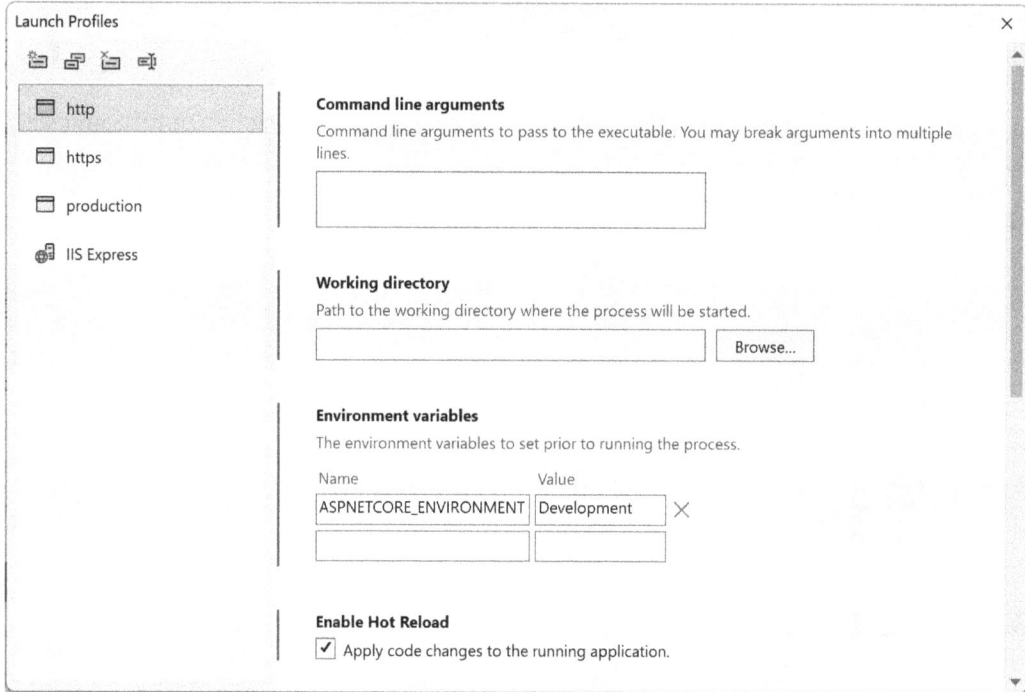

Figure 3.3 – The Launch Profiles dialog in Visual Studio 2022

If you make changes here, you need to restart the application to apply the changes.

We have learned how to set the environment. With the environment set, we can use different configurations for different environments.

Understanding the priorities of configuration and environment variables

We introduced quite a few different ways to read configuration values and environment variables. Let us see how configuration and environment variables are prioritized.

The following table shows the priorities of the configuration sources (the lowest number has the highest priority):

Source	Priority
Command-line arguments	1
Non-prefixed environment variables	2

Source	Priority
User secrets (`Development` environment only)	3
`appsettings.{Environment}.json`	4
`appsettings.json`	5

Table 3.2 – Priorities of configuration sources

If other configuration providers are registered in the `Program.cs` file, the later registered providers have higher priority than the earlier registered providers.

In terms of environment variables such as `ASPNETCORE_ENVIRONMENT`, the following table shows the priorities:

Source	Priority
Command-line arguments	1
`launchSettings.json` (development purposes only)	2
Environment variable in the current process	3
System environment variable	4

Table 3.3 – Priorities of environment variables

Note that there are some other ways to configure the environment that are not listed in the preceding table. For example, if you deploy the ASP.NET Core application to Azure App Service, you can set the `ASPNETCORE_ENVIRONMENT` environment variable in the App Service configuration. For Linux apps and container apps, Azure App Service passes these settings to the container using the `--env` flag to set the environment variable in the container.

Checking the environment in the code

Keep running the application in the `Production` environment. Let's try to access the Swagger UI page: `http://localhost:5161/swagger`. You will find it shows a `404 Not Found` error. Why?

This is because the Swagger UI page is only enabled in the `Development` environment. You can see the code in the `Program.cs` file:

```
if (app.Environment.IsDevelopment())
{
    app.UseSwagger();
    app.UseSwaggerUI();
}
```

The preceding code means that the Swagger UI page is enabled for the development environment only.

The three environment names, `Development`, `Staging`, and `Production`, are predefined in the ASP.NET Core framework, as shown next:

```
public static class Environments
{
    public static readonly string Development = "Development";
    public static readonly string Staging = "Staging";
    public static readonly string Production = "Production";
}
```

The `app.Environment.IsDevelopment()` method checks the current environment. If the current environment is `Development`, the Swagger UI page is enabled. Otherwise, it is disabled.

To set the environment in code, use the following code when creating `WebApplicationBuilder`:

```
var builder = WebApplication.CreateBuilder(new WebApplicationOptions
{
    EnvironmentName = Environments.Staging
});
```

The environment name is stored in the `IHostEnvironment.EnvironmentName` property. You can define your own environment names. For example, you can define an environment name such as `Test`. But the framework provides built-in methods, such as `IsDevelopment()`, `IsStaging()`, and `IsProduction()`, to check the environment. If you define your own environment names, you can use the `IHostEnvironment.IsEnvironment(string environmentName)` method to check the environment.

We can use the `System.Environment` class to get the environment variable in the code, such as `ASPNETCORE_ENVIRONMENT`. Add the following code to the `Program.cs` file:

```
// Omitted for brevity
var app = builder.Build();
// Read the environment variable ASPNETCORE_ENVIRONMENT
var environmentName = Environment.GetEnvironmentVariable("ASPNETCORE_
ENVIRONMENT");
Console.WriteLine($"ASPNETCORE_ENVIRONMENT is {environmentName}");
```

Run the application, and you can see the `ASPNETCORE_ENVIRONMENT` environment variable in the console:

```
PS C:\dev\web-api-with-asp-net\example_code\chapter3\EnvironmentsDemo>
dotnet run
Building...
ASPNETCORE_ENVIRONMENT is Development
info: Microsoft.Hosting.Lifetime[14]
      Now listening on: http://localhost:5161
# Omitted for brevity
```

What should we do for different environments?

For the development environment, we can enable the Swagger UI page, show detailed error messages, output debugging information, and so on. For the production environment, we should configure the application for best performance and maximum security. Consider the following points for the production environment:

- Disable the Swagger UI page.
- Disable detailed error messages.
- Show a friendly error message.
- Do not output debugging information.
- Enable cache.
- Enable HTTPS.
- Enable the response compression.
- Enable monitoring and logging.

Summary

In this chapter, we explored three important components of ASP.NET Core: routing, configuration, and environments. We discussed how to configure routing for an ASP.NET Core web API application and how to read parameters from the request. Additionally, we learned how to read configuration values from various sources, such as `appsettings.json`, environment variables, and command-line arguments. We also explored how to read configurations based on the environment, allowing us to enable different features for different environments.

In the next chapter, we will learn about two more essential components of ASP.NET Core: logging and middleware.

4

ASP.NET Core Fundamentals (Part 2)

In *Chapter 3*, we learned about three important components in ASP.NET Core: routing, configuration, and environment. Next, let's continue to explore the other components in ASP.NET Core.

In this chapter, we will cover the following topics:

- Logging
- Middleware

ASP.NET Core provides a flexible logging API that works with many logging providers. We can send and store the logs in various destinations, such as the console, text files, Azure Application Insights, and so on. When the application has issues, logs can help us find out what is going on. Logging belongs to a bigger topic called *observability*, which is a set of practices that help us understand the state of the application. In this chapter, we will learn how to use the logging API in ASP.NET Core.

Middleware is a component that can be plugged into the request pipeline to handle requests and responses. It is one of the most important improvements in ASP.NET Core compared to the traditional ASP.NET framework. The request pipeline is a chain of middleware components, and each of these components can do something with the request or response, or pass it to the next middleware component in the pipeline. In this chapter, we will learn how to use built-in middleware components and how to develop custom middleware components.

By the end of this chapter, you will be able to use the logging API to log messages, set up log levels, and configure log providers. Additionally, you will be able to develop custom middleware components to process requests and responses.

Technical requirements

The code examples in this chapter can be found at `https://github.com/PacktPublishing/Web-API-Development-with-ASP.NET-Core-8/tree/main/samples/chapter4`. You can use VS 2022 or VS Code to open the solutions.

Logging

Logging is an important part of any application. A well-designed logging system can capture data that helps you diagnose problems and monitor the application in production. Logging gives you insight into how, when, where, and why significant system events occurred so that you know how the application is performing and how users are interacting with it. Logging may help you to identify security weaknesses or potential attacks. Logging can also help you audit users activities.

Logging should not impact the performance of the application. It should be fast and efficient. It should not affect any logic of the application. When you add logging to an application, you should consider the following points:

- What information should be logged?
- What format should log messages be in?
- Where should log messages be sent?
- How long should log messages be kept?
- How to make sure log messages won't impact the performance of the application?

In this section, we will discuss how to use the logging system in ASP.NET Core.

Let's create a new project to learn how to use the logging API. Create a new ASP.NET Core web API project named `LoggingDemo` using the following command:

```
dotnet new webapi -n LoggingDemo -controllers
```

You can also download the source code named `LoggingDemo` from the `samples/chapter4` folder in the chapter's GitHub repository.

Using built-in logging providers

ASP.NET Core supports a logging API that works with various logging providers, including built-in logging providers and third-party logging providers. The default ASP.NET Core web API template has the following logging providers registered:

- Console logging provider
- Debug logging provider

- EventSource logging provider

- EventLog logging provider (Windows only)

To clearly see how these logging providers work, let's remove all the pre-registered logging providers and then add the console logging provider. Open the Program.cs file and add the following code:

```
var builder = WebApplication.CreateBuilder(args);
builder.Logging.ClearProviders();
builder.Logging.AddConsole();
```

Now, only the console logging provider is enabled. Let's use the console logging provider to output log messages. Open the WeatherForecastController.cs file; you can see the ILogger<WeatherForecastController> interface is injected into the constructor already:

```
private readonly ILogger<WeatherForecastController> _logger;
public WeatherForecastController(ILogger<WeatherForecastController> logger
{
    _logger = logger;
}
```

Open the WeatherForecastController.cs file in the project. Add the following code to the Get() method:

```
[HttpGet(Name = "GetWeatherForecast")]
public IEnumerable<WeatherForecast> Get()
{
    _logger.Log(LogLevel.Information, "This is a logging message.");
    // Omitted for brevity
}
```

Run the application using the dotnet run command. Request the /WeatherForecast endpoint using the browser. You can see the log message in the console:

```
info: LoggingDemo.WeatherForecastController[0]
      This is a logging message.
```

If you run the application in VS 2022 and run the application using the *F5* key, you can see a log message in the console window, but you cannot see it in the **Output** window of VS 2022:

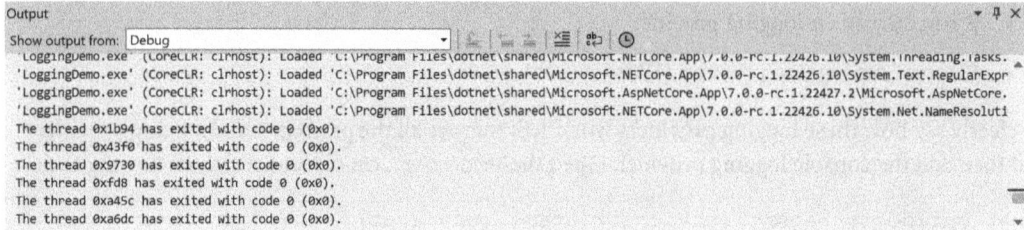

Figure 4.1 – The Output window for debug messages in VS 2022

To send logging messages to the **Output** window, we need to enable the `Debug` logging provider. Open the `Program.cs` file and add the following code:

```
builder.Logging.ClearProviders();
builder.Logging.AddConsole();
builder.Logging.AddDebug();
```

Press *F5* to run the application in VS 2022 again. Now, you can see the log message in the **Output** window:

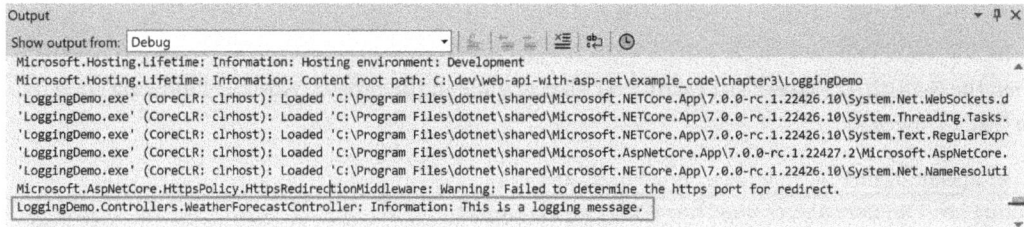

Figure 4.2 – Debug logging messages in VS 2022

So, if want to add more other logging providers, we can call the extension methods of the `ILoggingBuilder` interface. Some third-party logging providers also provide the extension methods of the `ILoggingBuilder` interface.

For example, if we need to write the log messages to the Windows event log, we can add the `EventLog` logging provider. Add the following code to the `Program.cs` file:

```
builder.Logging.AddEventLog();
```

Test the application and we should be able to see the log messages in the Windows event log.

Wait – why can't we see it in the event log?

This is a specific scenario for the `EventLog` logging provider. Because it is a Windows-only logging provider, it does not inherit the default logging provider settings. We need to specify the logging level in the `appsettings.json` file. Open the `appsettings.Development.json` file and update the `Logging` section:

```
{
    "Logging": {
        "LogLevel": {
            "Default": "Trace",
            "Microsoft.AspNetCore": "Warning"
        },
        "EventLog": {
            "LogLevel": {
                "Default": "Information"
            }
        }
    }
}
```

We need to add an `EventLog` section to specify the logging level for the `EventLog` logging provider. If it is not specified, the default logging level is `Warning`, which is higher than `Information`. This would result in us not being able to see `Information` logging messages. Run the application again, and now we can see the log messages in the event log:

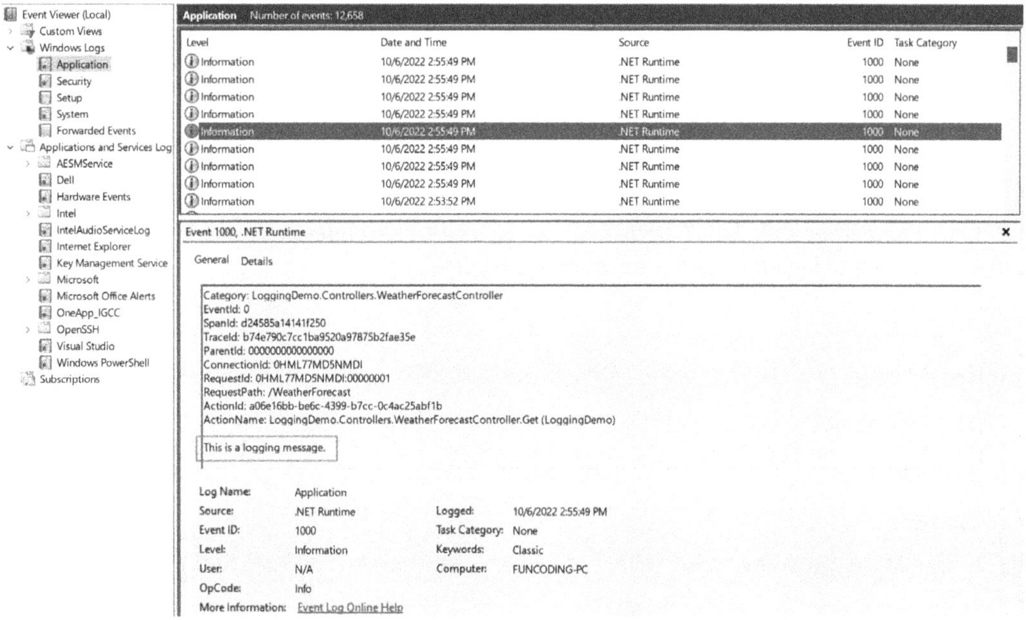

Figure 4.3 – Event log on Windows

We just introduced a new term – **logging level**. What is it?

Logging levels

In the preceding example, we used a `Log` method that accepts a `LogLevel` parameter. The `LogLevel` parameter indicates the severity of the log message. The `LogLevel` parameter can be one of the following values:

Level	Value	Description
Trace	0	Used for the most detailed messages. These messages may contain sensitive application data. These messages are disabled by default and should never be enabled in a production environment.
Debug	1	Used for debugging information and development. Use with caution in production because of the high volume. Normally, these logs should not have a long-term value.
Information	2	Used for tracking the general flow of the application. These logs should have a long-term value.
Warning	3	Used to indicate potential problems or unexpected events. These issues typically do not cause the application to fail.
Error	4	Used to indicate failures in the current operation or request and not an application-wide failure. These errors and exceptions cannot be handled.
Critical	5	Used to indicate critical failures that require immediate attention; for example, data loss scenarios.
None	6	Used to specify a logging category that should not write messages.

Table 4.1 – Logging levels

To simplify the method call, the `ILogger<TCategoryName>` interface provides the following extension methods to log messages for different logging levels:

- `LogTrace()`
- `LogDebug()`
- `LogInformation()`
- `LogWarning()`
- `LogError()`
- `LogCritical()`

You can use the `LogInformation()` method to replace the `Log()` method in the preceding example:

```
_logger.LogInformation("This is a logging message.");
```

You will see the same log message in the console window.

Let's add a `LogTrace()` method in the `WeatherForecastController.cs` file, which will send a `Trace` log:

```
_logger.LogTrace("This is a trace message");
```

Run the application using `dotnet run` and request the `WeatherForecast` endpoint again. You will not see the trace message in the console window. Why? Because the trace message is disabled by default. Open the `appsettings.json` file; we can find the following configuration:

```
"Logging": {
  "LogLevel": {
    "Default": "Information",
    "Microsoft.AspNetCore": "Warning"
  }
},
```

Based on the configuration, the default logging level is `Information`. Look back at the logging level table we introduced before. The `Trace` logging level is 0, which is less than the `Information` logging level. So, the `Trace` logging level will not output by default. To enable the `Trace` logging level, we need to change the `Default` logging level to `Trace`. But there is another question – should we enable `Trace` logging for all environments?

The answer is *it depends*. The `Trace` logging level is used for the most detailed messages, which means it may contain sensitive application data. We can enable `Trace` logging in the development environment, but we may not want to enable it in the production environment. To achieve this, we can use the `appsettings.Development.json` file to override the `appsettings.json` file. That is what we learned in *Chapter 3*. Open the `appsettings.Development.json` file and update the following configuration to enable the `Trace` log:

```
"Logging": {
  "LogLevel": {
    "Default": "Trace",
    "Microsoft.AspNetCore": "Warning"
  }
}
```

Now, run the application again. You should be able to see a trace message in the console window for the development environment.

> **Important note**
> To specify the logging level for the production environment, we can add an `appsettings.Production.json` file, and then override the settings for the `Logging` section.

Keep in mind that the logging levels such as `Trace`, `Debug`, and `Information` will produce a lot of log messages. If we need to enable them in the production environment for troubleshooting, we need to be careful. Think about where we want to store logging messages.

You may notice that in the `appsettings.json` file, there is a `Microsoft.AspNetCore` logging section. It is used to control the logging level for the ASP.NET Core framework. ASP.NET Core uses the category name to differentiate logging messages produced from the framework and the application. Check the code where we inject the `ILogger` service into the controller:

```
public WeatherForecastController(ILogger<WeatherForecastController>
logger)
{
    _logger = logger;
}
```

The `ILogger<TCategoryName>` interface is defined in the `Microsoft.Extensions.Logging` namespace. The `TCategoryName` type parameter is used to categorize log messages. You can use any string values for the category name, but using the class name as the logging category name is a common practice.

Logging parameters

These `Log{LOG LEVEL}()` methods have some overloads, such as the following:

- `Log{LOG LEVEL}(string? message, params object?[] args)`
- `Log{LOG LEVEL}(EventId eventId, string? message, params object?[] args)`
- `Log{LOG LEVEL}(Exception exception, string message, params object[] args)`
- `Log{LOG LEVEL}(EventId eventId, Exception? exception, string? message, params object?[] args)`

The parameters of these methods are set out here:

- The `eventId` parameter is used to identify the log message
- The `message` parameter is used as a format string
- The `args` parameter is used to pass arguments for the format string
- The `exception` parameter is used to pass the exception object

For example, we can define an `EventIds` class to identify log messages, like so:

```
 public class EventIds
 {
     public const int LoginEvent = 2000;
     public const int LogoutEvent = 2001;
     public const int FileUploadEvent = 2002;
     public const int FileDownloadEvent = 2003;
     public const int UserRegistrationEvent = 2004;
     public const int PasswordChangeEvent = 2005;
     // Omitted for brevity
 }
```

Then, we can use the `eventId` parameter to identify log messages:

```
 _logger.LogInformation(EventIds.LoginEvent, "This is a logging message
 with event id.");
```

Some logging providers can use the `eventId` parameter to filter log messages.

We have introduced how to use the `message` parameter. You can use a plain string as the message, or you can use a format string and use the `args` parameter to pass arguments for the format string. Here is an example that uses the `message` and `args` parameters:

```
 _logger.LogInformation("This is a logging message with args: Today
 is {Week}. It is {Time}.", DateTime.Now.DayOfWeek, DateTime.Now.
 ToLongTimeString());
```

If an exception occurs, you can use the `LogError()` method with the `exception` parameter to log the exception:

```
 try
 {
     // Omitted for brevity
 }
 catch (Exception ex)
 {
     _logger.LogError(ex, "This is a logging message with exception.");
 }
```

When using the `LogError()` method to log an exception, it is important to pass the exception object to the `exception` parameter. This is essential in order to preserve stack trace information; simply logging the exception message is not sufficient.

Using third-party logging providers

The logging system in ASP.NET Core is designed to be extensible. The default logging providers, including the console logging provider and the `Debug` logging provider, can output logging messages in the console window or debug window, which is convenient for development. But in the production environment, we may want to send log messages to a file, a database, or a remote logging service. We can use third-party logging providers to achieve this.

There are many third-party logging frameworks or libraries that work with ASP.NET Core, such as the following:

Logging provider	Website	GitHub repo
`Serilog`	`https://serilog.net/`	`https://github.com/serilog/serilog-aspnetcore`
`NLog`	`https://nlog-project.org/`	`https://github.com/NLog/NLog.Extensions.Logging`
`log4net`	`https://logging.apache.org/log4net/`	`https://github.com/huorswords/Microsoft.Extensions.Logging.Log4Net.AspNetCore`

Table 4.2 – Third-party logging providers

Some other platforms provide rich features for collecting and analyzing the log messages, such as the following:

- **Exceptionless** (`https://exceptionless.com/`)
- **ELK Stack** (`https://www.elastic.co/elastic-stack/`)
- **Sumo Logic** (`https://www.sumologic.com/`)
- **Seq** (`https://datalust.co/seq`)
- **Sentry** (`https://sentry.io`)

These platforms can provide a dashboard to view log messages. Technically, they are not just logging providers but also platforms for observability that contain logging, tracing, metrics, and so on.

Let's start with a simple example. How can we print logging messages to a file?

We can use **Serilog** to write the log messages to a file. Serilog is a popular logging framework that works with .NET. It provides a standard logging API and a rich set of sinks that write log events to storage in various formats. These sinks target a variety of destinations, such as the following:

- File
- Azure Application Insights

- Azure Blob Storage

- Azure Cosmos DB

- Amazon CloudWatch

- Amazon DynamoDB

- Amazon Kinesis

- Exceptionless

- Elasticsearch

- Sumo Logic

- Email

- PostgreSQL

- RabbitMQ

Serilog provides a `Serilog.AspNetCore` NuGet package that integrates with ASP.NET Core. It has a set of extension methods to configure the logging system. To use it, install the `Serilog.AspNetCore` NuGet package by running this command:

```
dotnet add package Serilog.AspNetCore
```

Next, let's use the `Serilog.Sinks.File` sink to write log messages to a file. Install the `Serilog.Sinks.File` NuGet package with this command:

```
dotnet add package Serilog.Sinks.File
```

Then, update the `Program.cs` file to configure the logging system:

```
using Serilog;

var builder = WebApplication.CreateBuilder(args);
builder.Logging.ClearProviders();
var logger = new LoggerConfiguration().WriteTo.File(Path.
Combine(AppDomain.CurrentDomain.BaseDirectory, "logs/log.txt"),
rollingInterval: RollingInterval.Day, retainedFileCountLimit: 90).
CreateLogger();
builder.Logging.AddSerilog(logger);
```

In the preceding code, we first clear the default logging providers. Then, we create a `Serilog.ILogger` instance and add it to the logging system. The `WriteTo.File` method is used to configure the `Serilog.Sinks.File` sink. It will write log messages to a file named `log.txt` in the `logs` folder. The `rollingInterval` parameter is used to specify the rolling interval. In the current example, we set it up as daily. The `retainedFileCountLimit` parameter is used to specify the maximum number of log files to keep. In this case, we have kept the number of files to 90. Then, we call the `CreateLogger` method to create a `Serilog.ILogger` instance. Finally, we call the `AddSerilog()` method to add the `Serilog.ILogger` instance to the logging system.

There's no need to change the code that uses the `ILogger` service. The `ILogger` service is still injected into the controller. The only difference is that log messages will be written to a file instead of the console window.

Run the application again and request the `/WeatherForecast` endpoint. You should be able to see log messages in the `logs/log.txt` file.

> **Important note**
>
> If you write logging messages to a file, please make sure that there is enough disk space and that proper retention policies are set. Otherwise, the disk space may get exhausted. Also, it is recommended to use some professional logging systems for monitoring in the production environment. For example, if your application is deployed in Azure, you can easily integrate it with Azure Application Insights. Storing logging messages in a text file is not easy to manage and analyze.
>
> If the amount of logging messages is huge, consider sending them to a message queue, such as RabbitMQ, and then have a separate process to consume the messages and write them to a database. Keep in mind that the logging system should not be a bottleneck for the application. Do not use asynchronous methods for logging because logging should be fast, and frequently switching between threads may cause performance issues.

In the `LoggingDemo` project, we have already configured the logging system to write log messages to a file. You can check the `Program.cs` file to see how it works. Serilog provides a rich set of sinks, and it supports various configurations as well. You can find more provided sinks here: `https://github.com/serilog/serilog/wiki/Provided-Sinks`.

Structured logging

In the *Logging parameters* section, we showed how we can use the `args` parameter to format a message string. You may wonder if we can use string concatenation to achieve the same outcome; for example, like this:

```
logger.LogInformation($"This is a logging message with string
concatenation: Today is {DateTime.Now.DayOfWeek}. It is {DateTime.Now.
ToLongTimeString()}.");
```

The answer is yes and no. If you do not care about the values of the parameters, it does seem to work. However, a modern way to handle logs is to use **structured logging** instead of a plain string message. Structured logging is a way to log messages in a structured format. Parameters such as the day of the week are identified and structured so that a system can process them, which means we can perform special operations on them, such as filtering, searching, and so on. Let's dive into the details.

Serilog efficiently supports structured logging when you use `args` parameters instead of string concatenation. Let's update the `Program.cs` file to send logs to the console window in a structured format:

```
var logger = new LoggerConfiguration()
    .WriteTo.File(Path.Combine(AppDomain.CurrentDomain.
BaseDirectory, "logs/log.txt"), rollingInterval: RollingInterval.Day,
retainedFileCountLimit: 90)
    .WriteTo.Console(new JsonFormatter())
    .CreateLogger();
builder.Logging.AddSerilog(logger);
```

Serilog supports multiple sinks in the same logging pipeline. In the preceding code, we add the console sink after the file sink. The console sink is automatically installed with the `Serilog.AspNetCore` NuGet package, so you do not need to manually install it. `JsonFormatter` is used to format log messages in JSON format. You can also specify the `formatter` type for the file sink.

In the `LoggingController` file, add a new action method to compare the structured logs and string concatenation logs:

```
[HttpGet]
[Route("structured-logging")]
public ActionResult StructuredLoggingSample()
{
    logger.LogInformation("This is a logging message with args: Today
is {Week}. It is {Time}.", DateTime.Now.DayOfWeek, DateTime.Now.
ToLongTimeString());
    logger.LogInformation($"This is a logging message with string
concatenation: Today is {DateTime.Now.DayOfWeek}. It is {DateTime.Now.
ToLongTimeString()}.");
    return Ok("This is to test the difference between structured
logging and string concatenation.");
}
```

Run the application again and request the `api/Logging/structured-logging` endpoint. You should be able to see log messages in the console window in JSON format.

Log messages with structured logging look like this:

```
{
    "Timestamp":"2022-11-22T09:59:44.6590391+13:00",
    "Level":"Information",
    "MessageTemplate":"This is a logging message with args: Today is
{Week}. It is {Time}.",
    "Properties":{
        "Week":"Tuesday",
```

```
        "Time":"9:59:44 AM",
        "SourceContext":"LoggingDemo.Controllers.LoggingController",
        "ActionId":"9fdba8d6-8997-4cba-a9e1-0cefe36cabd1",
        "ActionName":"LoggingDemo.Controllers.LoggingController.
StructuredLoggingSample (LoggingDemo)",
        "RequestId":"0HMMC0D2M1GC4:00000001",
        "RequestPath":"/api/Logging/structured-logging",
        "ConnectionId":"0HMMC0D2M1GC4"
    }
}
```

Log messages with string concatenation look like this:

```
{
    "Timestamp":"2022-11-22T09:59:44.6597035+13:00",
    "Level":"Information",
    "MessageTemplate":"This is a logging message with string
concatenation: Today is Tuesday. It is 9:59:44 AM.",
    "Properties":{
        "SourceContext":"LoggingDemo.Controllers.LoggingController",
        "ActionId":"9fdba8d6-8997-4cba-a9e1-0cefe36cabd1",
        "ActionName":"LoggingDemo.Controllers.LoggingController.
StructuredLoggingSample (LoggingDemo)",
        "RequestId":"0HMMC0D2M1GC4:00000001",
        "RequestPath":"/api/Logging/structured-logging",
        "ConnectionId":"0HMMC0D2M1GC4"
    }
}
```

Note that the structured logging has `Week` and `Time` properties, while the string concatenation does not. Structured logging is more flexible and easier to process. Therefore, structured logging is recommended instead of string concatenation.

A great tool to analyze structured logging messages is Seq (`https://datalust.co/seq`). Seq is a powerful log management tool that creates the visibility you need to quickly identify and diagnose problems in your applications. It is a commercial product, but it provides a free trial. You can download it here: `https://datalust.co/download`.

Install it on your local machine. When you see the following window, please take note of the **Listen URI** value. We will configure it for Serilog later. It uses the default port `5341`:

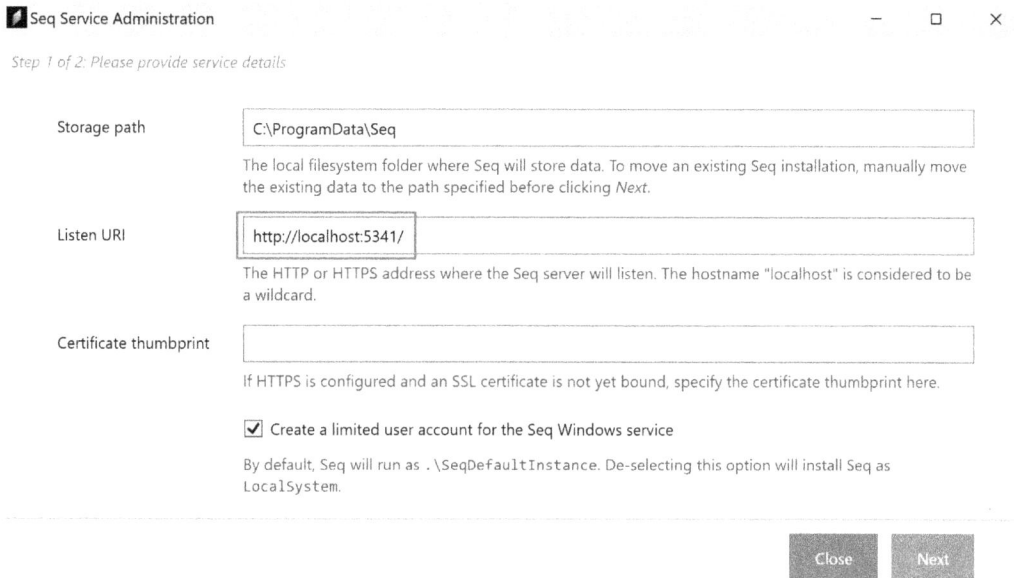

Figure 4.4 – Installing Seq

Next, we need to configure Serilog to send log messages to Seq. Serilog has a sink for Seq, so we can easily install it using the following command:

```
dotnet add package Serilog.Sinks.Seq
```

In the `Program.cs` file, update the logging configuration to send log messages to Seq:

```
var logger = new LoggerConfiguration()
    .WriteTo.File(formatter: new JsonFormatter(), Path.
Combine(AppDomain.CurrentDomain.BaseDirectory, "logs/log.txt"),
rollingInterval: RollingInterval.Day, retainedFileCountLimit: 90)
    .WriteTo.Console(new JsonFormatter())
    .WriteTo.Seq("http://localhost:5341")
    .CreateLogger();
```

Run the application again and request the `api/Logging/structured-logging` endpoint. You should be able to see log messages in Seq:

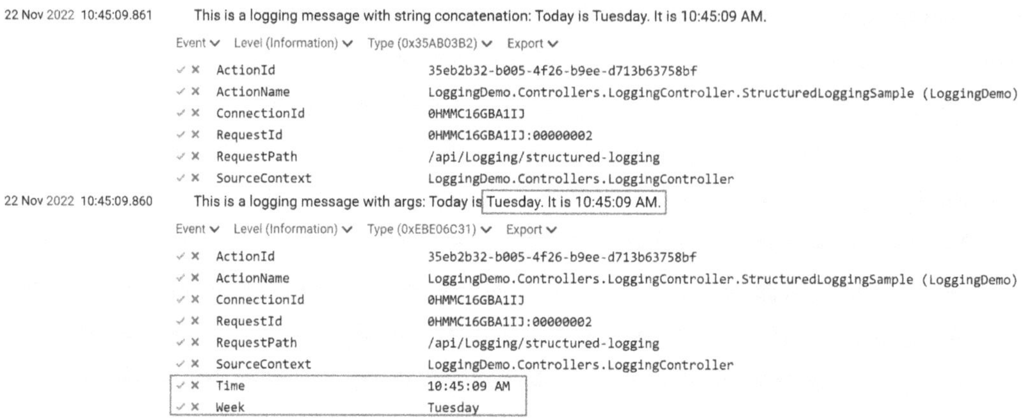
22 Nov 2022 10:45:09.861
This is a logging message with string concatenation: Today is Tuesday. It is 10:45:09 AM.
Event Level (Information) Type (0x35AB03B2) Export
ActionId 35eb2b32-b005-4f26-b9ee-d713b63758bf
ActionName LoggingDemo.Controllers.LoggingController.StructuredLoggingSample (LoggingDemo)
ConnectionId 0HMMC16GBA1IJ
RequestId 0HMMC16GBA1IJ:00000002
RequestPath /api/Logging/structured-logging
SourceContext LoggingDemo.Controllers.LoggingController
22 Nov 2022 10:45:09.860
This is a logging message with args: Today is Tuesday. It is 10:45:09 AM.
Event Level (Information) Type (0xEBE06C31) Export
ActionId 35eb2b32-b005-4f26-b9ee-d713b63758bf
ActionName LoggingDemo.Controllers.LoggingController.StructuredLoggingSample (LoggingDemo)
ConnectionId 0HMMC16GBA1IJ
RequestId 0HMMC16GBA1IJ:00000002
RequestPath /api/Logging/structured-logging
SourceContext LoggingDemo.Controllers.LoggingController
Time 10:45:09 AM
Week Tuesday

Figure 4.5 – Structured logging in Seq

We can search the `Week` property to filter log messages, as shown in the following figure:

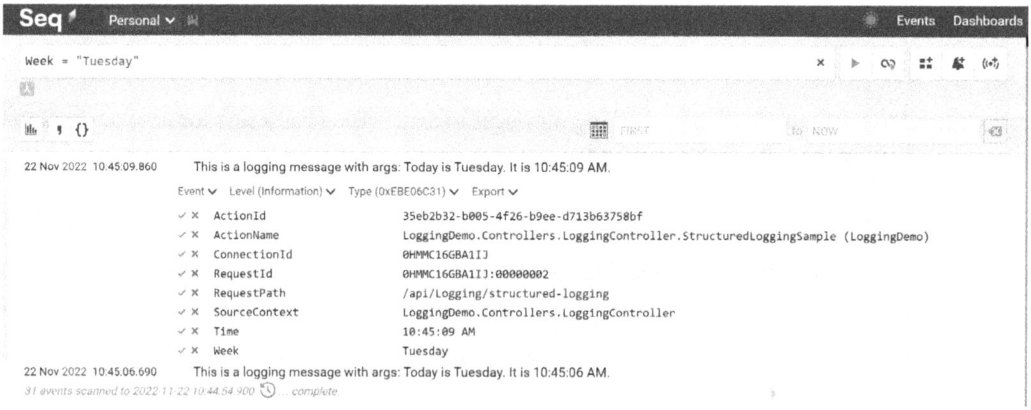
Seq
Personal
Events
Dashboards
Week = "Tuesday"
22 Nov 2022 10:45:09.860
This is a logging message with args: Today is Tuesday. It is 10:45:09 AM.
Event Level (Information) Type (0xEBE06C31) Export
ActionId 35eb2b32-b005-4f26-b9ee-d713b63758bf
ActionName LoggingDemo.Controllers.LoggingController.StructuredLoggingSample (LoggingDemo)
ConnectionId 0HMMC16GBA1IJ
RequestId 0HMMC16GBA1IJ:00000002
RequestPath /api/Logging/structured-logging
SourceContext LoggingDemo.Controllers.LoggingController
Time 10:45:09 AM
Week Tuesday
22 Nov 2022 10:45:06.690
This is a logging message with args: Today is Tuesday. It is 10:45:06 AM.

Figure 4.6 – Filtering structured logs in Seq

Now, we understand why structured logging is more powerful than string concatenation when you send logs. Serilog provides many powerful features for logging, such as enrichers. You can find more information about Serilog on its official website: `https://serilog.net/`.

> **Important note**
>
> Seq is a good choice for local development. However, you may need to purchase a license for your critical applications and services for better support, or you may want to use another logging system. Please carefully consider your requirements before you choose a logging system. Don't forget that we can use configurations to switch between different logging systems for different environments.

What should/shouldn't we log?

Logging is a powerful tool to help us diagnose problems, monitor an application, audit a system, and so on. But logging is not free. It consumes resources and may slow down applications. Depending on the purpose of logging, we may need to log different information. Generally, there are some scenarios that we should log:

- Input/output validation errors, such as invalid input parameters or invalid response data

- Authentication and authorization failures

- Application errors, exceptions, and warnings, such as database connection errors, network errors, and so on

- Application startup and shutdown

- High-risk operations, such as deleting a record from the database, changing a user's password, transferring money, and so on

- Legal compliance, such as auditing, terms of service, personal data consent, and so on

- Critical business events, such as a new order being placed, a new user being registered, and so on

- Any suspicious activities, such as brute-force attacks, account lockouts, and so on

What information should we log? If the information in the log message is not sufficient to diagnose the problem, it would be useless. Generally, we should log the following information:

- **When**: What time did the event happen?

- **Where**: What is the application name and version? What is the hostname or IP address? What is the module or component name?

- **Who**: Which user or client is involved in the event? What is the username, request ID, or client ID?

- **What**: What is the event? What severity level does the event have? What is the error message or stack trace? Any other descriptive information?

We may need to include more information based on the requirements. For example, for web API applications, we also need to log the request path, HTTP method, headers, status code, and so on. For database applications, we may need to log the SQL statement, parameters, and so on.

There is some information that we should not log:

- Application source code

- Sensitive application information, such as application secrets, passwords, encryption keys, database connection strings, and so on

- Sensitive user information, such as **personally identifiable information** (**PII**); for example, health status, government identification, and so on

- Bank account information, credit card information, and so on

- Information that users do not consent to share

- Any other information that may violate the law or regulations

We have explained the basics of logging in ASP.NET Core. Next, we will learn another important component of ASP.NET Core: middleware.

Middleware

In this section, we will introduce middleware, which is one of the most important improvements in ASP.NET Core.

To follow this section, you can run the following command to create a new ASP.NET Core web API project:

```
dotnet new webapi -n MiddlewareDemo -controllers
```

You can download the example project named `MiddlewareDemo` from the chapter's GitHub repository.

What is middleware?

ASP.NET Core is a middleware-based framework. An ASP.NET Core application is built upon a set of middleware components. Middleware is a software component that is responsible for handling requests and responses. Multiple middleware components form a pipeline to process requests and generate responses. In this pipeline, each middleware component can perform a specific task, such as authentication, authorization, logging, and so on. Then, it passes the request to the next middleware component in the pipeline. It is a huge improvement over the traditional ASP.NET framework, which is based on the HTTP module and HTTP handler. In this way, the ASP.NET Core framework is more flexible and extensible. You can add or remove middleware components as needed. You can also write your own middleware components.

The following diagram shows a middleware pipeline:

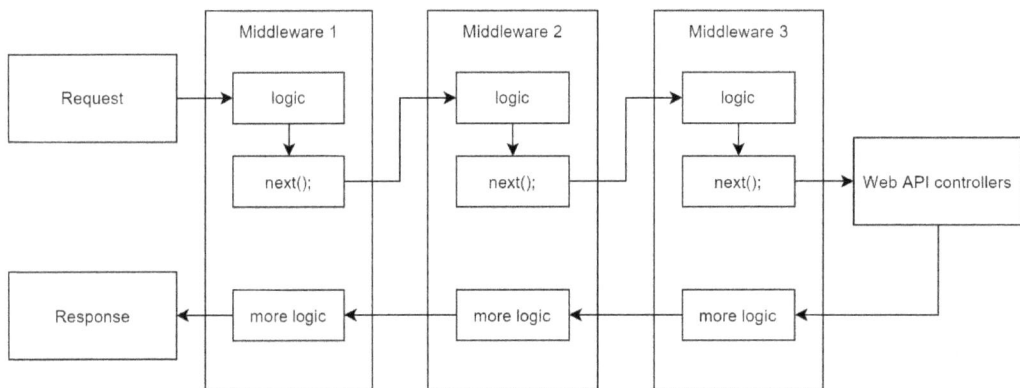

Figure 4.7 – Middleware pipeline

A middleware component can perform tasks before and after the next middleware component in the pipeline. It can also choose whether to pass the request to the next middleware component or stop processing the request and generate a response directly.

Creating simple middleware

Let's see an example. Open the `MiddlewareDemo` project in VS Code. Add the following code to the `Program.cs` file:

```
var app = builder.Build();

app.Run(async context =>
{
    await context.Response.WriteAsync("Hello world!");
});
```

> **Important note**
>
> In the preceding code, `builder.Build()` returns a `WebApplication` instance. It is the host for the web API project, which is responsible for app startup and lifetime management. It also manages logging, DI, configuration, middleware, and so on.

Run the application with `dotnet run` or `dotnet watch`, and you will see all requests will return a `Hello world!` response, regardless of the URL. This is because the `app.Run()` method handles all requests. In this case, the middleware short-circuits the pipeline and returns the response directly. We can also call it *terminal middleware*.

To use multiple middleware components, we can use the `app.Use()` method. The `app.Use()` method adds a middleware component to the pipeline. Update the `Program.cs` file as follows:

```
var app = builder.Build();

app.Use(async (context, next) =>
{
    var logger = app.Services.GetRequiredService<ILogger<Program>>();
    logger.LogInformation($"Request Host: {context.Request.Host}");
    logger.LogInformation("My Middleware - Before");
    await next(context);
    logger.LogInformation("My Middleware - After");
    logger.LogInformation($"Response StatusCode: {context.Response.
StatusCode}");
});
```

Run the application and request the /weatherforecast endpoint. You can see the following output in the console:

```
info: Program[0]
      Request Host: localhost:5170
info: Program[0]
      My Middleware - Before
info: Program[0]
      My Middleware - After
info: Program[0]
      Response StatusCode: 200
```

The WebApplication instance uses an app.Use() method to add a simple middleware component to the pipeline. The middleware component is an anonymous function that takes two parameters: context and next. Let's take a closer look at these:

- The context parameter is an instance of the HttpContext class that contains request and response information

- The next parameter is a delegate that is used to pass the request to the next middleware component in the pipeline

- The await next() statement passes the request to the next middleware component in the pipeline

This middleware does not do anything with the request and response. It only outputs some information to the console. Let us add another middleware component to the pipeline. Update the Program.cs file as follows:

```
var app = builder.Build();

app.Use(async (context, next) =>
{
    var logger = app.Services.GetRequiredService<ILogger<Program>>();
    logger.LogInformation($"ClientName HttpHeader in Middleware 1:
{context.Request.Headers["ClientName"]}");
    logger.LogInformation($"Add a ClientName HttpHeader in Middleware
1");
    context.Request.Headers.TryAdd("ClientName", "Windows");
    logger.LogInformation("My Middleware 1 - Before");
    await next(context);
    logger.LogInformation("My Middleware 1 - After");
    logger.LogInformation($"Response StatusCode in Middleware 1:
{context.Response.StatusCode}");
});
```

```
app.Use(async (context, next) =>
{
    var logger = app.Services.GetRequiredService<ILogger<Program>>();
    logger.LogInformation($"ClientName HttpHeader in Middleware 2:
{context.Request.Headers["ClientName"]}");
    logger.LogInformation("My Middleware 2 - Before");
    context.Response.StatusCode = StatusCodes.Status202Accepted;
    await next(context);
    logger.LogInformation("My Middleware 2 - After");
    logger.LogInformation($"Response StatusCode in Middleware 2:
{context.Response.StatusCode}");
});
```

In this example, we add two middleware components to the pipeline. The first middleware component adds a ClientName HTTP header to the request. The second middleware component sets the response status code to 202 Accepted. Run the application and request the /weatherforecast URL. You will see the following output in the console:

```
info: Program[0]
      ClientName HttpHeader in Middleware 1:
info: Program[0]
      Add a ClientName HttpHeader in Middleware 1
info: Program[0]
      My Middleware 1 - Before
info: Program[0]
      ClientName HttpHeader in Middleware 2: Windows
info: Program[0]
      My Middleware 2 - Before
info: Program[0]
      My Middleware 2 - After
info: Program[0]
      Response StatusCode in Middleware 2: 202
info: Program[0]
      My Middleware 1 - After
info: Program[0]
      Response StatusCode in Middleware 1: 202
```

Note the following points:

- The original request does not contain the ClientName HTTP header. So, the value of the ClientName HTTP header in Middleware 1 is empty.

- Before Middleware 1 passes the request to Middleware 2, Middleware 1 adds the ClientName HTTP header to the request.

- `Middleware 1` does not output `My Middleware 1 - After` to the console after its `await next(context);` code. Instead, `Middleware 2` outputs the `ClientName` HTTP header value to the console.

- `Middleware 2` changes the response status code to `202 Accepted`. The response will be passed to `Middleware 1`. Then, `Middleware 1` outputs `My Middleware 1 - After` with the new response status code.

That indicates how middleware components work in the pipeline.

How to assemble middleware components

Besides the `app.Use()` method, the `WebApplication` class also provides `Map()`, `MapWhen()`, `UseWhen()`, and `Run()` methods to add middleware components to a pipeline. The following table shows the differences between these methods:

Method	Description
`app.Map()`	Maps a request path to a sub-request pipeline. The middleware component is only executed when the request path matches the specified path.
`app.MapWhen()`	Runs a sub-request pipeline when a given predicate is matched.
`app.Use()`	Adds an inline delegate to the application's request pipeline.
`app.UseWhen()`	Adds an inline delegate to the application's request pipeline when a given predicate is matched. It is rejoined to the main pipeline if it does not short-circuit or contain terminal middleware.
`app.Run()`	Adds a terminal middleware component to the pipeline. It prevents further middleware components from processing requests.

Table 4.3 – The differences between the app.Map(), app.MapWhen(),
app.Use(), app.UseWhen(), and app.Run() methods

To handle a request, we use `app.Map()` or `app.MapWhen()` to configure which request path (or predicate) should be handled by which middleware component. The `app.Use()` and `app.Run()` methods are used to add middleware components to the pipeline. One pipeline can have multiple `app.Use()` methods. But only one `app.Run()` method is allowed in one pipeline, and the `app.Run()` method must be the last method in the pipeline. The following diagram illustrates the process:

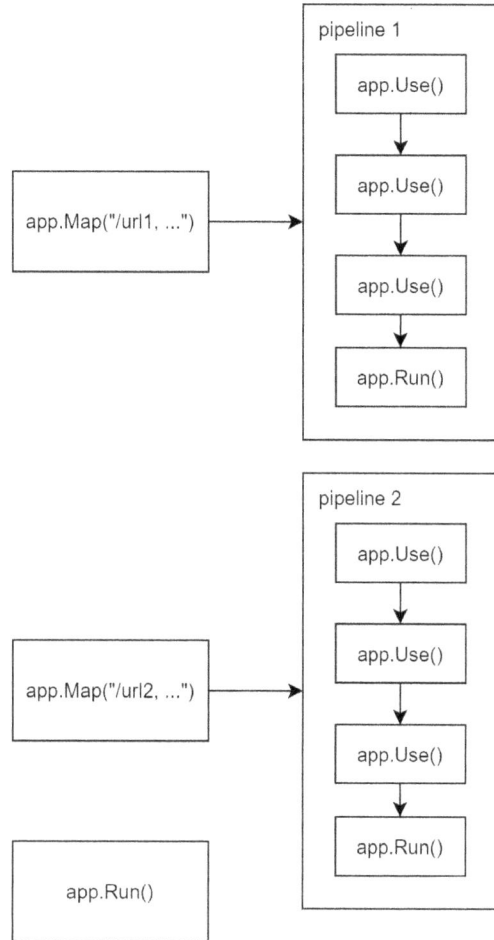

Figure 4.8 – The relationship between the app.Map(), app.Use(), and app.Run() methods

Let's see an example. We will develop a lottery application that allows users to call an API to check whether they get the lucky number. Lucky numbers are generated randomly. We will add a couple of middleware components to the pipeline.

Update the Program.cs file as follows:

```
app.Map("/lottery", app =>
{
    var random = new Random();
    var luckyNumber = random.Next(1, 6);
    app.UseWhen(context => context.Request.QueryString.Value ==
$"?{luckyNumber.ToString()}", app =>
    {
```

```
        app.Run(async context =>
        {
            await context.Response.WriteAsync($"You win! You got the
lucky number {luckyNumber}!");
        });
    });
    app.UseWhen(context => string.IsNullOrWhiteSpace(context.Request.
QueryString.Value), app =>
    {
        app.Use(async (context, next) =>
        {
            var number = random.Next(1, 6);
            context.Request.Headers.TryAdd("number", number.
ToString());
            await next(context);
        });
        app.UseWhen(context => context.Request.Headers["number"] ==
luckyNumber.ToString(), app =>
        {
            app.Run(async context =>
            {
                await context.Response.WriteAsync($"You win! You got
the lucky number {luckyNumber}!");
            });
        });
    });
    app.Run(async context =>
    {
        var number = "";
        if (context.Request.QueryString.HasValue)
        {
            number = context.Request.QueryString.Value?.Replace("?",
"");
        }
        else
        {
            number = context.Request.Headers["number"];
        }
        await context.Response.WriteAsync($"Your number is {number}.
Try again!");
    });
});
app.Run(async context =>
{
```

```
    await context.Response.WriteAsync($"Use the /lottery URL to play.
You can choose your number with the format /lottery?1.");
});
```

This is a fun lottery program. Let's see how it is configured:

1. First, we use the `app.Map()` method to map the `/lottery` request path to a sub-request pipeline. In this part, there are a couple of things to do:

   I.   We use a `Random` instance to generate a lucky number. Note that this number is generated only once when the application starts.

   II. Next, we add a middleware component to the pipeline using the `app.UseWhen()` method. This middleware works only when the request has a query string. If the query string is the same as the lucky number, it uses `app.Run()` to write the response. This branch is done.

   III. Next, we add another middleware component when the request does not have a query string. This middleware consists of two sub-middleware components:

   • The first one generates a random number and adds it to the HTTP header, then passes it to the second sub-middleware.

   • The second one uses `app.UseWhen()` to check the HTTP headers of the request. If the HTTP header contains the lucky number, it uses `app.Run()` to write the response. This branch is done. This part shows how we use the `app.UseWhen()` method to rejoin a middleware component to the main pipeline.

   IV. Next, we add a middleware component to the pipeline using the `app.Run()` method. This middleware component is used to handle all other requests for the `/lottery` URL. It writes the response to the client and shows the number that the client has chosen. Note that if the user already got the lucky number, this part will not be executed.

2. At the end of the program, we have another middleware component using the `app.Run()` method. This middleware component is used to handle all other requests. It shows how to play the game.

Run the application and request the `/lottery` endpoint a couple of times. Sometimes, you will see you win the lottery. Or, you can include a query string in the URL, such as `/lottery?1`. You should be able to notice that the lucky number is the same for all requests. If you restart the application, the lucky number may change, because it is generated randomly when the application starts.

There are a couple of things to note:

- A middleware component is initialized only once when the application starts.
- You can mix the `app.Map()`, `app.MapWhen()`, `app.Use()`, `app.UseWhen()`, and `app.Run()` methods in one pipeline. But be careful to use them in the right order.
- `app.Run()` must be the last method in a pipeline.
- You cannot change the response after `await next();` or `await next.Invoke();` is called because it may cause a protocol violation or corrupt the response body format. For example, if the response has been sent to the client, if you change the headers or the status code, it will throw an exception. If you want to change the response, please do so before `await next();` or `await next.Invoke();` is called.

You may wonder what the difference between `app.MapWhen()` and `app.UseWhen()` is. Both of them are used to configure a conditional middleware execution. The differences are as follows:

- `app.MapWhen()`: Used to branch the request pipeline based on the given predicate
- `app.UseWhen()`: Used to conditionally add a branch in the request pipeline that is rejoined to the main pipeline if it does not short-circuit or contain a terminal middleware

To clarify the difference, update the `Program.cs` file as follows:

```
app.UseWhen(context => context.Request.Query.ContainsKey("branch"),
app =>
{
    app.Use(async (context, next) =>
    {
        var logger = app.ApplicationServices.
GetRequiredService<ILogger<Program>>();
        logger.LogInformation($"From UseWhen(): Branch used =
{context.Request.Query["branch"]}");
        await next();
    });
});

app.Run(async context =>
{
    await context.Response.WriteAsync("Hello world!");
});
```

Run the application using `dotnet run` and request the `http://localhost:5170/?branch=1` URL. You will see the console window outputs the following message:

```
info: Program[0]
      From UseWhen(): Branch used = 1
```

And the response is `Hello world!`. If you request any other URL, such as `http://localhost:5170/test`, you will still see a `Hello world!` response. But you will not see the console window that outputs the log message. That says `app.UseWhen()` only works when the predicate is `true`. If the predicate is `false`, the pipeline will continue to execute the next middleware component.

Next, let us try to change `app.UseWhen()` to `app.MapWhen()`. Update the `Program.cs` file as follows:

```
app.MapWhen(context => context.Request.Query.ContainsKey("branch"),
app =>
{
    app.Use(async (context, next) =>
    {
        var logger = app.ApplicationServices.
GetRequiredService<ILogger<Program>>();
        logger.LogInformation($"From MapWhen(): Branch used =
{context.Request.Query["branch"]}");
        await next();
    });
});
```

Run the application and request the `http://localhost:5170/?branch=1` URL again. You will see a log message in the console window, but it returns a `404` error! Why?

That is because the `app.MapWhen()` method is used to branch the request pipeline based on the given predicate. If the predicate is `true`, the request pipeline will be branched to the sub-pipeline defined in this `app.MapWhen()` method. But when the `next()` method is called, it does not have a next middleware component to execute, even though there is an `app.Run()` method defined at the end of the program. So, it returns a `404` error.

To make it work, we need to add another `app.Run()` method to the sub-pipeline. Update the `Program.cs` file as follows:

```
app.MapWhen(context => context.Request.Query.ContainsKey("branch"),
app =>
{
    app.Use(async (context, next) =>
    {
        var logger = app.ApplicationServices.
GetRequiredService<ILogger<Program>>();
        logger.LogInformation($"From MapWhen(): Branch used =
{context.Request.Query["branch"]}");
        await next();
    });
    app.Run(async context =>
```

```
    {
        var branchVer = context.Request.Query["branch"];
        await context.Response.WriteAsync($"Branch used =
{branchVer}");
    });
});
```

Now, run the application again and request the `http://localhost:5170/?branch=1` URL. You can see the logging message is `From MapWhen(): Branch used = 1`, and the response is returned as expected.

The middleware mechanism is very powerful. It makes the ASP.NET Core application incredibly flexible. But it may cause issues if you make incorrect orders. Please use it wisely.

In this section, we introduced how to apply middleware components using delegate methods. ASP.NET Core provides a lot of built-in middleware components to simplify development. In the next section, we will explore some of the built-in middleware components.

Built-in middleware

ASP.NET Core framework provides a lot of built-in middleware components. Check the code of the `Program.cs` file. You can find some code like this:

```
if (app.Environment.IsDevelopment())
{
    app.UseSwagger();
    app.UseSwaggerUI();
}
app.UseHttpsRedirection();
app.UseAuthorization();
```

There are some middleware components to enable Swagger, HTTPS redirection, authorization, and so on. You can find a full list of built-in middleware components here: `https://learn.microsoft.com/en-us/aspnet/core/fundamentals/middleware/?view=aspnetcore-8.0#built-in-middleware`.

Here are some common built-in middleware components. Note the **Order** column. Some middleware components must be called in a specific order. Some middleware may terminate the requests:

Middleware	Description	Order
Authentication	Enables authentication support.	Before `HttpContext.User` is needed. Terminal for OAuth callbacks.
Authorization	Enables authorization support.	Immediately after the Authentication middleware.

Middleware	Description	Order
CORS	Configures **cross-origin resource sharing (CORS)**.	Before components that use CORS. `UseCors` currently must go before `UseResponseCaching`.
Health Check	Checks the health status of the application and its dependencies.	Terminal if a request matches a health check endpoint.
HTTPS Redirection	Redirects all HTTP requests to HTTPS.	Before components that consume the URL.
Response Caching	Enables response cache.	Before components that require caching. `UseCors` must come before `UseResponseCaching`.
Endpoint Routing	Defines and constrains request routes.	Terminal for matching routes.

Table 4.4 – Common built-in middleware components

We will not cover all the built-in middleware components in this book. But we will introduce some of them in the following sub-sections, such as rate-limiting, request timeouts, short-circuits, and so on.

Using the rate-limiting middleware

The rate-limiting middleware is a new built-in middleware provided in ASP.NET Core 7. It is used to limit the number of requests that a client can make in a given time window. It is very useful to prevent **distributed denial-of-service (DDoS)** attacks.

The rate-limiting middleware defines four policies:

- Fixed window
- Sliding window
- Token bucket
- Concurrency

This section is just to introduce how to use the middleware, so we will not cover details of the policies. We will use the fixed window policy in this section. This policy uses a fixed time window to limit the number of requests. For example, we can limit the number of requests to 10 per 10 seconds. When the time window expires, a new time window starts, and the counter is reset to 0.

Update the `Program.cs` file as follows:

```
builder.Services.AddRateLimiter(_ =>
    _.AddFixedWindowLimiter(policyName: "fixed", options =>
```

```
        {
            options.PermitLimit = 5;
            options.Window = TimeSpan.FromSeconds(10);
            options.QueueProcessingOrder = QueueProcessingOrder.
OldestFirst;
            options.QueueLimit = 2;
        }));
// Omitted for brevity
app.UseRateLimiter();
app.MapGet("/rate-limiting-mini", () => Results.Ok($"Hello {DateTime.
Now.Ticks.ToString()}")).RequireRateLimiting("fixed");
```

The preceding code adds the rate-limiting middleware to a minimal API request pipeline. It creates a fixed window policy named `fixed`. The `options` property means a maximum of 5 requests per each 10-second window are allowed.

Run the application and request the `http://localhost:5170/rate-limiting-mini` URL 10 times. You will see the response is `Hello 638005....` But the sixth request will be pending until the time window expires. You can try other policies if you want. To practice more, you can move the configuration for the policy to the `appsettings.json` file.

To apply this rate-limiting middleware to a controller-based API, we need to add the `EnableRateLimiting` attribute to the controller or action, as follows:

```
[HttpGet("rate-limiting")]
[EnableRateLimiting(policyName: "fixed")]
public ActionResult RateLimitingDemo()
{
    return Ok($"Hello {DateTime.Now.Ticks.ToString()}");
}
```

Next, we will introduce another built-in middleware component: the request timeout middleware.

Using the request timeouts middleware

ASP.NET Core 8 introduces the request timeout middleware, which allows developers to set a timeout for an endpoint. If the request is not completed within the allotted time, a `HttpContext.RequestAborted` cancellation token is triggered, allowing the application to handle the timeout request. This feature helps to prevent the application from being blocked by long-running requests.

To apply the request timeout middleware to the ASP.NET Core web API project, update the `Program.cs` file as follows:

```
builder.Services.AddRequestTimeouts();
// Omitted for brevity
app.UseRequestTimeouts();
```

The request timeout middleware can be used for a specific endpoint. To do this, we need to add the `EnableRequestTimeout` attribute to the controller or action, as follows:

```
[HttpGet("request-timeout")]
[RequestTimeout(5000)]
public async Task<ActionResult> RequestTimeoutDemo()
{
    var delay = _random.Next(1, 10);
    logger.LogInformation($"Delaying for {delay} seconds");
    try
    {
        await Task.Delay(TimeSpan.FromSeconds(delay), Request.
HttpContext.RequestAborted);
    }
    catch
    {
        logger.LogWarning("The request timed out");
        return StatusCode(StatusCodes.Status503ServiceUnavailable,
"The request timed out");
    }
    return Ok($"Hello! The task is complete in {delay} seconds");
}
```

In the preceding code, we use the `RequestTimeout` attribute to set the timeout to 5 seconds. In the action method, we use a `Task.Delay()` method to simulate a long-running task. The delay time is generated randomly. If the request is not completed within 5 seconds, a `Request.HttpContext.RequestAborted` cancellation token is triggered. Then, we can handle the timeout request in the `catch` block.

Run the application using the `dotnet run` command and request the `/api/request-timeout` endpoint a few times. Sometimes, you will get a `503` response. Note that the request timeout middleware does not work in the debug mode. To test this middleware, please ensure that the debugger is not attached to the application.

Similarly, if you want to apply this middleware to a minimal API, you can use the `WithRequestTimeout` method, as follows:

```
app.MapGet("/request-timeout-mini", async (HttpContext context,
ILogger<Program> logger) =>
  {
      // Omited for brevity
  }).WithRequestTimeout(TimeSpan.FromSeconds(5));
```

The timeout configuration can be configured with a policy. Then, we can apply the policy to the controller or action. Update the `Program.cs` file as follows:

```
builder.Services.AddRequestTimeouts(option =>
  {
      option.DefaultPolicy = new RequestTimeoutPolicy { Timeout =
TimeSpan.FromSeconds(5) };
      option.AddPolicy("ShortTimeoutPolicy", TimeSpan.FromSeconds(2));
      option.AddPolicy("LongTimeoutPolicy", TimeSpan.FromSeconds(10));
  });
```

The preceding code defines three timeout policies. `DefaultPolicy` is used when no policy is specified. `ShortTimeoutPolicy` is used for short-running requests with a timeout of 2 seconds, while `LongTimeoutPolicy` is used for long-running requests with a timeout of 10 seconds. To apply the policy to the controller or action, the `EnableRequestTimeout` attribute can be used as follows:

```
[HttpGet("request-timeout-short")]
  [RequestTimeout("ShortTimeoutPolicy")]
  public async Task<ActionResult> RequestTimeoutShortDemo()
```

If the action method does not specify a policy, `DefaultPolicy` will be used.

Using the short-circuit middleware

The short-circuit middleware is another new middleware component introduced in ASP.NET Core 8. This middleware is used to short-circuit a request when it is not necessary to continue processing the request. For example, web robots may request the `/robots.txt` file to check if the website allows crawling. As a web API application, we do not need to process this request. However, the execution of the request pipeline will still continue. The short-circuit middleware can be used to short-circuit the request and return a response directly.

To apply the short-circuit middleware to a specific endpoint, add the following code to the `Program.cs` file:

```
app.MapGet("robots.txt", () => Results.Content("User-agent: *\
nDisallow: /", "text/plain")).ShortCircuit();
```

The preceding code uses the `ShortCircuit()` method to short-circuit the request. If the request path is `/robots.txt`, it will return a text/plain response directly.

Another way to use the short-circuit middleware is to use `MapShortCircuit` as follows:

```
app.MapShortCircuit((int)HttpStatusCode.NotFound, "robots.txt",
"favicon.ico");
```

In this example, when a request is made to `/robots.txt` or `/favicon.ico`, a `404 Not Found` response will be returned directly. This ensures that the server is not burdened with unnecessary requests.

> **Important note**
>
> The short-circuit middleware should be placed at the start of the pipeline to prevent other middleware components from processing the request unnecessarily. This will ensure that the request is handled in the most efficient manner.

ASP.NET Core provides a wide range of built-in middleware components. These components use extension methods such as `AddXXX()` and `UseXXX()` to add middleware components to the pipeline. In the following section, we will explore how to create a custom middleware component and apply it to the pipeline using an extension method.

Creating a custom middleware component

If the built-in middleware cannot meet your requirements, you can create your own middleware components. A custom middleware component does not have to derive from a base class or an interface. But a middleware class does need to follow some conventions:

- It must have a public constructor that accepts a `RequestDelegate` parameter.
- It must have a public method named `Invoke()` or `InvokeAsync()` that accepts a `HttpContext` parameter and returns a `Task`. The `HttpContent` parameter must be the first parameter.
- It can use DI to inject additional dependencies.
- An extension method is needed to add the middleware to the `IApplicationBuilder` instance.

Consider this scenario. For better tracking, we want to use the concept of correlation ID in the application. The correlation ID is a unique identifier for each request. It is used to track the request through the application, especially in a microservice architecture. ASP.NET Core provides a `HttpContext.TraceIdentifier` property to store the unique identifier. By default, Kestrel generates the ID using the `{ConnectionId}:{Request number}` format; for example, `0HML6LNF87PBV:00000001`.

If we have multiple services, such as *Service A* and *Service B*, when the client calls *Service A*, *Service A* will generate a `TraceIdentifier` instance for the current request, then *Service A* will call *Service B*, and *Service B* will generate another `TraceIdentifier` instance. It is hard to track the request through multiple services. We need to use the same `TraceIdentifier` instance for the request to be tracked.

The idea is to generate a correlation ID for each request chain. Then, set it in the `X-Correlation-Id` header for request/response as well. When we call *Service B* from *Service A*, attach the `X-Correlation-Id` header to the HTTP request headers so that we can attach the `X-Correlation-Id` value in the logs for future diagnostics. To do this, we need to create a custom middleware component. Create a new class named `CorrelationIdMiddleware` in the project folder:

```
public class CorrelationIdMiddleware(RequestDelegate next,
ILogger<CorrelationIdMiddleware> logger)
{    private const string CorrelationIdHeaderName = "X-Correlation-
Id";

    public async Task InvokeAsync(HttpContext context)
    {
        var correlationId = context.Request.
Headers[CorrelationIdHeaderName].FirstOrDefault();
        if (string.IsNullOrEmpty(correlationId))
        {
            correlationId = Guid.NewGuid().ToString();
        }
        context.Request.Headers.TryAdd(CorrelationIdHeaderName,
correlationId);
        // Log the correlation ID
        logger.LogInformation("Request path: {RequestPath}.
CorrelationId: {CorrelationId}", context.Request.Path, correlationId);
        context.Response.Headers.TryAdd(CorrelationIdHeaderName,
correlationId);
        await next(context);
    }
}
```

In the preceding code, the `CorrelationIdMiddleware` class has a public constructor that accepts a `RequestDelegate` parameter. It also has a public method named `InvokeAsync()` that accepts a `HttpContext` parameter and returns a `Task` instance. The `InvokeAsync()` method is the entry point of the middleware. It gets the correlation ID from the request header. If it is not found, it generates a new one. Then, it sets the `HttpContext.TraceIdentifier` property and adds the correlation ID to the response header. Finally, it calls the next middleware component in the pipeline. It also uses a logger via DI to log the correlation ID.

Next, add a new extension method to the `IApplicationBuilder` instance:

```
public static class CorrelationIdMiddlewareExtensions
{
    public static IApplicationBuilder UseCorrelationId(this
IApplicationBuilder builder)
    {
        return builder.UseMiddleware<CorrelationIdMiddleware>();
    }
}
```

Now, we can apply the correlation ID middleware in the `Program.cs` file:

```
app.UseCorrelationId();
```

Open the `WeatherForecastController.cs` file and add the following code to the `Get()` method:

```
[HttpGet(Name = "GetWeatherForecast")]
public IEnumerable<WeatherForecast> Get()
{
    // Get the "X-Correlation-Id" header from the request
    var correlationId = Request.Headers["X-Correlation-Id"].
FirstOrDefault();
    // Log the correlation ID
    _logger.LogInformation("Handling the request. CorrelationId:
{CorrelationId}", correlationId);
    // Call another service with the same "X-Correlation-Id" header
when you set up the HttpClient
    //var httpContent = new StringContent("Hello world!");
    //httpContent.Headers.Add("X-Correlation-Id", correlationId);
    // Omitted for brevity
```

Run the application and request the `http://localhost:5170/WeatherForecast` URL. You will see the response header contains the `X-Correlation-Id` property, as shown next:

```
content-type: application/json; charset=utf-8
date: Wed,05 Oct 2022 08:17:55 GMT
server: Kestrel
transfer-encoding: chunked
x-correlation-id: de67a42b-fd95-4ba1-bd2a-28a54c878d4a
```

You can also see the correlation ID in the log, such as the following:

```
MiddlewareDemo.CorrelationIdMiddleware: Information: Request path: /
WeatherForecast. CorrelationId: 795bf955-50a1-4d71-a90d-f859e636775a
...
```

```
MiddlewareDemo.Controllers.WeatherForecastController: Information:
Handling the request. CorrelationId: 795bf955-50a1-4d71-a90d-
f859e636775a
```

In this way, we can use the correlation ID to track the request through multiple services, especially in a microservice architecture.

We did not call any other services in this example. You can have a try yourself. Create another service and call it from the current service. Apply the same correlation ID middleware in another service. It can get the X-Correlation-Id header from the request headers and continue to use it. Then, you will see the same correlation ID is used for each request chain in both services.

> **Important note**
>
> The preceding example is purely for demonstration purposes. Microsoft provides a NuGet package called Microsoft.AspNetCore.HeaderPropagation that can be used to propagate headers to downstream services. You can find the sample code here: https://github.com/ dotnet/aspnetcore/tree/main/src/Middleware/HeaderPropagation/ samples/HeaderPropagationSample. You can also check *Chapter 16* to learn more about the distributed tracing using OpenTelemetry.

Summary

In this chapter, we learned about the logging framework in ASP.NET Core, and introduced a third-party logging framework, Serilog, to help us write logs to different sinks, such as files, console, and Seq, which is a tool that analyzes logs using structured logging. We also learned what middleware is, how to use the built-in middleware components, and how to create a custom middleware component.

It is time to implement some real business logic in the next chapter. We will introduce **Entity Framework Core** (**EF Core**), a powerful **object-relational mapper** (**ORM**) framework, to help us access the database.

5

Data Access in ASP.NET Core (Part 1: Entity Framework Core Fundamentals)

In *Chapter 2*, we introduced a simple ASP.NET Core application to manage blog posts, which uses a static field to store the data in memory. In many real-world applications, the data is persisted in databases – such as SQL Server, MySQL, SQLite, PostgreSQL, and so on – so we will need to access the database to implement the CRUD operations.

In this chapter, we will learn about data access in ASP.NET Core. There are many ways to access the database in ASP.NET Core, such as through ADO.NET, Entity Framework Core, and Dapper, among others. In this chapter, we will focus on Entity Framework Core, which is the most popular **object-relational mapping (ORM)** framework in .NET Core.

Entity Framework Core, or **EF Core** for short, is an open-source ORM framework that allows us to create and manage mapping configurations between the database schema and the object models. It provides a set of APIs to perform CRUD operations using LINQ methods, which is like operating the objects in memory. EF Core supports many database providers, such as SQL Server, SQLite, PostgreSQL, MySQL, and so on. It also supports many other features, such as migrations, change tracking, and so on.

In this chapter, we will cover the following topics:

- Why use ORM?
- Configuring the DbContext class
- Implementing CRUD controllers
- Basic LINQ queries
- Configuring the mapping between models and database tables

By the end of this chapter, you will be able to use EF Core to access the database in ASP.NET Core applications and perform basic CRUD operations.

Technical requirements

The code example in this chapter can be found at `https://github.com/PacktPublishing/Web-API-Development-with-ASP.NET-Core-8/tree/main/samples/chapter5/`. You can use VS 2022 or VS Code to open the solution.

You are expected to have basic knowledge of SQL queries and LINQ. If you are not familiar with them, you can refer to the following resources:

- SQL queries: `https://www.w3schools.com/sql/`
- LINQ: `https://learn.microsoft.com/en-us/dotnet/csharp/linq/`

Why use ORM?

To operate the data in relational databases, we need to write SQL statements. However, SQL statements are not easy to maintain and are not type-safe. Every time you update the database schema, you need to update the SQL statements as well, which is error-prone. In many traditional applications, the logic is tightly coupled with the database. For example, the logic could be defined in a SQL database directly, such as stored procedures, triggers, and so on. This makes the application hard to maintain and extend.

ORM helps us to map the database schema to the object model, so we can operate the data in the database just like we operate the objects in memory. ORM can translate the CRUD operations to SQL statements, which means it is like an abstract layer between the application and the database. The data access logic is decoupled from the database, so we can easily change the database without changing the code. Also, it provides strong type safety, so we can avoid runtime errors caused by type mismatch.

Keep in mind that we are not saying that ORM is the best solution for all scenarios. Sometimes, we need to write SQL statements directly to achieve the best performance. For example, if we need to generate a complex data report, we may need to write SQL statements to optimize the performance of the query. However, for most scenarios, ORM provides more benefits than drawbacks.

There are many ORM frameworks in .NET. In this book, we will use EF Core, which is the most popular ORM framework in .NET Core. The following are the reasons why we chose EF Core:

- **Open-source**: EF Core is an open-source project and is mainly maintained by Microsoft, so it is well-supported. The contribution is also very active.

- **Multiple database support**: EF Core supports many database providers, such as SQL Server, SQLite, PostgreSQL, MySQL and so on. Developers can use the same APIs to access different databases.

- **Migration**: EF Core supports database migrations, which allows us to update the database schema easily.

- **LINQ support**: EF Core provides support for LINQ, which allows us to use a familiar syntax to query the database.

- **Code-first approach**: EF Core supports the code-first approach, which means we can define the database schema using C# code, and EF Core will generate the database schema automatically.

- **Performance**: EF Core is designed to be lightweight and performant. It supports query caching and lazy loading to help improve performance. Also, EF Core provides asynchronous APIs, which allows us to perform database operations asynchronously to improve the scalability of the application. In addition, EF Core supports raw SQL queries, enabling us to write SQL statements directly to achieve the best performance.

Overall, EF Core is a good choice for most scenarios if you are using .NET Core. So, in this book, we will be using EF Core as the ORM framework.

To use the .NET Core CLI to perform EF Core-related tasks, we first need to install the `dotnet-ef` tool. You can install it using the following command:

```
dotnet tool install --global dotnet-ef
```

It is recommended to install the tool as a global tool, so you can use it in any project for convenience.

Next, create a new web API project using the following command:

```
dotnet new webapi -n BasicEfCoreDemo -controllers
```

Then, navigate to the project folder and run the following command to install EF Core packages:

```
dotnet add package Microsoft.EntityFrameworkCore.SqlServer
dotnet add package Microsoft.EntityFrameworkCore.Design
```

The first package is the database provider, which is used to connect the application to a SQL Server database. For this demo application, we will use **LocalDB**, which is a lightweight version of SQL Server. The second package contains shared design-time components for EF Core tools, which are required to perform database migrations.

> **What is LocalDB?**
>
> LocalDB is designed to be used as a substitute for the full version of SQL Server; it is suitable for development and testing, but not for production use. We can use LocalDB for development and replace the connection string when we deploy the application to production. LocalDB is installed with VS 2022. If you do not have VS 2022 by default, you can find the installation package at `https://learn.microsoft.com/en-us/sql/database-engine/configure-windows/sql-server-express-localdb`.
>
> LocalDB is supported by Windows only. If you use macOS or Linux, you can use SQLite instead of LocalDB, or use a Docker container to run SQL Server. For more information about SQLite, please refer to `https://docs.microsoft.com/en-us/ef/core/providers/sqlite/`.
>
> For more information about SQL Server on Docker, please refer to `https://learn.microsoft.com/en-us/sql/linux/quickstart-install-connect-docker`. Note that there are many other database providers, such as SQLite, PostgreSQL, MySQL, and so on. You can find the full list of database providers at `https://docs.microsoft.com/en-us/ef/core/providers/`. Some providers are not maintained by Microsoft.

Next, let's explore how to use EF Core to access the database.

Configuring the DbContext class

To represent the database, EF Core uses the `DbContext` class, which allows us to query and save data. An instance of the `DbContext` class maintains the database connection and maps the database schema to the object model. It also tracks the changes in objects and manages the transactions. If you are familiar with OOP, you can think of the `DbContext` class as a bridge between the database and the object model, just like an interface. When you query or save data, you operate the objects through the `DbContext` class, and EF Core will translate the operations to the corresponding SQL statements.

In this chapter, we will develop a simple application to manage invoices. This application will be used to demonstrate how to use EF Core to access the database, including how to define the database schema, how to perform CRUD operations, and how to use migrations to update the database schema.

You can follow *Chapter 1* to define the API contract first. The API contract defines the endpoints and the request/response models. When we define the API contract, note that we need to consult stakeholders to understand the requirements. For example, we need to know the fields of the invoice, the data types of the fields, and so on. We also need to understand the business rules, such as *the invoice number should be unique*, and *the invoice amount should be greater than 0*, for example. That means we will spend lots of time on the API design phase. Here, we assume that we have already defined the API contract, and we can start to develop the application.

Creating models

The first step is to define the models. A model, also known as an entity, is a class that represents an object in the real world, which will be mapped to a table (or multiple tables) in the database. In this demo application, we need to define the `Invoice` model.

An invoice can be defined as the following class:

```
namespace BasicEfCoreDemo.Models;
public class Invoice{
    public Guid Id { get; set; }
    public string InvoiceNumber { get; set; } = string.Empty;
    public string ContactName { get; set; } = string.Empty;
    public string? Description { get; set; }
    public decimal Amount { get; set; }
    public DateTimeOffset InvoiceDate { get; set; }
    public DateTimeOffset DueDate { get; set; }
    public InvoiceStatus Status { get; set; }
}
```

InvoiceStatus is a custom enum type, which is defined as the following code:

```
public enum InvoiceStatus
{
    Draft,
    AwaitPayment,
    Paid,
    Overdue,
    Cancelled
}
```

You can create a file named Invoice.cs in the Models folder and copy the Invoice class code into the file.

> **Important note**
>
> We use the Guid type for the Id property, which is the unique identifier for the invoice. You can also use int or long as the identifier. Either way has its pros and cons. For example, int is more efficient than Guid, but it is not unique across databases. When the database grows, you may need to split the data into multiple databases, which means the int identifier may not be unique anymore. On the other hand, Guid is unique no matter how many databases you have, but it is more expensive to store, insert, query, and sort the records than using int or long. The Guid primary key with the cluster index may cause poor performance in some scenarios. In this demo application, we use Guid as the identifier for now. We will discuss more about the techniques to optimize the application performance in future chapters.
>
> We also use the DateTimeOffset type for the InvoiceDate and DueDate properties, which is the recommended type for date and time in .NET Core. You can also use the DateTime type if you do not care about the time zone. DateTimeOffset includes a time zone offset from UTC time, and it is supported by both .NET type and SQL Server. This is helpful if you want to avoid the time zone issues.

We may need more properties in the future, such as contact information, invoice items, and so on, but we will add them later. Let's focus on only the model for now.

Creating and configuring the DbContext class

Next, we will create a `DbContext` class to represent the database. Create a file named `InvoiceDbContext.cs` in the `Data` folder and add the following code:

```
using BasicEfCoreDemo.Models;
using Microsoft.EntityFrameworkCore;

namespace BasicEfCoreDemo.Data;

public class InvoiceDbContext(DbContextOptions<InvoiceDbContext>
options) : DbContext(options)

{
    public DbSet<Invoice> Invoices => Set<Invoice>();
}
```

In the preceding code, we have done the following:

- Inherited the `DbContext` class and defined the `InvoiceDbContext` class, which represents the database.

- Defined the `Invoices` property, which is a `DbSet<Invoice>` type. It is used to represent the `Invoices` table in the database.

> **Important note**
>
> Why we do not use `public DbSet<Invoice> Invoices { get; set; }` here? The reason is that if the `DbSet<T>` properties are not initialized, the compiler will emit warnings from them because the nullable reference type feature is enabled by default. So we can use a `Set<TEntity>()` method to initialize the property to eliminate the warning. Another way to fix that is to use the null-forgiving operator, `!`, which forces the silencing of the compiler warnings. The `DbContext` base constructor will initialize the `DbSet<T>` properties for us, so it is safe to use `!` for this case. If you do not mind seeing the warnings, using `public DbSet<Invoice> Invoices { get; set; }` also works. You can use either method.

Next, let's configure the database connection string. Open the `appsettings.json` file and add the following code to the `ConnectionStrings` section:

```
"ConnectionStrings": {
    "DefaultConnection": "Server=(localdb)\\
mssqllocaldb;Database=BasicEfCoreDemoDb;Trusted_
Connection=True;MultipleActiveResultSets=true"
  }
```

> **Important note**
>
> You can use other databases, such as SQLite or PostgreSQL, but you need to install the corresponding database provider and change the connection string accordingly. To learn more about connection strings, please refer to `https://learn.microsoft.com/en-us/dotnet/framework/data/adonet/connection-string-syntax`. There is a website called `https://connectionstrings.com/` that can generate connection strings for different database providers.

In the preceding connection string, we use `Server=(localdb)\\mssqllocaldb` to specify the server as a LocalDB instance, and `Database=BasicEfCoreDemoDb` to specify the name of the database. You can change the database name to whatever you want. The `Trusted_Connection=True` option specifies that the connection is trusted, which means you do not need to provide the username and password. The `MultipleActiveResultSets=true` option specifies that the connection can have **Multiple Active Result Sets** (**MARS**). This means that you can have multiple independent queries executing on the same connection. This option is required for the `Include()` method in EF Core.

Open the `Program.cs` file and add the following code after `builder` is created:

```
builder.Services.AddDbContext<InvoiceDbContext>(options =>
    options.UseSqlServer(builder.Configuration.
GetConnectionString("DefaultConnection")));
```

The preceding code registers the `InvoiceDbContext` class to the dependency injection container. The `AddDbContext<TContext>()` method is an extension method that accepts a `DbContextOptionsBuilder` parameter, which calls the `UseSqlServer()` method to configure the database provider to use SQL Server or LocalDB. Note that we use the `UseSqlServer()` method for both SQL Server and LocalDB. The difference is that LocalDB has a `(localdb)\\mssqllocaldb` server name by default. We also pass the database connection string to the `UseSqlServer()` method, which should be the same as the name we defined in the `appsettings.json` file.

Currently, this code just registers the `InvoiceDbContext` class to the dependency injection container, but we have not created the database yet. Next, we will create the database using the `dotnet ef` command.

Creating the database

We have defined the `InvoiceDbContext` class, and the instance of `InvoiceDbContext` is added to the dependency injection container. Next, we need to create the database and the `Invoices` table before we can use it. To create the database and the `Invoices` table, we need to run the following command to apply the database migration:

```
dotnet ef migrations add InitialDb
```

The `InitialDb` parameter is the migration name. You can use any name you like as long as it is a valid C# identifier. It is recommended to use a meaningful name, such as `InitialDb`, `AddInvoiceTable`, and so on.

The preceding command creates a couple of migration files, such as `<timestamp>_InitialDb.cs` and `<timestamp>_InitialDb.Designer.cs`, which are stored in the `Migrations` folder. The `<timestamp>_InitialDb.cs` migration file contains an `Up()` method to create the database and the tables. It also has a `Down()` method to roll back the changes. Note that this command does not create the database; it just creates the migration files. Please do not manually modify or delete the migration files as they are required to apply or roll back the database changes.

Here is a sample of the migration files:

```
protected override void Up(MigrationBuilder migrationBuilder)
{
    migrationBuilder.CreateTable(
        name: "Invoices",
        columns: table => new
        {
            Id = table.Column<Guid>(type: "uniqueidentifier",
nullable: false),
            InvoiceNumber = table.Column<string>(type:
"nvarchar(max)", nullable: false),
            ContactName = table.Column<string>(type: "nvarchar(max)",
nullable: false),
            Description = table.Column<string>(type: "nvarchar(max)",
nullable: true),
            Amount = table.Column<decimal>(type: "decimal(18,2)",
nullable: false),
            InvoiceDate = table.Column<DateTimeOffset>(type:
"datetimeoffset", nullable: false),
            DueDate = table.Column<DateTimeOffset>(type:
"datetimeoffset", nullable: false),
            Status = table.Column<int>(type: "int", nullable: false)
        },
```

```
        constraints: table =>
        {
            table.PrimaryKey("PK_Invoices", x => x.Id);
        });
}
// Omitted for brevity
```

As you can see, the `Up()` method creates the table, columns, and constraints. The `Down()` method drops the table. You can use `dotnet ef migrations remove` to remove the migration files.

Important note

You may see a warning message like this:

Microsoft.EntityFrameworkCore.Model.Validation[30000]

No store type was specified for the decimal property 'Amount' on entity type 'Invoice'. This will cause > values to be silently truncated if they do not fit in the default precision and scale. Explicitly > specify the SQL server column type that can accommodate all the values in 'OnModelCreating' using > 'HasColumnType', specify precision and scale using 'HasPrecision', or configure a value converter using > 'HasConversion'.

This is because we did not specify the precision and scale for the `Amount` property. We will fix it later. Currently, EF Core will use the default precision and scale for the `decimal` type, which is `decimal(18,2)`.

The migration file has been created, but it has not been applied to the database yet. Next, run the following command to create the database and the `Invoices` table:

```
dotnet ef database update
```

If the command is successful, we should find the database file in your user folder, such as `C:\Users\{username}\BasicEfCoreDemoDb.mdf` if you use Windows. You can use `%USERPROFILE%` to get the user folder path.

Important note

You may encounter an error `System.Globalization.CultureNotFoundException:` Only the invariant culture is supported in globalization-invariant mode. See `https://aka.ms/GlobalizationInvariantMode` for more information. `(Parameter 'name')`. This is because starting in .NET 6, the globalization invariant mode is enabled by default. You can disable it by setting the `InvariantGlobalization` property to `false` in the `csproj` file.

There are several tools you can use to open the LocalDB database file – for example, **SQL Server Management Studio** (**SSMS**), which is supported by Microsoft. You can download it here: `https://learn.microsoft.com/en-us/sql/ssms/download-sql-server-management-studio-ssms`. You can also use other tools, such as Dbeaver (`https://dbeaver.io/`), a free, universal database tool, or JetBrains DataGrip (`https://www.jetbrains.com/datagrip/`), a powerful database IDE. We'll be using SSMS.

Open the database file in SSMS, and you will see that the `BasicEfCoreDemoDb` database has been created. It will have two tables – `Invoices` and `__EFMigrationsHistory`:

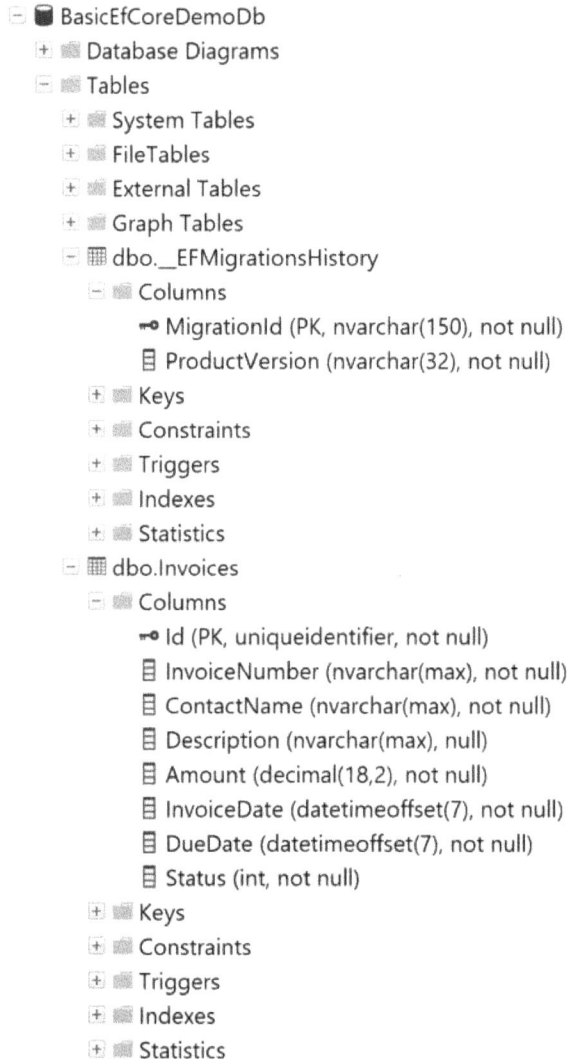

```
BasicEfCoreDemoDb
  Database Diagrams
  Tables
    System Tables
    FileTables
    External Tables
    Graph Tables
    dbo.__EFMigrationsHistory
      Columns
        MigrationId (PK, nvarchar(150), not null)
        ProductVersion (nvarchar(32), not null)
      Keys
      Constraints
      Triggers
      Indexes
      Statistics
    dbo.Invoices
      Columns
        Id (PK, uniqueidentifier, not null)
        InvoiceNumber (nvarchar(max), not null)
        ContactName (nvarchar(max), not null)
        Description (nvarchar(max), null)
        Amount (decimal(18,2), not null)
        InvoiceDate (datetimeoffset(7), not null)
        DueDate (datetimeoffset(7), not null)
        Status (int, not null)
      Keys
      Constraints
      Triggers
      Indexes
      Statistics
```

Figure 5.1 – The database created by EF Core migration

The `__EFMigrationsHistory` table is used to track the migrations. It is created automatically by EF Core. Do not manually modify it.

Now that we have created the database and the `Invoices` table. Next,let's add some seed data to the table.

Adding seed data

Open the `InvoiceDbContext.cs` file and add the following code to the `OnModelCreating()` method:

```
protected override void OnModelCreating(ModelBuilder modelBuilder)
{
    modelBuilder.Entity<Invoice>().HasData(
        new Invoice
        {
            Id = Guid.NewGuid(),
            InvoiceNumber = "INV-001",
            ContactName = "Iron Man",
            Description = "Invoice for the first month",
            Amount = 100,
            InvoiceDate = new DateTimeOffset(2023, 1, 1, 0, 0, 0,
TimeSpan.Zero),
            DueDate = new DateTimeOffset(2023, 1, 15, 0, 0, 0,
TimeSpan.Zero),
            Status = InvoiceStatus.AwaitPayment
        },
        // Omitted for brevity. You can check the full code in the
sample project.
}
```

We need to create a new database migration to apply the changes to the database. Run the following command:

```
dotnet ef migrations add AddSeedData
dotnet ef database update
```

If you check the database in SSMS, you will see that the seed data is added to the `Invoices` table.

The data is ready. Next, we will create the controllers to handle the HTTP requests and operate the data with the database.

Implementing CRUD controllers

In this section, we will implement the controllers to handle the HTTP requests, which are the GET, POST, PUT, and DELETE operations that are used to retrieve, create, update, and delete data, respectively.

Creating the controller

If you have installed the `dotnet aspnet-codegenerator` tool following *Chapter 2*, you can use the following command to create a controller with the specific `DbContext`. Do not forget to install the `Microsoft.VisualStudio.Web.CodeGeneration.Design` NuGet package, which is required by the `dotnet aspnet-codegenerator` tool:

```
# Install the tool if you have not installed it yet.
#dotnet tool install -g dotnet-aspnet-codegenerator
dotnet add package Microsoft.VisualStudio.Web.CodeGeneration.Design
dotnet-aspnet-codegenerator controller -name InvoicesController -api
-outDir Controllers --model Invoice --dataContext InvoiceDbContext
-async -actions
```

The preceding command has some parameters as shown here:

- `-name`: The name of the controller.

- `-api`: Indicates that the controller is an API controller.

- `-outDir`: The output directory of the controller.

- `--model`: The model class name. In this case, it is the `Invoice` class.

- `--dataContext`: The `DbContext` class name. In this case, it is the `InvoiceDbContext` class.

- `-async`: Indicates that the actions of the controller are asynchronous.

For more information about the `dotnet aspnet-codegenerator` tool, see `https://learn.microsoft.com/en-us/aspnet/core/fundamentals/tools/dotnet-aspnet-codegenerator`.

The `dotnet aspnet-codegenerator` tool will create a controller with the following actions:

```
using BasicEfCoreDemo.Data;
using BasicEfCoreDemo.Models;
using Microsoft.AspNetCore.Mvc;
using Microsoft.EntityFrameworkCore;

namespace BasicEfCoreDemo.Controllers
{
    [Route("api/[controller]")]
    [ApiController]
    public class InvoicesController : ControllerBase
    {
        private readonly InvoiceDbContext _context;

        public InvoicesController(InvoiceDbContext context)
```

```csharp
        {
            _context = context;
        }


        // GET: api/Invoices
        [HttpGet]
        public async Task<ActionResult<IEnumerable<Invoice>>>
GetInvoices()
        {
            if (_context.Invoices == null)
            {
                return NotFound();
            }
            return await _context.Invoices.ToListAsync();
        }


        // GET: api/Invoices/5
        [HttpGet("{id}")]
        public async Task<ActionResult<Invoice>> GetInvoice(Guid id)
        {
            if (_context.Invoices == null)
            {
                return NotFound();
            }
            var invoice = await _context.Invoices.FindAsync(id);

            if (invoice == null)
            {
                return NotFound();
            }

            return invoice;
        }


        // PUT: api/Invoices/5
        // To protect from overposting attacks, see https://
go.microsoft.com/fwlink/?linkid=2123754
        [HttpPut("{id}")]
        public async Task<IActionResult> PutInvoice(Guid id, Invoice
invoice)
        {
            if (id != invoice.Id)
            {
                return BadRequest();
```

```
            }

            _context.Entry(invoice).State = EntityState.Modified;

            try
            {
                await _context.SaveChangesAsync();
            }
            catch (DbUpdateConcurrencyException)
            {
                if (!InvoiceExists(id))
                {
                    return NotFound();
                }
                else
                {
                    throw;
                }
            }

            return NoContent();
        }


        // POST: api/Invoices
        // To protect from overposting attacks, see https://
go.microsoft.com/fwlink/?linkid=2123754
        [HttpPost]
        public async Task<ActionResult<Invoice>> PostInvoice(Invoice
invoice)
        {
            if (_context.Invoices == null)
            {
                return Problem("Entity set 'InvoiceDbContext.
Invoices' is null.");
            }
            _context.Invoices.Add(invoice);
            await _context.SaveChangesAsync();

            return CreatedAtAction("GetInvoice", new { id = invoice.Id
}, invoice);
        }
```

```
        // DELETE: api/Invoices/5
        [HttpDelete("{id}")]
        public async Task<IActionResult> DeleteInvoice(Guid id)
        {
            if (_context.Invoices == null)
            {
                return NotFound();
            }
            var invoice = await _context.Invoices.FindAsync(id);
            if (invoice == null)
            {
                return NotFound();
            }

            _context.Invoices.Remove(invoice);
            await _context.SaveChangesAsync();

            return NoContent();
        }

        private bool InvoiceExists(Guid id)
        {
            return (_context.Invoices?.Any(e => e.Id == id)).
  GetValueOrDefault();
        }
    }
}
```

It is so easy! The dotnet aspnet-codegenerator tool has generated the controller with basic CRUD operations. You can run the application and test the API endpoints with the Swagger UI. We will explain the code of the controller in detail.

How controllers work

In *Chapter 2* and *Chapter 3*, we introduced how HTTP requests are mapped to the controller actions. In this chapter, we focus on data access and database operations.

First, we use DI to inject the InvoiceDbContext instance into the controller, which handles the database operations. As developers, normally, we do not need to worry about the database connection. InvoiceDbContext is registered as scoped, which means that each HTTP request will create a new InvoiceDbContext instance, and the instance will be disposed of after the request is completed.

Once we get the `InvoiceDbContext` instance, we can use the `DbSet` property to access the entity set. The `DbSet<Invoice>` property represents a collection of the `Invoice` model class, which is mapped to the `Invoices` table in the database. We can use **LINQ** methods, such as `FindAsync()`, `Add()`, `Remove()`, and `Update()`, to retrieve, add, remove, and update the entity in the database, respectively. The `SaveChangesAsync()` method is used to save the changes to the database. In this way, we operate the database through .NET objects, which is much easier than using SQL statements. That is the power of ORMs.

> **What is LINQ?**
>
> **Language-Integrated Query (LINQ)** is a set of features in .NET that provide a consistent and expressive way to query and manipulate data from various data sources, such as a database, XML, and in-memory collections. With LINQ, you can write queries in a declarative way, which is much easier than using SQL statements. We will show you some basic LINQ queries in the next section. For more information about LINQ, see `https://learn.microsoft.com/en-us/dotnet/csharp/programming-guide/concepts/linq/`.

Let us look at the generated SQL statements. Use `dotnet run` to start the application and test the `api/Invoices` API endpoint with the Swagger UI or any tool you like. You can see the following SQL statements in the **Debug** window:

```
info: Microsoft.EntityFrameworkCore.Database.Command[20101]
      Executed DbCommand (26ms) [Parameters=[], CommandType='Text',
CommandTimeout='30']
      SELECT [i].[Id], [i].[Amount], [i].[ContactName], [i].
[Description], [i].[DueDate], [i].[InvoiceDate], [i].[InvoiceNumber],
[i].[Status]
      FROM [Invoices] AS [i]
```

The logs are helpful for understanding the SQL statements generated by EF Core. EF Core executes the SQL query and then maps the result to the models. It significantly simplifies data access and database operations.

Next, let us learn how to query data using LINQ in controllers.

Basic LINQ queries

This book is not intended to be a LINQ handbook. However, we will show you some basic LINQ queries in this section:

- Querying the data
- Filtering the data
- Sorting the data

- Paging the data
- Creating the data
- Updating the data
- Deleting the data

Querying the data

The `DbSet<Invoice> Invoices` property in the `InvoiceDbContext` class represents a collection of the `Invoice` entity. We can use LINQ methods to query the data. For example, we can use the `ToListAsync()` method to retrieve all the invoices from the database:

```
var invoices = await _context.Invoices.ToListAsync();
```

That is how the `GetInvoices` action method works.

To find a specific invoice, we can use the `FindAsync()` method, as shown in the `GetInvoice()` action method:

```
var invoice = await _context.Invoices.FindAsync(id);
```

The `FindAsync()` method accepts the primary key value as the parameter. EF Core will translate the `FindAsync()` method to the SQL `SELECT` statement, as shown here:

```
Executed DbCommand (15ms) [Parameters=[@__get_Item_0='?' (DbType =
Guid)], CommandType='Text', CommandTimeout='30']
      SELECT TOP(1) [i].[Id], [i].[Amount], [i].[ContactName], [i].
[Description], [i].[DueDate], [i].[InvoiceDate], [i].[InvoiceNumber],
[i].[Status]
      FROM [Invoices] AS [i]
      WHERE [i].[Id] = @__get_Item_0
```

We can also use `Single()` or `SingleOrDefault()` methods to find a specific entity. For example, we can use the `SingleAsync()` method to find the invoice with the specified ID:

```
var invoice = await _context.Invoices.SingleAsync(i => i.Id == id);
```

> **Important note**
>
> You may notice that we use `SingleAsync()` instead of `Single()` in the code. Many methods of EF Core have both synchronous and asynchronous versions. The asynchronous versions are suffixed with `Async`. It is recommended to use the asynchronous versions in the controller actions because they are non-blocking and can improve the performance of the application.

If you have LINQ experience, you may know there are other methods – such as `First()`, `FirstOr-Default()`, and so on – that can be used to find a specific entity. The differences are listed as follows:

- `Find()` or `FindAsync()` is used to find an entity by the primary key value. If the entity is not found, it returns `null`. Note that these two methods are related to the tracking state of the entity. If the entity is already tracked by `DbContext`, the `Find()` and `FindAsync()` methods will return the tracked entity immediately without querying the database. Otherwise, they will execute the SQL `SELECT` statement to retrieve the entity from the database.

- `Single()` or `SingleAsync()` can accept a predicate as the parameter. It returns the *single* entity that satisfies the predicate and throws an exception if the entity is not found or more than one entity satisfies the condition. If it is called without a predicate, it returns the only entity of the collection and throws an exception if more than one entity exists in the collection.

- `SingleOrDefault()` or `SingleOrDefaultAsync()` can accept a predicate as the parameter. It also returns the *single* entity that satisfies the predicate and throws an exception if more than one entity satisfies the condition but returns a default value if the entity is not found. If it is called without a predicate, it returns a default value (or a specified default value) if the collection is empty and throws an exception if more than one entity exists in the collection.

- `First()` or `FirstAsync()` can accept a predicate as the parameter. It returns the *first* entity that satisfies the predicate and throws an exception if the entity is not found or the collection is null or empty. If it is called without a predicate, it returns the first entity of the collection and throws an exception if the collection is null or empty.

- `FirstOrDefault()` or `FirstOrDefaultAsync()` can accept a predicate as the parameter. It also returns the *first* entity that satisfies the predicate. If the entity is not found or the collection is empty, it returns a default value (or a specified default value). If it is called without a predicate, it returns the first entity if the collection is not empty; otherwise, it returns a default value (or a specified default value). If the collection is null, it throws an exception.

These methods are kind of confusing. The recommended practice is as follows:

- If you want to find an entity by the primary key value and leverage the tracking state to improve the performance, use `Find()` or `FindAsync()`.

- If you are sure that the entity exists and only one entity satisfies the condition, use `Single()` or `SingleAsync()`. If you would like to specify a default value when the entity is not found, use `SingleOrDefault()` or `SingleOrDefaultAsync()`.

- If you are not sure whether the entity exists, or there may be more than one entity that satisfies the condition, use `First()` or `FirstAsync()`. If you would like to specify a default value when the entity is not found, use `FirstOrDefault()` or `FirstOrDefaultAsync()`.

- Do not forget to check whether the result is `null` if you use `Find()`, `FindAsync()`, `Single-OrDefault()`, `SingleOrDefaultAsync()`, `FirstOrDefault()`, and `FirstOr-DefaultAsync()`.

Filtering the data

If the table contains a lot of records, we may want to filter the data based on some conditions, instead of returning all of them. We can use the `Where()` method to filter the invoices by status. Update the `GetInvoices` action method as shown here:

```
[HttpGet]
public async Task<ActionResult<IEnumerable<Invoice>>>
GetInvoices(InvoiceStatus? status)
{
    // Omitted for brevity
    return await _context.Invoices.Where(x => status == null ||
x.Status == status).ToListAsync();
}
```

The `Where()` method accepts a lambda expression as the parameter. A lambda expression is a concise way to define a delegate method inline, which is widely used in LINQ queries to define filtering, sorting, and projection operations. In the preceding example, the `x => status == null ||` `x.Status == status` lambda expression means that the `Status` property of the `Invoice` entity is equal to the `status` parameter if the `status` parameter is not `null`. EF Core will translate the lambda expression to the SQL `WHERE` clause.

Run the application and check the Swagger UI. You will find the `/api/Invoices` endpoint now has a `status` parameter. You can use the parameter to filter the invoices by status:

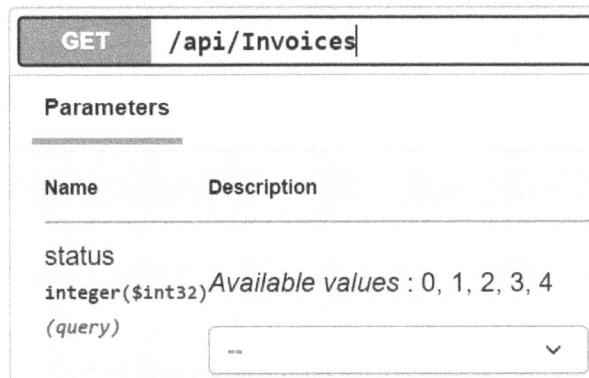

Figure 5.2 – Filtering the invoices by status

Send a request with a status parameter to the `/api/Invoices` endpoint. You will get the invoices with the specified status. The SQL query is shown here:

```
info: Microsoft.EntityFrameworkCore.Database.Command[20101]
      Executed DbCommand (44ms) [Parameters=[@__status_0='?' (Size =
16)], CommandType='Text', CommandTimeout='30']
```

```
    SELECT [i].[Id], [i].[Amount], [i].[ContactName], [i].
[Description], [i].[DueDate], [i].[InvoiceDate], [i].[InvoiceNumber],
[i].[Status]
    FROM [Invoices] AS [i]
    WHERE [i].[Status] = @__status_0
```

You can see that the `Where()` method is translated to the SQL `WHERE` clause.

Sorting and paging

It is not always enough to filter the data. We may also want to sort the data based on some properties and return a subset of the data using paging. We can use some methods, such as `OrderBy()`, `OrderByDescending()`, `Skip()`, `Take()`, and so on, to sort and page the data. Update the `GetInvoices` action method as shown here:

```
[HttpGet]
public async Task<ActionResult<IEnumerable<Invoice>>> GetInvoices(int
page = 1, int pageSize = 10, InvoiceStatus? status = null)
{
    // Omitted for brevity
    return await _context.Invoices.AsQueryable().Where(x => status ==
null || x.Status == status)
                .OrderByDescending(x => x.InvoiceDate)
                .Skip((page - 1) * pageSize)
                .Take(pageSize)
                .ToListAsync();
}
```

In the preceding code, we use the `AsQueryable()` method to convert the `DbSet<Invoice>` to `IQueryable<Invoice>`. We can use `IQueryable` to build a query. The `Where()` and `OrderByDescending()` methods return a new `IQueryable` object. So, we can chain the LINQ methods to build a new query. The `Where()` method is used to filter the data, the `OrderByDescending()` method is used to sort the data based on the `InvoiceDate` property in descending order, and the `Skip()` and `Take()` methods are used to page the data. The `Skip()` method skips the first `pageSize * (page - 1)` records, and the `Take()` method returns the next `pageSize` records. At the end of the statement, the `ToListAsync()` method executes the query and returns the result.

In fact, the `AsQueryable()` method is not required here because the `DbSet<TEntity>` class implements the `IQueryble<TEntity>` interface, which means the `DbSet<Invoice>` property is already an `IQueryable` object. We can chain the LINQ methods directly.

What is IQueryable?

When we use some LINQ methods, such as `Where()`, `OrderBy()`, `Skip()`, and `Take()`, EF Core will not execute the query immediately. It will build a query and return a new `IQueryable` object. `IQueryable` is an interface in the `System.Linq` namespace that represents a queryable collection of entities that can be used to compose a query against a specific data source, such as a database. It allows us to build complex queries by chaining the LINQ methods, but it postpones the query execution until the very last moment when the result is needed. Generally, when we call the `ToListAsync()` method, the query will be translated to a server-specific query language, such as SQL, and executed against the database. This can improve the performance of the application because we do not need to fetch all the data from the database before we can filter and sort the data.

Run the application using `dotnet run` and check the Swagger UI, you will see the `page` and `pageSize` parameters are added to the `/api/Invoices` endpoint. You can use the parameters to page the invoices as follows:

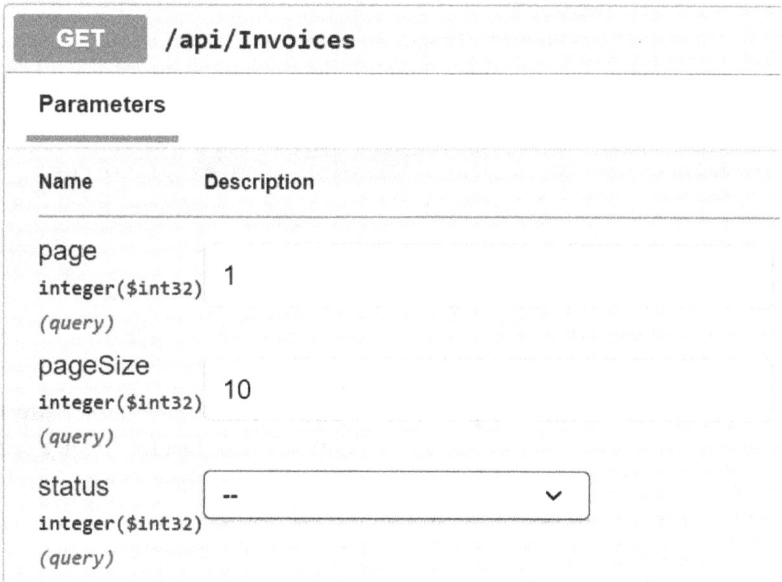

Figure 5.3 – Sorting and paging the invoices

The generated SQL query is shown here:

```
info: Microsoft.EntityFrameworkCore.Database.Command[20101]
      Executed DbCommand (40ms) [Parameters=[@__status_0='?' (Size =
16) (DbType = AnsiString), @__p_1='?' (DbType = Int32), @__p_2='?'
(DbType = Int32)], CommandType='Text', CommandTimeout='30']
      SELECT [i].[Id], [i].[Amount], [i].[ContactName], [i].
[Description], [i].[DueDate], [i].[InvoiceDate], [i].[InvoiceNumber],
```

```
[i].[Status]
    FROM [Invoices] AS [i]
    WHERE [i].[Status] = @__status_0
    ORDER BY [i].[InvoiceDate] DESC
    OFFSET @__p_1 ROWS FETCH NEXT @__p_2 ROWS ONLY
```

Note that the SQL statement uses the OFFSET/FETCH clause to page the data. These keywords are supported by SQL Server but may not be supported by other databases. For example, MySQL uses the LIMIT clause to page the data. EF Core can eliminate the differences between different databases. It will translate the LINQ query to the correct SQL statement for the database. In this way, developers can focus on writing the LINQ query in a database-agnostic way. That is the beauty of EF Core.

Creating an entity

Next, let's see how to create a new invoice. Check the code of the PostInvoice action method:

```
[HttpPost]
public async Task<ActionResult<Invoice>> PostInvoice(Invoice invoice)
{
    if (_context.Invoices == null)
    {
        return Problem("Entity set 'InvoiceDbContext.Invoices' is
null.");
    }
    _context.Invoices.Add(invoice);
    await _context.SaveChangesAsync();
    return CreatedAtAction("GetInvoice", new { id = invoice.Id },
invoice);
```

}The PostInvoice action method accepts an Invoice object as the request body. It uses the Add() method to add the invoice to the Invoices entity set. Note that this change occurs in the memory. The data will not be added to the database until the SaveChangesAsync() method is called to save the changes to the database. The CreatedAtAction() method returns a 201 Created response with the location of the newly created invoice. You can return a 200 OK response instead but it is recommended to return a 201 Created response for the creation of a new resource.

You can send a POST request to the /api/invoices endpoint to create a new invoice and see how the SQL statement is generated from the logs. It should be similar to the following:

```
info: Microsoft.EntityFrameworkCore.Database.Command[20101]
    Executed DbCommand (3ms) [Parameters=[@p0='?' (DbType =
Guid), @p1='?' (Precision = 18) (Scale = 2) (DbType = Decimal), @
p2='?' (Size = 32), @p3='?' (Size = 256), @p4='?' (DbType =
DateTimeOffset), @p5='?' (DbType = DateTimeOffset), @p6='?' (Size =
32) (DbType = AnsiString), @p7='?' (Size = 16) (DbType = AnsiString)],
CommandType='Text', CommandTimeout='30']
```

```
    SET IMPLICIT_TRANSACTIONS OFF;
    SET NOCOUNT ON;
    INSERT INTO [Invoices] ([Id], [Amount], [ContactName],
[Description], [DueDate], [InvoiceDate], [InvoiceNumber], [Status])
    VALUES (@p0, @p1, @p2, @p3, @p4, @p5, @p6, @p7);
```

> **Important note**
>
> The JSON body for the POST action does not need to contain the Id property. EF Core will generate a new Guid value for the Id property.

Updating an entity

To update an entity, we use the Put request. The code of the PutInvoice action method is shown here:

```
[HttpPut("{id}")]
public async Task<IActionResult> PutInvoice(Guid id, Invoice invoice)
{
    if (id != invoice.Id)
    {
        return BadRequest();
    }
    _context.Entry(invoice).State = EntityState.Modified;
    try
    {
        await _context.SaveChangesAsync();
    }
    catch (DbUpdateConcurrencyException)
    {
        if (!InvoiceExists(id))
        {
            return NotFound();
        }
        else
        {
            throw;
        }
    }
    return NoContent();
}
```

The `PutInvoice` action method accepts the `id` parameter and the `Invoice` object as the request body. If you check the Swagger UI, you will see the `id` parameter is defined in the URL but the `Invoice` object is defined in the request body. This is because `Invoice` is not a primitive type, so ASP.NET Core can only get it from the request body. We discussed this in the *Binding source attributes* section in *Chapter 3*.

Next, we use the `_context.Entry()` method to get the `EntityEntry` object of the invoice. Then, we set the `State` property to `EntityState.Modified`. It seems that the `EntityState` enum plays an important role here. What is the `EntityState` enum?

In EF Core, each instance of `DbContext` has a `ChangeTracker` to track the changes in the entities, which is a powerful feature of EF Core. In other words, EF Core knows the state of each entity – whether it is added, deleted, or modified. When we update the entity, we just update the entity in the memory. EF Core can track the changes. When the `SaveChangesAsync()` method is called, it will generate the SQL statement to update the data in the database.

The `EntityState` enum can have the following values:

- `Detached`: The entity is not being tracked by the context.
- `Unchanged`: The entity is being tracked by the context, but the values are not changed.
- `Deleted`: The entity is being tracked and exists in the database, but it has been marked for deletion. So, when the `SaveChangesAsync()` method is called, EF Core will generate the SQL statement to delete the entity from the database.
- `Modified`: The entity is being tracked and exists in the database, and the values have been modified in `DbContext`. When the `SaveChangesAsync()` method is called, EF Core will generate the SQL statement to update the entity in the database.
- `Added`: The entity is being tracked but it does not exist in the database. When the `SaveChangesAsync()` method is called, EF Core will generate the SQL statement to insert the entity into the database.

In the *Creating an entity* section, we used the `Add()` method to add the entity to the entity set. This is equivalent to setting the `State` property to `Added`, such as in the following code:

```
// _context.Invoices.Add(invoice); This is equivalent to the following
code
_context.Entry(invoice).State = EntityState.Added;
await _context.SaveChangesAsync();
```

Similar to the `Add()` method, changing the state of the entity does not modify the data in the database. You must call the `SaveChangesAsync()` method to save the changes to the database.

Let's try to call the `PutInvoice` action method to update an invoice. Send a PUT request to the `/api/invoices/{id}` endpoint in the Swagger UI. The request body is like this:

```
{
  "id": "0d501380-83d9-44f4-9087-27c8f09082f9",
  "invoiceNumber": "INV-001",
  "contactName": "Spider Man",
  "description": "Invoice for the first month",
  "amount": 100,
  "invoiceDate": "2023-01-01T00:00:00+00:00",
  "dueDate": "2023-01-15T00:00:00+00:00",
  "status": 1
}
```

Please update the JSON body to change the `contactName` property only. The SQL statement generated by EF Core is shown here:

```
info: Microsoft.EntityFrameworkCore.Database.Command[20101]
      Executed DbCommand (39ms) [Parameters=[@p7='?' (DbType =
Guid), @p0='?' (Precision = 18) (Scale = 2) (DbType = Decimal), @
p1='?' (Size = 32), @p2='?' (Size = 4000), @p3='?' (DbType =
DateTimeOffset), @p4='?' (DbType = DateTimeOffset), @p5='?' (Size =
32) (DbType = AnsiString), @p6='?' (Size = 16) (DbType = AnsiString)],
CommandType='Text', CommandTimeout='30']
      SET IMPLICIT_TRANSACTIONS OFF;
      SET NOCOUNT ON;
      UPDATE [Invoices] SET [Amount] = @p0, [ContactName] = @
p1, [Description] = @p2, [DueDate] = @p3, [InvoiceDate] = @p4,
[InvoiceNumber] = @p5, [Status] = @p6
      OUTPUT 1
      WHERE [Id] = @p7;
```

You can see EF Core omits the `Id` column in the UPDATE statement. This is because the `Id` column is the primary key of the `Invoices` table. EF Core knows that it does not need to update the `Id` column. But EF Core will update the other properties irrespective of whether the values are changed or not because the `EntityState` of the entity is `Modified`.

Sometimes we want to update only the changed properties. For example, if we just want to update the `Status` property, the SQL statement should not have to update other columns. To do this, we can find the entity that needs to be updated, and then update the properties explicitly. Let us update the `PutInvoice` action method to do this:

```
var invoiceToUpdate = await _context.Invoices.FindAsync(id);
if (invoiceToUpdate == null)
{
    return NotFound();
```

```
}
invoiceToUpdate.Status = invoice.Status;
await _context.SaveChangesAsync();
```

In this example, we first find the entity by the `FindAsync()` method. Then, we update the `Status` property. EF Core will mark the `Status` property as modified. Finally, we call the `SaveChangesAsync()` method to save the changes to the database. You can see the generated SQL statement only updates the `Status` property, which is like this:

```
info: Microsoft.EntityFrameworkCore.Database.Command[20101]
      Executed DbCommand (2ms) [Parameters=[@__get_Item_0='?' (DbType
= Guid)], CommandType='Text', CommandTimeout='30']
      SELECT TOP(1) [i].[Id], [i].[Amount], [i].[ContactName], [i].
[Description], [i].[DueDate], [i].[InvoiceDate], [i].[InvoiceNumber],
[i].[Status]
      FROM [Invoices] AS [i]
      WHERE [i].[Id] = @__get_Item_0
info: Microsoft.EntityFrameworkCore.Database.Command[20101]
      Executed DbCommand (2ms) [Parameters=[@p1='?' (DbType = Guid),
@p0='?' (Size = 16) (DbType = AnsiString)], CommandType='Text',
CommandTimeout='30']
      SET IMPLICIT_TRANSACTIONS OFF;
      SET NOCOUNT ON;
      UPDATE [Invoices] SET [Status] = @p0
      OUTPUT 1
      WHERE [Id] = @p1;
```

However, in an actual scenario, normally, the endpoint will receive the whole entity, not just the changed properties. We may not know which properties are changed. In this case, we can just update all the properties in the code. EF Core can track the state of the entity, so it is smart enough to determine which properties are changed. Let us update the `PutInvoice` action method to explicitly update all the properties:

```
// Omitted for brevity
var invoiceToUpdate = await _context.Invoices.FindAsync(id);
if (invoiceToUpdate == null)
{
    return NotFound();
}
invoiceToUpdate.InvoiceNumber = invoice.InvoiceNumber;
invoiceToUpdate.ContactName = invoice.ContactName;
invoiceToUpdate.Description = invoice.Description;
invoiceToUpdate.Amount = invoice.Amount;
invoiceToUpdate.InvoiceDate = invoice.InvoiceDate;
invoiceToUpdate.DueDate = invoice.DueDate;
```

```
invoiceToUpdate.Status = invoice.Status;
await _context.SaveChangesAsync();
// Omitted for brevity
```

Send a PUT request to the /api/Invoices/{id} endpoint and attach a JSON body to the request. If you just update the Status and Description properties, the SQL statement will be like this:

```
info: Microsoft.EntityFrameworkCore.Database.Command[20101]
      Executed DbCommand (17ms) [Parameters=[@p2='?' (DbType = Guid),
@p0='?' (Size = 256), @p1='?' (Size = 16) (DbType = AnsiString)],
CommandType='Text', CommandTimeout='30']
      SET IMPLICIT_TRANSACTIONS OFF;
      SET NOCOUNT ON;
      UPDATE [Invoices] SET [Description] = @p0, [Status] = @p1
      OUTPUT 1
      WHERE [Id] = @p2;
```

The preceding SQL statement slightly improves the performance because it only updates the changed properties. It might not be a big deal for a small table, but if you have a large table that has many columns, it will be a good practice. However, it needs a SELECT statement to get the entity first. Choose the way that suits your scenario.

There is an issue in the preceding code. If the entity has many properties, it will be tedious to update all the properties one by one. In this case, we can use the Entry method to get the EntityEntry object and then set the CurrentValues property to the new values. Let us update the PutInvoice action method to use the Entry method:

```
// Update only the properties that have changed _context.
Entry(invoiceToUpdate).CurrentValues.SetValues(invoice);
```

The SetValues() method will set all the properties of the entity to the new values. EF Core can detect the changes and mark the properties that have changed as Modified. So, we do not need to manually set each property. This way is a good practice when updating an entity that has many properties. Also, the object used to update the properties does not have to be the same type as the entity. It is useful in a layered application. For example, the entity received from the client is a **Data Transfer Object (DTO)** object, and the entity in the database is a domain object. In this case, EF Core will update the properties that match the names of the properties in the DTO object.

Note that the SetValues() method only updates the simple properties, such as string, int, decimal, DateTime, and so on. If the entity has a navigation property, the SetValues() method will not update the navigation property. In this case, we need to update the properties explicitly.

Test the /api/Invoices/{id} endpoint by sending a PUT request again. You can see the generated SQL statement is similar to the previous one.

Deleting an entity

In the generated code of the DeleteInvoice action method, we can see the following code:

```
[HttpDelete("{id}")]
public async Task<IActionResult> DeleteInvoice(Guid id)
{
    if (_context.Invoices == null)
    {
        return NotFound();
    }
    var invoice = await _context.Invoices.FindAsync(id);
    if (invoice == null)
    {
        return NotFound();
    }
    _context.Invoices.Remove(invoice);
    await _context.SaveChangesAsync();
    return NoContent();
}
```

The logic is to find the entity first, and then remove it from DbSet using the Remove() method. Finally, we call the SaveChangesAsync() method to save the changes to the database. If you understand EntityState already, you might know that the Remove() method is equivalent to setting EntityState to Deleted, as follows:

```
_context.Entry(invoice).State = EntityState.Deleted;
```

The generated SQL statement is like this:

```
info: Microsoft.EntityFrameworkCore.Database.Command[20101]
      Executed DbCommand (2ms) [Parameters=[@__get_Item_0='?' (DbType
= Guid)], CommandType='Text', CommandTimeout='30']
      SELECT TOP(1) [i].[Id], [i].[Amount], [i].[ContactName], [i].
[Description], [i].[DueDate], [i].[InvoiceDate], [i].[InvoiceNumber],
[i].[Status]
      FROM [Invoices] AS [i]
      WHERE [i].[Id] = @__get_Item_0
info: Microsoft.EntityFrameworkCore.Database.Command[20101]
      Executed DbCommand (3ms) [Parameters=[@p0='?' (DbType = Guid)],
CommandType='Text', CommandTimeout='30']
      SET IMPLICIT_TRANSACTIONS OFF;
      SET NOCOUNT ON;
      DELETE FROM [Invoices]
      OUTPUT 1
      WHERE [Id] = @p0;
```

As you can see, EF Core generates two SQL statements, which seems a little bit unnecessary to find the entity first. When we delete an entity, the only thing we need is the primary key. So, we can update the `DeleteInvoice()` action like this:

```
// Omitted for brevity
var invoice = new Invoice { Id = id };
_context.Invoices.Remove(invoice);
await _context.SaveChangesAsync();
// Omitted for brevity
```

Now, the SQL statement is as follows:

```
info: Microsoft.EntityFrameworkCore.Database.Command[20101]
      Executed DbCommand (2ms) [Parameters=[@p0='?' (DbType = Guid)],
CommandType='Text', CommandTimeout='30']
      SET IMPLICIT_TRANSACTIONS OFF;
      SET NOCOUNT ON;
      DELETE FROM [Invoices]
      OUTPUT 1
      WHERE [Id] = @p0;
```

It is much simpler than the previous one.

From EF Core 7.0, we have a new method called `ExecuteDeleteAsync()` that can be used to delete an entity without loading it first. The code is as follows:

```
await _context.Invoices.Where(x => x.Id == id).ExecuteDeleteAsync();
```

> **Important note**
>
> The `ExecuteDeleteAsync()` method does not involve the change tracker, so it will execute the SQL statement immediately to the database. It does not need to call the `SaveChangesAsync()` method at the end. It is a recommended way to delete one entity (or more) from EF Core 7.0 and later versions. However, if the entity already exists in `DbContext` and is tracked by the change tracker, executing the SQL statement directly may cause the data in `DbContext` and the database to be inconsistent. In this case, you may need to use the `Remove()` method or set the `EntityState` property to `Deleted` to delete the entity from `DbContext`. Please consider your scenario carefully before using the `ExecuteDeleteAsync()` method.

You must be wondering how EF Core knows the names of columns and tables in the database. We will talk about the configuration and see how EF Core maps the models to the database next.

Configuring the mapping between models and database

ORM, as the name suggests, is used to map the objects to the relational database. EF Core uses the mapping configuration to map the models to the database. In the previous section, we saw that we did not configure any mappings; however, EF Core could still map the models to the database automatically. This is because EF Core has a set of built-in conventions to configure the mappings. We can also explicitly customize the configuration to meet our needs. In this section, we will discuss the configuration in EF Core, including the following:

- Mapping conventions
- Data annotations
- Fluent API

Mapping conventions

There are some conventions in EF Core for mapping the models to the database:

- The database uses the `dbo` schema by default.
- The table name is the plural form of the model name. For example, we have a `DbSet<Invoice>` `Invoices` property in the `InvoiceDbContext` class, so the table name is `Invoices`.
- The column name is the property name.
- The data type of the column is based on the property type and the database providers. Here is a list of the default mapping for some common C# types in SQL Server:

.NET type	SQL Server data type
int	int
long	bigint
string	nvarchar(max)
bool	bit
datetime	datetime
double	float
decimal	decimal(18,2)
byte	tinyint
short	smallint
byte[]	varbinary(max)

Table 5.1 – Default mapping for some common C# types in SQL Server

- If a property is named `Id` or `<entity name>Id`, EF Core will map it to the primary key.

- If EF Core detects that the relationship between two models is one-to-many, it will map the navigation property to a foreign key column in the database automatically.

- If a column is a primary key, EF Core will create a clustered index for it automatically.

- If a column is a foreign key, EF Core will create a non-clustered index for it automatically.

- An enum type is mapped to the underlying type of the enum. For example, the `InvoiceStatus` enum is mapped to the `int` type in the database.

However, sometimes we need to refine the mapping. For example, we may want to use the `varchar(100)` column instead of the `nvarchar(max)` column for a `string` property. We may also want to save enums as strings in the database instead of `int` values. In such cases, we can override the default conventions to customize the mapping based on our needs.

There are two ways to explicitly configure the mapping between the models and the database:

- Data annotations
- Fluent API

Let us see how to use data annotations and Fluent API to customize the mapping.

Data annotations

Data annotations are attributes that you can apply to the model classes to customize the mapping. For example, you can use the `Table` attribute to specify the table name and use the `Column` attribute to specify the column name. The following code shows how to use data annotations to customize the mapping:

```
[Table("Invoices")]
public class Invoice
{
    [Column("Id")]
    [Key]
    public Guid Id { get; set; }

    [Column(name: "InvoiceNumber", TypeName = "varchar(32)")]
    [Required]
    public string InvoiceNumber { get; set; } = string.Empty;

    [Column(name: "ContactName")]
    [Required]
    [MaxLength(32)]
    public string ContactName { get; set; } = string.Empty;
```

```
        [Column(name: "Description")]
        [MaxLength(256)]
        public string? Description { get; set; }

        [Column("Amount")]
        [Precision(18, 2)]
        [Range(0, 999999999999999.99)]
        public decimal Amount { get; set; }

        [Column(name: "InvoiceDate", TypeName = "datetimeoffset")]
        public DateTimeOffset InvoiceDate { get; set; }

        [Column(name: "DueDate", TypeName = "datetimeoffset")]
        public DateTimeOffset DueDate { get; set; }

        [Column(name: "Status", TypeName = "varchar(16)")]
        public InvoiceStatus Status { get; set; }
    }
```

In the preceding code, each property has one or more data annotations. These data annotations are attributes that you can apply to the model classes to customize the mapping. You can specify some mapping information, such as the table name, column name, column data type, and so on.

Here is a list of the most commonly used data annotations:

Attribute	Description
Table	The table name that the model class is mapped to.
Column	The column name in the table that a property is mapped to. You can also specify the column data type using the TypeName property.
Key	Specifies the property as a key.
ForeignKey	Specifies the property as a foreign key.
NotMapped	The model or property is not mapped to the database.
Required	The value of the property is required.
MaxLength	Specifies the maximum length of the value in the database. Applied to string or array values only.
Index	Creates an index on the column that the property is mapped to.
Precision	Specifies the precision and scale of the property if the database supports precision and scale facets.

Attribute	Description
DatabaseGenerated	Specifies how the database should generate the values for the property. If you use `int` or `long` as the primary key for an entity, you can use this attribute and set `DatabaseGeneratedOption` as `Identity`, such as `[DatabaseGenerated(DatabaseGeneratedOption.Identity)]`.
TimeStamp	Specifies the property is used for concurrency management. The property will map to a `rowversion` type in SQL Server. The implementation may vary in different database providers.

Table 5.2 – Commonly used data annotations

In this way, the mapping configuration is embedded in the model classes. It is easy to understand, but it is a little bit intrusive, which means the model classes are polluted with the database-related configuration. To decouple the model classes from the database mapping configuration, we can use Fluent API.

> **Important note**
>
> Every time the mapping is changed, you need to run the `dotnet ef migrations add <migration name>` command to generate a new migration. Then, run the `dotnet ef database update` command to update the database.

Fluent API

Fluent API is a set of extension methods that you can use to configure the mappings gracefully. It is the most flexible and powerful way to apply the mapping configuration without polluting the model classes. Another important thing to note is that Fluent API has a higher priority than data annotations. If you configure the same property in both data annotations and Fluent API, Fluent API will override the data annotations. So, Fluent API is the recommended way to configure the mapping.

Fluent API is applied in the order of the method calls. If there are two calls to configure the same property, the latest call will override the previous configuration.

To use Fluent API, we need to override the `OnModelCreating()` method in the derived `DbContext` class. The following code shows how to use Fluent API to configure the mapping:

```
protected override void OnModelCreating(ModelBuilder modelBuilder)
{
    // Seed data is omitted for brevity
```

```
        modelBuilder.Entity<Invoice>(b =>
    {
        b.ToTable("Invoices");
        b.HasKey(i => i.Id);
        b.Property(p => p.Id).HasColumnName("Id");
        b.Property(p => p.InvoiceNumber).
HasColumnName("InvoiceNumber").HasColumnType("varchar(32)").
IsRequired();
        b.Property(p => p.ContactName).HasColumnName("ContactName").
HasMaxLength(32).IsRequired();
        b.Property(p => p.Description).HasColumnName("Description").
HasMaxLength(256);
        // b.Property(p => p.Amount).HasColumnName("Amount").
HasColumnType("decimal(18,2)").IsRequired();
        b.Property(p => p.Amount).HasColumnName("Amount").
HasPrecision(18, 2);
        b.Property(p => p.InvoiceDate).HasColumnName("InvoiceDate").
HasColumnType("datetimeoffset").IsRequired();
        b.Property(p => p.DueDate).HasColumnName("DueDate").
HasColumnType("datetimeoffset").IsRequired();
        b.Property(p => p.Status).HasColumnName("Status").
HasMaxLength(16).HasConversion(
            v => v.ToString(),
            v => (InvoiceStatus)Enum.Parse(typeof(InvoiceStatus),
v));
    });
}
```

In the preceding code, we use the `Entity()` method to configure the `Invoice` entity. This method accepts an `Action<EntityTypeBuilder<TEntity>>` parameter to specify the mappings. The `EntityTypeBuilder<TEntity>` class has a lot of methods to configure the entity, such as table name, column name, column data type, and so on. You can chain these methods in a fluent way to configure the entity, so it is called Fluent API.

Here is a list of the most commonly used Fluent API methods:

Method	Description	Equivalent data annotation
HasDefaultSchema()	Specifies the database schema. The default schema is dbo.	N/A
ToTable()	The table name that the model class is mapped to.	Table
HasColumnName()	The column name in the table that a property is mapped to.	Column

Method	Description	Equivalent data annotation
HasKey()	Specifies the property as a key.	Key
Ignore()	Ignores a model or a property from the mapping. This method can be applied on an entity level or a property level.	NotMapped
IsRequired()	The value of the property is required.	Required
HasColumnType()	Specifies the data type of the column that the property is mapped to.	Column with the TypeName property
HasMaxLength()	Specifies the maximum length of the value in the database. Applied to a string or array only.	MaxLength
HasIndex()	Creates an index on the specific property.	Index
IsRowVersion()	Specifies the property is used for concurrency management. The property will map to a rowversion type in SQL Server. The implementation may vary in different database providers.	TimeStamp
HasDefaultValue()	Specifies a default value for a column. The value must be a constant.	N/A
HasDefaultValueSql()	Specifies a SQL expression to generate the default value for a column, such as GetUtcDate().	N/A
HasConversion()	Defines a value converter to map the property to the column data type. It contains two Func expressions to convert the values.	N/A
ValueGeneratedOnAdd()	Specifies the value of the property to be generated by the database when a new entity is added. EF Core will ignore this property when inserting a record.	DatabaseGenerated with the DatabaseGeneratedOption. Identity option

Table 5.3 – Commonly used Fluent API methods

There are some other methods to configure the relationships using Fluent API, such as `HasOne()`, `HasMany()`, `WithOne()`, `WithMany()`, and so on. We will cover them in the next chapter.

Separating the mapping configurations

In a large project, there may be a lot of model classes. If we put all the mapping configurations in the `OnModelCreating()` method, the method will be very long and hard to maintain. To make the code more readable and maintainable, we can extract the mapping configuration to one class or a couple of separate classes.

One way to do this is to create an extension method for the `ModelBuilder` class. Create a new class named `InvoiceModelCreatingExtensions` in the `Data` folder. Then, add the following code to the class:

```
public static class InvoiceModelCreatingExtensions
{
    public static void ConfigureInvoice(this ModelBuilder builder)
    {
        builder.Entity<Invoice>(b =>
        {
            b.ToTable("Invoices");
            // Other mapping configurations are omitted for brevity
        });
    }
}

// You can continue to create the mapping for other entities, or
create separate files for each entity.
```

Then, in the `OnModelCreating()` method, call the extension method:

```
modelBuilder.ConfigureInvoice();
// You can continue to call the extension methods for other entities.
such as
// modelBuilder.ConfigureInvoiceItem();
```

Now, the `OnModelCreating()` method is much cleaner and easier to read.

Another way to separate the mapping configurations is to implement the `IEntityTypeConfiguration<TEntity>` interface. Create a new class named `InvoiceConfiguration` in the `Data` folder. Then, add the following code to the class:

```
public class InvoiceConfiguration : IEntityTypeConfiguration<Invoice>
{
    public void Configure(EntityTypeBuilder<Invoice> builder)
```

```
    {
        builder.ToTable("Invoices");
        // Other mapping configurations are omitted for brevity
    }
}
```

Then, in the OnModelCreating() method, there are two ways to apply the configuration:

- If you use the ApplyConfiguration() method, add the following code to the OnModelCreating() method:

  ```
  modelBuilder.ApplyConfiguration(new InvoiceConfiguration());
  // You can continue to call the ApplyConfiguration method for
  other entities. such as
  // modelBuilder.ApplyConfiguration(new
  InvoiceItemConfiguration());
  ```

- Or you can call the Configure() method directly:

  ```
  new InvoiceConfiguration().Configure(modelBuilder.
  Entity<Invoice>());
  // You can continue to call the Configure method for other
  entities. such as
  // new InvoiceItemConfiguration().Configure(modelBuilder.
  Entity<InvoiceItem>());
  ```

As the project grows, it might be a little bit tedious to call the mapping configuration for each entity. In this case, EF Core has a method called ApplyConfigurationsFromAssembly() to apply all the configurations in the assembly, as shown in the following code:

```
modelBuilder.ApplyConfigurationsFromAssembly(typeof(InvoiceDbContext).
Assembly);
```

You can choose the one that fits your project best. One reminder is that if you use the ApplyConfigurationsFromAssembly() method, you need to make sure that all the configuration classes are in the same assembly as the DbContext class. Also, you cannot control the order of the configurations. If the order matters, you need to call each configuration explicitly in the correct order.

After you run the dotnet ef migrations add <migration name> command, you can find that the generated migration file has the following code:

```
migrationBuilder.AlterColumn<string>(
    name: "Status",
    table: "Invoices",
    type: "varchar(16)",
    nullable: false,
    oldClrType: typeof(int),
```

```
        oldType: "int");

migrationBuilder.AlterColumn<string>(
    name: "InvoiceNumber",
    table: "Invoices",
    type: "varchar(32)",
    nullable: false,
    oldClrType: typeof(string),
    oldType: "nvarchar(max)");
```

The preceding snippet shows that the Status property is changed from int to varchar(16), and the InvoiceNumber property is changed from nvarchar(max) to varchar(32). Then, you can run the dotnet ef database update command to update the database. You will see the Status column is stored as strings.

> **Important note**
>
> During the migration, the data may get lost if the data type is changed. For example, if the data type is changed from nvarchar(max) to varchar(32), the original data will be truncated to 32 characters. Please make sure you understand the data type change before you run the migration.

It is recommended to explicitly configure the mapping for each entity in order to ensure optimal performance. For example, nvarchar(max) requires more storage space than varchar, so the default mapping configuration may not be the most efficient. Additionally, the default dbo database schema may not be suitable for your particular scenario. Therefore, explicitly configuring the mapping is a recommended practice.

Summary

In this chapter, we learned how to access the database using EF Core. We implemented CRUD operations using the DbContext class. We introduced some basic LINQ queries, such as query, filter, sort, create, update, and delete. We also learned how to configure the mapping using data annotations and Fluent API. With the knowledge gained in this chapter, you can build a simple application to access the database.

However, the application we built in this chapter is quite basic and only has one entity. In a real-world project, there are usually multiple entities and relationships between them.

In the next chapter, we will learn how to configure relationshiseps between entities using EF Core.

Data Access in ASP.NET Core (Part 2 – Entity Relationships)

In *Chapter 5*, we introduced the fundamentals of **Entity Framework Core (EF Core)**, including how to create a `DbContext` class and how to use it to access data.

You can recap the basic concepts of relationships in *Chapter 1*, in the *Defining the Relationships between Resources* section, where we introduced relationships between resources. For example, in a blog system, a post has a collection of comments, and a user has a collection of posts. In an invoice system, an invoice has a collection of invoice items, and an invoice item belongs to an invoice. An invoice also has a contact, which can have one or more contact persons and can have one address.

In this chapter, we will continue to explore the features of EF Core. We will learn how to manage relationships between entities using Fluent APIs. Finally, we will discuss how to implement CRUD operations for entities with relationships.

We will cover the following topics in this chapter:

- Understanding one-to-many relationships
- Understanding one-to-one relationships
- Understanding many-to-many relationships
- Understanding owned entities

After reading this chapter, you should be able to configure relationships between entities using Fluent APIs in EF Core and implement CRUD operations for entities with relationships in your ASP.NET Core applications.

Technical requirements

The code examples in this chapter can be found at `https://github.com/PacktPublishing/Web-API-Development-with-ASP.NET-Core-8/tree/main/samples/chapter6`. You can use VS 2022 or VS Code to open the solutions.

You are expected to have basic knowledge of **Structured Query Language** (**SQL**) queries and **Language-Integrated Query** (**LINQ**). If you are not familiar with them, you can refer to the following resources:

- **SQL queries**: `https://www.w3schools.com/sql/`

- **LINQ**: `https://learn.microsoft.com/en-us/dotnet/csharp/linq/`

Understanding one-to-many relationships

One-to-many relationships are the most common relationships in a relational database. They are also called **parent-child (children)** relationships. For example, an invoice has a collection of invoice items. In this section, we will learn how to configure a one-to-many relationship in EF Core and how to implement CRUD operations for entities with a one-to-many relationship.

Let us continue to use the invoice sample application. You can find the sample code of the `EfCoreRelationshipsDemo` project in the `chapter6` folder. If you would like to test the code following the book, you can continue to work on the `BasicEfCoreDemo` project. Note that the `InvoiceDbContext` class has been renamed `SampleDbContext` in the sample code.

Next, let us update the `Invoice` class and create an `InvoiceItem` class, then define the one-to-many relationship between them.

One-to-many configuration

To demonstrate a one-to-many relationship, we need to add a new class named `InvoiceItem` in the `Models` folder and add some additional properties to the `Invoice` class to represent the relationship between them.

The code of the `InvoiceItem` class is as follows:

```
public class InvoiceItem
{
    public Guid Id { get; set; }
    public string Name { get; set; } = string.Empty;
    public string? Description { get; set; }
    public decimal UnitPrice { get; set; }
    public decimal Quantity { get; set; }
    public decimal Amount { get; set; }
```

```
    public Guid InvoiceId { get; set; }
    public Invoice? Invoice { get; set; }
}
```

The `InvoiceItem` class has a set of properties to store the invoice item data, such as `Name`, `Description`, `UnitPrice`, and so on. It also has an `InvoiceId` property to store the ID of the invoice that the invoice item belongs to, and an `Invoice` property to reference the invoice. To get started with the configuration process, follow these steps:

1. Update the `Invoice` class as follows:

    ```
    public class Invoice
    {
        public Guid Id { get; set; }
        // Omitted for brevity
        // Add a collection of invoice items
        public List<InvoiceItem> InvoiceItems { get; set; } = new
    ();
    }
    ```

    In the preceding code, we defined a relationship between `Invoice` and `InvoiceItem`. An invoice has a collection of invoice items, and an invoice item belongs to an invoice. It is a one-to-many relationship, where we can identify these terms:

    - **Principal entity**: This is the *one* entity in a one-to-many relationship. It is also called the parent entity. In the preceding case, `Invoice` is the principal entity.

    - **Dependent entity**: This is the *many* entity in a one-to-many relationship. It is also called the child entity. In the preceding case, `InvoiceItem` is the dependent entity. It has an `InvoiceId` foreign key property to identify the parent entity.

    - **Principal key**: This is the primary key of the principal entity, which uniquely identifies the principal entity. In the preceding case, the `Id` property of the `Invoice` class is the principal key.

    - **Foreign key**: The `InvoiceId` property of the `InvoiceItem` class is the foreign key, which is used to store the principal key value of the parent entity.

    - **Navigation property**: This is used to reference the related entity. It can be defined on the principal or dependent entity. There are two types of navigation properties:

        - **Collection navigation property**: This is defined on the principal entity, which is used to reference a collection of related entities. For example, the `InvoiceItems` property of the `Invoice` class is a collection navigation property.

        - **Reference navigation property**: This is defined on the dependent entity, which is used to reference a single related entity. For example, the `Invoice` property of the `InvoiceItem` class is a reference navigation property.

2. Because we added a new model, we need to update the `DbContext` class. Open the `SampleDbContext` class and add the following code:

```
public DbSet<InvoiceItem> InvoiceItems => Set<InvoiceItem>();
```

3. Also, it is a good practice to configure the mapping for the new model. Add a new class in the `Data` folder and name it `InvoiceItemConfiguration`:

```
public class InvoiceItemConfiguration :
IEntityTypeConfiguration<InvoiceItem>
{
    public void Configure(EntityTypeBuilder<InvoiceItem>
builder)
    {
        builder.ToTable("InvoiceItems");
        builder.Property(p => p.Id).
HasColumnName(nameof(InvoiceItem.Id));
        builder.Property(p => p.Name).
HasColumnName(nameof(InvoiceItem.Name)).HasMaxLength(64).
IsRequired();
        builder.Property(p => p.Description).
HasColumnName(nameof(InvoiceItem.Description)).
HasMaxLength(256);
        builder.Property(p => p.UnitPrice).
HasColumnName(nameof(InvoiceItem.UnitPrice)).HasPrecision(8, 2);
        builder.Property(p => p.Quantity).
HasColumnName(nameof(InvoiceItem.Quantity)).HasPrecision(8, 2);
        builder.Property(p => p.Amount).
HasColumnName(nameof(InvoiceItem.Amount)).HasPrecision(18, 2);
        builder.Property(p => p.InvoiceId).
HasColumnName(nameof(InvoiceItem.InvoiceId));
    }
}
```

4. Once we define navigation properties for `Invoice` and `InvoiceItem`, EF Core can discover the relationship between these two entities. Let us create a migration using the `dotnet ef migrations add AddInvoiceItem` command. Then, check the generated migration file. You will find that EF Core has added the following code:

```
migrationBuilder.CreateTable(
    name: "InvoiceItems",
    columns: table => new
    {
        Id = table.Column<Guid>(type: "uniqueidentifier",
nullable: false),
        // Omitted for brevity
        InvoiceId = table.Column<Guid>(type: "uniqueidentifier",
nullable: false)
```

```
        },
        constraints: table =>
        {
            table.PrimaryKey("PK_InvoiceItems", x => x.Id);
            table.ForeignKey(
                name: "FK_InvoiceItems_Invoices_InvoiceId",
                column: x => x.InvoiceId,
                principalTable: "Invoices",
                principalColumn: "Id",
                onDelete: ReferentialAction.Cascade);
        });

    // Omitted for brevity
    migrationBuilder.CreateIndex(
        name: "IX_InvoiceItems_InvoiceId",
        table: "InvoiceItems",
        column: "InvoiceId");
```

EF Core will create a new `InvoiceItems` table and add a foreign key constraint to the `InvoiceId` column. The name of the foreign key constraint is `FK_<dependent type name>_<principal type name>_<foreign key property name>`. It will also create an index on the `InvoiceId` column.

Another thing you need to be aware of is that the `onDelete` action is set to `ReferentialAction.Cascade`, which means that if the parent entity is deleted, all related child entities will also be deleted.

Let's think about a question – what if we do not have the `InvoiceId` property in the `InvoiceItem` class? Can EF Core still discover the relationship between these two entities? You can use the `dotnet ef migrations remove` command to remove the last migration, delete the `InvoiceId` property in the `InvoiceItem` class, and then add a migration again. You will see that EF Core can still create a column named `InvoiceId` in the `InvoiceItems` table, and apply the foreign key constraint to it, which is called **shadow foreign key** property. This is because EF Core has its built-in convention to do this. There are a few scenarios where EF Core can discover one-to-many relationships between entities:

- The dependent entity has a reference navigation property to the principal entity
- The principal entity has a collection navigation property to the dependent entity
- The reference navigation property and the collection navigation property are included at both ends
- The reference navigation property and the collection navigation property are included at both ends, and the foreign key property is included in the dependent entity

We can explicitly configure the relationship between entities to change the default behavior of EF Core if the convention does not work for us. Follow these steps:

1. To explicitly configure the one-to-many relationship between entities, we can use the `HasOne()`, `WithMany()`, and `HasMany()`, `WithOne()` methods. Add the following code to the `InvoiceConfiguration` class:

    ```
    builder.HasMany(x => x.InvoiceItems)
        .WithOne(x => x.Invoice)
        .HasForeignKey(x => x.InvoiceId);
    ```

    The `HasMany()` method is used to configure the collection navigation property, and the `WithOne()` method is used to configure the reference navigation property. The `HasForeignKey()` method is used to configure the foreign key property. So, the preceding code explicitly configures that one invoice can have many invoice items, and the `InvoiceId` property of the `InvoiceItem` class is the foreign key. If you add a migration now, you will find that EF Core will generate the same code as that generated by the convention.

2. A relationship can be also defined for the `InvoiceItem` class. Remove the preceding configuration code for the `Invoice` class, and add the following code to the `InvoiceItemConfiguration` class:

    ```
    builder.HasOne(i => i.Invoice)
        .WithMany(i => i.InvoiceItems)
        .HasForeignKey(i => i.InvoiceId)
        .OnDelete(DeleteBehavior.Cascade);
    ```

    It should be easy to understand now. The `HasOne()` method is used to configure the reference navigation property, and the `WithMany` method is used to configure the collection navigation property.

    Note that we also explicitly configured the `OnDelete()` action to `Cascade`, which is the same as the one generated by the convention. But we can change it to other options if needed. That said, the Fluent API is more flexible than the convention.

3. We just need to configure the relationship on one side of the relationship. So, please clean up the test code before you add the migration file and apply the migration to the database. After the migration is applied, you can check the database schema to see if the foreign key constraint is created, as shown in the following figure:

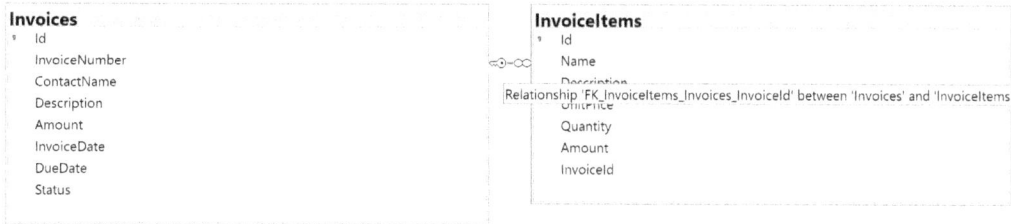

Invoices
* Id
  InvoiceNumber
  ContactName
  Description
  Amount
  InvoiceDate
  DueDate
  Status

InvoiceItems
* Id
  Name
  Description
  UnitPrice
  Quantity
  Amount
  InvoiceId

Relationship 'FK_InvoiceItems_Invoices_InvoiceId' between 'Invoices' and 'InvoiceItems'

Figure 6.1 – A foreign key constraint is created in the database

As a one-to-many relationship can be defined in either direction, which side should we configure the relationship on? It depends on the scenario. If the two entities have a strong one-to-many relationship, it does not really matter which side we configure the relationship on. But if the two entities are loosely coupled, we had better configure the relationship on the dependent entity.

For example, a `User` entity is shared by many other entities, such as `Post`, `Comment`, `Invoice`, and so on. Each `Post` entity can have an `Author` property that is a reference navigation property to the `User` entity, and `Comment` and `Invoice` also do the same. However, the `User` entity does not need to have the collection navigation properties to the `Post`, `Comment`, and `Invoice` entities. In this case, we should configure the relationship on the `Post`, `Comment`, and `Invoice` entities.

To configure this kind of relationship, we can ignore the parameter for the `WithMany` method, because the `User` entity does not have the collection navigation property to the `Post` entity, as shown in the following code:

```
public void Configure(EntityTypeBuilder<Post> builder)
{
    // Omitted for brevity
    builder.HasOne(x => x.Author)
        .WithMany()
        .HasForeignKey(x => x.AuthorId);
    // Omitted for brevity
}
```

Next, let's see how we implement CRUD operations for entities with a one-to-many relationship.

One-to-many CRUD operations

CRUD operations for entities with a one-to-many relationship differ from those without a relationship. For example, when retrieving an invoice, we may need to query both the `Invoices` table and the `InvoiceItems` table in order to also retrieve the associated invoice items. Additionally, when deleting an invoice, we must consider whether to delete the related invoice items as well.

EF Core can assist us in managing various scenarios. For instance, when we need to retrieve an invoice and its invoice items, EF Core can generate a `LEFT JOIN` query to join the two tables. To implement CRUD operations for entities with a one-to-many relationship, let us explore the following sections.

Creating data

First, let us create a new invoice with a few invoice items. You do not need to update the code of the `PostInvoice` action:

1. Run the application using `dotnet run`. Send a `POST` request to the `/api/Invoices` endpoint. The JSON body is like this:

```
{
    "invoiceNumber": "INV-004",
    "contactName": "Hulk",
    "description": "Invoice for the first month",
    "amount": 300,
    "invoiceDate": "2022-12-28T01:39:42.915Z",
    "dueDate": "2022-12-28T01:39:42.915Z",
    "status": 1,
    "invoiceItems": [
      {
        "name": "Invoice Item 1",
        "description": "",
        "unitPrice": 100,
        "quantity": 2,
        "amount": 200
      },
      {
        "name": "Invoice Item 2",
        "description": "",
        "unitPrice": 50,
        "quantity": 2,
        "amount": 100
      }
    ]
}
```

Note that it does not need to contain the `Id` property for the `Invoice` class, and the `Id` property for the `InvoiceItem` class is also not included. EF Core will generate them automatically.

After you send the `POST` request, you will see an exception in the log:

```
An unhandled exception has occurred while executing the request.
System.Text.Json.JsonException: A possible object cycle was
detected. This can either be due to a cycle or if the object
```

```
depth is larger than the maximum allowed depth of 32. Consider
using ReferenceHandler.Preserve on JsonSerializerOptions to
support cycles. Path: $.InvoiceItems.Invoice.InvoiceItems.
Invoice.InvoiceItems.Invoice.InvoiceItems.Invoice.InvoiceItems.
Invoice.InvoiceItems.Invoice.InvoiceItems.Invoice.InvoiceItems.
Invoice.InvoiceItems.Invoice.InvoiceItems.Invoice.InvoiceItems.
```

The exception is thrown because the `Invoice` class has a collection navigation property to the `InvoiceItem` class, and the `InvoiceItem` class has a reference navigation property to the `Invoice` class. So, there is a cycle in the JSON serialization. Some serialization frameworks, such as `Newtonsoft.Json`, `System.Text.Json`, do not allow such cycles. ASP.NET Core uses `System.Text.Json` for JSON serialization by default. So, we need to configure the `System.Text.Json` framework to ignore the cycle.

2. Open the `Program.cs` file and add the following code to `builder.Services.AddControllers()`:

```
builder.Services
    .AddControllers()
    .AddJsonOptions(options =>
    {
        options.JsonSerializerOptions.ReferenceHandler =
ReferenceHandler.IgnoreCycles;
    });
```

Another way to fix the exception is to use the `[JsonIgnore]` attribute to decorate the `Invoice` property in the `InvoiceItem` class. But if you have many entities with such a relationship, it is tedious to decorate all of them. Choose the way you prefer.

This exception occurs after data is saved to the database. So, if you check the database, you will find that the invoice and the invoice items are saved to the database:

	Id	Name	Description	UnitPrice	Quantity	Amount	InvoiceId
1	9092D563-D1C5-4271-8B67-08DAE9EDC256	Invoice Item 1		100.00	2.00	200.00	6F4BA1B0-6296-4B1B-5C54-08DAE9EDC252
2	F597E344-B527-4551-8B68-08DAE9EDC256	Invoice Item 2		50.00	2.00	100.00	6F4BA1B0-6296-4B1B-5C54-08DAE9EDC252

Figure 6.2 – The invoice items are saved to the database with the invoice ID

What is System.Text.Json?

`System.Text.Json` is a new JSON serialization framework provided since .NET Core 3.0. It is faster and more efficient than `Newtonsoft.Json`. It is also the default JSON serialization framework in ASP.NET Core 3.0 and later versions. It is recommended to use `System.Text.Json` instead of `Newtonsoft.Json` in new projects.

From the preceding example, you can see these points:

- EF Core generates an `Id` property of the principal entity if it is not defined in the model.

- EF Core generates an `Id` property of the dependent entity if it is not defined in the model.

- EF Core generates a foreign key property of the dependent entity, which is `InvoiceId` in this case, if it is not defined in the model.

- When the principal entity is added to the database, dependent entities are also added to the database automatically. You do not need to add dependent entities explicitly.

So, what if you want to add a new invoice item to an existing invoice? You can do it in two ways:

- Get the invoice first, then add the new invoice item to the `InvoiceItems` collection of the invoice, and then call the `SaveChanges()` method to save the changes to the database. This is an `Update` operation for the invoice, which means it should be a `PUT` action.

- Create a new invoice item, set the `InvoiceId` property to the `Id` property of the invoice, and then call the `SaveChanges()` method to save the changes to the database. This is a `Create` operation for the invoice item, which means it should be a `POST` action. Also, you need to provide an endpoint for the invoice item separately.

An invoice item cannot exist without an invoice. So, typically, you interact with the invoice item through the invoice. From a practical point of view, the first way is more common if the dependent entity count is not large. However, it depends on your scenario. If the principal entity has a large number of dependent entities, updating the entire principal entity may be inefficient and expensive. In this case, you can expose a separate endpoint to operate the dependent entity. For example, one blog post may have a large number of comments. It is common to add a new comment to a blog post, but it is not necessary to update the entire blog post and other comments. This is related to another concept, **domain-driven design** (**DDD**), which is to model domain objects and their relationships. We will talk about it in later chapters.

Querying data

Now we have an invoice and some invoice items in the database, we can send a `GET` request to the `/api/Invoices` endpoint. You can see the following response:

```
[
  {
    "id": "a224e90a-c01c-499b-7a9b-08dae9f04218",
    "invoiceNumber": "INV-004",
    "contactName": "Hulk",
    "description": "Invoice for the first month",
    "amount": 300,
    "invoiceDate": "2022-12-28T01:39:42.915+00:00",
```

```
      "dueDate": "2022-12-28T01:39:42.915+00:00",
      "status": 1,
      "invoiceItems": []
    },
    ...
  ]
```

The response contains a list of invoices. But the `InvoiceItems` property is empty. This is because the `InvoiceItems` property is a collection navigation property. By default, EF Core does not include dependent entities in the query result, so you need to explicitly include these in the query result. Follow these steps to query the invoice and invoice items from the database:

1. Open the `InvoicesController.cs` file, and update the code of the `GetInvoices()` method to this:

    ```
    // GET: api/Invoices
    [HttpGet]
    public async Task<ActionResult<IEnumerable<Invoice>>>
    GetInvoices(int page = 1, int pageSize = 10,
        InvoiceStatus? status = null)
    {
        // Omitted for brevity
        return await context.Invoices
            .Include(x => x.InvoiceItems)
            .Where(x => status == null || x.Status == status)
            .OrderByDescending(x => x.InvoiceDate)
            .Skip((page - 1) * pageSize)
            .Take(pageSize)
            .ToListAsync();
    }
    ```

    In the preceding code, we use the `Include` method to include dependent entities in the query result.

2. Restart the application and send the same request again. Now, you will see the result includes invoice items, as shown here:

    ```
    [
      {
        "id": "a224e90a-c01c-499b-7a9b-08dae9f04218",
        "invoiceNumber": "INV-004",
        "contactName": "Hulk",
        "description": "Invoice for the first month",
        "amount": 300,
        "invoiceDate": "2022-12-28T01:39:42.915+00:00",
        "dueDate": "2022-12-28T01:39:42.915+00:00",
    ```

```
    "status": 1,
    "invoiceItems": [
      {
        "id": "8cc52722-5b99-4d0c-07ef-08dae9f04223",
        "name": "Invoice Item 1",
        "description": "",
        "unitPrice": 100,
        "quantity": 2,
        "amount": 200,
        "invoiceId": "a224e90a-c01c-499b-7a9b-08dae9f04218",
        "invoice": null
      },
      {
        "id": "2d3f739a-2280-424b-07f0-08dae9f04223",
        "name": "Invoice Item 2",
        "description": "",
        "unitPrice": 50,
        "quantity": 2,
        "amount": 100,
        "invoiceId": "a224e90a-c01c-499b-7a9b-08dae9f04218",
        "invoice": null
      }
    ]
  },
  ...
]
```

In the response, the Invoice property of the invoice item is null. It is not required to include the principal entity in the dependent entity. As we explained in the previous section, this is a cyclic reference. If you do not want to include a null value for the Invoice property in the response, you can use the [JsonIgnore] attribute to decorate the Invoice property of the InvoiceItem model so that System.Text.Json will not serialize the Invoice property anymore.

The generated SQL query is as follows:

```
info: Microsoft.EntityFrameworkCore.Database.Command[20101]
      Executed DbCommand (35ms) [Parameters=[@__p_0='?' (DbType
= Int32), @__p_1='?' (DbType = Int32)], CommandType='Text',
CommandTimeout='30']
      SELECT [t].[Id], [t].[Amount], [t].[ContactName],
[t].[Description], [t].[DueDate], [t].[InvoiceDate], [t].
[InvoiceNumber], [t].[Status], [i0].[Id], [i0].[Amount], [i0].
[Description], [i0].[InvoiceId], [i0].[Name], [i0].[Quantity],
[i0].[UnitPrice]
      FROM (
```

```
        SELECT [i].[Id], [i].[Amount], [i].[ContactName],
[i].[Description], [i].[DueDate], [i].[InvoiceDate], [i].
[InvoiceNumber], [i].[Status]
        FROM [Invoices] AS [i]
        ORDER BY [i].[InvoiceDate] DESC
        OFFSET @__p_0 ROWS FETCH NEXT @__p_1 ROWS ONLY
    ) AS [t]
    LEFT JOIN [InvoiceItems] AS [i0] ON [t].[Id] = [i0].
[InvoiceId]
        ORDER BY [t].[InvoiceDate] DESC, [t].[Id]
```

As you see, when the LINQ query uses the `Include()` method to include dependent entities, EF Core will generate a `LEFT JOIN` query.

Important note

The `Include()` method is a convenient way to include dependent entities. However, it may cause performance issues when the collection of dependent entities is large. For example, a post may have hundreds or thousands of comments. It is not a good idea to include all comments in the query result for a list page. In this case, it is not necessary to include dependent entities in the query.

---

3. Note that the query includes `Invoice` data in each row of the result. For some scenarios, it may cause a so-called **Cartesian explosion** problem, which means the amount of duplicated data in the result is too large and may cause performance issues. In this case, we can split the queries into two steps. First, we query the invoices, and then we query the invoice items. You need to use the `AsSplitQuery()` method as follows:

```
[HttpGet]
public async Task<ActionResult<IEnumerable<Invoice>>>
GetInvoices(int page = 1, int pageSize = 10,
    InvoiceStatus? status = null)
{
    // Omitted for brevity
    return await context.Invoices
        .Include(x => x.InvoiceItems)
        .Where(x => status == null || x.Status == status)
        .OrderByDescending(x => x.InvoiceDate)
        .Skip((page - 1) * pageSize)
        .Take(pageSize)
        .AsSplitQuery()
        .ToListAsync();
}
```

The generated SQL query is as follows:

```
info: Microsoft.EntityFrameworkCore.Database.Command[20101]
      FROM (          SELECT [i].[Id], [i].[InvoiceDate]
      Executed DbCommand (2ms) [Parameters=[@__p_0='?' (DbType
= Int32), @__p_1='?' (DbType = Int32)], CommandType='Text',
CommandTimeout='30']
      SELECT [i].[Id], [i].[Amount], [i].[ContactName],
[i].[Description], [i].[DueDate], [i].[InvoiceDate], [i].
[InvoiceNumber], [i].[Status]
      FROM [Invoices] AS [i]
      ORDER BY [i].[InvoiceDate] DESC, [i].[Id]
      OFFSET @__p_0 ROWS FETCH NEXT @__p_1 ROWS ONLY
info: Microsoft.EntityFrameworkCore.Database.Command[20101]
      Executed DbCommand (2ms) [Parameters=[@__p_0='?' (DbType
= Int32), @__p_1='?' (DbType = Int32)], CommandType='Text',
CommandTimeout='30']
      SELECT [i0].[Id], [i0].[Amount], [i0].[Description], [i0].
[InvoiceId], [i0].[Name], [i0].[Quantity], [i0].[UnitPrice],
[t].[Id]      FROM (
          SELECT [i].[Id], [i].[InvoiceDate]
          FROM [Invoices] AS [i]
          ORDER BY [i].[InvoiceDate] DESC
          OFFSET @__p_0 ROWS FETCH NEXT @__p_1 ROWS ONLY
      ) AS [t]         INNER JOIN [InvoiceItems] AS [i0] ON [t].
[Id] = [i0].[InvoiceId]
      ORDER BY [t].[InvoiceDate] DESC, [t].[Id]
```

The query contains two SELECT statements. The first SELECT statement is used to query the invoices. The second SELECT statement is used to query the invoice items. The INNER JOIN query is used to join the two queries.

4. You can also configure the default query-splitting behavior globally by using the UseQuerySplittingBehavior() method in the OnConfiguring() method of your DbContext class. The following code shows how to configure the default query splitting behavior to SplitQuery in the SampleDbContext class:

```
protected override void OnConfiguring(DbContextOptionsBuilder
optionsBuilder)
{
    base.OnConfiguring(optionsBuilder);
    optionsBuilder.UseSqlServer(_configuration.
GetConnectionString("DefaultConnection"),
        b => b.UseQuerySplittingBehavior(QuerySplittingBehavior.
SplitQuery));
}
```

In this case, you don't need to use the `AsSplitQuery()` method in your LINQ queries. If you want to execute a specific query in a single query, you can use the `AsSingleQuery()` method like this:

```
[HttpGet]
public async Task<ActionResult<IEnumerable<Invoice>>>
GetInvoices(int page = 1, int pageSize = 10,
    InvoiceStatus? status = null)
{
    // Omitted for brevity

    return await _context.Invoices
        .Include(x => x.InvoiceItems)
        .Where(x => status == null || x.Status == status)
        .OrderByDescending(x => x.InvoiceDate)
        .Skip((page - 1) * pageSize)
        .Take(pageSize)
        .AsSingleQuery()
        .ToListAsync();
}
```

However, split queries may cause other issues. For example, multiple queries increase the number of round trips to the database. In addition, if another thread modifies the data between the two queries, the result may be inconsistent. Therefore, you should consider the pros and cons of split queries to fit your scenarios.

Retrieving data

Next, let's see how to retrieve data by ID. In the `GetInvoice` action, we use `await _context. Invoices.FindAsync(id)` to find the invoice by its ID. Send a `Get` request to the `/api/ Invoices/{id}` endpoint with a valid ID. You will see the response contains an empty `InvoiceItems` array. This is because the `InvoiceItems` property is not included in the query. To include the `InvoiceItems` property in the query, you can use the `Include` method in the LINQ query. The following code shows how to use the `Include` method to include the `InvoiceItems` property in the query:

```
[HttpGet("{id}")]
public async Task<ActionResult<Invoice>> GetInvoice(int id)
{
    var invoice = await context.Invoices
        .Include(x => x.InvoiceItems)
        .SingleOrDefaultAsync(x => x.Id == id);

    if (invoice == null)
    {
```

```
        return NotFound();
    }

    return invoice;
}
```

The generated SQL query is as follows:

```
info: Microsoft.EntityFrameworkCore.Database.Command[20101]
      Executed DbCommand (4ms) [Parameters=[@__id_0='?' (DbType =
Guid)], CommandType='Text', CommandTimeout='30']
      SELECT [i0].[Id], [i0].[Amount], [i0].[Description], [i0].
[InvoiceId], [i0].[Name], [i0].[Quantity], [i0].[UnitPrice], [t].[Id]
      FROM (
          SELECT TOP(1) [i].[Id]
          FROM [Invoices] AS [i]
          WHERE [i].[Id] = @__id_0
      ) AS [t]
      INNER JOIN [InvoiceItems] AS [i0] ON [t].[Id] = [i0].[InvoiceId]
      ORDER BY [t].[Id]
```

The query contains two `SELECT` statements, and the `INNER JOIN` query is used to join the two statements. In this way, you can retrieve the invoice and invoice items in a single query.

Deleting data

In the *One-to-many configuration* section, we introduced how to configure the `OnDelete` action to set the `DeleteBehavior` enum to `Cascade`. There are other options for the `DeleteBehavior` enum. Think about the following scenario in a one-to-many relationship:

- An invoice has a list of invoice items
- A user deletes an invoice

In this case, you may want to delete the related invoice items when the invoice is deleted because an invoice item cannot exist without an invoice. This behavior is called **cascade delete**. To delete the data, follow these steps:

1. Run the application and send a `Delete` request to the `/api/Invoices/{id}` endpoint with a valid ID. You will see the invoice and the related invoice items are deleted from the database. Note that if the `OnDelete()` method is configured as `Cascade` or `ClientCascade`, loading related entities using the `Include()` method in the LINQ query is not required. The cascade delete behavior is applied at the database level. You can see the generated SQL query, which just deletes the `Invoice` entity, here:

    ```
    info: Microsoft.EntityFrameworkCore.Database.Command[20101]
    ```

```
      Executed DbCommand (9ms) [Parameters=[@__id_0='?' (DbType
= Guid)], CommandType='Text', CommandTimeout='30']
      DELETE FROM [i]
      FROM [Invoices] AS [i]
      WHERE [i].[Id] = @__id_0
```

However, in some scenarios, you may want to keep dependent entities when the principal entity is deleted, for example:

- A category has a list of blog posts

- A user deletes a category

It is not necessary to delete blog posts when a category is deleted, because the blog posts can still exist without a category and can be assigned to another category. However, if a category is deleted, the `CategoryId` property of a blog post, which is a foreign key, will no longer match the primary key of any category. Therefore, you may want to set the `CategoryId` property to `null` when a category is deleted. This behavior is called **nullification**. To allow this, the requirement is that the `CategoryId` property is nullable. If the `CategoryId` property of a blog post entity is not nullable, EF Core will throw an exception when you try to delete a category because it will violate the foreign key constraint.

2. In the sample code, there is an example of this case. You can find the `Category` and `Post` classes in the `Models` folder. Similar to the `Invoice` and `InvoiceItem` classes, they have a one-to-many relationship. However, the `CategoryId` property in the `Post` class is nullable. Therefore, you can set `DeleteBehavior` to `ClientSetNull` to nullify the `CategoryId` property when a category is deleted.

The following code shows how to configure `DeleteBehavior` to `ClientSetNull`:

```
public class PostConfiguration : IEntityTypeConfiguration<Post>
{
    public void Configure(EntityTypeBuilder<Post> builder)
    {
        builder.ToTable("Posts");
        // Omitted for brevity
        builder.Property(p => p.CategoryId).
HasColumnName("CategoryId");
        builder.HasOne(p => p.Category)
            .WithMany(c => c.Posts)
            .HasForeignKey(p => p.CategoryId)
            .OnDelete(DeleteBehavior.ClientSetNull);
    }
}
```

In the `OnDelete()` method, you can pass the `DeleteBehavior` enum to set `DeleteBehavior` to `ClientSetNull`. The `ClientSetNull` value means that the foreign key property will be set to `null` when the principal entity is deleted.

3. In the `CategoriesController` class, you can find the `DeleteCategory()` method. It is similar to the `DeleteInvoice()` method in the `InvoicesController` class. The only difference is that we need to remove the relationship between the category and the blog posts before deleting the category. The following code shows how to remove the relationship between the category and the blog posts:

```
var category = await context.Categories.Include(x => x.Posts).
SingleOrDefaultAsync(x => x.Id == id);
if (category == null)
{
    return NotFound();
}
category.Posts.Clear();
// Or you can update the posts to set the category to null
// foreach (var post in category.Posts)
// {
//     post.Category = null;
// }
context.Categories.Remove(category);
await context.SaveChangesAsync();
```

You can clear the `Posts` property of the category entity, or you can update the `Category` property of the blog posts to set it to `null`. In this way, the `CategoryId` property of the blog posts will be set to `null` when the category is deleted. Also, it is required to load related entities using the `Include` method, because EF Core needs to track the changes of related entities as well.

4. Run the application and send a `Delete` request to the `/api/Categories/{id}` endpoint with a valid ID. Check the database, and you will see the category is deleted, but the blog posts are not deleted. Instead, the **CategoryId** property of the blog posts is set to **NULL**:

	Id	Title	Content	CategoryId
1	3FA85F64-5717-4562-B3FC-2C963F66AFA6	My first post	Post content	NULL

Figure 6.3 – The CategoryId property of the blog posts is set to NULL when the category is deleted

5. Check the generated SQL query, and you will see that EF Core executes two SQL queries. The first query is to update the `CategoryId` property of the blog posts to `null`. The second query is to delete the category. The generated SQL query is as follows:

```
info: Microsoft.EntityFrameworkCore.Database.Command[20101]
      Executed DbCommand (6ms) [Parameters=[@p1='?' (DbType
```

```
    = Guid), @p0='?' (DbType = Guid), @p2='?' (DbType = Guid)],
CommandType='Text', CommandTimeout='30']
        SET NOCOUNT ON;
        UPDATE [Posts] SET [CategoryId] = @p0
        OUTPUT 1
        WHERE [Id] = @p1;
        DELETE FROM [Categories]
        OUTPUT 1
        WHERE [Id] = @p2;
```

That says the posts now have no category, which means the relationship between the category and the blog posts is removed. You can assign the blog posts to another category to recreate the relationship following your business logic.

It is important to understand the consequences of deleting an entity when it has relationships. Keep in mind that some databases may not support cascading deletes. So, the `DeleteBehavior` enum contains quite a few values to allow you to fine-tune the behavior when deleting an entity. Generally, it is recommended to use `ClientCascade` or `ClientSetNull` because EF Core can perform cascading deletes or nullification if the database does not support cascading deletes.

So far, we have learned how to configure a one-to-many relationship and how to implement CRUD operations for entities with a one-to-many relationship. Next, let's move on to another type of relationship: a one-to-one relationship.

Understanding one-to-one relationships

A one-to-one relationship means that one entity has a relationship with only one entity of another type. For example, a bicycle requires one lock, which can only be used for that particular bicycle. Similarly, a person is only allowed to possess one driver's license, which is designated for their use only. In our sample code, a `Contact` entity has only one `Address` entity, and an `Address` entity belongs to only one `Contact` entity. In the previous section, you learned how to configure a one-to-many relationship using the `HasOne()`/`WithMany()` and `HasMany()`/`WithOne()` methods. In this section, you will learn how to configure a one-to-one relationship using the `HasOne()` and `WithOne()` methods.

One-to-one configuration

In a one-to-one relationship, both sides have a reference navigation property. Technically, both sides have equal positions. However, to explicitly configure the relationship, we need to specify which side is the dependent side and which side is the principal side. The foreign key property is normally defined on the dependent side. In the following example, we will configure a one-to-one relationship between the `Contact` class and the `Address` class:

```
public class Contact
{
```

```
    public Guid Id { get; set; }
    public string FirstName { get; set; } = string.Empty;
    public string LastName { get; set; } = string.Empty;
    public string? Title { get; set; }
    public string Email { get; set; } = string.Empty;
    public string Phone { get; set; } = string.Empty;
    public Address Address { get; set; }
}

public class Address
{
    public Guid Id { get; set; }
    public string Street { get; set; } = string.Empty;
    public string City { get; set; } = string.Empty;
    public string State { get; set; } = string.Empty;
    public string ZipCode { get; set; } = string.Empty;
    public string Country { get; set; } = string.Empty;
    public Guid ContactId { get; set; }
    public Contact Contact { get; set; }
}
```

In the preceding code, a `ContactId` foreign key is defined in the `Address` class, which implies that the `Address` class is the dependent entity, and the `Contact` class is the principal entity. If you do not define a foreign key property here, EF Core will automatically choose one of the entities to be the dependent entity. However, because `Contact` and `Address` are equal in a one-to-one relationship, EF Core may not choose the correct entity as we expect. So, we need to explicitly define a foreign key property in the dependent entity.

The configuration of a one-to-one relationship is as follows:

```
public class ContactConfiguration : IEntityTypeConfiguration<Contact>
{
    public void Configure(EntityTypeBuilder<Contact> builder)
    {
        builder.ToTable("Contacts");
        builder.HasKey(c => c.Id);
        // Omitted for brevity
        builder.Property(c => c.Phone).IsRequired();
    }
}

public class AddressConfiguration : IEntityTypeConfiguration<Address>
{
    public void Configure(EntityTypeBuilder<Address> builder)
```

```
    {
        builder.ToTable("Addresses");
        builder.HasKey(a => a.Id);
        // Omitted for brevity
        builder.Ignore(a => a.Contact);
        builder.HasOne(a => a.Contact)
            .WithOne(c => c.Address)
            .HasForeignKey<Address>(a => a.ContactId);
    }
}
```

The preceding code uses `HasOne`/`WithOne` to define the one-to-one relationship. This can be defined in either the `Contact` configuration or `Address` configuration. The `HasForeignKey` method is used to specify the foreign key property. If you want to define the relationship in the `Contact` configuration, the code may look like this:

```
builder.HasOne(c => c.Address)
    .WithOne(a => a.Contact)
    .HasForeignKey<Address>(a => a.ContactId);
```

Run the following code to add the migration and update the database:

```
dotnet ef migrations add AddContactAndAddress
dotnet ef database update
```

You will see the following code creates a `ContactId` foreign key on the `Addresses` table:

```
migrationBuilder.CreateTable(
    name: "Addresses",
    columns: table => new
    {
        Id = table.Column<Guid>(type: "uniqueidentifier", nullable:
false),
        // Omitted for brevity
    },
    constraints: table =>
    {
        table.PrimaryKey("PK_Addresses", x => x.Id);
        table.ForeignKey(
            name: "FK_Addresses_Contacts_ContactId",
            column: x => x.ContactId,
            principalTable: "Contacts",
            principalColumn: "Id",
            onDelete: ReferentialAction.Cascade);
    });
```

After the migration is applied, the relationship between **Contacts** and **Addresses** is configured successfully, as shown in *Figure 6.4*:

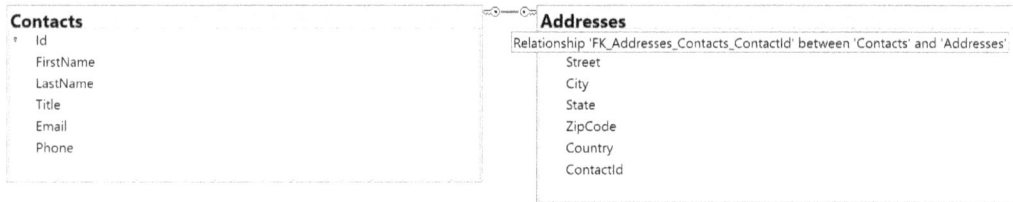

Contacts	Addresses
Id	Relationship 'FK_Addresses_Contacts_ContactId' between 'Contacts' and 'Addresses'
FirstName	Street
LastName	City
Title	State
Email	ZipCode
Phone	Country
	ContactId

Figure 6.4 – A ContactId foreign key is created on the Addresses table

Next, let's see how to implement CRUD operations for entities with a one-to-one relationship.

One-to-one CRUD operations

The CRUD operations of a one-to-one relationship are similar to those of a one-to-many relationship. EF Core can simplify CRUD operations for you. So, in this section, we will not explain all CRUD operations in detail. You will find a controller named ContactsController.cs in the sample repo, which implements CRUD operations for the Contact entity. You can check the code for the details.

The Contact and Address entities have a one-to-one relationship, meaning each Contact entity has one Address property and each Address entity belongs to only one Contact property. To illustrate how to query a contact with its address, we will use this example:

1. To create a new contact with its address, you can send a POST request to the api/contacts endpoint. The request body is as follows:

```
{
    "firstName": "John",
    "lastName": "Doe",
    "email": "john.doe@example.com",
    "phone": "1234567890",
    "address": {
        "street": "123 Main St",
        "city": "Wellington",
        "state": "Wellington",
        "zipCode": "6011",
        "country": "New Zealand"
    }
}
```

In the JSON request body, the address object is a property of the Contact object. It is not required to send the ContactId property in the request body. EF Core will automatically set the ContactId property to the Id property of the Contact object.

2. Similarly, when you query the contacts through the `api/contacts` endpoint, the `Address` object will not be included in the response body by default. You need to explicitly use the `Include` method to include the `Address` object in the query, as shown in the following code:

```
// GET: api/Contacts
[HttpGet]
public async Task<ActionResult<IEnumerable<Contact>>>
GetContacts()
{
    if (context.Contacts == null)
    {
        return NotFound();
    }
    return await context.Contacts.Include(x => x.Address).
ToListAsync();
}
```

You can check the `ContactsController.cs` file for other CRUD operations and test them in Postman or any other REST client you like.

We have explored two types of relationships: one-to-many and one-to-one. Now, let us delve into another type of relationship: many-to-many.

Understanding many-to-many relationships

A many-to-many relationship is when an entity can be associated with multiple entities and vice versa. For example, a movie can have many actors, and an actor can act in many movies; a post can have many tags, and a tag can have many posts; a student can enroll in many courses, and a course can have many students, and so on. In this section, we will introduce how to configure a many-to-many relationship in EF Core.

Many-to-many configuration

In a many-to-many relationship, we need to define a collection navigation property on both sides. Here is an example of a many-to-many relationship between a `Movie` entity and an `Actor` entity:

```
public class Movie
{
    public Guid Id { get; set; }
    public string Title { get; set; } = string.Empty;
    public string? Description { get; set; }
    public int ReleaseYear { get; set; }
    public List<Actor> Actors { get; set; } = new List<Actor>();
}
```

```
public class Actor
{
    public Guid Id { get; set; }
    public string Name { get; set; } = string.Empty;
    public List<Movie> Movies { get; set; } = new List<Movie>();
}
```

EF Core can detect the many-to-many relationship automatically following the convention. If you run the `dotnet ef migrations add AddMovieAndActor` command to add a migration, you will see the following code in the migration file:

```
migrationBuilder.CreateTable(
    name: "ActorMovie",
    columns: table => new
    {
        ActorsId = table.Column<Guid>(type: "uniqueidentifier",
nullable: false),
        MoviesId = table.Column<Guid>(type: "uniqueidentifier",
nullable: false)
    },
    constraints: table =>
    {
        table.PrimaryKey("PK_ActorMovie", x => new { x.ActorsId,
x.MoviesId });
        table.ForeignKey(
            name: "FK_ActorMovie_Actors_ActorsId",
            column: x => x.ActorsId,
            principalTable: "Actors",
            principalColumn: "Id",
            onDelete: ReferentialAction.Cascade);
        table.ForeignKey(
            name: "FK_ActorMovie_Movies_MoviesId",
            column: x => x.MoviesId,
            principalTable: "Movies",
            principalColumn: "Id",
            onDelete: ReferentialAction.Cascade);
    });
```

Besides the code that creates `Movies` and `Actors` tables, the migration file also creates a join table named `ActorMovie` to store the foreign keys for both sides. The `ActorMovie` table has two foreign key properties, `ActorsId` and `MoviesId`, to associate the `Actor` entity and the `Movie` entity.

However, sometimes the automatic detection of the many-to-many relationship may not meet our requirements. For example, we may want to call the table `MovieActor` instead of `ActorMovie`, we may want to specify the foreign key properties as `ActorId` and `MovieId` instead of `ActorsId` and `MoviesId`, or we may even want to add some additional properties to the join table. In these cases, we can explicitly configure the many-to-many relationship.

First, we need to define a join entity to store the foreign keys for both sides. Here is an example of a join entity named `MovieActor`:

```
public class MovieActor
{
    public Guid MovieId { get; set; }
    public Movie Movie { get; set; } = null!;
    public Guid ActorId { get; set; }
    public Actor Actor { get; set; } = null!;
    public DateTime UpdateTime { get; set; }
}
```

Also, we need to add a collection navigation property to the `Movie` and `Actor` entities:

```
public class Movie
{
    public Guid Id { get; set; }
    // Omitted other properties
    public List<MovieActor> MovieActors { get; set; } = new ();
}

public class Actor
{
    public Guid Id { get; set; }
    // Omited other properties
    public List<MovieActor> MovieActors { get; set; } = new ();
}
```

Then, we configure the many-to-many relationship in the `Movie` configuration using the `HasMany()` / `WithMany()` methods, as shown here:

```
public class MovieConfiguration : IEntityTypeConfiguration<Movie>
{
    public void Configure(EntityTypeBuilder<Movie> builder)
    {
        // Omitted for brevity
        builder.HasMany(m => m.Actors)
            .WithMany(a => a.Movies)
            .UsingEntity<MovieActor>(
                j => j
                    .HasOne(ma => ma.Actor)
                    .WithMany(a => a.MovieActors)
```

```
                        .HasForeignKey(ma => ma.ActorId),
            j => j
                .HasOne(ma => ma.Movie)
                .WithMany(m => m.MovieActors)
                .HasForeignKey(ma => ma.MovieId),
            j =>
            {
                // You can add more configuration here
                j.Property(ma => ma.UpdateTime).
HasColumnName("UpdateTime").HasDefaultValueSql("CURRENT_TIMESTAMP");
                j.HasKey(ma => new { ma.MovieId, ma.ActorId });
            }
        );
    }
}
```

Similarly, the configuration can be added to the `Actor` configuration as well. After adding the configuration, run the `dotnet ef migrations add AddMovieAndActor` command to add a migration; you will see the following code in the migration file:

```
migrationBuilder.CreateTable(
    name: "MovieActor",
    columns: table => new
    {
        MovieId = table.Column<Guid>(type: "uniqueidentifier", nullable:
false),
        ActorId = table.Column<Guid>(type: "uniqueidentifier", nullable:
false),
        UpdateTime = table.Column<DateTime>(type: "datetime2",
nullable: false, defaultValueSql: "CURRENT_TIMESTAMP")
    },
    constraints: table =>
    {
        table.PrimaryKey("PK_MovieActor", x => new { x.MovieId,
x.ActorId });
        table.ForeignKey(
            name: "FK_MovieActor_Actors_ActorId",
            column: x => x.ActorId,
            principalTable: "Actors",
            principalColumn: "Id",
            onDelete: ReferentialAction.Cascade);
        table.ForeignKey(
            name: "FK_MovieActor_Movies_MovieId",
            column: x => x.MovieId,
            principalTable: "Movies",
```

```
            principalColumn: "Id",
            onDelete: ReferentialAction.Cascade);
    });
```

You can see that the join table is renamed `MovieActor`, and the foreign key properties are renamed `MovieId` and `ActorId`. Also, the `UpdateTime` property is added to the join table.

After the migration is applied, you can see the join table in the database:

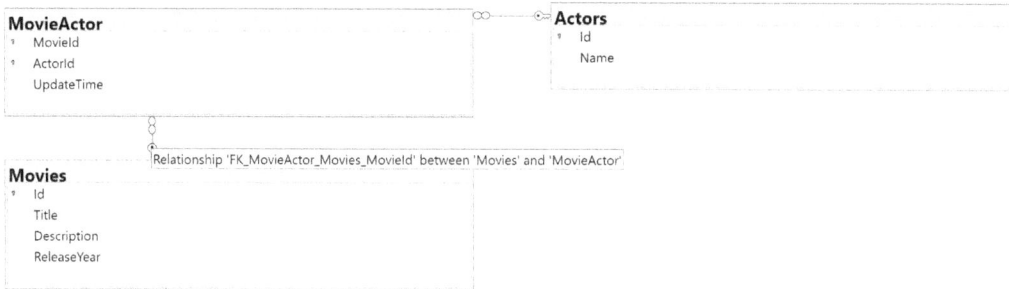

MovieActor
- MovieId
- ActorId
- UpdateTime

Actors
- Id
- Name

Relationship 'FK_MovieActor_Movies_MovieId' between 'Movies' and 'MovieActor'

Movies
- Id
- Title
- Description
- ReleaseYear

Figure 6.5 – The join table in the database

Another way (before EF Core 5.0) to configure a many-to-many relationship is to use the join entity to represent two separate one-to-many relationships. Here is an example of configuring a many-to-many relationship for the `Movie` and `Actor` entities:

```
public class MovieActorsConfiguration :
IEntityTypeConfiguration<MovieActor>
{
    public void Configure(EntityTypeBuilder<MovieActor> builder)
    {
        builder.ToTable("MovieActors");
        builder.HasKey(sc => new { sc.MovieId, sc.ActorId });
        builder.HasOne(sc => sc.Actor)
            .WithMany(s => s.MovieActors)
            .HasForeignKey(sc => sc.ActorId);
        builder.HasOne(sc => sc.Movie)
            .WithMany(c => c.MovieActors)
            .HasForeignKey(sc => sc.MovieId);
    }
}
```

In the preceding code, we configured two one-to-many relationships for the `Movie` and `Actor` entities on the `MovieActor` join entity. Each one-to-many relationship uses the `HasMany()`, `WithMany()`, and `HasForeignKey()` methods to configure the relationship. This combination of one-to-one relationships creates a many-to-many relationship.

You can use either way to configure a many-to-many relationship. The `HasMany()`/`WithMany()` methods is more convenient and easier to use.

Many-to-many CRUD operations

In a many-to-many relationship, such as `Movie` and `Actor`, we may need to get the actors of a movie or get the movies of an actor. So, we need to expose both entities through the REST API. You can create two controllers using the following commands:

```
dotnet-aspnet-codegenerator controller -name MoviesController -api
-outDir Controllers --model Movie --dataContext SampleDbContext -async
-actions
dotnet-aspnet-codegenerator controller -name ActorsController -api
-outDir Controllers --model Actor --dataContext SampleDbContext -async
-actions
```

Run the application and create some movies and actors.

We can include the actors of a movie when creating a movie. For example, we can create a movie with a couple of actors using the following JSON payload format:

```
{
    "title": "The Shawshank Redemption",
    "releaseYear": "1994",
    "actors": [
        {
            "name": "Tim Robbins"
        },
        {
            "name": "Morgan Freeman"
        },
        {
            "name": "Bob Gunton"
        },
        {
            "name": "William Sadler"
        }
    ]
}
```

You will see the following result in the database:

Results Messages

	Id	Title	Description	ReleaseYear
1	80000EED-1B24-4A0F-B1FC-08DAFD00E79A	The Shawshank Redemption	NULL	1994

	Id	Name
1	172770E2-D689-48B5-62B9-08DAFD01CE37	Tim Robbins
2	52567445-30D5-4D1F-62BA-08DAFD01CE37	Morgan Freeman
3	C577CC8F-3309-45C0-62BB-08DAFD01CE37	Bob Gunton
4	F7275337-3E60-4514-62BC-08DAFD01CE37	William Sadler

	MovieId	ActorId	UpdateTime
1	80000EED-1B24-4A0F-B1FC-08DAFD00E79A	172770E2-D689-48B5-62B9-08DAFD01CE37	2023-01-23 18:28:27.9866667
2	80000EED-1B24-4A0F-B1FC-08DAFD00E79A	52567445-30D5-4D1F-62BA-08DAFD01CE37	2023-01-23 18:28:27.9866667
3	80000EED-1B24-4A0F-B1FC-08DAFD00E79A	C577CC8F-3309-45C0-62BB-08DAFD01CE37	2023-01-23 18:28:27.9866667
4	80000EED-1B24-4A0F-B1FC-08DAFD00E79A	F7275337-3E60-4514-62BC-08DAFD01CE37	2023-01-23 18:28:27.9866667

Figure 6.6 – The join table has been populated

Similarly, you can also include the movies of an actor when creating an actor. However, if we include related entities arbitrarily, we may end up with duplicate entities. For example, we create an actor with a couple of movies using the following JSON payload format:

```
{
    "name": "Tim Robbins",
    "movies": [
        {
            "title": "The Shawshank Redemption",
            "releaseYear": "1994"
        },
        {
            "title": "Green Mile",
            "releaseYear": "1999"
        }
    ]
}
```

As a result, you will have two movies with the same title in the database. To avoid this, there are some options:

- Add a unique index to the `Title` property of the `Movie` entity to ensure that the title is unique. This is the simplest solution and can prevent duplicate entities from being added to the database.
- Check if the entity already exists in the database before adding it.
- Add movies and actors separately, and then update either the movie or the actor to include the other entity using the IDs of the other entity, instead of the whole entity.

You can use a combination of the preceding options to improve the implementation. To add a unique index to the `Title` property of the `Movie` entity, you can update the following code of the `MovieConfiguration` class:

```
public void Configure(EntityTypeBuilder<Movie> builder)
{
    builder.ToTable("Movies");
    builder.HasKey(m => m.Id);
    builder.Property(p => p.Title).HasColumnName("Title").
HasMaxLength(128).IsRequired();
    // Add a unique index to the Title property
    builder.HasIndex(p => p.Title).IsUnique();
    // Omitted for brevity
}
```

You can make the same change to the `Name` property of the `Actor` entity. After the change, you need to create a new migration and apply it to the database. This helps to prevent duplicate entities on the database level. If a request contains duplicate entities, the database will throw an exception.

It is possible to add or update related entities in the same request as the main entity. But sometimes, it might not be necessary. For example, an actor just acts in a new movie, and you want to create a new movie and add the actor to the movie. You can update the actor to include the new movie, but you have to send the whole actor entity in the request, including existing movies. It causes unnecessary data transfer.

To make the API easy to use, it is pragmatic to expose an additional API endpoint to update a collection of related entities only, rather than updating the whole entity. For example, we can create an `/api/actors/{id}/movies` endpoint to update the movies of an actor. It is a good practice to avoid updating a collection of related entities in the same request. We can just send the IDs of the related entities to the API endpoint. From the perspective of the API, the relationship is treated as a resource. In the `ActorsController.cs` file, you will find the following code:

```
[HttpPost("{id}/movies/{movieId}")]
public async Task<IActionResult> AddMovie(Guid id, Guid movieId)
{
    if (_context.Actors == null)
    {
        return NotFound("Actors is null.");
    }
    var actor = await _context.Actors.Include(x => x.Movies).
SingleOrDefaultAsync(x => x.Id == id);
    if (actor == null)
    {
        return NotFound($"Actor with id {id} not found.");
```

```
    }

    var movie = await _context.Movies.FindAsync(movieId);
    if (movie == null)
    {
        return NotFound($"Movie with id {movieId} not found.");
    }

    if (actor.Movies.Any(x => x.Id == movie.Id))
    {
        return Problem($"Movie with id {movieId} already exists for
Actor {id}.");
    }

    actor.Movies.Add(movie);
    await _context.SaveChangesAsync();

    return CreatedAtAction("GetActor", new { id = actor.Id }, actor);
}

[HttpGet("{id}/movies")]
public async Task<IActionResult> GetMovies(Guid id)
{
    if (_context.Actors == null)
    {
        return NotFound("Actors is null.");
    }

    var actor = await _context.Actors.Include(x => x.Movies).
SingleOrDefaultAsync();
    if (actor == null)
    {
        return NotFound($"Actor with id {id} not found.");
    }

    return Ok(actor.Movies);
}

[HttpDelete("{id}/movies/{movieId}")]
public async Task<IActionResult> DeleteMovie(Guid id, Guid movieId)
{
    if (_context.Actors == null)
    {
        return NotFound("Actors is null.");
```

```
    }

    var actor = await _context.Actors.Include(x => x.Movies).
SingleOrDefaultAsync();
    if (actor == null)
    {
        return NotFound($"Actor with id {id} not found.");
    }

    var movie = await _context.Movies.FindAsync(movieId);
    if (movie == null)
    {
        return NotFound($"Movie with id {movieId} not found.");
    }

    actor.Movies.Remove(movie);
    await _context.SaveChangesAsync();

    return NoContent();
}
```

The preceding code exposes a few endpoints to add, get, and delete the movies of an actor. You can test the endpoints using the JSON payload format, as shown in the preceding code snippet:

- To add a movie to an actor, send a POST request to the /api/actors/{id}/movies/
  {movieId} endpoint. The AddMovie action will check if the movie already exists in the
  database. If it does, it will then check if the movie already exists in the movies of the actor. If
  not, it adds the movie to the collection and then saves the changes to the database.

- To get the movies of an actor, send a GET request to the /api/actors/{id}/movies
  endpoint. The GetMovies action will return the movies of the actor. This endpoint can be
  updated to support pagination, sorting, and filtering.

- To delete a movie from an actor, send a DELETE request to the /api/actors/{id}/
  movies/{movieId} endpoint. The DeleteMovie action will remove the movie from
  the collection and then save the changes to the database. Note that it does not delete the movie
  from the database; it just deletes the relationship between the movie and the actor.

It is also possible to add a similar endpoint to the MoviesController.cs file to update the actors
of a movie. You can use the same approach to implement endpoints. Try it yourself!

> **Important note**
>
> When you call the `/api/actors` endpoint, you may find that the response contains `MovieActors` as well. This is not useful for the client. You can use the `JsonIgnore` attribute to ignore the property when serializing the response.

We have now discussed three common types of relationships: one-to-many, one-to-one, and many-to-many. You should now have a good understanding of how to configure relationships and implement CRUD operations for entities with relationships. Let's move on to our next topic: owned entity types.

Understanding owned entities

In the previous sections, we have learned some relationships are optional, but some are required. For example, a post can exist without a category, but a student ID card cannot exist without a student. For the latter, we can say a student owns an ID card. Similarly, a contact owns an address. We can also find some examples of one-to-many relationships. For example, an invoice owns many invoice items because an invoice item cannot exist without an invoice. In this section, we will introduce the concept of owned entities.

Owned entity types are entity types that are part of the owner and cannot exist without the owner. You can use common one-to-one or one-to-many relationships to model the owned entities, but EF Core provides a more convenient way called owned entity types. You can use the `OwnsOne()` or `OwnsMany()` method to define owned entity types, instead of using the `HasOne()` or `HasMany()` method. For example, to configure the `InvoiceItem` entity as an owned entity type of the `Invoice` entity, you can use the following code:

```
public class InvoiceConfiguration : IEntityTypeConfiguration<Invoice>
{
    public void Configure(EntityTypeBuilder<Invoice> builder)
    {
        // Omitted for brevity
        // Use the owned type to configure the InvoiceItems collection
        builder.OwnsMany(p => p.InvoiceItems, a =>
            {
                a.WithOwner( => x.Invoice).HasForeignKey(x =>
x.InvoiceId);
                a.ToTable("InvoiceItems");
                // Omitted for brevity
            }
        );
    }
}
```

As shown in the preceding code, you can use the `OwnsMany()`/`WithOwner()` methods to configure the owned entity type. The `OwnsMany()`/`WithOwner()` method specifies the owner of the owned entity type. The `HasForeignKey()` method specifies the foreign key property of the owned entity type. The configuration of the `InvoiceItem` entity is stored in the `InvoiceConfiguration` class.

Similarly, the configuration of the `Address` entity could be stored in the `ContactConfiguration` class like this:

```
public class ContactConfiguration : IEntityTypeConfiguration<Contact>
{
    public void Configure(EntityTypeBuilder<Contact> builder)
    {
        // Omitted for brevity
        // Use owned entity type
        builder.OwnsOne(c => c.Address, a =>
        {
            a.WithOwner(x => x.Contact);
            a.Property(a => a.Street).HasColumnName("Street").
HasMaxLength(64).IsRequired();
            // Omitted for brevity
        });
    }
}
```

When you use the `OwnsOne()`/`WithOwner()` methods, you do not need to specify the foreign key property because the owned entity type will be stored in the same table as the owner by default. You can use the `ToTable` method to specify the table name of the owned entity type.

So, what is the difference between normal one-to-one or one-to-many and owned entity types? There are some differences:

- You cannot create a `DbSet<T>` property for an owned entity type. You can only use the `DbSet<T>` property for the owner. That means you do not have any way to access the owned entity type directly. You must access the owned entity type through the owner.

- When you query the owner, the owned entity type will be included automatically. You do not need to use the `Include()` method to include the owned entity type explicitly. So, please be careful if the owner has many owned entities. It may cause performance issues.

If your entities have a simple one-to-one or one-to-many relationship and the data is not large, you can use owned entity types to simplify the configuration. However, if the relationship is complex and the data size is large, you would be better off using normal one-to-one or one-to-many relationships because you can decide which related entities to include explicitly.

Summary

In this comprehensive chapter, we delved into modeling relationships between entities in EF Core. We explored various common relationship types, including one-to-one, one-to-many, and many-to-many relationships. We learned how to configure these relationships using essential methods such as `HasOne()`/`WithMany()`, `HasMany()`/`WithOne()`, and `HasMany()`/`WithMany()`. To broaden our understanding, we also explored configuring owned entity types using the `OwnsOne()`/`WithOwner()` and `OwnsMany`/`WithOwner()` methods.

To effectively operate on entities with relationships, we explained how to implement CRUD operations for each type of relationship. Particularly, we explained cascading delete operations, ensuring data integrity and efficient management of related entities.

The concepts learned in this chapter will help you model relationships between entities in your ASP.NET Core applications. In the next chapter, we will learn about some advanced topics of EF Core, such as concurrency control, performance tuning, and more.

7

Data Access in ASP.NET Core (Part 3: Tips)

In *Chapter 6*, we learned how to manage relationships between entities using the EF Core Fluent API. We introduced three types of relationships: one-to-one, one-to-many, and many-to-many. We also learned how to perform **CRUD** operations on related entities. With the knowledge we gained from *Chapter 6*, we can now build a simple data access layer for most web API applications. However, there are still some scenarios that we need to handle properly. For example, how do we improve the performance of data access? And what should we do if there are concurrency conflicts?

In this chapter, we will cover some advanced topics related to data access in ASP.NET Core, including `DbContext` pooling, performance optimization, raw SQL queries, and concurrency conflicts. We will also discuss some tips and tricks that can help you write better code.

We will cover the following topics:

- `DbContext` pooling
- Tracking versus no-tracking queries
- IQueryable versus IEnumerable
- Client versus server evaluation
- Raw SQL queries
- Bulk operations
- Concurrency conflicts
- Reverse engineering
- Other ORM frameworks

After reading this chapter, you will have gained a deeper understanding of EF Core and be able to use it more effectively in your applications. You will have learned how to use no-tracking queries to improve query performance, as well as how to use raw SQL queries to execute complex queries. Additionally, you will understand how to use bulk operations to improve the performance of bulk data operations. Furthermore, you will be able to handle concurrency conflicts for large-scale applications and use reverse engineering to generate the entity classes and the DbContext class from an existing database.

Technical requirements

The code examples in this chapter can be found at `https://github.com/PacktPublishing/Web-API-Development-with-ASP.NET-Core-8`. You can use VS 2022 or VS Code to open the solutions.

Understanding DbContext pooling

In the previous chapter, we learned how to register the DbContext instance as a scoped service in the DI container using the AddDbContext() extension method. By default, a new DbContext instance is created for each request, which is generally not a problem since it is a lightweight object that does not consume many resources. However, in a high-throughput application, the cost of setting up various internal services and objects for each DbContext instance can add up. To address this, EF Core provides a feature called **DbContext pooling**, which allows the DbContext instance to be reused across multiple requests.

To enable DbContext pooling, you can replace the AddDbContext() method with the AddDbContextPool() method. This resets the state of the DbContext instance when it is disposed of, stores it in a pool, and reuses it when a new request comes in. By reducing the cost of setting up the DbContext instance, DbContext pooling can significantly improve the performance of your application for high-throughput scenarios.

You can find the sample code for this section from the /samples/chapter7/EfCoreDemo folder in this chapter's GitHub repository.

Open the Program.cs file in the EfCoreDemo project. The following code shows how to enable DbContext pooling:

```
services.AddDbContextPool<InvoiceDbContext>(options =>
{
    options.UseSqlServer(builder.Configuration.
GetConnectionString("DefaultConnection"));
});
```

The `AddDbContextPool()` method takes a `poolSize` parameter, which specifies the maximum number of `DbContext` instances that can be stored in the pool. The default value is `1024`, which is usually sufficient for most applications. If the pool is full, EF Core will start creating new `DbContext` instances as needed.

To validate whether `DbContext` pooling can improve the performance of the application, we can run a performance test. **Grafana k6** is an open-source load-testing tool that can be used to test the performance of web APIs. To use k6, you need to install NodeJS here: `https://nodejs.org/`. Then you can download it from `https://k6.io/docs/get-started/installation/`. k6 has packages for various platforms, including Windows, Linux, and macOS. Install k6 on your machine.

You can find a `script.js` file in the k6 folder in the project. The `script.js` file is a k6 script that contains the test scenarios. The following code shows the content of the `script.js` file:

```
import http from 'k6/http';
import { sleep } from 'k6';
export const options = {
    vus: 500,
    duration: '30s',
};
export default function () {
  http.get('http://localhost:5249/api/Invoices?page=1&pageSize=10');
  sleep(1);
}
```

This is a basic k6 test script that runs a 30-second, 500-VU load test. **Virtual users** (**VUs**) are the number of concurrent users that will be simulated. The test script will send GET requests to the `/api/Invoices?page=1&pageSize=10` endpoint for 30 seconds.

First, use the `AddDbContext()` method to register the `DbContext`, and run the application using the `dotnet run` command. Then, open a new terminal and run the following command to start the k6 test:

```
k6 run script.js
```

Next, use the `AddDbContextPool()` method to register the `DbContext`, and test the application again using the same k6 script. You can compare the results of the two tests to see whether `DbContext` pooling improves the performance of the application. For example, one test result of using the `AddDbContext()` method is as follows:

```
execution: local
   script: script.js
   output: -

scenarios: (100.00%) 1 scenario, 500 max VUs, 1m0s max duration (incl. graceful stop):
        * default: 500 looping VUs for 30s (gracefulStop: 30s)


     data_received..................: 6.5 MB 210 kB/s
     data_sent......................: 793 kB 26 kB/s
     http_req_blocked...............: avg=40.33ms  min=0s      med=0s      max=1.53s  p(90)=0s      p(95)=3.01ms
     http_req_connecting............: avg=40.29ms  min=0s      med=0s      max=1.53s  p(90)=0s      p(95)=1.89ms
     http_req_duration..............: avg=1.07s    min=2.75ms  med=1.14s max=2.6s     p(90)=1.62s   p(95)=1.76s
       { expected_response:true }...: avg=1.07s    min=2.75ms  med=1.14s max=2.6s     p(90)=1.62s   p(95)=1.76s
     http_req_failed................: 0.00%   ✓ 0             ✗ 7145
     http_req_receiving.............: avg=173.28µs min=0s      med=0s      max=16.67ms p(90)=552.4µs p(95)=999.95µs
     http_req_sending...............: avg=103.15µs min=0s      med=0s      max=8.36ms  p(90)=0s      p(95)=508µs
     http_req_tls_handshaking.......: avg=0s       min=0s      med=0s      max=0s      p(90)=0s      p(95)=0s
     http_req_waiting...............: avg=1.07s    min=507.8µs med=1.13s max=2.6s     p(90)=1.62s   p(95)=1.76s
     http_reqs......................: 7145    231.71879/s
     iteration_duration.............: avg=2.11s    min=1s      med=2.15s max=3.61s    p(90)=2.7s    p(95)=2.88s
     iterations.....................: 7145    231.71879/s
     vus............................: 500     min=500       max=500
     vus_max........................: 500     min=500       max=500


running (0m30.8s), 000/500 VUs, 7145 complete and 0 interrupted iterations
default ✓ [======================================] 500 VUs  30s
```

Figure 7.1 – The test result of using the AddDbContext() method

The following is the result of using the `AddDbContextPool` method:

```
execution: local
   script: script.js
   output: -

scenarios: (100.00%) 1 scenario, 500 max VUs, 1m0s max duration (incl. graceful stop):
        * default: 500 looping VUs for 30s (gracefulStop: 30s)


     data_received..................: 7.7 MB 245 kB/s
     data_sent......................: 947 kB 30 kB/s
     http_req_blocked...............: avg=23.73ms  min=0s      med=0s      max=1.02s  p(90)=0s      p(95)=508.95µs
     http_req_connecting............: avg=23.68ms  min=0s      med=0s      max=1.02s  p(90)=0s      p(95)=0s
     http_req_duration..............: avg=782.36ms min=505.49µs med=725.22ms max=2.54s p(90)=1.34s  p(95)=1.5s
       { expected_response:true }...: avg=782.36ms min=505.49µs med=725.22ms max=2.54s p(90)=1.34s  p(95)=1.5s
     http_req_failed................: 0.00%   ✓ 0             ✗ 8530
     http_req_receiving.............: avg=139.45µs min=0s      med=0s      max=4.92ms  p(90)=524.1µs p(95)=906.05µs
     http_req_sending...............: avg=64.05µs  min=0s      med=0s      max=7.79ms  p(90)=0s      p(95)=505.49µs
     http_req_tls_handshaking.......: avg=0s       min=0s      med=0s      max=0s      p(90)=0s      p(95)=0s
     http_req_waiting...............: avg=782.16ms min=505.49µs med=725.06ms max=2.54s p(90)=1.34s  p(95)=1.5s
     http_reqs......................: 8530    269.990737/s
     iteration_duration.............: avg=1.81s    min=1s      med=1.72s max=3.54s    p(90)=2.41s   p(95)=2.56s
     iterations.....................: 8530    269.990737/s
     vus............................: 248     min=248       max=500
     vus_max........................: 500     min=500       max=500


running (0m31.6s), 000/500 VUs, 8530 complete and 0 interrupted iterations
default ✓ [======================================] 500 VUs  30s
```

Figure 7.2 – The test result of using the AddDbContextPool() method

When using `AddDbContext()`, the average request duration is 1.07 s and 7,145 requests are completed, while when using `AddDbContextPool()`, the average request duration is 782.36 ms and 8,530 requests are completed. The results show that `DbContext` pooling can improve the performance of the application. Note that your results may vary depending on your machine's configuration. Also, the `dotnet run` command is used to run the application in development mode, which is not optimized for performance. So, this test is just for demonstration purposes and cannot reflect the real performance of the application. However, it can give you an idea of how `DbContext` pooling works.

> **Important note**
>
> For most applications, `DbContext` pooling is not necessary. You should enable `DbContext` pooling only if you have a high-throughput application. Therefore, before enabling `DbContext` pooling, it is important to test your application's performance with and without it to see whether there's any noticeable improvement.

In summary, while `DbContext` pooling can improve the performance of high-throughput applications, it's not a one-size-fits-all solution. Be sure to evaluate your application's specific needs before deciding whether to enable `DbContext` pooling or not.

Understanding the difference between tracking versus no-tracking queries

In this section, we will discuss the difference between tracking and no-tracking queries. What are tracking queries and no-tracking queries? Let us start from the beginning!

In the early days of .NET, the term **SqlHelper** was popular to refer to a static class that provided a set of methods to execute SQL queries. While SqlHelper simplified the process of executing SQL queries, developers still had to manage connection and transaction objects, write boilerplate code to map results to model objects, and work directly with the database.

ORM frameworks such as EF Core were created to solve these problems. They not only simplify the process of executing SQL queries and mapping the results to model objects but also provide the ability to track changes made to the entities returned by queries. When changes are saved, EF Core generates the appropriate SQL queries to update the database. This is called tracking and is a significant benefit of using an ORM framework such as EF Core.

However, tracking comes at a cost. This can add overhead and memory usage, especially when dealing with a large number of entities.

We introduced a little bit about tracking in *Chapter 5*. Let's see an example of tracking. You can find the sample code for this section from the `/samples/chapter7/EfCoreDemo` folder in this chapter's GitHub repository.

In the sample `EfCoreDemo` project, you can find the `GetInvoice` action in the `InvoicesController` class. The following code shows how the tracking works:

```
[HttpGet("{id}")]
public async Task<ActionResult<Invoice>> GetInvoice(Guid id)
{
    if (context.Invoices == null)
    {
        return NotFound();
    }
    logger.LogInformation($"Invoice {id} is loading from the
database.");
    var invoice = await context.Invoices.FindAsync(id);
    logger.LogInformation($"Invoice {invoice?.Id} is loaded from the
database."
    logger.LogInformation($"Invoice {id} is loading from the
context.");
    invoice = await context.Invoices.FindAsync(id);
    logger.LogInformation($"Invoice {invoice?.Id} is loaded from the
context.")
    if (invoice == null)
    {
        return NotFound();
    }
    return invoice;
}
```

In the preceding code, we added some logging statements to see how EF Core calls the database. Run the application and call the `GetInvoice` action. You will see an output in the console like this:

```
info: BasicEfCoreDemo.Controllers.InvoicesController[0]
      Invoice e61436dd-0dac-4e8b-7d61-08dae88bb288 is loading from the
database.
info: Microsoft.EntityFrameworkCore.Database.Command[20101]
      Executed DbCommand (30ms) [Parameters=[@__get_Item_0='?' (DbType
= Guid)], CommandType='Text', CommandTimeout='30']
      SELECT TOP(1) [i].[Id], [i].[Amount], [i].[ContactName], [i].
[Description], [i].[DueDate], [i].[InvoiceDate], [i].[InvoiceNumber],
[i].[Status]
      FROM [Invoices] AS [i]
      WHERE [i].[Id] = @__get_Item_0
info: BasicEfCoreDemo.Controllers.InvoicesController[0]
      Invoice e61436dd-0dac-4e8b-7d61-08dae88bb288 is loaded from the
database.
info: BasicEfCoreDemo.Controllers.InvoicesController[0]
```

```
        Invoice e61436dd-0dac-4e8b-7d61-08dae88bb288 is loading from the
context.
info: BasicEfCoreDemo.Controllers.InvoicesController[0]
        Invoice e61436dd-0dac-4e8b-7d61-08dae88bb288 is loaded from the
context.
```

When we call the `context.Invoices.FindAsync(id)` method for the first time, EF Core will query the database and return the `Invoice` entity. The second time, EF Core will return the `Invoice` entity from the context because the `Invoice` entity is already in the context.

Find() versus Single()

When we get an entity from the database by its primary key, we can use the `Find()` or `FindAsync()` methods. Also, we can use the `Single()` or `SingleOrDefault()` methods. They are similar, but they are not the same. The `Find()` and `FindAsync()` methods are methods of the `DbSet` class. If an entity with the given primary key values is being tracked by the context, the `Find()` or `FindAsync()` methods will return the tracked entity without making a request to the database. Otherwise, EF Core will make a query to the database to get the entity, attach it to the context, and return it. But if you use the `Single()` or `SingleOrDefault()` methods, EF Core will always make a query to the database to get the entity. The same is true for the `First()` and `FirstOrDefault()` methods. So, the `Find()` and `FindAsync()` methods are more efficient for getting an entity by its primary key. But in rare cases, `Find()` and `FindAsync()` may return outdated data if the entity is updated in the database after it is loaded into the context. For example, if you use the bulk-update `ExecuteUpdateAsync()` method, the update will not be tracked by `DbContext`. Then, if you use `Find()` or `FindAsync()` to get the entity from `DbContext`, you will get the outdated data. In this case, you should use `Single()` or `SingleOrDefault()` to get the entity from the database again. In most cases, you can use the `Find()` or `FindAsync()` methods to get an entity by its primary key when you are sure the entity is always tracked by the `DbContext`.

An entity has one of the following `EntityState` values: `Detached`, `Added`, `Unchanged`, `Modified`, or `Deleted`. We introduced the `EntityState` enum in *Chapter 5*. The following is how the states change:

- All the entities that are returned by the query (such as `Find()`, `Single()`, `First()`, `ToList()`, and their `async` overloads) are in the `Unchanged` state

- If you update the properties of the entity, EF Core will change the state to `Modified`

- If you call the `Remove()` method on the entity, EF Core will change the state to `Deleted`

- If you call the `Add()` method on the entity, EF Core will change the state to `Added`

- If you call the `Attach()` method on the untracked entity, EF Core will track the entity and set the state to `Unchanged`

- If you call the `Detach()` method on the tracked entity, EF Core will not track the entity and will change the state to `Detached`

Note that EF Core can track the changes at the property level, meaning that if you update a property of an entity, EF Core will only update the property when you call the `SaveChanges` method.

To retrieve the `EntityEntry` object for an entity, we can use the `Entry()` method, which contains the state of the entity and the changed properties. Use the sample `EfCoreDemo` project in the `/samples/chapter7/EfCoreDemo` folder. You can find the `PutInvoice` action within the `InvoicesController` class:

```
context.Entry(invoice).State = EntityState.Modified;
await context.SaveChangesAsync();
```

In the preceding code snippet, we obtained the `EntityEntry` object for the `Invoice` entity using the `Entry()` method and set its state to `Modified`. When `SaveChanges()` is called, EF Core persists the changes to the database.

By default, tracking is enabled in EF Core. However, there may be scenarios where you do not want EF Core to track changes to entities. For instance, in read-only queries within `Get` actions, where the `DbContext` only exists for the duration of the request, tracking is not necessary. Disabling tracking can enhance performance and save memory. If you don't intend to modify entities, you should disable tracking by calling the `AsNoTracking()` method on the query. Here's an example:

```
// To get the invoice without tracking
var invoice = await context.Invoices.AsNoTracking().
FirstOrDefaultAsync(x => x.Id == id);
// To return a list of invoices without tracking
var invoices = await context.Invoices.AsNoTracking().ToListAsync();
```

If you have lots of read-only queries and you feel it is tedious to call the `AsNoTracking()` method every time, you can disable tracking globally when you configure the `DbContext`. The following code shows how to do this:

```
protected override void OnConfiguring(DbContextOptionsBuilder
optionsBuilder)
{
    base.OnConfiguring(optionsBuilder);
    optionsBuilder.UseQueryTrackingBehavior(QueryTrackingBehavior.
NoTracking);
}
```

For any other queries that you want to track, you can call the `AsTracking()` method on the query, as shown in the following code:

```
// To get the invoice with tracking
var invoice = await context.Invoices.AsTracking().
FirstOrDefaultAsync(x => x.Id == id);
// To return a list of invoices with tracking
var invoices = await context.Invoices.AsTracking().ToListAsync();
```

In the preceding code, we explicitly call the `AsTracking()` method to enable tracking for the query, so that we can update the entity and save the changes to the database.

> **Important note**
>
> If an entity is a keyless entity, EF Core will never track it. Keyless entity types do not have keys defined on them. They are configured by a `[Keyless]` data annotation or a Fluent API `HasNoKey()` method. The keyless entity is often used for read-only queries or views. We will not discuss keyless entities in detail in this book. You can refer to the official documentation at `https://learn.microsoft.com/en-us/ef/core/modeling/keyless-entity-types` for more information.

Using no-tracking queries is a good way to improve performance for read-only scenarios. However, keep in mind that if you disable tracking, you will not be able to update the entities when you call the `SaveChanges()` method because EF Core cannot detect changes to untracked entities. So, it is important to consider the implications of using no-tracking queries before implementing them.

In addition to non-tracking queries, there are other factors that can affect the performance of data queries in EF Core. We will explore the differences between `IQueryable` and `IEnumerable` and how they impact query performance in the next section.

Understanding the difference between IQueryable and IEnumerable

When working with EF Core, you have two interfaces available to query the database: `IQueryable` and `IEnumerable`. Although these interfaces may seem similar at first glance, they have important differences that can affect your application's performance. In this section, we will discuss the differences between `IQueryable` and `IEnumerable`, how they work, and when to use each of them.

You might be familiar with the `IEnumerable` interface. The `IEnumerable` interface is a standard .NET interface that is used to represent a collection of objects. It is used to iterate through the collection. Many .NET collections implement the `IEnumerable` interface, such as `List`, `Array`, `Dictionary`, and so on. The `IEnumerable` interface has a single method called `GetEnumerator`, which returns an `IEnumerator` object. The `IEnumerator` object is used to iterate through the collection.

The first difference between `IQueryable` and `IEnumerable` is that `IQueryable` is in the `System.Linq` namespace, while `IEnumerable` is in the `System.Collections` namespace. The `IQueryable` interface inherits from the `IEnumerable` interface, so `IQueryable` can do everything that `IEnumerable` does. But why do we need the `IQueryable` interface?

One of the key differences between `IQueryable` and `IEnumerable` is that `IQueryable` is used to query data from a specific data source, such as a database. `IEnumerable` is used to iterate through a collection in memory. When we use `IQueryable`, the query will be translated into a specific query language, such as SQL, and executed against the data source to get the results when we call the `ToList()` (or `ToAway()`) method or iterate the items in the collection.

Download the sample code from the `/samples/chapter7/EfCoreDemo` folder in the chapter's GitHub repository. You can find a `GetInvoices` action in the `InvoicesController` class.

First, let's use the `IQueryable` interface to query the database:

```
// Use IQueryable
logger.LogInformation($"Creating the IQueryable...");
var list1 = context.Invoices.Where(x => status == null || x.Status ==
status);
logger.LogInformation($"IQueryable created");
logger.LogInformation($"Query the result using IQueryable...");
var query1 = list1.OrderByDescending(x => x.InvoiceDate)
    .Skip((page - 1) * pageSize)
    .Take(pageSize);
logger.LogInformation($"Execute the query using IQueryable");
var result1 = await query1.ToListAsync();
logger.LogInformation($"Result created using IQueryable");
```

In the preceding code, `context.Invoices` is a `DbSet<TEntity>` object, which implements the `IQueryable` interface. The `Where()` method is used to filter the invoices by status, and returns an `IQueryable` object. Then, we use some other methods to sort and paginate the invoices. When we call the `ToListAsync()` method, the query will be translated into a SQL query and executed against the database to get the results. The logs show the execution order of the code:

```
info: BasicEfCoreDemo.Controllers.InvoicesController[0]
      Creating the IQueryable...
info: BasicEfCoreDemo.Controllers.InvoicesController[0]
      IQueryable created
info: BasicEfCoreDemo.Controllers.InvoicesController[0]
      Query the result using IQueryable...
info: BasicEfCoreDemo.Controllers.InvoicesController[0]
      Execute the query using IQueryable
info: Microsoft.EntityFrameworkCore.Database.Command[20101]
```

```
      Executed DbCommand (49ms) [Parameters=[@__p_0='?' (DbType
= Int32), @__p_1='?' (DbType = Int32)], CommandType='Text',
CommandTimeout='30']
      SELECT [i].[Id], [i].[Amount], [i].[ContactName], [i].
[Description], [i].[DueDate], [i].[InvoiceDate], [i].[InvoiceNumber],
[i].[Status]
      FROM [Invoices] AS [i]
      ORDER BY [i].[InvoiceDate] DESC
      OFFSET @__p_0 ROWS FETCH NEXT @__p_1 ROWS ONLY
info: BasicEfCoreDemo.Controllers.InvoicesController[0]
      Result created using IQueryable
```

From the logs, we can see that the query is executed against the database when we call the
`ToListAsync()` method. The query contains the `ORDER BY`, `OFFSET`, and `FETCH NEXT`
clauses, which means the query is executed on the database server.

Next, let us use the `IEnumerable` interface to query the database:

```
// Use IEnumerable
logger.LogInformation($"Creating the IEnumerable...");
var list2 = context.Invoices.Where(x => status == null || x.Status ==
status).AsEnumerable();
logger.LogInformation($"IEnumerable created");
logger.LogInformation($"Query the result using IEnumerable...");
var query2 = list2.OrderByDescending(x => x.InvoiceDate)
    .Skip((page - 1) * pageSize)
    .Take(pageSize);
logger.LogInformation($"Execute the query using IEnumerable");
var result2 = query2.ToList();
logger.LogInformation($"Result created using IEnumerable");
```

In the preceding code, we use the `AsEnumerable()` method to convert the `IQueryable` object
to an `IEnumerable` object. Then, we sort and paginate the invoices and call the `ToList()` method
to get the results. The logs show the execution order of the code:

```
info: BasicEfCoreDemo.Controllers.InvoicesController[0]
      Creating the IEnumerable...
info: BasicEfCoreDemo.Controllers.InvoicesController[0]
      IEnumerable created
info: BasicEfCoreDemo.Controllers.InvoicesController[0]
      Query the result  using IEnumerable...
info: BasicEfCoreDemo.Controllers.InvoicesController[0]
      Execute the query using IEnumerable
info: Microsoft.EntityFrameworkCore.Database.Command[20101]
      Executed DbCommand (5ms) [Parameters=[], CommandType='Text',
CommandTimeout='30']
```

```
      SELECT [i].[Id], [i].[Amount], [i].[ContactName], [i].
[Description], [i].[DueDate], [i].[InvoiceDate], [i].[InvoiceNumber],
[i].[Status]
      FROM [Invoices] AS [i]
info: BasicEfCoreDemo.Controllers.InvoicesController[0]
      Result created using IEnumerable
```

Look at the logs. The generated SQL query does not contain the ORDER BY, OFFSET, and FETCH NEXT clauses, which means the query fetched all the invoices from the database and then filtered, sorted, and paged the invoices in memory. If we have a large number of entities in the database, the second query will be very slow and inefficient.

Now, we can see the difference between the two interfaces. The IQueryable interface is a deferred execution query, which means the query is not executed when we add more conditions to the query. The query will be executed against the database when we call the ToList() or ToArray() methods or iterate the items in the collection. So, in complex and heavy queries, we should always use the IQueryable interface to avoid fetching all the data from the database. Be careful when you call the ToList() or ToArray() methods because ToList() or ToArray() (and their async overloads) will execute the query immediately.

> **What LINQ methods can cause the query to be executed immediately?**
>
> There are a couple of operations that result in the query being executed immediately:
>
> - Use the for or foreach loop to iterate the items in the collection
> - Use the ToList(), ToArray(), Single(), SingleOrDefault(), First(), FirstOrDefault(), or Count() methods, or the async overloads of these methods

In this section, we explored the differences between IQueryable and IEnumerable. It is important to understand why we should use IQueryable instead of IEnumerable when querying the database for complex and heavy queries. Loading all the data from the database can cause performance issues if there are a large number of entities in the database.

Moving on, we will discuss another factor that can affect performance: client evaluation.

Client evaluation versus server evaluation

In this section, we will discuss the difference between client evaluation and server evaluation. In the old versions of EF Core (earlier than EF Core 3.0), the wrong usage of LINQ queries that have client evaluation can cause significant performance issues. Let's see what client evaluation and server evaluation are.

When we use EF Core to query data from the database, we can just write LINQ queries, and EF Core will translate the LINQ queries into SQL queries and execute them against the database. However, sometimes, the LINQ operation must be executed on the client side. Check the following code in the `SearchInvoices` action method in the `InvoicesController` class:

```
var list = await context.Invoices
    .Where(x => x.ContactName.Contains(search) || x.InvoiceNumber.
Contains(search))
    .ToListAsync();
```

When we use the `Contains()` method, EF Core can translate the LINQ query into the following SQL query:

```
info: Microsoft.EntityFrameworkCore.Database.Command[20101]
      Executed DbCommand (4ms) [Parameters=[@__search_0='?' (Size
= 32), @__search_0_1='?' (Size = 32) (DbType = AnsiString)],
CommandType='Text', CommandTimeout='30']
      SELECT [i].[Id], [i].[Amount], [i].[ContactName], [i].
[Description], [i].[DueDate], [i].[InvoiceDate], [i].[InvoiceNumber],
[i].[Status]
      FROM [Invoices] AS [i]
      WHERE (@__search_0 LIKE N'') OR CHARINDEX(@__search_0, [i].
[ContactName]) > 0 OR (@__search_0_1 LIKE '') OR CHARINDEX(@__
search_0, [i].[InvoiceNumber]) > 0
```

You can see that the SQL query uses some native SQL functions to filter the data, which means that the SQL query is executed on the database server. This is called **server evaluation**. EF Core tries to run server evaluation as much as possible.

Now, let's say we want to return the GST tax amount for each invoice. We can transfer the entity to a new object with the GST tax amount. Of course, the better way is to add a property for the tax in the `Invoice` entity. The following is a demonstration of how to do this.

Add a `static` method to calculate the GST tax amount:

```
private static decimal CalculateTax(decimal amount)
{
    return amount * 0.15m;
}
```

Update the code as follows:

```
var list = await context.Invoices
    .Where(x => x.ContactName.Contains(search) || x.InvoiceNumber.
Contains(search))
    .Select(x => new Invoice
    {
```

```
        Id = x.Id,
        InvoiceNumber = x.InvoiceNumber,
        ContactName = x.ContactName,
        Description = $"Tax: ${CalculateTax(x.Amount)}.
{x.Description}",
        Amount = x.Amount,
        InvoiceDate = x.InvoiceDate,
        DueDate = x.DueDate,
        Status = x.Status
    })
    .ToListAsync();
```

We updated the `Description` property by adding the GST tax calculation. When we run the application and call the endpoint, we will see the generated SQL query is the same as the previous query. But the `Description` property has been updated in the result. This means the conversion is done on the client side. This is called **client evaluation**.

This kind of client evaluation is acceptable because the query does need to fetch the data from the database. The cost is very low. However, it might cause problems for some queries. For example, we want to query the invoices that have a GST tax amount greater than $10. Update the code as follows:

```
var list = await context.Invoices
    .Where(x => (x.ContactName.Contains(search) || x.InvoiceNumber.
Contains(search)) && CalculateTax(x.Amount) > 10)
    .ToListAsync();
```

When we call the endpoint, we will see the following error:

```
fail: Microsoft.AspNetCore.Diagnostics.
DeveloperExceptionPageMiddleware[1]
      An unhandled exception has occurred while executing the request.
      System.InvalidOperationException: The LINQ expression
'DbSet<Invoice>()
         .Where(i => i.ContactName.Contains(__search_0) ||
i.InvoiceNumber.Contains(__search_0) && InvoicesController.
CalculateTax(i.Amount) > 10)' could not be translated. Additional
information: Translation of method 'BasicEfCoreDemo.Controllers.
InvoicesController.CalculateTax' failed. If this method can be
mapped to your custom function, see https://go.microsoft.com/
fwlink/?linkid=2132413 for more information. Either rewrite the query
in a form that can be translated, or switch to client evaluation
explicitly by inserting a call to 'AsEnumerable', 'AsAsyncEnumerable',
'ToList', or 'ToListAsync'. See https://go.microsoft.com/
fwlink/?linkid=2101038 for more information.
```

The error message is very clear. This is because the `CalculateTax()` method is not supported by EF Core. In old versions of EF Core (earlier than EF Core 3.0), EF Core will fetch all the data from the database and then filter the data in memory. It could cause performance issues. After EF Core

3.0, EF Core will throw an exception if the query cannot be translated correctly, to avoid potential performance issues.

But if you are sure the client evaluation is safe, such as when dealing with a small data size, you can explicitly use the `AsEnumerable()` method (or `AsAsyncEnumerable()`, `ToList()`, or `ToListAsync()`) to force EF Core to fetch all the data and then execute the query on the client side. Make sure you know what you are doing.

> **Why must the CalculateTax() method be static?**
>
> EF Core caches the compiled query due to the expensive nature of compiling the query. If the `CalculateTax()` method is not static, EF Core will need to maintain a reference to a constant expression of the `InvoicesController` through the `CalculateTax()` instance method, which could potentially lead to memory leaks. To prevent this, EF Core throws an exception if the `CalculateTax()` method is not static. Making the method static will ensure that EF Core does not capture constant in the instance.

The latest version of EF Core offers the benefit of preventing potential performance issues caused by client evaluation. If you encounter an exception similar to a previous one, you can review the query to ensure it is being translated correctly.

Next, we will discuss how to use raw SQL queries in EF Core. For some scenarios, we need to write raw SQL queries to execute complex queries.

Using raw SQL queries

Although EF Core can translate most LINQ queries into SQL queries, which is very convenient, sometimes we need to write raw SQL queries if the required query cannot be written in LINQ, or the generated SQL query is not efficient. In this section, we will explore how to use raw SQL queries in EF Core.

EF Core provides several methods to execute raw SQL queries:

- `FromSql()`
- `FromSqlRaw()`
- `SqlQuery()`
- `SqlQueryRaw()`
- `ExecuteSql()`
- `ExecuteSqlRaw()`

When we execute raw SQL queries, we must be careful to avoid SQL injection attacks. Let's see when we should use raw SQL queries and how to use them properly. You can download the sample code from the /samples/chapter7/EfCoreDemo folder in the chapter's GitHub repository.

FromSql() and FromSqlRaw()

We can use the FromSql() method to create a LINQ query based on an interpolated string. The FromSql() method is available in EF Core 7.0 and later versions. There is a similar method called FromSqlInterpolated() in older versions.

To execute the raw SQL query, we just need to pass the interpolated string to the FromSql() method, as follows:

```
var list = await context.Invoices
    .FromSql($"SELECT * FROM Invoices WHERE Status = 2")
    .ToListAsync();
```

We can also pass parameters to the raw SQL query. For example, we want to query the invoices that have a specific status:

```
[HttpGet]
[Route("status")]
public async Task<ActionResult<IEnumerable<Invoice>>>
GetInvoices(string status)
{
    // Omitted for brevity
    var list = await context.Invoices
        .FromSql($"SELECT * FROM Invoices WHERE Status = {status}")
        .ToListAsync();
    return list;
}
```

Wait, is it safe to insert a string into the SQL query directly? What if the status parameter is ' ; DROP TABLE Invoices; --? Will it cause a SQL injection attack?

That is a good question. Let us see how EF Core handles the parameters. Run the application and call the /api/invoices/status?status=AwaitPayment endpoint. We will see the generated SQL query is as follows:

```
info: Microsoft.EntityFrameworkCore.Database.Command[20101]
      Executed DbCommand (41ms) [Parameters=[p0='?' (Size = 4000)],
CommandType='Text', CommandTimeout='30']
      SELECT * FROM Invoices WHERE Status = @p0
```

The parameter is not inserted into the SQL query directly. Instead, EF Core uses the `@p0` parameter placeholder and passes the parameter value to the SQL query. This is called a parameterized query. It is safe to use the parameterized query to avoid SQL injection attacks. So, we do not need to worry about the safety of the `FromSql` method.

> **Why FromSql() is safe to use**
>
> The `FromSql()` method expects a parameter as the `FormattableString` type. So, it is required to use the interpolated string syntax by using the `$` prefix. The syntax looks like regular C# string interpolation, but it is not the same thing. A `FormattableString` type can include interpolated parameter placeholders. The interpolated parameter values will be automatically converted to the `DbParameter` type. So, it is safe to use the `FromSql()` method to avoid SQL injection attacks.

For some scenarios, we might need to build dynamic SQL queries. For example, we want to query the invoices according to user input, which specifies the property name and property value. For this case, we cannot use `FromSql` because it is not allowed to parameterize the column names. We need to use `FromSqlRaw` instead. However, we must be careful to avoid SQL injection attacks. It is the developer's responsibility to make sure the SQL query is safe. Here is an example:

```
[HttpGet]
[Route("free-search")]
public async Task<ActionResult<IEnumerable<Invoice>>>
GetInvoices(string propertyName, string propertyValue)
{
    if (context.Invoices == null)
    {
        return NotFound();
    }
    // Do something to sanitize the propertyName value
    var value = new SqlParameter("value", propertyValue);
    var list = await context.Invoices
        .FromSqlRaw($"SELECT * FROM Invoices WHERE {propertyName} = @
value", value)
        .ToListAsync();
    return list;
}
```

In the preceding example, the column name is not parameterized. Therefore, we must be careful to avoid SQL injection attacks. It is required to sanitize the `propertyName` value to make sure it is safe. Maybe you can check whether the value contains any special characters, such as `;`, `--`, and so on. If the value contains any special characters, you can throw an exception or remove the special characters before executing the SQL query. Also, if you allow the user to specify the column name, it will increase the effort to validate the column name because you need to check whether the column name exists in the database or whether the column has the correct index. Make sure you know what you are doing.

The propertyValue is parameterized, so it is safe to use.

After you build the SQL query using FromSql(), you can then apply the LINQ query operators to filter the data as you want. Remember that it is better to use FromSql() than FromSqlRaw().

When we use the FromSql() or FromSqlRaw() methods, keep in mind that there are some limitations:

- The data returned from the SQL query must contain all the properties of the entity, otherwise, EF Core cannot map the data to the entity.

- The column names returned from the SQL query must match the column names that the entity properties are mapped to.

- The SQL query can only query one table. If you need to query multiple tables, you can build the raw query first and then use the Include() method to include the related entities.

SqlQuery() and SqlQueryRaw()

The FromSql() method is useful when we want to query entities from the database using a raw SQL query. For some cases, we want to execute the raw SQL query and return a scalar value or non-entity type. For example, we want to query the IDs of invoices that have a specific status. We can use the SqlQuery() method to execute the raw SQL query and return a list of IDs. Here is an example:

```
[HttpGet]
[Route("ids")]
public ActionResult<IEnumerable<Guid>> GetInvoicesIds(string status)
{
    var result = context.Database
        .SqlQuery<Guid>($"SELECT Id FROM Invoices WHERE Status =
{status}")
        .ToList();
    return result;
}
```

The translated SQL query is as follows:

```
info: Microsoft.EntityFrameworkCore.Database.Command[20101]
      Executed DbCommand (22ms) [Parameters=[p0='?' (Size = 4000)],
CommandType='Text', CommandTimeout='30']
      SELECT Id FROM Invoices WHERE Status = @p0
```

Note that the SqlQuery() method is used on the Database property of the DbContext object. It is not available on the DbSet object.

The `SqlQueryRaw()` method is similar to the `SqlQuery()` method, but it allows us to build dynamic SQL queries like the `FromSqlRaw()` method. Similarly, you must take responsibility to avoid SQL injection attacks.

ExecuteSql() and ExecuteSqlRaw()

For some scenarios, where we do not need return values, we can use the `ExecuteSql` method to execute a raw SQL query. Normally, it is used to update or delete data or call a **stored procedure**. For example, when we need to delete all invoices that have a specific status, we can use the `ExecuteSql()` method to execute the raw SQL query. Here is an example:

```
[HttpDelete]
[Route("status")]
public async Task<ActionResult> DeleteInvoices(string status)
{
    var result = await context.Database
        .ExecuteSqlAsync($"DELETE FROM Invoices WHERE Status =
{status}");
    return Ok();
}
```

In this way, we do not need to load the entities from the database and then delete them one by one. It is much more efficient to use the `ExecuteSql()` method to execute the raw SQL query.

The `ExecuteSqlRaw()` method is similar to the `ExecuteSql()` method, but it allows us to build dynamic SQL queries like the `FromSqlRaw()` method. Similarly, you must be very careful to sanitize the SQL query to avoid SQL injection attacks.

In this section, we introduced how to use raw SQL queries in EF Core. We discussed the differences between `FromSql()` and `FromSqlRaw()`, `SqlQuery()` and `SqlQueryRaw()`, and `ExecuteSql()` and `ExecuteSqlRaw()`. We also discussed the limitations of these methods. Again, we must be very careful to avoid SQL injection attacks when we use raw SQL queries.

In one of the examples in this section, we showed you how to run a raw SQL query to delete a set of entities. EF Core 7.0 introduces a bulk operations feature that can make this easier. There are two new methods available for bulk operations, `ExecuteUpdate()` and `ExecuteDelete()`, which provide a more efficient way to update or delete data. In the following section, we will discuss this feature in more detail.

Using bulk operations

In this section, we will explore how to effectively update/delete data using EF Core. EF Core 7.0 or later offers the ability of bulk operations, which are easy to use and can improve the performance of update/delete operations. To take advantage of this feature, ensure you are using the most recent version of EF Core.

As we mentioned in the previous section, EF Core tracks the changes in entities. To update an entity, normally, we need to load the entity from the database, update the entity properties, and then call the `SaveChanges()` method to save the changes to the database. This is a very common scenario. Deleting an entity is similar. However, if we want to update or delete a large number of entities, it is not efficient to load the entities one by one and then update or delete them. For these scenarios, it is not required to track the changes in the entities. So, it would be better to use the bulk operations feature to update or delete data.

We can use a raw SQL query to update or delete data using the `ExecuteSql()` method. However, it lacks strong type support. Hardcoding the column names in the SQL query is not a good practice. From EF Core 7.0, we can use the `ExecuteUpdate()` and `ExecuteDelete()` methods to update or delete data. Note that these two methods do not involve the entity tracking feature. So, once you call these two methods, the changes will be executed immediately. There is no need to call the `SaveChanges()` method.

Next, let us see how to use these two methods. We will show you how to use the `ExecuteUpdate()` method and what SQL query is generated. The `ExecuteDelete()` method is similar. The sample code is located at the `/samples/chapter7/EfCoreDemo` folder in the chapter's GitHub repository.

ExecuteUpdate()

The `ExecuteUpdate()` method is used to update data without loading the entities from the database. You can use it to update one or more entities by adding the `Where()` clause.

For example, we want to update the status of the invoices that were created before a specific date. The code is as follows:

```
[HttpPut]
[Route("status/overdue")]
public async Task<ActionResult> UpdateInvoicesStatusAsOverdue(DateTime
date)
{
    var result = await context.Invoices
        .Where(i => i.InvoiceDate < date && i.Status == InvoiceStatus.
AwaitPayment)
        .ExecuteUpdateAsync(s => s.SetProperty(x => x.Status,
InvoiceStatus.Overdue));
```

```
        return Ok();
    }
```

The generated SQL query is as follows:

```
info: Microsoft.EntityFrameworkCore.Database.Command[20101]
      Executed DbCommand (46ms) [Parameters=[@__p_0='?' (DbType =
DateTimeOffset)], CommandType='Text', CommandTimeout='30']
      UPDATE [i]
      SET [i].[Status] = 'Overdue'
      FROM [Invoices] AS [i]
      WHERE [i].[InvoiceDate] < @__p_0 AND [i].[Status] =
'AwaitPayment'
```

This query can update multiple invoices at the same time. It does benefit from the strong type support but has the same efficiency as the raw SQL query. If you need to update more than one property, you can use the SetProperty() method multiple times, as seen in the following code:

```
var result = await context.Invoices
        .Where(i => i.InvoiceDate < date && i.Status == InvoiceStatus.
AwaitPayment)
        .ExecuteUpdateAsync(s =>
            s.SetProperty(x => x.Status, InvoiceStatus.Overdue)
             .SetProperty(x => x.LastUpdatedDate, DateTime.Now));
```

In addition, the Where() clause can reference the other entities. So, the ExecuteUpdate() method is always recommended to update multiple entities, instead of using the raw SQL query.

ExecuteDelete()

Similarly, we can use the ExecuteDelete() method to delete data without loading the entities from the database. This method can be used to delete one or more entities by adding the Where clause. For example, we want to delete the invoices that were created before a specific date. The code is as follows:

```
await context.Invoices.Where(x => x.InvoiceDate < date).
ExecuteDeleteAsync();
```

Again, these bulk operations do not track the changes in the entities. If a DbContext instance already loaded the entities, after the bulk update or delete, the entities in the context will still keep the old values. So, be careful when using these bulk operations.

In this section, we discussed how to use the bulk operations feature in EF Core. We introduced the ExecuteUpdate() and ExecuteDelete() methods, which can be used to update or delete data without loading the entities from the database. Compared to a raw SQL query, these two methods have strong type support. It is recommended to use these two methods to update or delete multiple entities.

Next, we will learn how to manage concurrency conflicts when updating data.

Understanding concurrency conflicts

An API endpoint can be called by multiple clients at the same time. If the endpoint updates data, the data may be updated by another client before the current client completes the update. When the same entity is updated by multiple clients, it can cause a concurrency conflict, which may result in data loss or inconsistency, or even cause data corruption. In this section, we will discuss how to handle concurrency conflicts in EF Core. You can download the sample project `ConcurrencyConflictDemo` from the `/samples/chapter7/ConcurrencyConflictDemo` folder in the chapter's GitHub repository.

There are two ways to handle concurrency conflicts:

- **Pessimistic concurrency control**: This uses database locks to prevent multiple clients from updating the same entity at the same time. When a client tries to update an entity, it will first acquire a lock on the entity. If the lock is acquired successfully, only this client can update the entity, and all other clients will be blocked from updating the entity until the lock is released. However, this approach may result in performance issues when the number of concurrent clients is large because managing locks is expensive. EF Core does not have built-in support for pessimistic concurrency control.

- **Optimistic concurrency control**: This way does not involve locks; instead, a version column is used to detect concurrency conflicts. When a client tries to update an entity, it will first get the value of the version column, and then compare this value with the old value when updating the entity. If the value of the version column is the same, it means that no other client has updated the entity. In this case, the client can update the entity. But if the value of the version column is different from the old value, it means that another client has updated the entity. In this case, EF Core will throw an exception to indicate the concurrency conflict. The client can then handle the exception and retry the update operation.

Let's see an example of concurrency conflicts. In the `ConcurrencyConflictDemo` project, we have a `Product` entity with an `Inventory` property, which is used to store the number of products in stock. We want to create an API endpoint to sell a product. When a client calls this endpoint, it will pass the product ID and the number of products to sell. The endpoint will then update the `Inventory` property by subtracting the number of products to sell. The logic is as follows:

- The client calls the API endpoint to sell a product.
- The application gets the product from the database.
- The application checks the `Inventory` property to make sure that the number of products in stock is enough for the sale:
  - If the number of products in stock is enough, the application subtracts the number of products being sold from the `Inventory` property and then calls the `SaveChanges()` method to save the changes to the database

- If the number of products in stock is not enough, the application returns an error message to the client

> **Important note**
>
> The example project uses the following code to reset the database in the `Program.cs` file when the application starts:
>
> `dbContext.Database.EnsureDeleted();`
>
> `dbContext.Database.EnsureCreated();`
>
> So, when you run the application, the database will be reset, and the `Inventory` property of product 1 will be set to `15`.

The following code shows the first version of the implementation of the API endpoint:

```
[HttpPost("{id}/sell/{quantity}")]
public async Task<ActionResult<Product>> SellProduct(int id, int
quantity)
{
    if (context.Products == null)
    {
        return Problem("Entity set 'SampleDbContext.Products' is
null.");
    }
    var product = await context.Products.FindAsync(id);
    if (product == null)
    {
        return NotFound();
    }
    if (product.Inventory < quantity)
    {
        return Problem("Not enough inventory.");
    }
    product.Inventory -= quantity;
    await context.SaveChangesAsync();
    return product;
}
```

There should be some other logic that handles the order creation and payment, and so on. We will not discuss that here; instead, we will focus on the concurrency conflicts caused by the product inventory update:

1. To simulate this concurrent scenario, we can pass a `delay` parameter to add a delay before saving the changes to the database. The following code shows how to add a delay:

    ```
    [HttpPost("{id}/sell/{quantity}")]
    public async Task<ActionResult<Product>> SellProduct(int id, int
    ```

```
quantity, int delay = 0)
{
    // Omitted code for brevity
    await Task.Delay(TimeSpan.FromSeconds(delay));
    product.Inventory -= quantity;
    await context.SaveChangesAsync();
    return product;
}
```

2. Now, let us try to call the API endpoint twice in a short time. The first POST request will pass a delay parameter with a value of 2 seconds:

    ```
    http://localhost:5273/api/Products/1/sell/10?delay=2
    ```

    The second POST request will pass the delay parameter with a value of 3 seconds:

    ```
    http://localhost:5273/api/Products/1/sell/10?delay=3
    ```

3. Send the first request and then send the second request in 2 seconds. The expected result should be that the first request will succeed and the second request will fail. But actually, both requests will succeed. The responses show that the Inventory property of the product is updated to 5, which is incorrect. The initial value of the Inventory property is 15, and we sold 20 products, so how can the Inventory property be updated to 5?

Let us see what happens in the application:

1. Client A calls the API endpoint to sell a product and wants to sell 10 products.

2. Client A checks the Inventory property and finds that the number of products in stock is 15, which is enough to sell.

3. Almost at the same time, client B calls the API endpoint to sell a product and wants to sell 10 products.

4. Client B checks the Inventory property and finds that the number of products in stock is 15 because client A has not updated the Inventory property yet.

5. Client A subtracts 10 from the Inventory property, which results in a value of 5, and saves the changes to the database. Now, the number of products in stock is 5.

6. Client B also subtracts 10 from the Inventory property and saves the changes to the database. The problem is that the number of products in stock has been updated to 5 by client A, but client B does not know this. So, client B also updates the Inventory property to 5, which is incorrect.

This is an example of concurrency conflict. Multiple clients try to update the same entity at the same time, and the result is not what we expected. In this case, client B should not be able to update the Inventory property because the number of products in stock is not enough. However, if the application does not handle concurrency conflicts, we may end up with incorrect data in the database.

To solve this problem, EF Core provides optimistic concurrency control. There are two ways to use optimistic concurrency control:

- Native database-generated concurrency token
- Application-managed concurrency token

Let us see how to use these two ways to handle concurrency conflicts.

Native database-generated concurrency token

Some databases, such as SQL Server, provide a native mechanism to handle concurrency conflicts. To use the native database-generated concurrency token in SQL Server, we need to create a new property for the `Product` class and add a `[Timestamp]` attribute to it. The following code shows the updated `Product` class:

```
public class Product
{
    public int Id { get; set; }
    public string Name { get; set; }
    public int Inventory { get; set; }
    // Add a new property as the concurrency token
    public byte[] RowVersion { get; set; }
}
```

In the Fluent API configuration, we need to add the following code to map the RowVersion property to the rowversion column in the database:

```
modelBuilder.Entity<Product>()
    .Property(p => p.RowVersion)
    .IsRowVersion();
```

If you prefer to use the data annotation configuration, you can add the `[Timestamp]` attribute to the RowVersion property, and EF Core will automatically map it to the rowversion column in the database, as shown in the following code:

```
public class Product
{
    public int Id { get; set; }
    public string Name { get; set; }
    public int Inventory { get; set; }
    // Add a new property as the concurrency token
    [Timestamp]
    public byte[] RowVersion { get; set; }
}
```

Do not forget to run the `dotnet ef migrations add AddConcurrencyControl` command to create a new migration. There is no need to run the `dotnet ef database update` command this time because we have the code to reset the database when the application starts.

> **Important note**
>
> If you want to configure the mapping in Fluent API, you can use the following code:
>
> ```
> modelBuilder.Entity<Product>()
>     .Property(p => p.RowVersion)
>     .IsRowVersion();
> ```
>
> This will generate the following migration:
>
> ```
> migrationBuilder.AddColumn<byte[]>(
> name: "RowVersion",
> table: "Products",
> type: "rowversion",
> rowVersion: true,
> nullable: false,
> defaultValue: new byte[0]);
> ```

Now, let's try to call the API endpoint again. Use the same requests as before, one with a `delay` parameter of 2 seconds, and the other with a `delay` parameter of 3 seconds. This time, we should see that the first request will succeed, but the second request will fail with a `DbUpdateConcurrencyException` exception:

```
fail: Microsoft.AspNetCore.Diagnostics.
DeveloperExceptionPageMiddleware[1]
      An unhandled exception has occurred while executing the request.
      Microsoft.EntityFrameworkCore.DbUpdateConcurrencyException:
The database operation was expected to affect 1 row(s), but
actually affected 0 row(s); data may have been modified or
deleted since entities were loaded. See http://go.microsoft.com/
fwlink/?LinkId=527962 for information on understanding and handling
optimistic concurrency exceptions.
```

Check the database. The `Inventory` column of product 1 has been updated to 5, which is correct.

If you check the SQL statement generated by EF Core, you will find that the `rowversion` column is included in the `WHERE` clause of the `UPDATE` statement:

```
info: Microsoft.EntityFrameworkCore.Database.Command[20101]
      Executed DbCommand (1ms) [Parameters=[@p1='?' (DbType = Int32),
@p0='?' (DbType = Int32), @p2='?' (Size = 8) (DbType = Binary)],
CommandType='Text', CommandTimeout='30']
```

```
SET IMPLICIT_TRANSACTIONS OFF;
SET NOCOUNT ON;
UPDATE [Products] SET [Inventory] = @p0
OUTPUT INSERTED.[RowVersion]
WHERE [Id] = @p1 AND [RowVersion] = @p2;
```

By using concurrency control, EF Core not only checks the ID of the entity but also checks the value of the `rowversion` column. If the value of the `rowversion` column is not the same as the value in the database, it means that the entity has been updated by another client, and the current update operation should be aborted.

Note that the `rowversion` column type is available for SQL Server, but not for other databases, such as SQLite. Different databases may have different types of concurrency tokens, or may not support the concurrency token at all. Please check the documentation of the database you are using to see whether it supports the built-in concurrency token. If not, you need to use the application-managed concurrency token, as shown in the next section.

Application-managed concurrency token

If the database does not support the built-in concurrency token, we can manually manage the concurrency token in the application. Instead of using the `rowversion` column, which can be automatically updated by the database, we can use a property in the entity class to manage the concurrency token and assign a new value to it every time the entity is updated.

Here is an example of using the application-managed concurrency token:

1. First, we need to add a new property to the `Product` class, as shown in the following code:

```
public class Product
{
    public int Id { get; set; }
    public string Name { get; set; }
    public int Inventory { get; set; }
    // Add a new property as the concurrency token
    public Guid Version { get; set; }
}
```

2. Update the Fluent API configuration to specify the `Version` property as the concurrency token:

```
modelBuilder.Entity<Product>()
    .Property(p => p.Version)
    .IsConcurrencyToken();
```

3. The corresponding data annotation configuration is as follows:

```
public class Product
{
    public int Id { get; set; }
    public string Name { get; set; }
    public int Inventory { get; set; }
    // Add a new property as the concurrency token
    [ConcurrencyCheck]
    public Guid Version { get; set; }
}
```

4. Because this Version property is not managed by the database, we need to manually assign a new value whenever the entity is updated.

The following code shows how to update the Version property when the entity is updated:

```
[HttpPost("{id}/sell/{quantity}")]
public async Task<ActionResult<Product>> SellProduct(int id, int quantity)
{
    // Omitted for brevity.
    product.Inventory -= quantity;
    // Manually assign a new value to the Version property.
    product.Version = Guid.NewGuid();
    await context.SaveChangesAsync();
    return product;
}
```

You can use the application-managed concurrency token in SQL Server as well. The only difference is that you need to manually assign a new value to the concurrency token property every time the entity is updated. But if you use the built-in concurrency token in SQL Server, you do not need to do that.

In the event of a concurrency conflict, it is essential to take the necessary steps to resolve the issue. This will be addressed in the following section.

Handling concurrency conflicts

When a concurrency conflict occurs, EF Core will throw a DbUpdateConcurrencyException exception. We can catch this exception and handle it in the application. For example, we can return a 409 Conflict status code to the client, and let the client decide what to do next:

```
[HttpPost("{id}/sell/{quantity}")]
public async Task<ActionResult<Product>> SellProduct(int id, int quantity)
{
```

```
    // Omitted for brevity.
    product.Inventory -= quantity;
    try
    {
        await context.SaveChangesAsync();
    }
    catch (DbUpdateConcurrencyException)
    {
        // Do not forget to log the error
        return Conflict($"Concurrency conflict for Product {product.
Id}.");
    }
    return product;
}
```

The preceding code returns a `409 Conflict` status code to the client when a concurrency conflict occurs. The client can then handle the exception and retry the update operation.

> **Important note**
>
> Some databases provide different isolation levels to handle concurrency conflicts. For example, SQL Server provides four isolation levels: `ReadUncommitted`, `ReadCommitted`, `RepeatableRead`, and `Serializable`. The default isolation level is `ReadCommitted`. Each isolation level has different behaviors when a concurrency conflict occurs and has its own pros and cons. Higher levels of isolation provide more consistency but also reduce concurrency. For more information, see *Isolation Levels* at `https://learn.microsoft.com/en-gb/sql/t-sql/statements/set-transaction-isolation-level-transact-sql`.

In this section, we discussed how to handle concurrency conflicts in EF Core. We introduced two ways to handle concurrency conflicts: native database-generated concurrency tokens and application-managed concurrency tokens. We also discussed how to handle exceptions when concurrency conflicts occur. Concurrency conflicts are a common issue in a highly concurrent environment. It is important to handle them properly to avoid data loss or inconsistency.

Reverse engineering

So far, we have learned how to use EF Core to create a database schema from the entity classes. This is called *code-first*. However, sometimes we need to work with an existing database. In this case, we need to create the entity classes and `DbContext` from the existing database schema. This is called *database-first* or *reverse engineering*. In this section, we will discuss how to use EF Core to reverse engineer the entity classes and `DbContext` from an existing database schema. This is useful when we want to migrate an existing application to EF Core.

Let's use the `EfCoreRelationshipsDemoDb` database as an example. If you have not created this database, please follow the steps in *Chapter 6* to create it. The sample code is located at the `/samples/chapter7/EfCoreReverseEngineeringDemo` folder in the chapter's GitHub repository.

1. First, let us create a new web API project. Run the following command in the terminal:

    ```
    dotnet new webapi -n EfCoreReverseEngineeringDemo
    ```

2. To use reverse engineering, we need to install the `Microsoft.EntityFrameworkCore.Design` NuGet package. Navigate to the `EfCoreReverseEngineeringDemo` folder, and run the following command in the terminal to install it:

    ```
    dotnet add package Microsoft.EntityFrameworkCore.Design
    ```

3. Then, add the `Microsoft.EntityFrameworkCore.SqlServer` NuGet package to the project:

    ```
    dotnet add package Microsoft.EntityFrameworkCore.SqlServer
    ```

    If you use other database providers, such as SQLite, you need to install the corresponding NuGet package. For example, to use SQLite, you need to install the `Microsoft.EntityFrameworkCore.Sqlite` NuGet package. You can find the list of supported database providers at `https://learn.microsoft.com/en-us/ef/core/providers/`.

4. Next, we will use the `dbcontext scaffold` command to generate the entity classes and `DbContext` from the database schema. This command needs the connection string of the database and the name of the database provider. We can run the following command in the terminal:

    ```
    dotnet ef dbcontext scaffold "Server=(localdb)\
    mssqllocaldb;Initial Catalog=EfCoreRelationshipsDemoDb;Trusted_
    Connection=True;" Microsoft.EntityFrameworkCore.SqlServer
    ```

    By default, the generated files will be placed in the current directory. We can use the `--context-dir` and `--output-dir` options to specify the output directory of the `DbContext` and entity classes. For example, we can run the following command to generate the `DbContext` and entity classes in the `Data` and `Models` folders, respectively:

    ```
    dotnet ef dbcontext scaffold "Server=(localdb)\
    mssqllocaldb;Initial Catalog=EfCoreRelationshipsDemoDb;Trusted_
    Connection=True;" Microsoft.EntityFrameworkCore.SqlServer
    --context-dir Data --output-dir Models
    ```

    The default name of the `DbContext` class will be the same as the database name, such as `EfCoreRelationshipsDemoDbContext.cs`. We can also change the name of the `DbContext` class by using the `--context` option. For example, we can run the following command to change the name of the `DbContext` class to `AppDbContext`:

    ```
    dotnet ef dbcontext scaffold "Server=(localdb)\
    ```

```
mssqllocaldb;Initial Catalog=EfCoreRelationshipsDemoDb;Trusted_
Connection=True;" Microsoft.EntityFrameworkCore.SqlServer
--context-dir Data --output-dir Models --context AppDbContext
```

If the command executes successfully, we will see the generated files in the `Data` and `Models` folders.

5. Open the `AppDbContext.cs` file, and we will see a warning in the following code:

```
protected override void OnConfiguring(DbContextOptionsBuilder
optionsBuilder)

#warning To protect potentially sensitive information in your
connection string, you should move it out of source code.
You can avoid scaffolding the connection string by using the
Name= syntax to read it from configuration - see https://
go.microsoft.com/fwlink/?linkid=2131148. For more guidance
on storing connection strings, see http://go.microsoft.com/
fwlink/?LinkId=723263.
        => optionsBuilder.UseSqlServer("Server=(localdb)\\
mssqllocaldb;Initial Catalog=EfCoreRelationshipsDemoDb;Trusted_
Connection=True;");
```

This warning tells us that we should not store the connection string in the source code. Instead, we should store it in a configuration file, such as `appsettings.json`.

In the `OnModelCreating` method, we can see the entity classes and their relationships have been configured in Fluent API style. If you prefer to use data annotations, you can use the `--data-annotations` option when you run the `dbcontext scaffold` command. But as we mentioned in *Chapter 5*, Fluent API is more powerful than data annotations, and it is recommended to use Fluent API.

EF Core is smart enough to detect the relationships between the entity classes if your database schema follows the conventions. However, if this is not the case, you may get unexpected results. Please review the generated code carefully to make sure the relationships are configured correctly.

Keep in mind that the generated code is just a starting point. Some models or properties may not be represented correctly in the database. For example, if your models have inheritance, the generated code will not include the base class because the base class is not represented in the database. Also, some column types may not be able to be mapped to the corresponding CLR types. For example, the `Status` column in the `Invoice` table is of the `nvarchar(16)` type, which will be mapped to the `string` type in the generated code, instead of the `Status` enum type.

You can update the generated code to suit your needs, but be aware that the next time you run the `dbcontext scaffold` command, the changes will be overwritten. You can use partial classes to add your own code to the generated classes, as the generated classes are declared as `partial`.

In this section, we discussed how to use EF Core to reverse engineer the entity classes and `DbContext` from an existing database schema. It should be noted that EF Core strongly prefers the code-first approach. Unless you are working with an existing database, it is recommended to use the code-first approach to take advantage of the EF Core migrations feature.

Other ORM frameworks

In addition to EF Core, there are numerous other ORM frameworks available for .NET. Some of the most popular include the following:

- **Dapper** (`https://dapperlib.github.io/Dapper/`): Dapper is a micro-ORM framework that is designed to be fast and lightweight. Dapper does not support change tracking, but it is easy to use, and it is very fast. As the official documentation says, "*Dapper's simplicity means that many features that ORMs ship with are stripped out. It worries about the 95% scenario and gives you the tools you need most of the time. It doesn't attempt to solve every problem.*" The performance is one of the most important features of Dapper. Maybe it is not fair to compare Dapper's performance with EF Core, because EF Core provides many more features than Dapper. If you are looking for a simple ORM framework that is fast and easy to use, Dapper is a good choice. In some projects, Dapper is used with EF Core to provide the best of both worlds. Dapper is open-source and originally developed by Stack Overflow.

- **NHibernate** (`https://nhibernate.info/`): Like NUnit, NHibernate is a .NET implementation of the Hibernate ORM framework in Java. It is a mature, open-source ORM framework that has been around for a long time. It is very powerful and flexible. NHibernate is maintained by a community of developers.

- **PetaPoco** (`https://github.com/CollaboratingPlatypus/PetaPoco`): PetaPoco is a tiny, fast, easy-to-use micro-ORM framework, which only had 1,000+ lines of code in the original version. PetaPoco has a similar performance to Dapper because it uses dynamic method generation (MSIL) to assign column values to properties. PetaPoco now supports SQL Server, SQL Server CE, MS Access, SQLite, MySQL, MariaDB, PostgreSQL, Firebird DB, and Oracle. It uses T4 templates to generate the code. PetaPoco is open-source and currently maintained by a few core developers.

It is hard to say which one is the best. It depends on your needs. Dapper is known for its speed and performance, while EF Core is more feature-rich and provides better support for complex queries and relationships. When deciding which framework to use for a particular task, consider the performance implications of each approach, and also the trade-offs between the features and flexibility of the framework.

Summary

In this chapter, we delved into some advanced topics of Entity Framework. We started by exploring how to improve the performance of our application by using `DbContext` pooling and no-tracking queries. We then learned how to execute raw SQL queries safely and efficiently using parameterized queries, and how to leverage the new bulk operations feature in EF Core for faster data manipulation.

Next, we looked at how to handle concurrency scenarios using optimistic concurrency control, which allows multiple users to access and modify the same data simultaneously without conflicts. We also covered reverse engineering, a technique for generating entity classes and `DbContext` classes from an existing database schema, which can save time and effort in creating a data access layer.

To broaden our horizons beyond EF Core, we briefly introduced some other popular ORM frameworks, such as Dapper, NHibernate, and PetaPoco, and discussed their strengths and weaknesses. By the end of this chapter, you should have a solid understanding of how to leverage EF Core in a web API project to efficiently access and manipulate data, as well as some insights into other ORM options available to you.

However, EF Core is a very large topic, and we cannot cover everything in this book. For more information about EF Core, please refer to the official documentation at `https://learn.microsoft.com/en-us/ef/`.

In the next chapter, we will learn how to secure our web API project using authentication and authorization.

8

Security and Identity in ASP.NET Core

In *Chapter 7*, we discussed some more advanced topics of EF Core, such as `DbContext` pooling, performance optimization, and concurrency control. At this point, you should have the skills to create a web API application that accesses the database using EF Core. However, the application is not secure. Without any authentication, anyone who knows the URL can access the API, potentially exposing sensitive data to the public. To ensure the security of the web API application, we must take additional steps.

Security is a broad topic, and it is a crucial aspect of any application. In this chapter, we will explore some of the security features that ASP.NET Core provides, including authentication, authorization, and some best practices for securing your web API application. We will cover the following topics:

- Getting started with authentication and authorization
- Delving deeper into authorization
- Managing users and roles
- New Identity API endpoints in ASP.NET Core 8
- Understanding OAuth 2.0 and OpenID Connect
- Other security topics

After reading this chapter, you will have a basic understanding of the security features in ASP.NET Core. You will also know how to implement authentication and various authorization types in your web API applications, such as role-based authorization, claim-based authorization, and policy-based authorization.

Technical requirements

The code examples in this chapter can be found at `https://github.com/PacktPublishing/Web-API-Development-with-ASP.NET-Core-8`. You can use VS Code or VS 2022 to open the solutions.

Getting started with authentication and authorization

Authentication and authorization are two important aspects of security. Although these two terms are often used together, they are distinct concepts. Before we dive into the code, it is important to gain an understanding of the differences between authentication and authorization.

We have already built some web API applications. However, these APIs will be publicly available to anyone who knows the URL. For some resources, we want to restrict access to only authenticated users. For example, we have a resource that contains some sensitive information that should not be available to everyone. In this case, the application should be able to identify the user who is making the request. If the user is anonymous, the application should not allow the user to access the resource. This is where authentication comes into play.

For some scenarios, we also want to restrict access to some specific users. For example, we want to allow authenticated users to read the resource, but only admin users to update or delete the resource. In this case, the application should be able to check whether the user has the required permissions to execute the operation. This is where authorization is used.

Long story short, authentication is used to know who the user is, while authorization is used to know what the user can do. Together, these processes are used to ensure that the user is who they claim to be and that they have the required permissions to access the resource.

ASP.NET Core provides the Identity framework, which has a rich set of features for authentication and authorization. In this chapter, we will explore how to use the Identity framework to implement authentication and authorization in ASP.NET Core. We will also introduce some third-party authentication providers.

Think about a scenario where we want to build a web API application that allows users to register and log in. For a specific endpoint, we only want to allow authenticated users to access the resource. In this section, we will explore how to implement this scenario. From this example, you will learn how to implement a basic authentication and authorization system in ASP.NET Core; this will help you prepare for the next section.

We will use the following resources in this example:

- `POST /account/register`: This resource will be used to register a new user. The user should send the username and password in the request body. After validating the username and password, the application will create a new user in the database and return a JWT token to the user. This JWT token will be used to authenticate the user in subsequent requests.

- POST /account/login: This resource will be used to log into an existing user. After the user sends the username and password, the application will validate the credentials and return a JWT token to the user if the credentials are valid. The JWT token will be used to authenticate the user in subsequent requests.

- GET /WeatherForecast: This resource will be used to get the weather forecast. It only allows authenticated users to access the resource. The user should send the JWT token in the Authorization header to authenticate the user.

There should be more endpoints to manage users, such as updating the user profile, deleting the user, resetting the password, and more. However, we are not building a complete application in this chapter. To keep things simple, we will only focus on the minimal features required to demonstrate the authentication and authorization features in ASP.NET Core.

What is JWT?

JWT stands for *JSON Web Token*. It is an industry standard for representing claims securely between two parties. The RFC for JWT is RFC 7519: `https://www.rfc-editor.org/rfc/rfc7519`. A JWT token consists of three parts: header, payload, and signature. So, typically, a JWT token looks like `xxxxx.yyyyy.zzzzz`. The header contains the algorithm used to sign the token, the payload contains the claims, and the signature is used to verify the integrity of the token. For more information about JWT, see `https://jwt.io/introduction`.

Creating a sample project with authentication and authorization

To begin, we must prepare the project and add any necessary NuGet packages. Additionally, we need to configure the database context to enable us to store user information in the database. Follow these steps:

1. Create a new ASP.NET Core web API project by running the following command:

   ```
   dotnet new webapi -n AuthenticationDemo -controllers
   ```

   This command will create a new ASP.NET Core web API project named `AuthenticationDemo`. Open the project in VS Code. You can find the start project in the `/samples/chapter8/AuthenticationDemo/BasicAuthenticationDemo/start` folder.

2. Now, it's time to add the required NuGet packages. We will use ASP.NET Core Identity to implement the authentication. ASP.NET Core Identity is a membership system that provides authentication and authorization features. It is a part of the ASP.NET Core framework. We need to install the following NuGet packages:

   - `Microsoft.AspNetCore.Identity.EntityFrameworkCore`: This package is used for the EF Core implementation of ASP.NET Core Identity.

   - `Microsoft.EntityFrameworkCore.SqlServer`: This package is used to connect to SQL Server.

- `Microsoft.EntityFrameworkCore.Tools`: This package is used to enable the necessary EF Core tools.

- `Microsoft.AspNetCore.Authentication.JwtBearer`: This package is used to enable JWT authentication.

The ASP.NET Core Identity package already comes with the default project template, so we do not need to install it.

3. Next, we will add the database context. We will use EF Core to access the database. But first, we need an entity model to represent the user. Create a new folder named `Authentication` and add a new class named `AppUser` to it. The `AppUser` class inherits from the `IdentityUser` class, which is provided by ASP.NET Core Identity, as shown in the following code:

```
public class AppUser : IdentityUser
{
}
```

The `IdentityUser` class already contains the properties that we need to represent a user for most of the scenarios, such as `UserName`, `Email`, `PasswordHash`, `PhoneNumber`, and others.

4. Next, we need to create a database context to access the database. Add a new class named `AppDbContext` to the `Authentication` folder. The `AppDbContext` class inherits from the `IdentityDbContext` class, which is provided by ASP.NET Core Identity, as shown in the following code:

```
public class AppDbContext(DbContextOptions<AppDbContext>
options, IConfiguration configuration)
  : IdentityDbContext<AppUser>(options)
{

    protected override void
OnConfiguring(DbContextOptionsBuilder optionsBuilder)
    {
        base.OnConfiguring(optionsBuilder);
        optionsBuilder.UseSqlServer(_configuration.
GetConnectionString("DefaultConnection"));
    }
}
```

To provide the database connection string, we need to add the following configuration to the `appsettings.json` file:

```
"ConnectionStrings": {
    "DefaultConnection": "Server=(localdb)\\
mssqllocaldb;Database=AuthenticationDemo;Trusted_
Connection=True;MultipleActiveResultSets=true"
}
```

As we can see, this `AppDbContext` is purely for ASP.NET Core Identity. If you have other entities in your application, you can create a separate `DbContext` for them if you want to. You can use the same connection string for both `DbContext`s.

5. Next, we will need to create a few models for registering and logging in users because, when we register a user, we need to send the username, password, and email address. When we log a user in, we need to send the username and password. It would be good if we had separate models for these different scenarios.

6. Create a new class named `AddOrUpdateAppUserModel` in the `Authentication` folder. This class will be used to represent the user when we register a new user. The `AddOrUpdateAppUserModel` class should contain the following properties:

```
public class AddOrUpdateAppUserModel
{
    [Required(ErrorMessage = "User name is required")]
    public string UserName { get; set; } = string.Empty;

    [EmailAddress]
    [Required(ErrorMessage = "Email is required")]
    public string Email { get; set; } = string.Empty;

    [Required(ErrorMessage = "Password is required")]
    public string Password { get; set; } = string.Empty;
}
```

7. Similarly, create a new class named `LoginModel` in the `Authentication` folder, as shown in the following code:

```
public class LoginModel
{
    [Required(ErrorMessage = "User name is required")]
    public string UserName { get; set; } = string.Empty;

    [Required(ErrorMessage = "Password is required")]
    public string Password { get; set; } = string.Empty;
}
```

We can add more properties to the `AppUser` class if we need to. For example, you can add `FirstName`, `LastName`, and also a `ProfilePicture` property to store the user's profile picture, as shown in the following code:

```
public class AppUser : IdentityUser
{
    public string FirstName { get; set; }
    public string LastName { get; set; }
```

```
        public string ProfilePicture { get; set; }
    }
```

If you add additional properties to the `AppUser` class, you need to add the corresponding properties for the `AddOrUpdateAppUserModel` as well.

8. Next, we need to configure the authentication service. First, let's update the `appsettings.json` file to provide the configurations for JWT tokens:

```
"JwtConfig": {
    "ValidAudiences": "http://localhost:5056",
    "ValidIssuer": "http://localhost:5056",
    "Secret": "c1708c6d-7c94-466e-aca3-e09dcd1c2042"
}
```

Update the configurations as per your requirements. Because we use the same web API to issue and validate the JWT token, we use the same URL for the `ValidAudiences` and `ValidIssuer` properties. The `Secret` property is used to sign the JWT token. You can use any string as the secret. In this case, we can use a GUID value. Also, please note that this is for demo purposes only. In a real-world application, you should store the secret in a secure location, such as Azure Key Vault.

9. Update the code in the `Program.cs` file, as shown in the following code:

```
// Omitted for brevity
builder.Services.AddControllers();

builder.Services.AddDbContext<AppDbContext>();
builder.Services.AddIdentityCore<AppUser>()
    .AddEntityFrameworkStores<AppDbContext>()
    .AddDefaultTokenProviders();

builder.Services.AddAuthentication(options =>
{
    options.DefaultAuthenticateScheme = JwtBearerDefaults.
AuthenticationScheme;
    options.DefaultChallengeScheme = JwtBearerDefaults.
AuthenticationScheme;
    options.DefaultScheme = JwtBearerDefaults.
AuthenticationScheme;
}).AddJwtBearer(options =>
{
    var secret = builder.Configuration["JwtConfig:Secret"];
    var issuer = builder.Configuration["JwtConfig:ValidIssuer"];
    var audience = builder.
```

```
Configuration["JwtConfig:ValidAudiences"];
    if (secret is null || issuer is null || audience is null)
    {
        throw new ApplicationException("Jwt is not set in the
configuration");
    }
    options.SaveToken = true;
    options.RequireHttpsMetadata = false;
    options.TokenValidationParameters = new
TokenValidationParameters()
    {
        ValidateIssuer = true,
        ValidateAudience = true,
        ValidAudience = audience,
        ValidIssuer = issuer,
        IssuerSigningKey = new SymmetricSecurityKey(Encoding.
UTF8.GetBytes(secret))
    };
});
// Omitted for brevity
app.UseHttpsRedirection();

app.UseAuthentication();
app.UseAuthorization();
// Omitted for brevity
```

In the preceding code, we configured the authentication service to use JWT tokens. The `AddIdentityCore()` method adds and configures the identity system for the specified `User` type. We also added `AppDbContext` and `AppUser` to the service collection and specified that we want to use EF Core to store the user data. The `AddDefaultTokenProviders()` method adds the default token providers for the application, which are used to generate tokens. The `Services.AddAuthentication()` method configures the authentication service to use JWT tokens. The `AddJwtBearer()` method configures the JWT bearer authentication handler, including the token validation parameters. We use some configurations from the `appsettings.json` file to configure the token validation parameters.

Finally, we need to call the `UseAuthentication()` and `UseAuthorization()` methods to enable authentication and authorization in the application.

10. Now, it's time to create and update the database. We have already created the database context and the user entity. So, now, we need to create the database. To do that, just run the following command:

```
dotnet ef migrations add InitialDb
dotnet ef database update
```

If the commands are executed successfully, you should see the database created with the following tables:

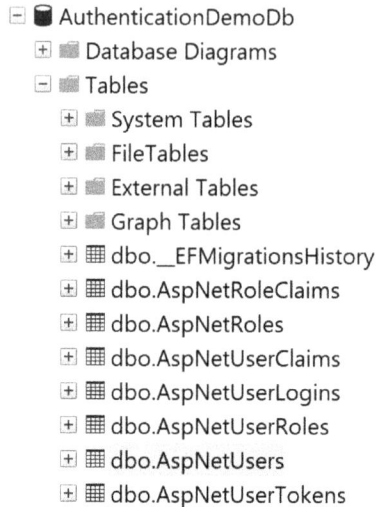

```
⊟ 🗄 AuthenticationDemoDb
    ⊞ 📁 Database Diagrams
    ⊟ 📁 Tables
        ⊞ 📁 System Tables
        ⊞ 📁 FileTables
        ⊞ 📁 External Tables
        ⊞ 📁 Graph Tables
        ⊞ ▦ dbo.__EFMigrationsHistory
        ⊞ ▦ dbo.AspNetRoleClaims
        ⊞ ▦ dbo.AspNetRoles
        ⊞ ▦ dbo.AspNetUserClaims
        ⊞ ▦ dbo.AspNetUserLogins
        ⊞ ▦ dbo.AspNetUserRoles
        ⊞ ▦ dbo.AspNetUsers
        ⊞ ▦ dbo.AspNetUserTokens
```

Figure 8.1 – The database tables created by ASP.NET Core Identity

11. Another way to check whether the database has been created is to add the following code to the `Program.cs` file:

```
using (var serviceScope = app.Services.CreateScope())
{
    var services = serviceScope.ServiceProvider;
    // Ensure the database is created.
    var dbContext = services.GetRequiredService<AppDbContext>();
    dbContext.Database.EnsureCreated();
}
```

You can use either of the methods to check whether the database is created in the development environment.

The data for the users will be stored in these tables, which is convenient when using the default tables provided by ASP.NET Core Identity.

Next, let's apply the `Authorize` attribute to enable authentication and authorization for `WeatherForecastController`:

1. Update `WeatherForecastController` by adding an `[Authorize]` attribute, as shown in the following code:

```
[Authorize]
[ApiController]
```

```
[Route("[controller]")]
public class WeatherForecastController : ControllerBase
{
    // ...
}
```

This attribute will ensure that the user is authenticated before accessing the controller. If the user is not authenticated, the controller will return a `401 Unauthorized` response. Test this by running the application and calling the `/WeatherForecast` endpoint. You should see a `401` response:

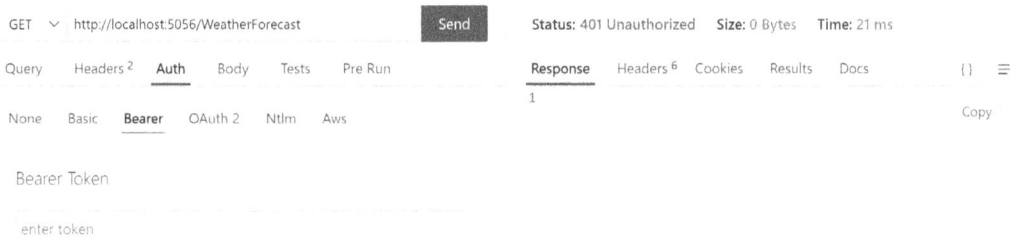

Figure 8.2 – When the user is not authenticated, the controller returns a 401 response

The `Authorize` attribute can be applied to the controller or the action method. If the attribute is applied to the controller, all the action methods in the controller will be protected. If the attribute is applied to the action method, only that action method will be protected.

You can also use the `AllowAnonymous` attribute to allow anonymous access to the controller or action method. Note that the `AllowAnonymous` attribute overrides the `Authorize` attribute. So, if you apply both attributes to the controller or action method, the `AllowAnonymous` attribute will take precedence, which means that the controller or action method will be accessible to all the users.

2. Next, let's add `AccountController` to handle the authentication requests. For example, we need to provide a `/account/register` endpoint. When the user sends the username and password, the application will create a record of the user in the database and generate a JWT token.

To generate a JWT token, we need to provide the following information:

- **The issuer of the token**: This is the server that issues the token.

- **The audience of the token**: This is the server that consumes the token. It can be the same as the issuer or include multiple servers.

- **The secret key to sign the token**: This is used to validate the token.

We already defined these values in the `appsettings.json` file as described in the previous steps.

3. Next, create a new controller named AccountController to handle the authentication requests. Create a new class named AccountController in the Controllers folder. The AccountController class should inherit from the ControllerBase class, as shown in the following code:

```
[ApiController]
[Route("[controller]")]
public class AccountController(UserManager<AppUser> userManager,
IConfiguration configuration)
  : ControllerBase
{
}
```

We use the UserManager class to manage the users. The UserManager class is provided by ASP.NET Core Identity. We also need to inject the IConfiguration interface to get the configuration values from the appsettings.json file.

4. Create a new method named Register() in the AccountController class. This method will be used to register a new user. The Register() method should accept an AddOrUpdateAppUserModel object as a parameter, as shown in the following code:

```
[HttpPost("register")]
public async Task<IActionResult> Register([FromBody]
AddOrUpdateAppUserModel model)
{
    // Check if the model is valid
    if (ModelState.IsValid)
    {
        var existedUser = await userManager.
FindByNameAsync(model.UserName);
        if (existedUser != null)
        {
            ModelState.AddModelError("", "User name is already
taken");
            return BadRequest(ModelState);
        }
        // Create a new user object
        var user = new AppUser()
        {
            UserName = model.UserName,
            Email = model.Email,
            SecurityStamp = Guid.NewGuid().ToString()
        };
        // Try to save the user
        var result = await userManager.CreateAsync(user, model.
Password);
```

```
        // If the user is successfully created, return Ok
        if (result.Succeeded)
        {
            var token = GenerateToken(model.UserName);
            return Ok(new { token });
        }
        // If there are any errors, add them to the ModelState
object
        // and return the error to the client
        foreach (var error in result.Errors)
        {
            ModelState.AddModelError("", error.Description);
        }
    }
    // If we got this far, something failed, redisplay form
    return BadRequest(ModelState);
}
```

The `UserManager` class provides a set of methods to manage the users, such as `FindByNameAsync`, `FindByNameAsync()`, `CreateAsync()`, and others. The `GenerateToken()` method is a private method that generates a JWT token for the user, as shown in the following code:

```
private string? GenerateToken(string userName)
{
    var secret = _configuration["JwtConfig:Secret"];
    var issuer = _configuration["JwtConfig:ValidIssuer"];
    var audience = _configuration["JwtConfig:ValidAudiences"];
    if (secret is null || issuer is null || audience is null)
    {
        throw new ApplicationException("Jwt is not set in the
configuration");
    }
    var signingKey = new SymmetricSecurityKey(Encoding.UTF8.
GetBytes(secret));
    var tokenHandler = new JwtSecurityTokenHandler();
    var tokenDescriptor = new SecurityTokenDescriptor
    {
        Subject = new ClaimsIdentity(new[]
        {
            new Claim(ClaimTypes.Name, userName)
        }),
        Expires = DateTime.UtcNow.AddDays(1),
        Issuer = issuer,
        Audience = audience,
```

```
        SigningCredentials = new SigningCredentials(signingKey,
    SecurityAlgorithms.HmacSha256Signature)
        };
        var securityToken = tokenHandler.
    CreateToken(tokenDescriptor);
        var token = tokenHandler.WriteToken(securityToken);

        return token;
    }
```

In the preceding code, we are using the `JwtSecurityTokenHandler` class to generate the JWT token. The `JwtSecurityTokenHandler` class is provided by the `System.IdentityModel.Tokens.Jwt` NuGet package. First, we get the configuration values from the `appsettings.json` file. Then, we create a `SymmetricSecurityKey` object using the secret key. The `SymmetricSecurityKey` object is used to sign the token.

Next, we created a `SecurityTokenDescriptor` object, which contains the following properties:

- `Subject`: The subject of the token. The subject can be any value, such as the username, email address, and so on.

- `Expires`: The expiration date of the token.

- `Issuer`: The issuer of the token.

- `Audience`: The audience of the token.

- `SigningCredentials`: The credentials to sign the token. Note that we use the `HmacSha256Signature` algorithm to sign the token. It is a 256-bit HMAC cryptographic algorithm for digital signatures. If you encounter an error such as `IDX10603: The algorithm: 'HS256' requires the SecurityKey.KeySize to be greater than '128' bits.`, please check the secret key in the `appsettings.json` file. The secret key should be at least 16 characters long (16 * 8 = 128).

Finally, we used the `JwtSecurityTokenHandler` class to create and write the token to a string value.

5. Now, we can test the `Register()` method. Use `dotnet run` to run the application. You can use the Swagger UI or any other tools to test the API. Send a `POST` request with the following JSON data to the `http://localhost:5056/account/register` endpoint:

```
{
    "userName": "admin",
    "email": "admin@example.com",
    "password": "Passw0rd!"
}
```

You will see the response similar to the following:

POST ∨ http://localhost:5056/Account/register   **Send**   Status: 200 OK   **Size:** 272 Bytes   **Time:** 176 ms

Query   Headers ²   Auth   **Body** ¹   Tests   Pre Run

Json   Xml   Text   Form   Form-encode   Graphql   Binary

Json Content   Format

```
1  {
2    "userName": "admin",
3    "email": "admin@example.com",
4    "password": "Passw0rd!"
5  }
```

Response   Headers ⁶   Cookies   Results   Docs

```
1  {
2    "token": "eyJhbGciOiJIUzI1NiIsInR5cCI6IkpXVCJ9
       .eyJ1bmlxdWVfbmFtZSI6ImFkbWluIiwibmJmIjoxNjgwNDMyOT
       c1LCJleHAiOjE2ODA1MTkzNzUsImlhdCI6MTY4MDQzMjk3NSwia
       XNzIjoiaHR0cDovL2xvY2FsaG9zdDo1MDU2IiwiYXVkIjoiaHR0
       cDovL2xvY2FsaG9zdDo1MDU2In0.NXmtnHMPDJoniWNjLtVHA
       -QrT2w4jYIXC8pBSppHgH4"
3  }
```

Figure 8.3 – Registering a new user

As we can see, the `Register()` method returns a JWT token. The token is valid for 1 day. We can use this token to authenticate the user in the future. If you check the database, you will see that a new user has been created in the `AspNetUsers` table, and the password is hashed, as shown in the following screenshot:

	Id	UserName	NormalizedUserName	Email	NormalizedEmail	EmailConfirmed	PasswordHash
1	7af60d22-9811-4314-bbfa-1ae460ec0d68	admin	ADMIN	admin@example.com	ADMIN@EXAMPLE.COM	0	AQAAAAIAAYagAAAAECK9e1G04YMLv3h5SSIP5cbt/UvqxRwm...

Figure 8.4 – The new user has been created in the database

6. Copy the token value and send a `GET` request to the `/WeatherForecast` endpoint. You need to attach the `Bearer` token to the request header, as shown in the following screenshot:

GET ∨ http://localhost:5056/WeatherForecast   **Send**   Status: 200 OK   **Size:** 386 Bytes   **Time:** 81 ms

Query   Headers ²   **Auth** ¹   Body   Tests   Pre Run

None   Basic   **Bearer**   OAuth 2   Ntlm   Aws

Bearer Token

```
eyJhbGciOiJodHRwOi8vd3d3LnczLm9yZy8yMDAxLzA0L3htbGRzaWct
bW9yZSNobWFjLXNoYTI1NiIsInR5cCI6IkpXVCJ9.eyJodHRwOi8vc2NoZ
W1hcy54bWxzb2FwLm9yZy93cy8yMDA1LzA1L2lkZW50aXR5L2NsYWl
tcy9uYW1lIjoiYWRtaW4iLCJleHAiOjE2NzkxOTI4OTAsImlzcyI6Imh0dHA
6Ly9sb2NhbGhvc3Q6NTA1NiIsImF1ZCI6Imh0dHA6Ly9sb2NhbGhvc3Q
6NTA1NiJ9.2Yt07TGx8M6m8IXMEjJndat0fPjtOr99ny3Wo8lSkes
```

Token Prefix   Bearer

Response   Headers ⁶   Cookies   Results   Docs

```
1  [
2    {
3      "date": "2023-03-21",
4      "temperatureC": 35,
5      "temperatureF": 94,
6      "summary": "Sweltering"
7    },
8    {
9      "date": "2023-03-22",
10     "temperatureC": 50,
11     "temperatureF": 121,
12     "summary": "Balmy"
13   },
14   {
15     "date": "2023-03-23",
16     "temperatureC": 5,
17     "temperatureF": 40,
18     "summary": "Mild"
19   },
20   {
21     "date": "2023-03-24",
22     "temperatureC": 42,
23     "temperatureF": 107,
24     "summary": "Chilly"
25   },
```

Figure 8.5 – Sending a request with the Bearer token

> **Important note**
>
> When you attach the bearer token to the request, please note that there is a prefix of `Bearer` before the token value. So, the actual format should be
>
> `Authorization: Bearer <token>.`

OK, it works! Your API is now secured. The next step is to create a login method to authenticate the user. It is quite straightforward. Create a new method named `Login` in the `AccountController` class. The `Login()` method should accept an `AddOrUpdateAppUserModel` object as a parameter, as shown in the following code:

```
[HttpPost("login")]
public async Task<IActionResult> Login([FromBody] LoginModel model)
{
    // Get the secret in the configuration

    // Check if the model is valid
    if (ModelState.IsValid)
    {
        var user = await _userManager.FindByNameAsync(model.UserName);
        if (user != null)
        {
            if (await _userManager.CheckPasswordAsync(user, model.
Password))
            {
                var token = GenerateToken(model.UserName);
                return Ok(new { token });
            }
        }
        // If the user is not found, display an error message
        ModelState.AddModelError("", "Invalid username or password");
    }
    return BadRequest(ModelState);
}
```

We use the `UserManager` class to find the user by the username. If the user is found, we use the `CheckPasswordAsync()` method to check the password. If the password is correct, we generate a new token and return it to the client. If the user is not found or the password is incorrect, we return an error message to the client.

So far, we have created a web API project with basic authentication and authorization. We also created a controller to handle account-related operations. Note that in this example, we have not implemented any specific authorization rules. All authenticated users can access the `WeatherForecast` endpoint.

Next, we will discuss the details of the JWT token.

Understanding the JWT token structure

A JWT token is a string value. It is composed of three parts, separated by a dot (.):

- Header

- Payload

- Signature

The header and payload are encoded using the `Base64Url` algorithm. We can use `jwt.io` to decode the token. Copy the token in the response body and paste it into the `Encoded` field on the `jwt.io` website. You will see the decoded token, as shown in the following screenshot:

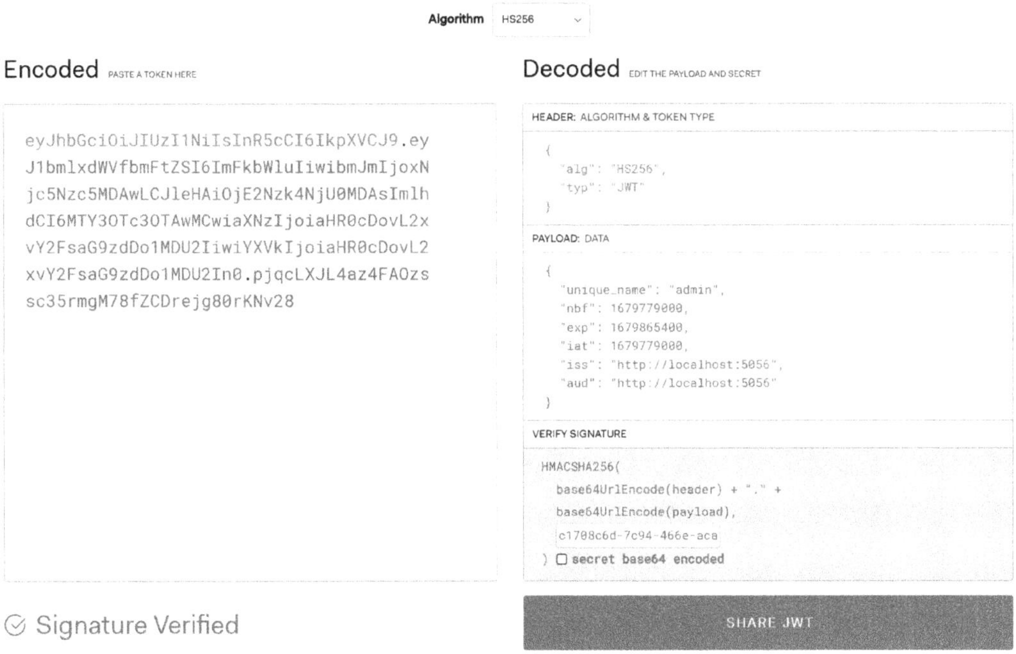

Figure 8.6 – Decoding the JWT token

The header contains the algorithm used to sign the token. In our case, we use the `HmacSha256Signature` algorithm. So, the decoded header is as follows:

```
{
  "alg": "HS256",
  "typ": "JWT"
}
```

The payload contains the claims of the token and some other additional data. In our case, the decoded payload is as follows:

```
{
  "unique_name": "admin",
  "nbf": 1679779000,
  "exp": 1679865400,
  "iat": 1679779000,
  "iss": "http://localhost:5056",
  "aud": "http://localhost:5056"
}
```

There are some recommended (but not mandatory) registered claim names defined in RFC7519:

- sub: The sub (subject) claim identifies the principal that is the subject of the token

- nbf: The nbf (not before) claim identifies the time before which the token *must not* be accepted for processing

- exp: The exp (expiration time) claim identifies the expiration time on or after which the token *must not* be accepted for processing

- iat: The iat (issued at) claim identifies the time at which the token was issued

- iss: The iss (issuer) claim identifies the principal that issued the token

- aud: The aud (audience) claim identifies the recipients that the token is intended for

Note that in our case, we use the same value for the iss and aud claims because we use the same web API to issue and validate the token. In a real-world application, normally, there is a separate authentication server to issue the token so that the iss and aud claims have different values.

The signature is used to verify the token to make sure the token does not tamper. There are various algorithms to generate the signature. In our case, we use the HmacSha256Signature algorithm, so the signature is generated using the following formula:

HMACSHA256(base64UrlEncode(header) + "." + base64UrlEncode(payload), secret)

Therefore, the token typically looks like xxxxx.yyyyy.zzzzz, which can be easily passed in the HTTP request header, or stored in the local storage of the browser.

Consuming the API

At this point, we have a secured API. You can find a sample client application named AuthenticationDemoClient in the samples\chapter8\AuthenticationDemo\ BasicAuthenticationDemo\end folder. The client application is a simple console application. It uses the HttpClient class to send HTTP requests to the API. The main code is like this:

Login:

```
var httpClient = new HttpClient();
// Create a post request with the user name and password
var request = new HttpRequestMessage(HttpMethod.Post, "http://
localhost:5056/authentication/login");
request.Content = new StringContent(JsonSerializer.Serialize(new
LoginModel()
{
    UserName = userName,
    Password = password
}), Encoding.UTF8, "application/json");
var response = await httpClient.SendAsync(request);
var token = string.Empty;
if (response.IsSuccessStatusCode)
{
    var content = await response.Content.ReadAsStringAsync();
    var jwtToken = JsonSerializer.Deserialize<JwtToken>(content);
    Console.WriteLine(jwtToken.token);
    token = jwtToken.token;
}
```

Get weather forecast:

```
request = new HttpRequestMessage(HttpMethod.Get, "http://
localhost:5056/WeatherForecast");
// Add the token to the request header
request.Headers.Authorization = new
AuthenticationHeaderValue("Bearer", token);
response = await httpClient.SendAsync(request);
if (response.IsSuccessStatusCode)
{
    var content = await response.Content.ReadAsStringAsync();
    var weatherForecasts = JsonSerializer.
Deserialize<IEnumerable<WeatherForecast>>(content);
    foreach (var weatherForecast in weatherForecasts)
    {
        Console.WriteLine("Date: {0:d}", weatherForecast.Date);
        Console.WriteLine($"Temperature (C): {weatherForecast.
TemperatureC}");
        Console.WriteLine($"Temperature (F): {weatherForecast.
TemperatureF}");
        Console.WriteLine($"Summary: {weatherForecast.Summary}");
    }
}
```

First, the client application sends a request to the login API to get the token. Then, it attaches the token to the request header and sends the request to the weather forecast API. If the token is valid, the API will return the data.

Configuring the Swagger UI to support authorization

You probably prefer to use the Swagger UI to test the APIs. The default configuration of the Swagger UI does not support authorization. We need to update the `AddSwaggerGen()` method in the `Program` class to support authorization. Update the `Program` class as follows:

```
builder.Services.AddSwaggerGen(c =>
{
    c.AddSecurityDefinition("Bearer", new OpenApiSecurityScheme
    {
        Description = "JWT Authorization header using the Bearer
scheme. Example: \"Authorization: Bearer {token}\"",
        Name = "Authorization",
        In = ParameterLocation.Header,
        Type = SecuritySchemeType.ApiKey,
        Scheme = "Bearer"
    });
    c.AddSecurityRequirement(new OpenApiSecurityRequirement
    {
        {
            new OpenApiSecurityScheme
            {
                Reference = new OpenApiReference
                {
                    Type = ReferenceType.SecurityScheme,
                    Id = "Bearer"
                }
            },
            new string[] { }
        }
    });
});
```

The preceding code adds the `Bearer` security definition to the Swagger UI. The `AddSecurityRequirement` method adds the `Authorization` header to the Swagger UI. Now, when you run the application, you will see the **Authorize** button in the Swagger UI. Click the **Authorize** button; you will see a pop-up window that allows you to enter the token, as shown in the following screenshot:

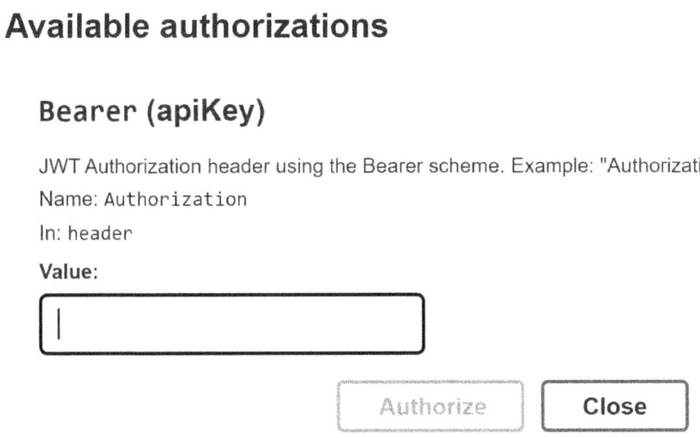

Available authorizations

Bearer (apiKey)

JWT Authorization header using the Bearer scheme. Example: "Authorization: Bearer {token}"

Name: `Authorization`

In: `header`

Value:

```
|
```

<button>Authorize</button> <button>Close</button>

Figure 8.7 – Entering the token in the Swagger UI

Enter the token in the **Value** field. Then, click the **Authorize** button. Now, you can test the APIs using the Swagger UI directly:

> **Caution**
> You need to add the `Bearer` prefix to the token with a space.

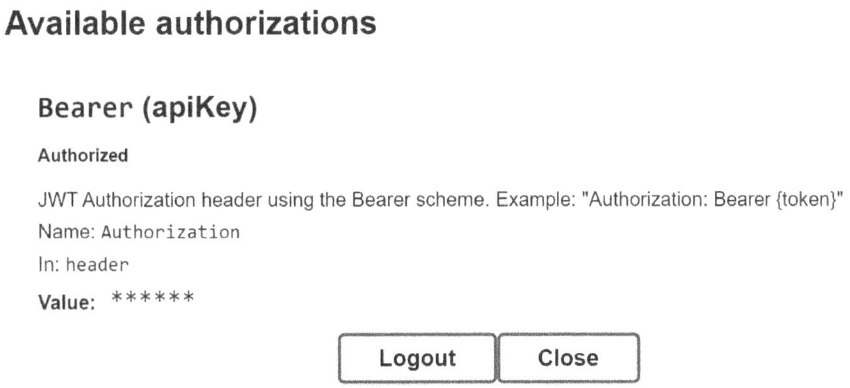

Available authorizations

Bearer (apiKey)

Authorized

JWT Authorization header using the Bearer scheme. Example: "Authorization: Bearer {token}"

Name: `Authorization`

In: `header`

Value: ******

<button>Logout</button> <button>Close</button>

Figure 8.8 – The Swagger UI is authorized

You can find more information about the configuration for the Swagger UI here: `https://github.com/domaindrivendev/Swashbuckle.AspNetCore`.

In this section, we discussed the implementation of a web API project that supports authentication and authorization, including the creation of a controller to handle the login request. Additionally, we explored how to generate a JWT token and validate it, as well as how to use a console application to access the project resource and how to configure the Swagger UI to test the APIs with authorization.

In the next section, we will learn more about authorization in ASP.NET Core. We will explore a couple of authorization types, including role-based authorization, claim-based authorization, and policy-based authorization.

Delving deeper into authorization

Authorization is the process of determining whether a user is allowed to perform a specific action. In the previous section, we implemented a web API project that enables simple authentication and authorization. By using the `Authorize` attribute, only authenticated users can access the API. However, in many scenarios, we need to implement granular authorization. For example, some resources are only accessible to the administrator, while some resources are accessible to normal users. In this section, we will explore how to implement granular authorization in ASP.NET Core, including role-based authorization, claim-based authorization, and policy-based authorization.

Role-based authorization

You can find the starter app and the completed app in this book's GitHub repository at `chapter8/AuthorizationDemo/RoleBasedAuthorizationDemo`. The starter app is similar to the application we created in the previous section:

1. We'll start with the starter app. Don't forget to create the database and run the migrations using the following commands:

    ```
    dotnet ef database update
    ```

    A role is a set of permissions that are assigned to a user. For example, the administrator role has permission to access all resources, while the normal user role has limited permissions. A user can be assigned to multiple roles. A role can be assigned to multiple users.

2. Next, let's define a few roles in the project. Create a class named `AppRoles` that is defined as follows:

    ```
    public static class AppRoles
    {
        public const string Administrator = "Administrator";
        public const string User = "User";
        public const string VipUser = "VipUser";
    }
    ```

3. In the `Program` class, we need to explicitly call the `AddRoles()` method after the `AddIdentityCore()` method. The updated code is as follows:

    ```
    // Use the `AddRoles()` method
    builder.Services.AddIdentityCore<AppUser>()
        .AddRoles<IdentityRole>()
    ```

```
.AddEntityFrameworkStores<AppDbContext>()
.AddDefaultTokenProviders();
```

If you use the `AddIdentity()` method, you do not need to call the `AddRoles()` method. The `AddIdentity()` method will call the `AddRoles()` method internally.

4. We also need to check whether the roles exist in the database. If not, we will create the roles. Add the code, as follows:

```
using (var serviceScope = app.Services.CreateScope())
{
    var services = serviceScope.ServiceProvider;
    var roleManager = app.Services.
GetRequiredService<RoleManager<IdentityRole>>();
    if (!await roleManager.RoleExistsAsync(AppRoles.User))
    {
        await roleManager.CreateAsync(new IdentityRole(AppRoles.
User));
    }

    if (!await roleManager.RoleExistsAsync(AppRoles.VipUser))
    {
        await roleManager.CreateAsync(new IdentityRole(AppRoles.
VipUser));
    }

    if (!await roleManager.RoleExistsAsync(AppRoles.
Administrator))
    {
        await roleManager.CreateAsync(new IdentityRole(AppRoles.
Administrator));
    }
}
```

5. Use `dotnet run` to run the application. You will see that the roles are created in the database:

▦ Results ▤ Messages

	Id	Name	NormalizedName	ConcurrencyStamp
1	5e040dfe-19d4-4c5f-904d-a06a907f8622	User	USER	NULL
2	b4db3767-6722-4a2f-8288-e52221597f87	VipUser	VIPUSER	NULL
3	c3f86a9d-5fcd-48e6-bd9c-624911c03d14	Administrator	ADMINISTRATOR	NULL

Figure 8.9 – Roles in the database

6. In the `AccountController` class, we have a `Register()` method that is used to register a new user. Let's update the `Register()` method to assign the `User` role to the new user. The updated code is as follows:

```
// Omitted for brevity
// Try to save the user
var userResult = await userManager.CreateAsync(user, model.
Password);
// Add the user to the "User" role
var roleResult = await userManager.AddToRoleAsync(user,
AppRoles.User);

// If the user is successfully created, return Ok
if (userResult.Succeeded && roleResult.Succeeded)
{
    var token = GenerateToken(model.UserName);
    return Ok(new { token });
}
```

Similarly, we can create a new action to register an administrator or a VIP user. You can check the code in the completed app.

7. You can register a new administrator using any HTTP client you like. After the users are created, you can view the users and their roles in the database, as shown in *Figure 8.10*:

	Id	UserName	NormalizedUserName	Email	NormalizedEmail	EmailConfirmed	PasswordHash
1	07c98d50-88fd-44f7-8719-590ff92e1f28	admin	ADMIN	admin@example.com	ADMIN@EXAMPLE.COM	0	AQAAAAIAAYagAAAAEEwClbtV1pVEW7/vT0dMv84Ob6c78iAD/...
2	9cf357be-0da4-4057-9452-1d2e4a3bb91e	user	USER	user@example.com	USER@EXAMPLE.COM	0	AQAAAAIAAYagAAAAEO7ko9kLwW5DncpxJcBtR2EKEI47FXHP...
3	d368507d-fa3f-4e9a-aeaf-d225e2fa625a	vipuser	VIPUSER	vipuser@example.com	VIPUSER@EXAMPLE.COM	0	AQAAAAIAAYagAAAAEOyJyv76Yin2e9/f9EKQuYnWx48p/8mhS...

	UserId	RoleId
1	9cf357be-0da4-4057-9452-1d2e4a3bb91e	5e040dfe-19d4-4c5f-904d-a06a907f8622
2	d368507d-fa3f-4e9a-aeaf-d225e2fa625a	b4db3767-6722-4a2f-8288-e52221597f87
3	07c98d50-88fd-44f7-8719-590ff92e1f28	c3f86a9d-5fcd-48e6-bd9c-624911c03d14

Figure 8.10 – Users and their roles in the database

The data of the `AspNetUserRoles` table is used to store the relationship between users and roles. The `UserId` column is the primary key of the `AspNetUsers` table, while the `RoleId` column is the primary key of the `AspNetRoles` table.

8. Next, we need to update the method that is used to generate the JWT token. When we generate the token, we need to include the roles of the user in the token. We can use the `GetRolesAsync()` method to get the roles and then add them to the claims. The updated code is as follows:

```
var userRoles = await userManager.GetRolesAsync(user);
var claims = new List<Claim>
{
```

```
            new(ClaimTypes.Name, userName)
    };
    claims.AddRange(userRoles.Select(role => new Claim(ClaimTypes.
    Role, role)));
    var tokenDescriptor = new SecurityTokenDescriptor
    {
        Subject = new ClaimsIdentity(claims),
        Expires = DateTime.UtcNow.AddDays(1),
        Issuer = issuer,
        Audience = audience,
        SigningCredentials = new SigningCredentials(signingKey,
    SecurityAlgorithms.HmacSha256Signature)
    };
```

9. Try to run the application and register a new user or log in with an existing user. Copy the token in the response and paste it to the jwt.io website to decode the payload. You will see that the roles are included in the token, as shown here:

```
{
    "unique_name": "admin",
    "role": "Administrator",
    "nbf": 1679815694,
    "exp": 1679902094,
    "iat": 1679815694,
    "iss": "http://localhost:5056",
    "aud": "http://localhost:5056"
}
```

10. Now, let's update the WeatherForecastController class to implement role-based authorization. Add a new action for administrators, as follows:

```
[HttpGet("admin", Name = "GetAdminWeatherForecast")]
[Authorize(Roles = AppRoles.Administrator)]
public IEnumerable<WeatherForecast> GetAdmin()
{
    return Enumerable.Range(1, 20).Select(index => new
WeatherForecast
    {
        Date = DateOnly.FromDateTime(DateTime.Now.
AddDays(index)),
        TemperatureC = Random.Shared.Next(-20, 55),
        Summary = Summaries[Random.Shared.Next(Summaries.
Length)]
    })
    .ToArray();
}
```

The Authorize attribute is used to specify the role that is allowed to access the API. In the preceding code, only authenticated users with the Administrator role can access the API.

Now, you can test the API. If you use the token of a normal user to access the /WeatherForecast/admin endpoint, you will get a 403 Forbidden response.

Generally, the administrator role should have permission to access all resources. But in our current application, the administrator user cannot access the /WeatherForecast endpoint. There are multiple ways to fix this.

The first way is that when we register a new administrator, we can assign the Administrator role to the user and also assign the User role (or any other roles) to the user. This way, the administrator user can access all resources.

We can also update the Authorize attribute to allow multiple roles, like so:

```
[HttpGet(Name = "GetWeatherForecast")]
[Authorize(Roles = $"{AppRoles.User},{AppRoles.VipUser},{AppRoles.
Administrator}")]
public IEnumerable<WeatherForecast> Get()
{
    // Omitted for brevity
}
```

The preceding code means that the user must have at least one of the specified roles to access the API.

Note that if you apply multiple Authorize attributes with specified roles to an action, the user must have all the roles to access the API. For example, consider the following code:

```
[HttpGet("vip", Name = "GetVipWeatherForecast")]
[Authorize(Roles = AppRoles.User)]
[Authorize(Roles = AppRoles.VipUser)]
public IEnumerable<WeatherForecast> GetVip()
{
    // Omitted for brevity
}
```

The preceding code states that the user must have both the User and VipUser roles to access the API. If the user has only one of the roles, the user will get a 403 Forbidden response.

Besides this, we can also define a policy to specify the roles that are allowed to access the API. For example, in the Program class, we can add the following code:

```
builder.Services.AddAuthorization(options =>
{
    options.AddPolicy("RequireAdministratorRole", policy => policy.
RequireRole(AppRoles.Administrator));
```

```
    options.AddPolicy("RequireVipUserRole", policy => policy.
RequireRole(AppRoles.VipUser));
    options.AddPolicy("RequireUserRole", policy => policy.
RequireRole(AppRoles.User));
    options.AddPolicy("RequireUserRoleOrVipUserRole", policy =>
policy.RequireRole(AppRoles.User, AppRoles.VipUser));
});
```

Then, we can update the `Authorize` attribute to use the policy like this:

```
[HttpGet("admin-with-policy", Name =
"GetAdminWeatherForecastWithPolicy")]
[Authorize(Policy = "RequireAdministratorRole")]
public IEnumerable<WeatherForecast> GetAdminWithPolicy()
{
    // Omitted for brevity
}
```

If the `policy.RequireRole()` method has multiple roles in parameters, the user must have at least one of the roles to access the API. You can check the code in the completed app.

With that, we've implemented role-based authorization in ASP.NET Core. In the next section, we will learn how to implement claim-based authorization.

Claim-based authorization

When a user is authenticated, the user will have a set of claims that are used to store the information about the user. For example, the user can have a claim that specifies the user's role. So, technically, roles are also claims, but they are special claims that are used to store the roles of the user. We can store other information in the claims, such as the user's name, email address, date of birth, driving license number, and more. Once we've done this, the authorization system can check the claims to determine whether the user is allowed to access the resource. Claim-based authorization provides more granular access control than role-based authorization, but it can be more complex to implement and manage.

You can find the starter app and the completed app in the `chapter8/AuthorizationDemo/ClaimBasedAuthorizationDemo` folder in this book's GitHub repository:

1. We'll start with the starter app. Don't forget to create the database and run the migrations using the following commands:

    ```
    dotnet ef database update
    ```

    A claim is a key-value pair. Note that the claim type and value are case-sensitive. You can store the claims in the database or the code. When the user logs in, the claims can be retrieved from the database and added to the token. In this demo, we will hard-code the claims in the code for simplicity.

2. ASP.NET Core provides a built-in `ClaimTypes` class that contains the common claim types, such as `NameIdentifier`, `DateOfBirth`, `Email`, `Gender`, `GivenName`, `Name`, `PostalCode`, and others, including `Role`. This is why we said that roles are also claims. You can also define your own claim types. For example, we can define the following claim types in the `AppClaimTypes` class:

```
public static class AppClaimTypes
{
    public const string DrivingLicenseNumber =
"DrivingLicenseNumber";
    public const string AccessNumber = "AccessNumber";
}
```

3. Also, create a new `AppAuthorizationPolicies` class to define the authorization policies:

```
public static class AppAuthorizationPolicies
{
    public const string RequireDrivingLicenseNumber =
"RequireDrivingLicenseNumber";
    public const string RequireAccessNumber =
"RequireAccessNumber";
}
```

4. Then, we can add the claims to the token when the user logs in. Update the `GenerateToken` method in the `AccountController` class, as follows:

```
// Omitted for brevity
var tokenDescriptor = new SecurityTokenDescriptor
{
    Subject = new ClaimsIdentity(new[]
    {
        new Claim(ClaimTypes.Name, userName),
        // Suppose the user's information is stored in the
database so that we can retrieve it from the database
        new Claim(ClaimTypes.Country, "New Zealand"),
        // Add our custom claims
        new Claim(AppClaimTypes.AccessNumber, "12345678"),
        new Claim(AppClaimTypes.DrivingLicenseNumber,
"123456789")
    }),
    Expires = DateTime.UtcNow.AddDays(1),
    Issuer = issuer,
    Audience = audience,
    SigningCredentials = new SigningCredentials(signingKey,
SecurityAlgorithms.HmacSha256Signature)
};
// Omitted for brevity
```

We can add any claims to the token. In the preceding code, we added the `Country`, `AccessNumber`, and `DrivingLicenseNumber` claims to the token.

5. Imagine that we have a requirement that only users who have their driving licenses can access the resource. We can implement this by adding the following code to the `Program` class:

```
builder.Services.AddAuthorization(options =>
{
    options.AddPolicy(AppAuthorizationPolicies.
RequireDrivingLicense, policy => policy.
RequireClaim(AppClaimTypes.DrivingLicenseNumber));
    options.AddPolicy(AppAuthorizationPolicies.
RequireAccessNumber, policy => policy.
RequireClaim(AppClaimTypes.AccessNumber));
});
```

So, the difference between role-based authorization and claim-based authorization is that claim-based authorization uses `policy.RequireClaim()` to check the claims, while role-based authorization uses `policy.RequireRole()` to check the roles.

6. At this point, we can update the `Authorize` attribute so that it uses the policy, like this:

```
[Authorize(Policy = AppAuthorizationPolicies.
RequireDrivingLicense)]
[HttpGet("driving-license")]
public IActionResult GetDrivingLicense()
{
    var drivingLicenseNumber = User.Claims.FirstOrDefault(c =>
c.Type == AppClaimTypes.DrivingLicenseNumber)?.Value;
    return Ok(new { drivingLicenseNumber });
}
```

7. Run the application and test the `/WeatherForecast/driving-license` endpoint. You will get a `401` Unauthorized response because the user does not have the `DrivingLicenseNumber` claim. Register a user or log in to get the token. Then, add the token to the `Authorization` header and call the `/WeatherForecast/driving-license` endpoint again. You will get a 200 OK response with `drivingLicenseNumber` in the response body.

The token now contains the claims, as shown in the following JSON response:

```
{
  "unique_name": "user",
  "http://schemas.xmlsoap.org/ws/2005/05/identity/claims/
country": "New Zealand",
  "AccessNumber": "12345678",
  "DrivingLicenseNumber": "123456789",
  "nbf": 1679824749,
  "exp": 1679911149,
```

```
    "iat": 1679824749,
    "iss": "http://localhost:5056",
    "aud": "http://localhost:5056"
}
```

8. This is the simplest way to implement claim-based authorization. The current approach only checks whether the token contains the claim; it does not check the value of the claim. We can check the values as well. The `RequireClaim()` method also has an overload that accepts `allowedValues` as a parameter. For example, we have a resource that can only be accessed by users based in New Zealand. We can update the `Program` class as follows:

```
builder.Services.AddAuthorization(options =>
{
    // Omitted for brevity
    options.AddPolicy(AppAuthorizationPolicies.RequireCountry,
policy => policy.RequireClaim(ClaimTypes.Country, "New
Zealand"));
});
```

The `allowedValues` parameter is an array of strings. So, we can pass multiple values to the `allowedValues` parameter. For example, we can allow users from New Zealand and Australia to access the resource:

```
options.AddPolicy(AppAuthorizationPolicies.RequireCountry,
policy => policy.RequireClaim(ClaimTypes.Country, "New Zealand",
"Australia"));
```

The action in the controller looks like this:

```
[Authorize(Policy = AppAuthorizationPolicies.RequireCountry)]
[HttpGet("country")]
public IActionResult GetCountry()
{
    var country = User.Claims.FirstOrDefault(c => c.Type ==
ClaimTypes.Country)?.Value;
    return Ok(new { country });
}
```

You can test the API by calling the `/WeatherForecast/country` endpoint. Now, only users who have the `Country` claim with the value of `New Zealand` can access the resource.

Similar to roles, we can apply multiple policies to a resource. For example, we can require the user to have both the `DrivingLicense` and `AccessNumber` claims to access the resource. Just like roles, you can add two policies to the `Authorize` attribute, which means that the user must have both the `DrivingLicense` and `AccessNumber` claims to access the resource. Here's an example:

```
[Authorize(Policy = AppAuthorizationPolicies.RequireDrivingLicense)]
[Authorize(Policy = AppAuthorizationPolicies.RequireAccessNumber)]
```

```
[HttpGet("driving-license-and-access-number")]
public IActionResult GetDrivingLicenseAndAccessNumber()
{
    var drivingLicenseNumber = User.Claims.FirstOrDefault(c => c.Type
== AppClaimTypes.DrivingLicenseNumber)?.Value;
    var accessNumber = User.Claims.FirstOrDefault(c => c.Type ==
AppClaimTypes.AccessNumber)?.Value;
    return Ok(new { drivingLicenseNumber, accessNumber });
}
```

Another way is to use the `RequireAssertion()` method, which allows us to execute a custom logic to check the claims. Update the `Program` class, as follows:

```
builder.Services.AddAuthorization(options =>
{
    // Omitted for brevity
    options.AddPolicy(AppAuthorizationPolicies.
RequireDrivingLicenseAndAccessNumber, policy => policy.
RequireAssertion(context =>
    {
        var hasDrivingLicenseNumber = context.User.HasClaim(c =>
c.Type == AppClaimTypes.DrivingLicenseNumber);
        var hasAccessNumber = context.User.HasClaim(c => c.Type ==
AppClaimTypes.AccessNumber);
        return hasDrivingLicenseNumber && hasAccessNumber;
    }));
});
```

In the preceding code, the `context` parameter contains the `User` property that contains the claims. We can use the `HasClaim()` method to check whether the user has the claim. Then, we can return `true` if the user has both the `DrivingLicenseNumber` and `AccessNumber` claims; otherwise, we return `false`. You can also use the `context.User.Claims` property to get the claims and check the values per your requirement.

The action in the controller looks like this:

```
[Authorize(Policy = AppAuthorizationPolicies.
RequireDrivingLicenseAndAccessNumber)]
[HttpGet("driving-license-and-access-number")]
public IActionResult GetDrivingLicenseAndAccessNumber()
{
    // Omitted for brevity
}
```

In this section, we learned how to implement claim-based authorization in ASP.NET Core. We also learned how to use the `RequireAssertion()` method to check the claims. If we need a more complex authorization logic, we can use policy-based authorization. But first, let's learn how authorization works in ASP.NET Core.

Understanding the authorization process

In the previous section, we learned how to implement role-based authorization and claim-based authorization. Let's delve deeper into the details. You may have noticed that when we use role-based authorization or claim-based authorization, we need to call the `AddPolicy()` method in the `AddAuthorization` method. The signature of the `AddPolicy()` method is as follows:

```
public void AddPolicy(string name, Action<AuthorizationPolicyBuilder>
configurePolicy)
{
    // Omitted for brevity
}
```

The `AddPolicy()` method accepts two parameters:

- A name parameter, which is the name of the policy
- A `configurePolicy` parameter, which is a delegate that accepts an `AuthorizationPolicyBuilder` parameter

You can press *F12* to check the source code of the `AuthorizationPolicyBuilder` class. You will find that it has some methods to configure the policy, such as `RequireRole()`, `RequireClaim()`, and others. The source code of the `RequireRole` method looks like this:

```
public AuthorizationPolicyBuilder RequireRole(IEnumerable<string>
roles)
{
    ArgumentNullThrowHelper.ThrowIfNull(roles);
    Requirements.Add(new RolesAuthorizationRequirement(roles));
    return this;
}
```

The source code of the `RequireClaim()` method is shown here:

```
public AuthorizationPolicyBuilder RequireClaim(string claimType)
{
    ArgumentNullThrowHelper.ThrowIfNull(claimType);
    Requirements.Add(new ClaimsAuthorizationRequirement(claimType,
allowedValues: null));
    return this;
}
```

Both the `RequireRole()` and `RequireClaim()` methods call the `Requirements.Add()` method under the hood. So, what is the `Requirements` object?

We are getting closer to the core of authorization in ASP.NET Core. The definition of the `Requirements` object is as follows:

```
public IList<IAuthorizationRequirement> Requirements { get; set; } =
new List<IAuthorizationRequirement>();
```

The `Requirements` object in the `AuthorizationPolicyBuilder` class is a list of `IAuthorizationRequirement` objects. The `IAuthorizationRequirement` interface is just a marker service, and it does not have any methods. Let's press *F12* on the `RolesAuthorizationRequirement` class and the `ClaimsAuthorizationRequirement` class. We will see their source code:

```
// RolesAuthorizationRequirement
public class RolesAuthorizationRequirement :
AuthorizationHandler<RolesAuthorizationRequirement>,
IAuthorizationRequirement
{
    // Omitted for brevity
}


// ClaimsAuthorizationRequirement
public class ClaimsAuthorizationRequirement :
AuthorizationHandler<ClaimsAuthorizationRequirement>,
IAuthorizationRequirement
{
    // Omitted for brevity
}
```

As we can see, both the `RolesAuthorizationRequirement` and `ClaimsAuthorizationRequirement` classes implement the `IAuthorizationRequirement` interface. They also implement the `AuthorizationHandler<TRequirement>` class, which is defined as follows:

```
public abstract class AuthorizationHandler<TRequirement> :
IAuthorizationHandler
        where TRequirement : IAuthorizationRequirement
{
    /// <summary>
    /// Makes a decision if authorization is allowed.
    /// </summary>
    /// <param name="context">The authorization context.</param>
    public virtual async Task HandleAsync(AuthorizationHandlerContext
context)
    {
```

```
        foreach (var req in context.Requirements.
OfType<TRequirement>())
        {
            await HandleRequirementAsync(context, req).
ConfigureAwait(false);
        }
    }


    /// <summary>
    /// Makes a decision if authorization is allowed based on a
specific requirement.
    /// </summary>
    /// <param name="context">The authorization context.</param>
    /// <param name="requirement">The requirement to evaluate.</param>
    protected abstract Task
HandleRequirementAsync(AuthorizationHandlerContext context,
TRequirement requirement);
}
```

So, each implementation of the `AuthorizationHandler<TRequirement>` class implements
the `HandleRequirementAsync()` method to check the requirements. For example, the
`RolesAuthorizationRequirement` class consists of the following code:

```
public RolesAuthorizationRequirement(IEnumerable<string> allowedRoles)
{
    ArgumentNullThrowHelper.ThrowIfNull(allowedRoles);

    if (!allowedRoles.Any())
    {
        throw new InvalidOperationException(Resources.Exception_
RoleRequirementEmpty);
    }
    AllowedRoles = allowedRoles;
}

/// <summary>
/// Gets the collection of allowed roles.
/// </summary>
public IEnumerable<string> AllowedRoles { get; }

/// <summary>
/// Makes a decision if authorization is allowed based on a specific
requirement.
/// </summary>
/// <param name="context">The authorization context.</param>
```

```
/// <param name="requirement">The requirement to evaluate.</param>
protected override Task
HandleRequirementAsync(AuthorizationHandlerContext context,
RolesAuthorizationRequirement requirement)
{
    if (context.User != null)
    {
        var found = false;

        foreach (var role in requirement.AllowedRoles)
        {
            if (context.User.IsInRole(role))
            {
                found = true;
                break;
            }
        }

        if (found)
        {
            context.Succeed(requirement);
        }
    }
    return Task.CompletedTask;
}
```

When a RolesAuthorizationRequirement instance is instantiated, it accepts a collection of roles from the constructor. Then, it uses the HandleRequirementAsync() method to check whether the user is in the role. If the user is in the role, it calls the context.Succeed() method to set the Succeeded property to true. Otherwise, it sets the Succeeded property to false.

If you check the implementation of the ClaimsAuthorizationRequirement class, you will find it is similar to the RolesAuthorizationRequirement class. It accepts claimType and a set of allowValues and checks whether the user has the claim, and whether the claim value is in the allowValues set.

The next question is – who is responsible for calling these methods?

Let's go back to the Program class to understand the middleware pipeline. We have the app.UseAuthorization() method in the Program file, which is used to add the authorization middleware. Press *F12* on the UseAuthorization method. We'll be able to view its source code:

```
public static IApplicationBuilder UseAuthorization(this
IApplicationBuilder app)
{
```

```
    // Omitted for brevity
    return app.UseMiddleware<AuthorizationMiddleware>();
}
```

Continue to press *F12* to check the source code of `AuthorizationMiddleware`. You will see the following code in the `Invoke()` method:

```
// Omitted for brevity
var authorizeData = endpoint?.Metadata.
GetOrderedMetadata<IAuthorizeData>() ?? Array.Empty<IAuthorizeData>();
var policies = endpoint?.Metadata.
GetOrderedMetadata<AuthorizationPolicy>() ?? Array.
Empty<AuthorizationPolicy>();
// Omitted for brevity
var policyEvaluator = context.RequestServices.
GetRequiredService<IPolicyEvaluator>();
var authenticateResult = await policyEvaluator.
AuthenticateAsync(policy, context);
// Omitted for brevity
var authorizeResult = await policyEvaluator.AuthorizeAsync(policy,
authenticateResult!, context, resource);
// Omitted for brevity
```

Now, we are closer. The `AuthorizationMiddleware` class gets the policies from the endpoint metadata and then calls the `IPolicyEvaluator.AuthenticateAsync()` method to check whether the user is authenticated, after which it calls the `IPolicyEvaluator.AuthorizeAsync()` method to check whether the user is authorized. The `IPolicyEvaluator` interface is defined as follows:

```
public interface IPolicyEvaluator
{
    Task<AuthenticateResult> AuthenticateAsync(AuthorizationPolicy
policy, HttpContext context);
    Task<PolicyAuthorizationResult> AuthorizeAsync(AuthorizationPolicy
policy, AuthenticateResult authenticationResult, HttpContext context,
object? resource);
}
```

The default implementation of `IPolicyEvaluator` has been injected into the DI container by the ASP.NET Core framework. You can find the source code of the `PolicyEvaluator` class here: `https://source.dot.net/#Microsoft.AspNetCore.Authorization.Policy/PolicyEvaluator.cs`. You will see it has an `IAuthorizationService` object injected into it, which is defined as follows:

```
public interface IAuthorizationService
{
    Task<AuthorizationResult> AuthorizeAsync(ClaimsPrincipal
```

```
user, object? resource, IEnumerable<IAuthorizationRequirement>
requirements);
    Task<AuthorizationResult> AuthorizeAsync(ClaimsPrincipal user,
object? resource, string policyName);
}
```

With that, we've found the IAuthorizationRequirement class we described earlier!

You can find the source code of the default implementation of IAuthorizationService here: https://source.dot.net/#Microsoft.AspNetCore.Authorization/ DefaultAuthorizationService.cs. It is also injected into the DI container by the framework. The core code is as follows:

```
public virtual async Task<AuthorizationResult>
AuthorizeAsync(ClaimsPrincipal user, object? resource,
IEnumerable<IAuthorizationRequirement> requirements)
{
    ArgumentNullThrowHelper.ThrowIfNull(requirements);

    var authContext = _contextFactory.CreateContext(requirements,
user, resource);
    var handlers = await _handlers.GetHandlersAsync(authContext).
ConfigureAwait(false);
    foreach (var handler in handlers)
    {
        await handler.HandleAsync(authContext).ConfigureAwait(false);
        if (!_options.InvokeHandlersAfterFailure && authContext.
HasFailed)
        {
            break;
        }
    }

    var result = _evaluator.Evaluate(authContext);
    if (result.Succeeded)
    {
        _logger.UserAuthorizationSucceeded();
    }
    else
    {
        _logger.UserAuthorizationFailed(result.Failure);
    }
    return result;
}
```

So, we end up with the following call stack:

1. Define the authorization policy (requirement) in the `Program` class.

2. Apply the authorization policy to the endpoint.

3. Apply the authorization middleware to the pipeline.

4. The request comes in with the `Authorization` header, which can be retrieved from the `HttpContext` object.

5. `AuthorizationMiddleware` calls the `IPolicyEvaluator.AuthorizeAsync()` method.

6. The `IPolicyEvaluator.AuthorizeAsync()` method calls the `IAuthorizationService.AuthorizeAsync()` method.

7. The `IAuthorizationService.AuthorizeAsync()` method calls the `IAuthorizationHandler.HandleAsync()` method to check whether the user is authorized.

Once we understand the call stack, we can easily implement an authorization policy by implementing the `IAuthorizationRequirement`, `IAuthorizationHandler`, and `IAuthorizationService` interfaces.

Policy-based authorization

In the previous section, we explained that both role-based authorization and claim-based authorization are implemented by the `IAuthorizationRequirement`, `IAuthorizationHandler`, and `IAuthorizationService` interfaces under the hood. If we have more complex authorization logic, we can use policy-based authorization directly, which allows us to define custom authorization policies to execute complex authorization logic.

For example, we have a scenario where we need to support the following authorization logic:

*   The special premium content can be accessed by the user who has a `Premium` subscription and is also based in New Zealand

*   Users who have a `Premium` subscription, but are not based in New Zealand, cannot access the special premium content

There may be other complex authorization logic in the real world. Let's implement the aforementioned authorization logic using policy-based authorization. You can find the sample code in the `/samples/chapter8/AuthorizationDemo/PolicyBasedAuthorization` folder:

1. First, add two classes to the `Authentication` folder, as follows:

    ```
    public static class AppClaimTypes
    {
        public const string Subscription = "Subscription";
    ```

```
    }

    public static class AppAuthorizationPolicies
    {
        public const string SpecialPremiumContent =
    "SpecialPremiumContent";
    }
```

These classes define the claim types and authorization policies we need. You can also use strings directly in the code, but it is recommended to use constants to avoid typos.

2. In the `AccountController` class, update the `GenerateToken()` method with a new claim, as follows:

```
    private string? GenerateToken(string userName, string country)
    {
        // Omitted for brevity
        var tokenDescriptor = new SecurityTokenDescriptor
        {
            Subject = new ClaimsIdentity(new[]
            {
                new Claim(ClaimTypes.Name, userName),
                new Claim(AppClaimTypes.Subscription, "Premium"),
                new Claim(ClaimTypes.Country, country)
            }),
            Expires = DateTime.UtcNow.AddDays(1),
            Issuer = issuer,
            Audience = audience,
            SigningCredentials = new SigningCredentials(signingKey,
    SecurityAlgorithms.HmacSha256Signature)
        };
        // Omitted for brevity
    }
```

We added a new claim, `AppClaimTypes.Subscription`, with a value of `Premium` to the token. This claim represents the user's subscription type. We also added a new claim, `ClaimTypes.Country`, to the token. This claim represents the user's country. In the real world, you can get the user's subscription type and country from the database. Let's assume we have the subscription type and country information in the token for simplicity.

3. Next, update the `Login()` method in the `AccountController` class to add the country to the claims and create another method for New Zealand users, as follows:

```
    [HttpPost("login-new-zealand")]
    public async Task<IActionResult> LoginNewZealand([FromBody]
    LoginModel model)
    {
```

```
    if (ModelState.IsValid)
    {
        var user = await userManager.FindByNameAsync(model.
UserName);
        if (user != null)
        {
            if (await userManager.CheckPasswordAsync(user,
model.Password))
            {
                var token = GenerateToken(model.UserName, "New
Zealand");
                return Ok(new { token });
            }
        }
        // If the user is not found, display an error message
        ModelState.AddModelError("", "Invalid username or
password");
    }
    return BadRequest(ModelState);
}


[HttpPost("login")]
public async Task<IActionResult> Login([FromBody] LoginModel
model)
{
    if (ModelState.IsValid)
    {
        var user = await userManager.FindByNameAsync(model.
UserName);
        if (user != null)
        {
            if (await userManager.CheckPasswordAsync(user,
model.Password))
            {
                var token = GenerateToken(model.UserName,
"Australia");
                return Ok(new { token });
            }
        }
        // If the user is not found, display an error message
        ModelState.AddModelError("", "Invalid username or
password");
    }
    return BadRequest(ModelState);
}
```

Again, this is a simplified implementation for demonstration purposes. In the real world, generally, there is only one login endpoint, and the country information is retrieved from the database or other sources, such as IP addresses.

4. Next, we need to implement the authorization policy. Create a new class named `SpecialPremiumContentRequirement` in the `Authorization` folder, as follows:

```
public class SpecialPremiumContentRequirement :
IAuthorizationRequirement
{
    public string Country { get; }

    public SpecialPremiumContentRequirement(string country)
    {
        Country = country;
    }
}
```

This class implements the `IAuthorizationRequirement` interface. The `Country` property represents the country where the premium content can be accessed. We can use this property to check whether the user is authorized to access the premium content.

5. Next, we need to implement the `AuthorizationHandler` interface. Create a `SpecialPremiumContentAuthorizationHandler` class in the `Authorization` folder, as follows:

```
public class SpecialPremiumContentAuthorizationHandler :
AuthorizationHandler<SpecialPremiumContentRequirement>
{
    protected override Task
HandleRequirementAsync(AuthorizationHandlerContext context,
SpecialPremiumContentRequirement requirement)
    {
        var hasPremiumSubscriptionClaim = context.User.
HasClaim(c => c.Type == "Subscription" && c.Value == "Premium");

        if (!hasPremiumSubscriptionClaim)
        {
            return Task.CompletedTask;
        }

        var countryClaim = context.User.FindFirst(c => c.Type ==
ClaimTypes.Country);
        if (countryClaim == null || string.
IsNullOrWhiteSpace(countryClaim.ToString()))
        {
            return Task.CompletedTask;
```

```
        }

        if (countryClaim.Value == requirement.Country)
        {
            context.Succeed(requirement);
        }

        return Task.CompletedTask;
    }
}
```

This handler is used to check whether the requirement is satisfied. If the user has a `Premium` subscription and is based in the country where the premium content can be accessed, the requirement is satisfied. Otherwise, the requirement is not satisfied.

6. Next, we need to register the authorization policy and the authorization handler. Update the `Program` class, as follows:

```
builder.Services.AddAuthorization(options =>
{
    options.AddPolicy(AppAuthorizationPolicies.
SpecialPremiumContent, policy =>
    {
        policy.Requirements.Add(new
SpecialPremiumContentRequirement("New Zealand"));
    });
});

builder.Services.AddSingleton<IAuthorizationHandler,
SpecialPremiumContentAuthorizationHandler>();
```

In the preceding code, we registered the authorization policy, `AppAuthorizationPolicies.SpecialPremiumContent`, with the `SpecialPremiumContentRequirement` requirement. The `SpecialPremiumContentRequirement` requirement is satisfied if the user has a `Premium` subscription and is based in New Zealand. We also registered the `SpecialPremiumContentAuthorizationHandler` handler as a singleton service.

7. Finally, we need to apply the authorization policy to the controller. Open the `WeatherForecastController` class and add a new action, as shown in the following code:

```
[Authorize(Policy = AppAuthorizationPolicies.
SpecialPremiumContent)]
[HttpGet("special-premium", Name = "GetPremiumWeatherForecast")]
public IEnumerable<WeatherForecast> GetPremium()
{
    // Omitted for brevity
}
```

This action can only be accessed by users who have a `Premium` subscription and are based in New Zealand. If the user does not have a `Premium` subscription or is not based in New Zealand, the authorization policy will not be satisfied, and the user will not be able to access the action.

You can test the application as we did in the previous section. The application has two login endpoints – one for New Zealand users and one for Australian users. If you log in as a New Zealand user, you can access the `WeatherForecast/special-premium` endpoint. Otherwise, you will get a `403` response.

There are some points to note for policy-based authorization:

- You can use one `AuthorizationHandler` instance to handle multiple requirements. in the `HandleAsync()` method, you can use `AuthorizationHandlerContext.PendingRequirements` to get all the pending requirements and then check them one by one.

- If you have multiple `AuthorizationHandler` instances, they will be invoked in any order, which means you cannot expect the order of the handlers.

- You need to call `context.Succeed(requirement)` to mark the requirement as satisfied.

What if the requirement is not satisfied? There are two options:

- Generally, you do not need to call `context.Fail()` to mark the failed requirement because there may be other handlers to handle the same requirement, which may be satisfied.

- If you want to make sure the requirement fails and indicate that the whole authorization process fails, you can call `context.Fail()` explicitly, and set the `InvokeHandlersAfterFailure` property to `false` in the `AddAuthorization()` method, like this:

```
builder.Services.AddAuthorization(options =>
  {
      options.AddPolicy(AppAuthorizationPolicies.PremiumContent,
policy =>
      {
          policy.Requirements.Add(new
PremiumContentRequirement("New Zealand"));
      });
      options.InvokeHandlersAfterFailure = false;
  });
```

In this section, we explored the three types of authorization available in ASP.NET Core: role-based, claim-based, and policy-based. We examined the source code to gain a deeper understanding of how authorization works. With this knowledge, you should now be able to confidently use the authorization features of ASP.NET Core. Next, we will learn how to manage users and roles.

Managing users and roles

In the previous sections, we implemented the authentication and authorization features. Generally, the application should also provide a way to manage users and roles. ASP.NET Core Identity provides a set of APIs to manage users and roles. In this section, we will introduce how to use these APIs.

Previously, we learned that the `IdentityDbContext` class is used to store the user and role information. So, we do not need to create a new database context class. Similarly, we can use `UserManager` and `RoleManager` to manage users and roles without having to write any code.

Here are some common operations for managing users by using the `UserManager` class:

Method	Description
`CreateAsync(TUser user, string password)`	Creates a user with the given password.
`UpdateUserAsync(TUser user)`	Updates a user.
`FindByNameAsync(string userName)`	Finds a user by name.
`FindByIdAsync(string userId)`	Finds a user by ID.
`FindByEmailAsync(string email)`	Finds a user by email.
`DeleteAsync(TUser user)`	Deletes a user.
`AddToRoleAsync(TUser user, string role)`	Adds the user to a role.
`GetRolesAsync(TUser user)`	Gets a list of roles for the user.
`IsInRoleAsync(TUser user, string role)`	Checks whether the user has a role.
`RemoveFromRoleAsync(TUser user, string role)`	Removes the user from a role.
`CheckPasswordAsync(TUser user, string password)`	Checks whether the password is correct for the user.
`ChangePasswordAsync(TUser user, string currentPassword, string newPassword)`	Changes the user's password. The user must provide the correct current password.
`GeneratePasswordReset-TokenAsync(TUser user)`	Generates a token for resetting the user's password. You need to specify `options.Token.PasswordResetTokenProvider` in the `AddIdentityCore()` method.

Method	Description
`GenerateEmailConfirma-tionTokenAsync(TUser user)`	Generates a token for confirming the user's email. You need to specify `options.Tokens.EmailConfir-mationTokenProvider` in the `AddIdentity-Core()` method.
`ConfirmEmailAsync(TUser user, string token)`	Checks whether the user has a valid email confirmation token. If the token matches, this method will set the `EmailConfirmed` property of the user to `true`.

Table 8.1 – Common operations for managing users

Here are some common operations for managing roles by using the `RoleManager` class:

Method	Description
`CreateAsync(TRole role)`	Creates a role
`RoleExistsAsync(string roleName)`	Checks whether the role exists
`UpdateAsync(TRole role)`	Updates a role
`DeleteAsync(TRole role)`	Deletes a role
`FindByNameAsync(string roleName)`	Finds a role by name

Table 8.2 – Common operations for managing roles

These APIs encapsulate the database operations, so we can use them to manage users and roles easily. Some of the methods return a `Task<IdentityResult>` object. The `IdentityResult` object contains a `Succeeded` property to indicate whether the operation is successful. If the operation is not successful, you can get the error messages by using the `Errors` property.

We will not cover all the APIs in this book. You can find more information in the ASP.NET Core documentation. Next, we will learn about the new built-in Identity API endpoints in ASP.NET Core 8.0.

New Identity API endpoints in ASP.NET Core 8

In the previous sections, we learned how to implement authentication and authorization using the built-in Identity APIs in ASP.NET Core. We developed a couple of endpoints to register, log in, and manage users and roles. ASP.NET Core 8.0 introduces a new set of features to simplify authentication for web APIs. In this section, we will introduce these new endpoints.

Note that this new feature is only for simple authentication scenarios. The token generated by the Identity API endpoints is opaque, not a JWT token, which means it is intended to be used by the same application only. However, it is still a choice for a quick start. In ASP.NET Core 8.0, we can use a new `MapIdentityApi()` method to map the Identity API endpoints without writing any implementation as we did in the previous sections. Let's learn how to use it:

1. First, follow *steps 1* to *5* in the *Creating a sample project with authentication and authorization* section to create a new web API project named `NewIdentityApiDemo`. Note that you do not need to install the `Microsoft.AspNetCore.Authentication.JwtBearer` package because we will not use JWT tokens in this sample project.

2. Add the authorization policy service and register `DbContext` in the `Program.cs` file, as follows:

   ```
   builder.Services.AddAuthorization();
   builder.Services.AddDbContext<AppDbContext>();
   ```

3. Run the following commands to create the database and the migration:

   ```
   dotnet ef migrations add InitialDb
   dotnet ef database update
   ```

4. Register the Identity API endpoints in the `Program.cs` file, as follows:

   ```
   builder.Services.AddIdentityApiEndpoints<AppUser>().
   AddEntityFrameworkStores<AppDbContext>();
   ```

   The `AddIdentityApiEndpoints()` method adds a set of common identity services to the application by calling the `AddIdentityCore<TUser>()` method under the hood. It also configures authentication to support identity bearer tokens and cookies, so we do not need to explicitly call the `AddIdentityCore<AppUser>()` method.

5. Map the Identity API endpoints in the `Program.cs` file, as follows:

   ```
   app.MapGroup("/identity").MapIdentityApi<AppUser>();
   ```

   The preceding code maps the Identity API endpoints to the `/identity` path. You can change it to any path you like, such as `api/accounts`, `/users`, and so on. Note that as we use an `AppUser` instead of the default `IdentityUser`, we must specify the `AppUser` type in the `MapIdentityApi()` method.

6. Apply the `[Authorize]` attribute to the `WeatherForecastController` class, as follows:

   ```
   [Authorize]
    [ApiController]
    [Route("[controller]")]
    public class WeatherForecastController : ControllerBase
   ```

```
{
    // Omitted for brevity
}
```

7. Run the application using `dotnet run`. You will see the new Identity API endpoints in the Swagger UI:

NewIdentityApiDemo   ︿

| POST | /identity/register | ⌄ |

| POST | /identity/login | ⌄ |

| POST | /identity/refresh | ⌄ |

| GET | /identity/confirmEmail | ⌄ |

| POST | /identity/resendConfirmationEmail | ⌄ |

| POST | /identity/forgotPassword | ⌄ |

| POST | /identity/resetPassword | ⌄ |

| POST | /identity/manage/2fa | ⌄ |

| GET | /identity/manage/info | ⌄ |

| POST | /identity/manage/info | ⌄ |

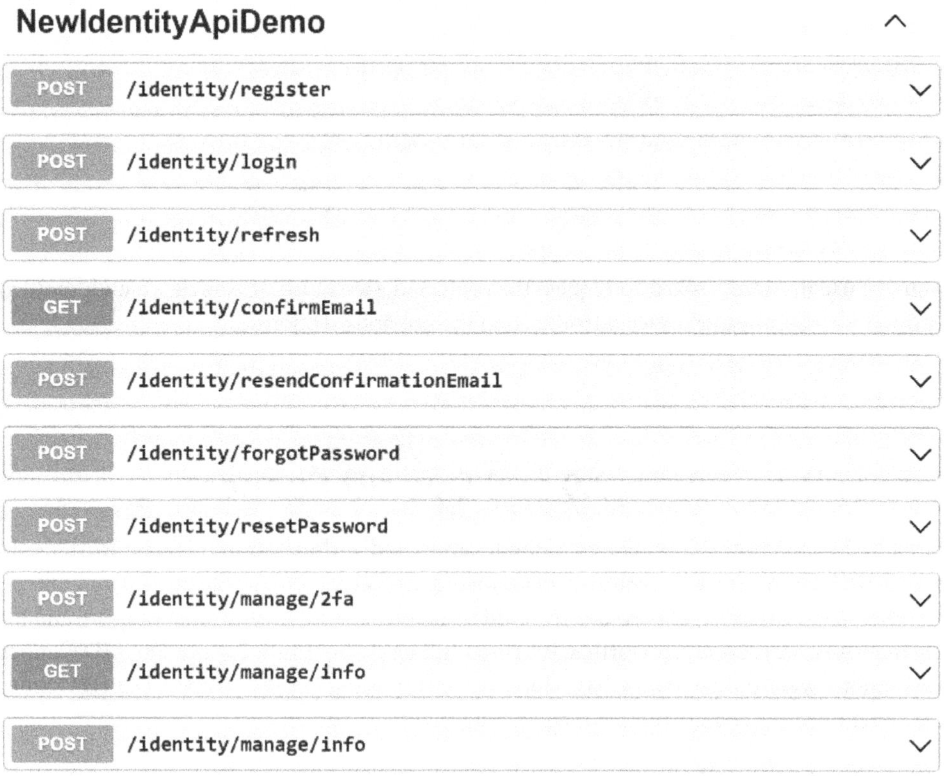

Figure 8.11 – Identity API endpoints in the Swagger UI

8. Now, you can explore the new Identity API endpoints. Send a POST request with the following body to the `/identity/register` endpoint to register a new user:

```
{
    "userName": "admin",
    "email": "admin@example.com",
    "password": "Passw0rd!"
}
```

You will see the request returns a 200 response. Then, send a POST request with the following body to the /identity/login endpoint to log in:

```
{
    "email": "admin@example.com",
    "password": "Passw0rd!"
}
```

You will get a response that contains the access token and refresh token:

```
{
    "tokenType": "Bearer",
    "accessToken": "CfDJ8L-NUxrCjhBJqmxaYaETqK0P0...",
    "expiresIn": 3600,
    "refreshToken": "CfDJ8L-NUxrCjhBJqmxaYaETqK2U..."
}
```

Then, you can use the access token to request the protected /weatherforecast endpoint with the Authorization header, as we introduced in the previous sections.

This new feature also provides endpoints such as refreshToken, confirmEmail, resetPassword, 2fa, and others. Feel free to explore them on your own.

Understanding OAuth 2.0 and OpenID Connect

Previously, we learned how to implement authentication and authorization using built-in Identity APIs in ASP.NET Core. However, you may encounter some terms such as OAuth 2.0 and OpenID Connect when you work on a real project. It would be helpful to understand what they are and how to use them in ASP.NET Core. It is worth authoring a full book on OAuth 2.0 and OpenID Connect. In this section, we will introduce some basic concepts surrounding OAuth 2.0 and OpenID Connect, as well as some third-party authentication and authorization providers.

What is OAuth 2.0?

Let's start with a real example. When you use LinkedIn, you may see a window that prompts you to sync your contacts from Outlook, Gmail, Yahoo, or other email services. This is because LinkedIn would like to know your contacts so that it can recommend you to invite your friends to join LinkedIn or to connect with them. This is a typical example where OAuth 2.0 is used:

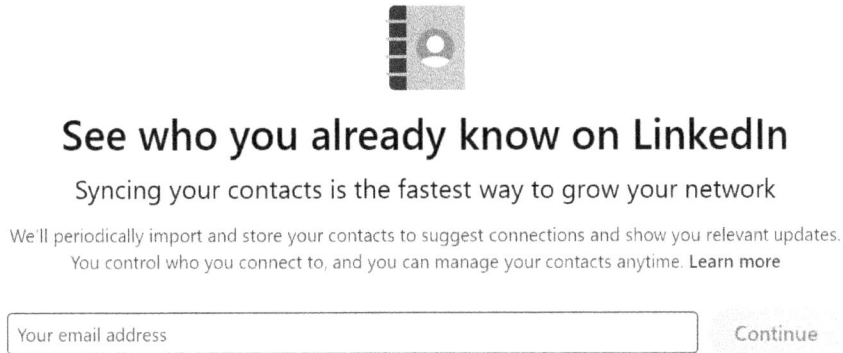

See who you already know on LinkedIn

Syncing your contacts is the fastest way to grow your network

We'll periodically import and store your contacts to suggest connections and show you relevant updates.
You control who you connect to, and you can manage your contacts anytime. **Learn more**

| Your email address | Continue |

Figure 8.12 – Syncing contacts on LinkedIn

If you fill in your email address and click the **Continue** button, you will be redirected to the email service provider's website. For example, I use Outlook, so I will see a window like this because I have multiple accounts:

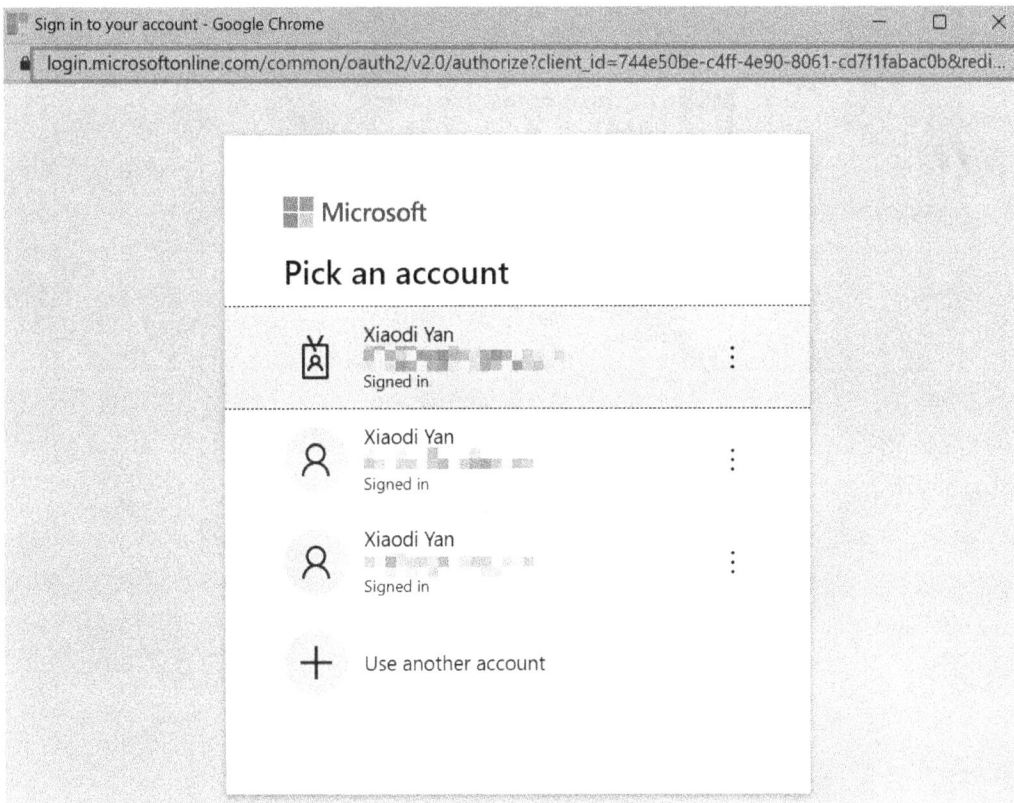

Figure 8.13 – Prompting to log into Outlook

Note the URL in the address bar. It will look something like this:

```
https://login.microsoftonline.com/common/oauth2/
v2.0/authorize?client_id=xxxxxxxx-xxxx-xxxx-xxxx-
xxxxxxxxxxxx&redirect_uri=https%3A%2F%2Fwww.linkedin.
com%2Fgenie%2Ffinishauth&scope=openid%20email%20People.Read&response_
type=code&state=xxxxxxxx-xxxx-xxxx-xxxx-xxxxxxxxxxxx
```

The URL contains the **client ID** of the application, which is used to identify the application. It also contains the **redirect URL** so that the authorization server can redirect the user back to the application after the user grants permission.

You need to log into the email service provider's website and authorize LinkedIn to access your contacts. If you have already logged in, you will see this window:

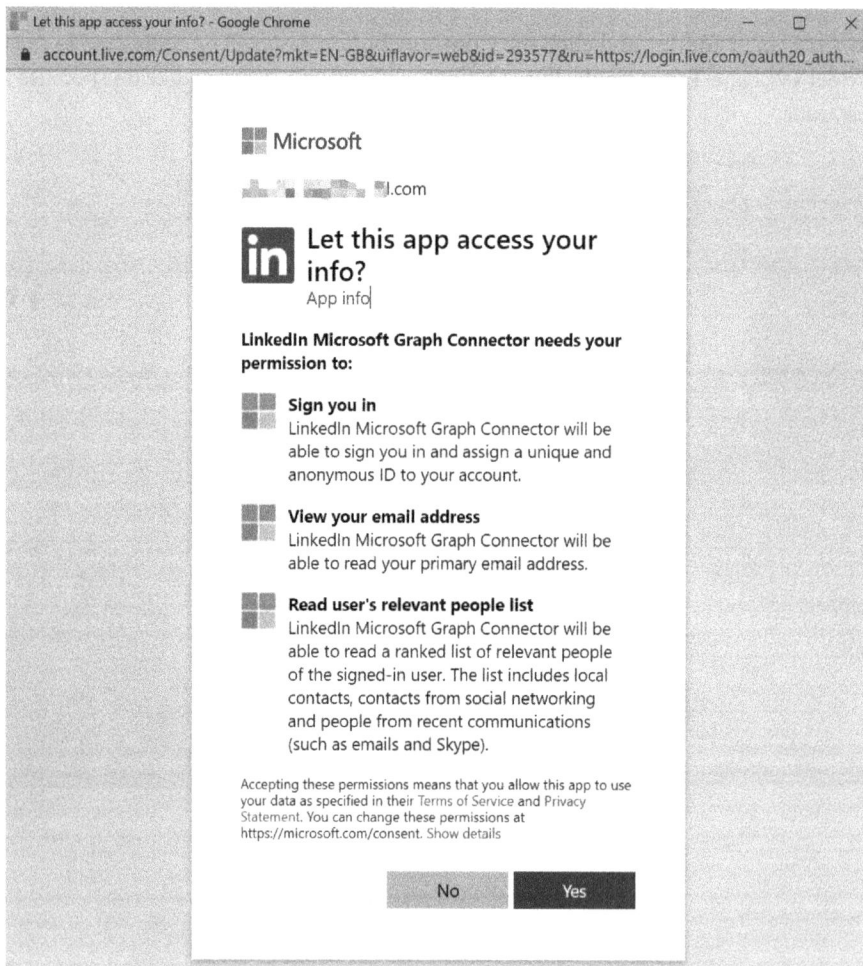

Figure 8.14 – Authorizing LinkedIn to access your contacts

After you authorize LinkedIn, you will be redirected back to LinkedIn. LinkedIn will get the contacts from the email service provider and show them to you.

We do not want to allow LinkedIn to know the password of our email address. In this case, OAuth 2.0 and OpenID Connect are used to authorize LinkedIn to access our contacts without knowing our password.

OAuth 2.0 implements a **delegated authorization** model. It allows a client to access a protected resource on behalf of a user. There are some entities involved in the OAuth 2.0 model:

- **Resource owner**: The user who owns the protected resource. In our example, the resource owner is the user who owns the email address.
- **Client**: The client application that wants to access the protected resource. In our example, the client is LinkedIn. Note that this client is not the user's browser.
- **Resource server**: The server that hosts the protected resource. In our example, the resource server is the email service provider – for example, Outlook.
- **Authorization server**: The server that handles the delegated authorization. In our example, the authorization server is Microsoft Identity Platform. An **Authorization** server has at least two endpoints:
  - The **authorization endpoint** is used to interact with the end user and obtain an authorization grant
  - The **token endpoint** is used with the client to exchange an authorization grant for an access token:

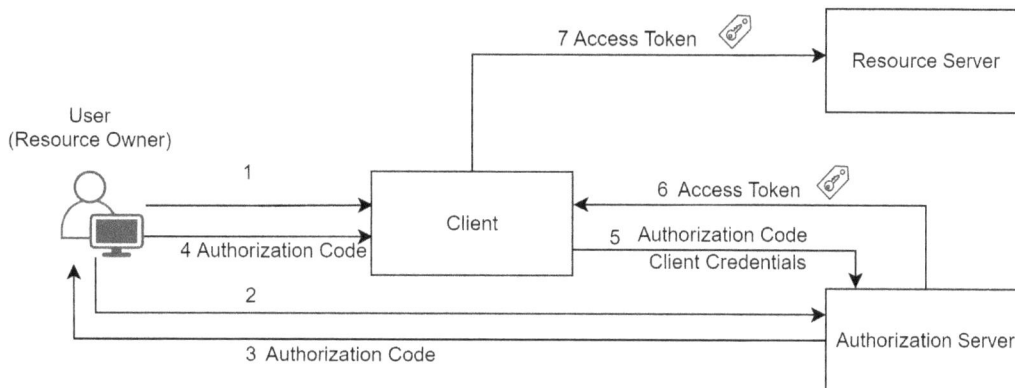

Figure 8.15 – OAuth 2.0 flow

Note that the client (LinkedIn) must register itself as a known client to the authorization server (Microsoft) before it can access the protected resource. The client must provide a **client ID** and a **client secret** to the authorization server to prove its identity. That is why we can see LinkedIn's Microsoft Graph Connector in *Figure 8.14*.

The common steps of OAuth 2.0 are as follows:

1. The client requests access to a protected resource.

2. The client redirects the user to the authorization server, such as Microsoft, Google, and so on. Specifically, it redirects to the authorization endpoint of the authorization server. After the user is authenticated, the authorization server will prompt the user, asking something like "Hi, I have a **known** client named LinkedIn, which wants to access my APIs using your privileges. Specifically, it wants to access your contacts so that it can send emails on your behalf. Do you want to grant access to LinkedIn?" This is what *Figure 8.14* shows.

3. Once the user accepts the request, the authorization server will generate an **authorization code**, which is just an opaque string that confirms the user did grant access to the client (LinkedIn). The authorization server will redirect the user back to the client (LinkedIn).

4. The authorization code is sent to the client (LinkedIn) as a query string parameter.

5. The client (LinkedIn) now has an authorization code. Next, it will use the authorization code, client ID, and client secret to request an **access token** from the token endpoint of the authorization server. It may ask something like "Hi, I am LinkedIn. This user has granted me access to the contacts of this email address. This is my client credentials (client id and client secret). I also have an authorization code. Can I get access to this?"

6. The authorization server will verify the client credentials and the authorization code. If all is good, it will generate an access token and send it back to the client (LinkedIn). The access token is a string that can be used to access the protected resource. It is usually a JWT token. It may also contain the **scope**, which is the permission that the client (LinkedIn) has been granted. For example, it may be **Contacts.Read**.

7. The client (LinkedIn) can now use this access token to access the protected resource. It may ask something like "Hi, I am LinkedIn. I have an access token. Can I access the contacts of this email address?" The resource server checks the access token and if it is valid, it will return the protected resource to the client (LinkedIn).

In this way, the client can access the protected resource without knowing the user's password. Because the access token has a scope, it can only access the protected resource within the scope. For example, if the scope is `Contacts.Read`, the client can only read the contacts, but it cannot modify the contacts. This mechanism provides a good balance between security and usability.

What is OpenID Connect?

OAuth was initially designed and released in 2006 and later revised and standardized in 2012 as OAuth 2.0. OAuth 2.0 solves the problem of delegated authorization. However, there are some other scenarios that OAuth 2.0 cannot solve. For example, your API may need to know the identity of the user who is accessing the API, but users may not want to create an account for your API. They may already have an account in some other services, such as Microsoft, Google, and others. In this case, it would be more convenient if the user could use their existing account to access your API. However, OAuth 2.0 was not designed to implement sign-in with an existing account. This is where a new specification named OpenID Connect comes in.

OpenID Connect is an authentication layer on top of OAuth 2.0 that was designed by the OpenID Foundation in 2014. OpenID Connect is like an extension of OAuth 2.0 that adds and defines some new features to retrieve the identity of the user, including profile information such as the user's name, email address, and so on. OpenID Connect uses similar terminology and concepts as OAuth 2.0, such as **client**, **resource owner**, **authorization server**, and others. However, keep in mind that OpenID Connect is not a replacement for OAuth 2.0. Instead, it is a specification that extends OAuth 2.0 to support authentication.

Many popular identity providers, such as Microsoft, Google, Facebook, and others, have implemented OpenID Connect so that you can integrate your API application with their identity provider. Then, users can use their existing account to sign into your API application. Here is an example of how OpenID Connect works on `Medium.com`:

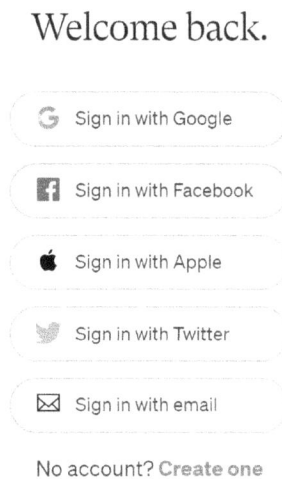

Welcome back.

G   Sign in with Google

f   Sign in with Facebook

   Sign in with Apple

   Sign in with Twitter

✉   Sign in with email

No account? Create one

Figure 8.16 – Medium.com sign-in with multiple identity providers

If you click **Sign in with Google**, you will be redirected to Google to sign in. Then, you will be redirected back to Medium.com so that you can use your existing Google account to sign in to Medium.com. This is what OpenID Connect does.

Similar to OAuth 2.0, OpenID Connect also generates an access token. It also introduces a new token called **ID token**, which is a JWT token that contains the identity of the user. The client application can inspect and validate the ID token to extract identity information about the user.

Integrating with other identity providers

Many identity providers support OpenID Connect so that you can integrate your API application with these platforms. Here are some popular identity providers:

- **Microsoft**: Microsoft provides Microsoft Identity Platform for authentication and authorization. For more information, see `https://learn.microsoft.com/zh-cn/azure/active-directory/develop/`.

- **Google**: Sign in with Google is a service that helps you quickly and easily manage user authentication and share the user's profile information with your application. For more information, see `https://developers.google.com/identity/gsi/web/guides/overview`.

- **Facebook**: Facebook Login is a convenient way for people to log into your application using their Facebook account. For more information, see `https://developers.facebook.com/products/facebook-login/`.

- **Auth0**: Auth0 is a cloud-based identity management platform that provides authentication, authorization, and related security services for web, mobile, and legacy applications. For more information, see `https://auth0.com/docs/quickstart/backend`.

- **Okta**: Okta is also a cloud-based identity platform that allows organizations to manage and secure user authentication and authorization across multiple applications and services. For more information, see `https://www.okta.com/`.

> **Important note**
>
> In March 2021, Okta acquired Auth0. However, the two companies will continue to operate separately. Generally, Auth0 targets smaller companies and is known for its developer-friendly features, but Okta is considered to be more focused on large enterprises and offers more advanced features such as network integration, single sign-on, and more.

If you need to build an identity provider yourself, there are also some open-source projects that you can use:

- `IdentityServer`: `IdentityServer` is one of the most flexible and standards-compliant OpenID Connect and OAuth 2.0 frameworks for ASP.NET Core. It is widely used by many companies to secure their applications and APIs. Note that `IdentityServer` is open-source, but it is not free now. The last free version is IdentityServer4, which was released in 2021, but it is no longer maintained. Duende Software now provides a commercial version of `IdentityServer`. For more information, see `https://duendesoftware.com/products/identityserver`.

- OpenIddict: OpenIddict is an open-source OpenID Connect stack for ASP.NET Core. It provides a versatile solution to implement OpenID Connect client, server, token validation, and more. However, it is not a turnkey solution. You need to write some custom code to implement some business logic, such as an authorization controller, and more. For more information, see https://github.com/openiddict/openiddict-core.

- KeyCloak: KeyCloak is an open-source identity and access management solution. It provides features such as single sign-on, user federation, strong authentication, user management, fine-grained authorization, and more. It is container-based, so it can easily be deployed in a containerized environment. For more information, see https://www.keycloak.org/.

We will not cover the details of how to integrate with these identity providers in this book. Please refer to the documentation instead.

Other security topics

As we mentioned at the beginning of this chapter, security is a very broad topic. In this section, we will briefly introduce some other security topics.

Always use Hypertext Transfer Protocol Secure (HTTPS)

HTTPS is a protocol that provides secure communication between a client and a server. It is a combination of the HTTP and **Secure Sockets Layer/Transport Layer Security** (**SSL/TLS**) protocols. HTTPS is used to encrypt communication between the client and the server, ensuring that sensitive data transmitted over the internet is secure and cannot be intercepted by unauthorized third parties. Google Chrome and other modern browsers will display a warning if you try to access a website that does not use HTTPS. Therefore, it is very important to use HTTPS for all your web applications.

The default ASP.NET Core web API template can use both HTTP and HTTPS. It is recommended to use HTTPS only. So, we need to configure the project to redirect all HTTP requests to HTTPS.

To do that, we need to add the following code to the `Program.cs` file:

```
app.UseHttpsRedirection();
```

This code applies the `UseHttpsRedirection` middleware to redirect HTTP requests to HTTPS.

When you run the application locally, ASP.NET Core will automatically generate a self-signed certificate and use it to encrypt communication. However, when you deploy the application to a production environment, you need to use a certificate issued by a trusted **certificate authority** (**CA**), such as DigiCert, Comodo, GeoTrust, and so on.

Using a strong password policy

The default password policy we implemented in previous sections is not secure enough. Users can use any password, which might be a security risk. It is important to force users to use strong, unique passwords that are difficult for others to guess or crack. Generally, a good password should be a combination of uppercase and lowercase letters, numbers, and special characters. The length of the password should be at least 8 characters. We can define a password policy to enforce these rules.

We can specify the password policy in the `Program` class. Add the following code after the `AddAuthentication()` method:

```
builder.Services.Configure<IdentityOptions>(options =>
{
    // Password settings
    options.Password.RequireDigit = true;
    options.Password.RequireLowercase = true;
    options.Password.RequireNonAlphanumeric = true;
    options.Password.RequireUppercase = true;
    options.Password.RequiredLength = 8;
    options.Password.RequiredUniqueChars = 1;
    // User settings
    options.User.AllowedUserNameCharacters =
    "abcdefghijklmnopqrstuvwxyzABCDEFGHIJKLMNOPQRSTUVWXYZ0123456789-.
_@+";
    options.User.RequireUniqueEmail = true;
});
```

The preceding code is easy to understand. In this example, we require the password to contain at least one uppercase letter, one lowercase letter, one number, and one special character, and the length of the password should be at least 8 characters. We also require the user's email to be unique. So, if a user tries to register with an email that is already in use, the registration will fail. Now, the user's password should be hard to guess.

We can also enforce the password policy when the user fails to log in. For example, if the user fails to log in three times, the account will be locked for 5 minutes. This can help prevent brute-force attacks. To enable this feature, add the following code after the `AddAuthentication()` method:

```
builder.Services.Configure<IdentityOptions>(options =>
{
    // Omitted for brevity
    // Lockout settings
    options.Lockout.DefaultLockoutTimeSpan = TimeSpan.FromMinutes(5);
    options.Lockout.MaxFailedAccessAttempts = 3;
    options.Lockout.AllowedForNewUsers = true;
});
```

This change works when we use `SignInManager.CheckPasswordSignInAsync()` method to sign in. In previous examples, we used `UserManager`. So, we need to update the `Login()` method in the `AuthenticationController` class. First, we need to inject `SignInManager` into the controller. Then, we must update the `AuthenticationController` class, as follows:

```
[HttpPost("login")]
public async Task<IActionResult> Login([FromBody] LoginModel model)
{
    // Check if the model is valid
    if (ModelState.IsValid)
    {
        var user = await _userManager.FindByNameAsync(model.UserName);
        if (user != null)
        {
            var result =
                await _signInManager.CheckPasswordSignInAsync(user,
model.Password, lockoutOnFailure: true);
            if (result.Succeeded)
            {
                var token = GenerateToken(model.UserName);
                return Ok(new { token });
            }
        }
        // If the user is not found, display an error message
        ModelState.AddModelError("", "Invalid username or password");
    }
    return BadRequest(ModelState);
}
```

The preceding code uses the `SignInManager.CheckPasswordSignInAsync()` method to sign in, which has a parameter named `lockoutOnFailure` that specifies whether the account should be locked out when the user fails to log in. The default value is `false`, so we need to use `true` to enable the lockout feature.

Note that if you use `AddIdentityCore<AppUser>()` in `Program.cs`, as we mentioned in the previous section, `SignInManager` is not available by default. In this case, you need to explicitly add the `SignInManager` service to the `ConfigureServices()` method, like this:

```
builder.Services.AddIdentityCore<AppUser>()
    .AddSignInManager()
    .AddEntityFrameworkStores<AppDbContext>()
    .AddDefaultTokenProviders();
```

Let's test the application. Run the application using `dotnet run` and create a new user using the `Register` API. You will find that if the password is too simple, you will get an error message. Here is a sample request:

```
{
  "userName": "user",
  "email": "user@example.com",
  "password": "123456"
}
```

You will get a `400` response with the following error message:

```
{
  "": [
    "Passwords must be at least 8 characters.",
    "Passwords must have at least one non alphanumeric character.",
    "Passwords must have at least one lowercase ('a'-'z').",
    "Passwords must have at least one uppercase ('A'-'Z')."
  ]
}
```

If you attempt to log in with an incorrect password more than three times, you will be locked out of the system for 5 minutes. During this period, you will not be able to access the system, even if you enter the correct password. After those 5 minutes have elapsed, you will be able to log in again.

Implementing two-factor authentication (2FA)

2FA is a security process that requires users to provide two different forms of authentication to verify their identity. Besides the common username and password, 2FA adds an extra layer of security by requiring users to provide a second authentication factor, such as a code sent to their mobile phone or authenticator app, fingerprint, face recognition, and so on. This makes it harder for hackers to gain access to user accounts. Even if the hacker gets the user's password, they still cannot get the second factor. 2FA is widely used in banking and financial services to protect users' sensitive information.

Multi-factor authentication (**MFA**) is a superset of 2FA. It requires users to provide more than two factors to verify their identity. There are two types of MFA:

- **MFA Time-based One-Time Password** (**TOTP**): MFA TOTP is a type of MFA that requires users to provide a code generated by an authenticator app (such as Google Authenticator or Microsoft Authenticator). The code is valid for a short period, usually 30 seconds. After the code expires, the user needs to generate a new code. This type of MFA is widely used in banking and financial services. If you use a bank app, you might have seen this type of MFA. It requires the server and authenticator app to have an accurate time.

- **MFA Fast Identity Online 2** (**FIDO2**): MFA FIDO2 is a type of MFA that requires users to authenticate using a hardware key, such as a USB key or a biometric device (such as a fingerprint scanner). It has become more popular in recent years. However, ASP.NET Core does not support FIDO2 directly yet.

- **MFA SMS**: MFA SMS is no longer recommended because there are many security issues with SMS.

To learn more about MFA, please refer to `https://learn.microsoft.com/en-us/aspnet/core/security/authentication/mfa`.

Implementing rate-limiting

Rate-limiting is a security mechanism that limits the number of requests a client can make to a server. It can prevent malicious clients from making too many requests, which can cause a **denial of service** (**DoS**) attack. ASP.NET Core provides a built-in rate-limiting middleware. We explained how to use it in *Chapter 4*.

Using model validation

Model validation is a security mechanism that prevents malicious users from sending invalid data to the server. We should always validate the data sent by the client. In other words, the client is not trusted. For example, we expect a property in the model to be an integer, but what if the client sends a string? The application should be able to handle this situation and reject the request directly before executing any business logic.

ASP.NET Core provides a built-in model binding and model validation mechanism. Model binding is used to convert the data sent by the client to the corresponding model. The data sent by the client can be in different formats, such as JSON, XML, form fields, or query strings. Model validation is used to check whether the data sent by the client is valid. We used model validation in the previous sections. For example, here is the code we used to register a new user:

```
// Create an action to register a new user
[HttpPost("register")]
public async Task<IActionResult> Register([FromBody]
AddOrUpdateAppUserModel model)
{
    // Check if the model is valid
    if (ModelState.IsValid)
    {
        // Omitted for brevity
    }
    return BadRequest(ModelState);
}
```

The `ModelState.IsValid` property represents whether the model is valid. So, how does ASP.NET Core validate the model? Look at the `AddOrUpdateAppUserModel` class:

```
public class AddOrUpdateAppUserModel
{
    [Required(ErrorMessage = "User name is required")]
    public string UserName { get; set; } = string.Empty;

    [EmailAddress]
    [Required(ErrorMessage = "Email is required")]
    public string Email { get; set; } = string.Empty;

    [Required(ErrorMessage = "Password is required")]
    public string Password { get; set; } = string.Empty;
}
```

We use the validation attributes to specify the validation rules. For example, `Required` is a built-in attribute annotation that specifies that the property is required. Here are some of the most commonly used ones besides `Required`:

- `CreditCard`: This specifies that the property must be a valid credit card number
- `EmailAddress`: This specifies that the property must be a valid email address
- `Phone`: This specifies that the property must be a valid phone number
- `Range`: This specifies that the property must be within a specified range
- `RegularExpression`: This specifies that the property must match a specified regular expression
- `StringLength`: This specifies that the property must be a string with a specified length
- `Url`: This specifies that the property must be a valid URL
- `Compare`: This specifies that the property must be the same as another property

If these built-in attributes cannot meet your requirements, you can also create custom attributes. For example, you can create an `Adult` attribute to validate the age of the user based on the user's birthday:

```
public class AdultAttribute : ValidationAttribute
{
    public string GetErrorMessage() => $"You must be at least 18 years old to register.";
    protected override ValidationResult IsValid(object value,
ValidationContext validationContext)
    {
        var birthDate = (DateTime)value;
        var age = DateTime.Now.Year - birthDate.Year;
```

```
        if (DateTime.Now.Month < birthDate.Month || (DateTime.Now.
Month == birthDate.Month && DateTime.Now.Day < birthDate.Day))
        {
            age--;
        }
        if (age < 18)
        {
            return new ValidationResult(GetErrorMessage());
        }
        return ValidationResult.Success;
    }
}
```

Then, you can use the `Adult` attribute in the model:

```
public class AddOrUpdateAppUserModel
{
    // Omitted for brevity
    [Required(ErrorMessage = "Birthday is required")]
    [Adult]
    public DateTime Birthday { get; set; }
}
```

You can also manually validate the model in the controller. For example, you can check whether the user's email is unique:

```
// Create an action to register a new user
[HttpPost("register")]
public async Task<IActionResult> Register([FromBody]
AddOrUpdateAppUserModel model)
{
    // Check if the email is unique
    if (await _userManager.FindByEmailAsync(model.Email) != null)
    {
        ModelState.AddModelError("Email", "Email already exists");
        return BadRequest(ModelState);
    }
    if (ModelState.IsValid)
    {
        // Omitted for brevity
    }
    return BadRequest(ModelState);
}
```

In the preceding code, we use the `AddModelError()` method to add a validation error to the model. The `ModelState.IsValid` property will return `false` if there is any validation error. In *Chapter 16*, we will discuss how to use the `ProblemDetails` class to return error information to the client and how to use `FluentValidation` to validate the model for more complex scenarios. You can refer to that chapter for more information.

Using parameterized queries

We explained how to use EF Core to execute SQL queries in previous chapters. Generally, if you use LINQ to query data, EF Core will generate parameterized queries for you. However, you need to take care of SQL injection attacks when you use the following methods:

- `FromSqlRaw()`
- `SqlQeuryRaw()`
- `ExecuteSqlRaw()`

These methods allow you to execute raw SQL queries without sanitizing the input. So, please make sure you sanitize the query statements before executing them.

Using data protection

Data protection is a security mechanism that prevents malicious users from accessing sensitive data. For example, if you store the user's password in the database, you should encrypt it before storing it. Another example is the user's credit card number, which should also be encrypted before it is stored.

The reason for this is that if the database is compromised, the attacker can easily access the user's sensitive data. In other words, the database is not trusted, just like the client. Data protection is another big topic, but it is beyond the scope of this book. ASP.NET Core provides a built-in data protection mechanism. If you would like to learn more about it, please refer to the official documentation: `https://learn.microsoft.com/en-us/aspnet/core/security/data-protection/introduction`.

Keeping secrets safe

Secrets are sensitive data that should not be exposed to the public. In our applications, we may have many secrets, such as the database connection string, the API keys, the client secrets, and so on. In the previous chapters, we often stored them in the `appsettings.json` file. However, we need to emphasize that this is not a good practice. These secrets should be stored in a safe place, such as Azure Key Vault, AWS Secrets Manager, or `kube-secrets`. Never upload them to the source code repository.

We will introduce **continuous integration/continuous deployment (CI/CD)** and explain how to store secrets safely in *Chapter 14*.

Keeping the framework up to date

The .NET Core framework is an open-source project. It is constantly being updated. We should always keep the framework up to date, including the NuGet packages. Note the life cycle of the .NET Core framework. Use the latest version of the framework as much as possible. If you are using an older version, you should consider upgrading it. You can find the life cycle of the .NET Core framework here: `https://dotnet.microsoft.com/en-us/platform/support/policy/dotnet-core`.

Checking the Open Web Application Security Project (OWASP) Top 10

OWASP is a nonprofit organization that provides information about web application security. It publishes a list of the most common web application security risks, which is called the OWASP Top 10. You can find the OWASP Top 10 here: `https://owasp.org/www-project-top-ten/`. You should check the list regularly to make sure your application is not vulnerable to any of the risks.

Also, OWASP provides a free resource called *DotNet Security Cheat Sheet*, where you can find the best practices for securing .NET Core applications. You can find it here: `https://cheatsheetseries.owasp.org/cheatsheets/DotNet_Security_Cheat_Sheet.html`.

Summary

In this chapter, we introduced the security and identity features of ASP.NET Core. We mainly learned how to use its built-in authentication and authorization mechanisms. We learned how to use the Identity framework to manage users and roles, and also explained role-based authorization, claim-based authorization, and policy-based authorization.

Then, we introduced OAuth 2.0 and OpenID Connect, which are the most popular authentication and authorization standards. After that, we explained several security practices, such as using HTTPS, strong passwords, parameterized queries, and more.

Again, security is a big topic, and we cannot cover all the details in one chapter. Please treat security as a continuous process, and always keep your application secure.

In the next chapter, we will get starssseted with testing, which is an important part of any software project. We will learn how to write unit tests for ASP.NET Core applications.

9

Testing in ASP.NET Core (Part 1 – Unit Testing)

Testing is an essential part of any software development process, including ASP.NET Core web API development. Testing helps to ensure that the application works as expected and meets the requirements. It also helps to ensure that any changes made to the code don't break existing functionality.

In this chapter, we'll look at the different types of testing that are available in ASP.NET Core and how to implement unit tests in ASP.NET Core web API applications.

We will cover the following topics in this chapter:

- Introduction to testing in ASP.NET Core
- Writing unit tests
- Testing the database access layer

By the end of this chapter, you will be able to write unit tests for your ASP.NET Core web API application to ensure that the code unit is functioning correctly. You will also learn how to use some libraries, such as `Moq` and `FluentAssertions`, to make your tests more readable and maintainable.

Technical requirements

The code examples in this chapter can be found at `https://github.com/PacktPublishing/Web-API-Development-with-ASP.NET-Core-8/tree/main/samples/chapter9`. You can use VS Code or VS 2022 to open the solutions.

Introduction to testing in ASP.NET Core

Different types of testing can be performed on an ASP.NET Core web API application, as follows:

- **Unit testing**: This is the process of testing individual units of code, such as methods and classes, to ensure that they work as expected. Unit tests should be small, fast, and isolated from other units of code. Mocking frameworks can be used to isolate units of code from their dependencies, such as databases and external services.

- **Integration testing**: This involves testing the integration between different components of the application to ensure that they work together as expected. This type of testing helps to identify issues that may arise when the application is deployed to a production environment. Generally, integration tests are slower than unit tests. Integration tests may use mock objects or real objects, depending on the scenario. For example, if the integration test is to test the integration between the application and a database, then a real database instance should be used. But if the integration test is to test the application's integration with an external service, such as a payment service, then we should use a mock object to simulate the external service. In the microservices architecture, integration tests are more complicated, as they may involve multiple services. Besides the integration tests for each service, there should also be integration tests for the entire system.

- **End-to-end testing**: This is the process of testing the application from the user's perspective to ensure that the entire system from start to finish, including the user interface, the web API, the database, and more, is working as expected. End-to-end testing typically involves simulating user interactions with the application, such as clicking buttons and entering data into forms.

- **Regression testing**: This involves testing whether the application still works as expected after new features are added or bugs are fixed. Regression testing is usually performed after the application is deployed to a production environment. It helps to ensure that the new features or bug fixes don't break existing functionality.

- **Load testing**: This involves testing whether the application can handle a normal load of users and requests. It helps to set the baseline for the performance of the application.

- **Stress testing**: This involves testing whether the application can handle extreme conditions, such as a sudden spike in the number of users and requests, or gradually increasing the load over a long period. It also determines whether the application can recover from failures and how long it takes to recover.

- **Performance testing**: This is a type of testing that evaluates the performance of the application under different workloads, including response time, throughput, resource usage, and more. Performance testing is a superset of load testing and stress testing. Generally, unit testing and integration testing are performed in the development environment and the staging environment, while performance testing is performed in a production-like environment, such as a **user acceptance testing** (**UAT**) environment, which closely mirrors the production environment in terms of infrastructure and configuration. This ensures that the performance tests are accurate and reliable. In some cases, limited performance testing can be performed in a development environment during scheduled maintenance windows to validate real-world performance scenarios.

Unit testing and integration testing are the most common types of testing that are written in .NET by developers. In this chapter, we will focus on unit testing; we will discuss integration testing in *Chapter 10*.

Writing unit tests

Unit tests are written to test individual units of code, such as methods and classes. Unit tests are typically written by developers who are familiar with the code. When developers develop new features or fix bugs, they should also write unit tests to ensure that the code works as expected. There are many unit testing frameworks available for .NET, including NUnit, xUnit, and MSTest. In this chapter, we will use xUnit to write unit tests since it is one of the most popular unit testing frameworks for modern .NET applications at present.

Preparing the sample application

The sample application, `InvoiceApp`, is a simple ASP.NET Core web API application that exposes a set of RESTful APIs for managing invoices. The sample application uses EF Core to store and retrieve data from a SQL Server database. It has the following endpoints:

- **Invoices API**: This endpoint is used to manage invoices. It supports the following operations:

  - `GET /api/invoices`: Retrieves a list of invoices

  - `GET /api/invoices/{id}`: Retrieves an invoice by ID

  - `POST /api/invoices`: Creates a new invoice

  - `PUT /api/invoices/{id}`: Updates an existing invoice

  - `DELETE /api/invoices/{id}`: Deletes an invoice

  - `PATCH /api/invoices/{id}/status`: Updates the status of an invoice

  - `POST /api/invoices/{id}/send`: Sends an invoice email to the contact

- **Contacts API**: This endpoint is used to manage contacts. Each invoice is associated with a contact. It supports the following operations:

  - `GET /api/contacts`: Retrieves a list of contacts

  - `GET /api/contacts/{id}`: Retrieves a contact by ID

  - `POST /api/contacts`: Creates a new contact

- `PUT /api/contacts/{id}`: Updates an existing contact

- `DELETE /api/contacts/{id}`: Deletes a contact

- `GET /api/contacts/{id}/invoices`: Retrieves a list of invoices for a contact

Note that the preceding endpoints are not enough to build a complete invoice management application. It is just a sample application to demonstrate how to write unit tests and integration tests for ASP. NET Core web API applications.

In addition, the sample application has a Swagger UI that can be used to test the APIs. *Figure 9.1* shows the Swagger UI for the sample application:

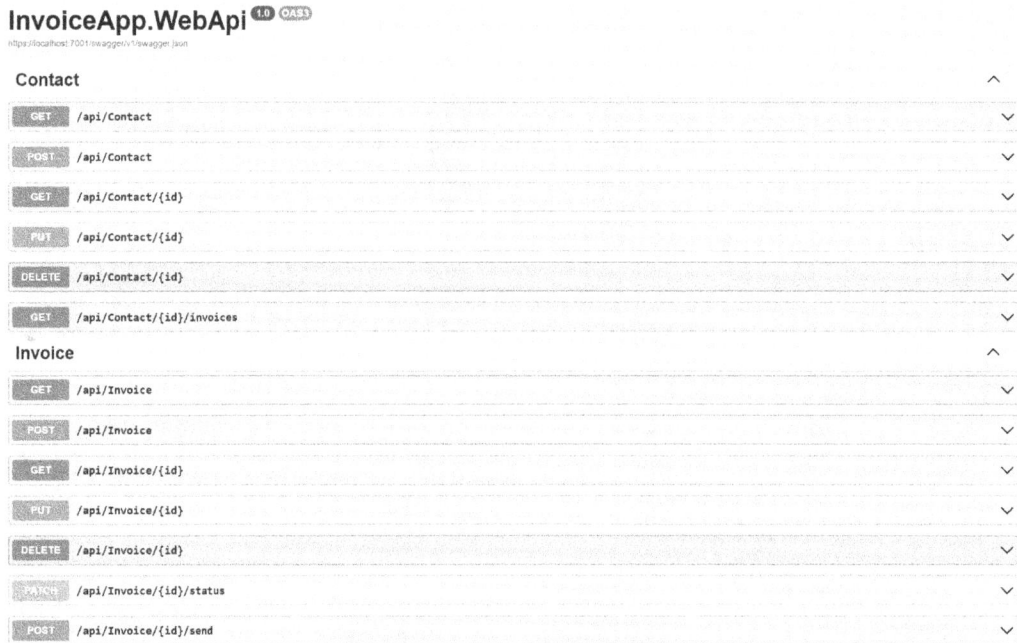

Figure 9.1 – The sample application API endpoints

After you run the sample application using the `dotnet run` command, you can access the Swagger UI at `http://localhost:5087/swagger/index.html`.

Now, we will use this sample application to demonstrate how to write unit tests for ASP.NET Core web API applications.

Setting up the unit tests project

We will use xUnit to write unit tests for the sample application. xUnit is a popular unit testing framework for .NET applications. It is a free, open-source project that has been around for many years. It is also the default unit testing framework for .NET Core and .NET 5+ applications. You can find more information about xUnit at `https://xunit.net/`.

To set up the test project, you can use VS 2022 or the .NET CLI. If you use VS 2022, you can create a new xUnit test project by right-clicking on the solution and selecting **Add | New Project**. Then, select **xUnit Test Project** from the list of project templates. You can name the project `InvoiceApp.UnitTests` and click **Create** to create the project:

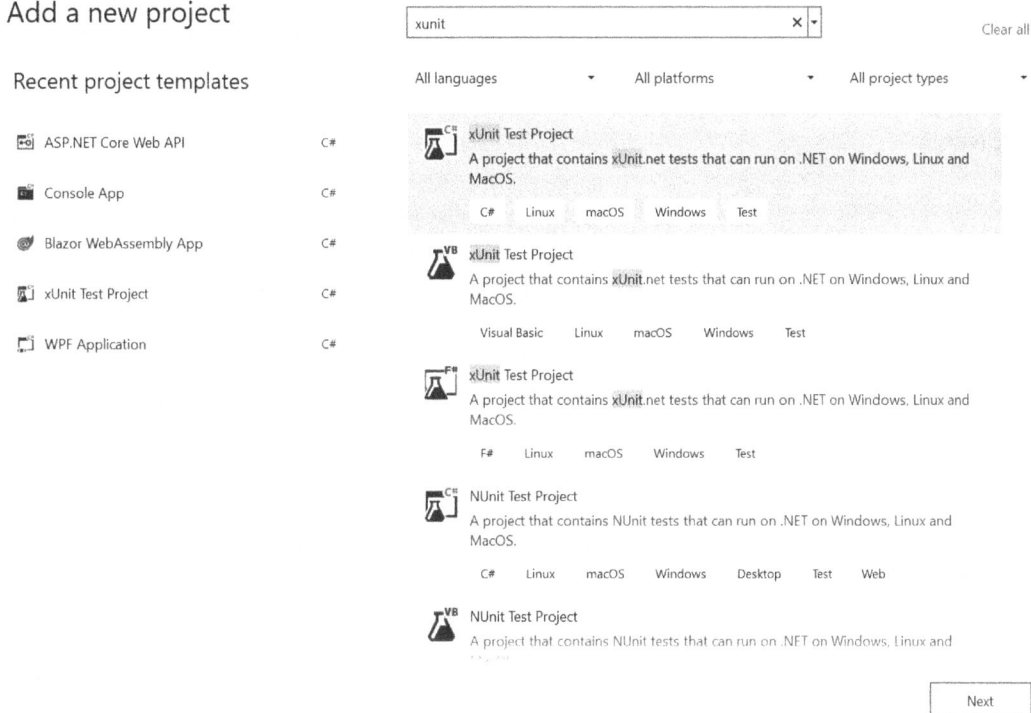

Figure 9.2 – Creating a new xUnit test project in VS 2022

After you create the project, add a project reference to the `InvoiceApp.WebApi` project so that the test project can access the classes in the main web API project. You can do this by right-clicking on the **Dependencies** node in the test project and selecting **Add Project Reference**. Then, select the `InvoiceApp.WebApi` project from the list of projects and click **OK** to add the project reference:

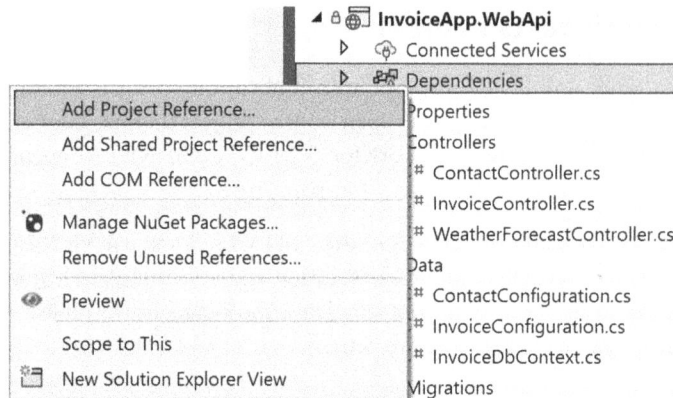

Figure 9.3 – Adding a project reference to the test project in VS 2022

If you use the .NET CLI, you can create a new xUnit test project by running the following commands in the terminal:

```
dotnet new xunit -n InvoiceApp.UnitTests
```

Then, you can add the test project to the solution by running the following commands:

```
dotnet sln InvoiceApp.sln add InvoiceApp.UnitTests/InvoiceApp.
UnitTests.csproj
```

You also need to add the reference to the main project by running the following command:

```
dotnet add InvoiceApp.UnitTests/InvoiceApp.UnitTests.csproj reference
InvoiceApp.WebApi/InvoiceApp.WebApi.csproj
```

The default xUnit test project template contains a sample unit test. You can delete the sample unit test named `UnitTest1.cs`; we will write our own unit tests in the next section.

If you create the test project starting from a blank .NET library project, you need to add the following packages to the test project:

- `Microsoft.NET.Test.Sdk`: This is required for running unit tests
- `xunit`: This is the xUnit framework that we will use to write unit tests
- `xunit.runner.visualstudio`: This is required for running unit tests in Visual Studio
- `coverlet.collector`: This is an open-source project that provides code coverage analysis for .NET applications

When we write unit tests, keep in mind that one unit test should test one unit of code, such as a method or a class. The unit test should be isolated from other units of code. If one method depends on another method, we should mock the other method to isolate the unit of code to ensure that we focus on the behavior of the unit of code that we are testing.

Writing unit tests without dependencies

Let's see the first example. In the sample application, you can find a `Services` folder, which includes the `IEmailService` interface and its implementation, `EmailService`. The `EmailService` class has a method named `GenerateInvoiceEmail()`. This method is a simple function that generates an email according to the `Invoice` entity. The following code shows the `GenerateInvoiceEmail()` method:

```
public (string to, string subject, string body)
GenerateInvoiceEmail(Invoice invoice)
{
    var to = invoice.Contact.Email;
    var subject = $"Invoice {invoice.InvoiceNumber} for {invoice.
Contact.FirstName} {invoice.Contact.LastName}";
    var body = $"""
        Dear {invoice.Contact.FirstName} {invoice.Contact.LastName},

        Thank you for your business. Here are your invoice details:
        Invoice Number: {invoice.InvoiceNumber}
        Invoice Date: {invoice.InvoiceDate.LocalDateTime.
ToShortDateString()}
        Invoice Amount: {invoice.Amount.ToString("C")}
        Invoice Items:
        {string.Join(Environment.NewLine, invoice.InvoiceItems.
Select(i => $"{i.Description} - {i.Quantity} x {i.UnitPrice.
ToString("C")}"))}

        Please pay by {invoice.DueDate.LocalDateTime.
ToShortDateString()}. Thank you!

        Regards,
        InvoiceApp
        """;
    return (to, subject, body);
}
```

> **Raw string literal**
>
> The body variable is a raw string literal, which is a new feature that was introduced in C# 11. Raw string literals are enclosed in triple quotes (`"""`). They can span multiple lines and can contain double quotes without escaping them. You can find more information about raw string literals at `https://learn.microsoft.com/en-us/dotnet/csharp/language-reference/tokens/raw-string`. Raw strings can also be used with interpolated strings, which is convenient for generating strings with variables.

There is no dependency in the `GenerateInvoiceEmail()` method, so we can write a unit test for this method without mocking any other methods. Create a class named `EmailServiceTests` in the `InvoiceApp.UnitTests` project. Then, add the following code to the class:

```
[Fact]
public void GenerateInvoiceEmail_Should_Return_Email()
{
    var invoiceDate = DateTimeOffset.Now;
    var dueDate = invoiceDate.AddDays(30);
    // Arrange
    var invoice = new Invoice
    {
        Id = Guid.NewGuid(),
        InvoiceNumber = "INV-001",
        Amount = 500,
        DueDate = dueDate,
        // Omit other properties for brevity
    };

    // Act
    var (to, subject, body) = new EmailService().
GenerateInvoiceEmail(invoice);

    // Assert
    Assert.Equal(invoice.Contact.Email, to);
    Assert.Equal($"Invoice INV-001 for John Doe", subject);
    Assert.Equal($"""
        Dear John Doe,

        Thank you for your business. Here are your invoice details:
        Invoice Number: INV-001
        Invoice Date: {invoiceDate.LocalDateTime.ToShortDateString()}
        Invoice Amount: {invoice.Amount.ToString("C")}
        Invoice Items:
        Item 1 - 1 x $100.00
        Item 2 - 2 x $200.00

        Please pay by {invoice.DueDate.LocalDateTime.
ToShortDateString()}. Thank you!

        Regards,
        InvoiceApp
        """, body);
}
```

The `Fact` attribute indicates that the `GenerateInvoiceEmail_Should_Return_Email()` method is a unit test so that xUnit can discover and run this method as a unit test. In the `GenerateInvoiceEmail_Should_Return_Email()` method, we created an `Invoice` object and passed it to the `GenerateInvoiceEmail()` method. Then, we used the `Assert` class to verify that the `GenerateInvoiceEmail()` method returns the expected email.

When writing unit tests, we follow the *Arrange-Act-Assert* pattern:

- **Arrange**: This is where we prepare the data and set up the environment for the unit test
- **Act**: This is where we call the method that we want to test
- **Assert**: This is where we verify that the method returns the expected result or that the method behaves as expected

To run the unit test in VS 2022, you can right-click on the `InvoiceApp.UnitTests` project or the `EmailServiceTest.cs` file and select **Run Tests**. You can also use the **Test Explorer** window to run the unit test by clicking the **Test** menu and selecting **Run all tests**:

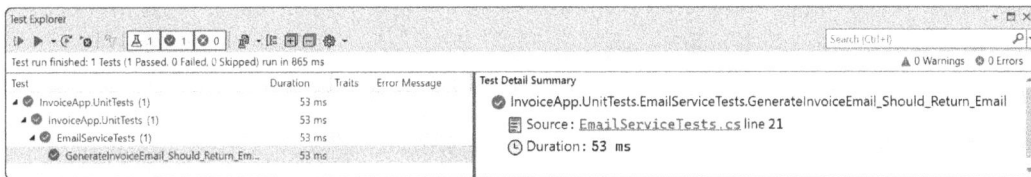

Figure 9.4 – Running unit tests in VS 2022

VS Code supports running unit tests as well. Click the **Testing** icon on the left-hand side of the VS Code window to open the **Test** view; you will see the unit tests, as shown in *Figure 9.5*:

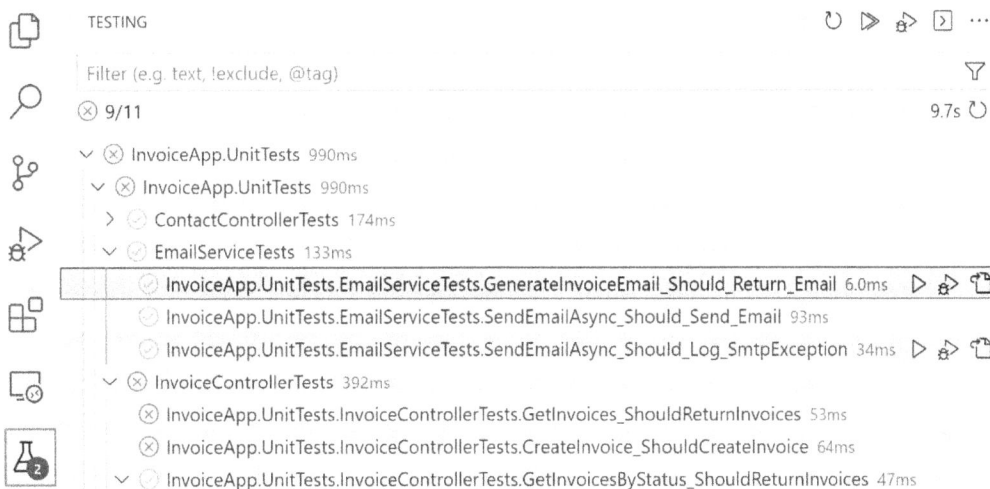

Figure 9.5 – Running unit tests in VS Code

If you use the .NET CLI, you can run the unit tests by running the following command in the terminal:

```
dotnet test
```

You will see the following output:

```
Starting test execution, please wait...
A total of 1 test files matched the specified pattern.

Passed!  - Failed:     0, Passed:     1, Skipped:     0, Total:     1,
Duration: < 1 ms - InvoiceApp.UnitTests.dll (net8.0)
```

The output shows that the unit test passed. If you want to see the detailed test results, you can run the following command:

```
dotnet test --verbosity normal
```

You will see detailed test results that show the test name, the result, the duration, and the output.

Writing unit tests with dependencies

Let's look at another example. In the `EmailService` class, there is a method named `SendEmailAsync()`, which sends an email to the recipient. In real-world applications, we usually use a third-party email service to send emails. To make the `EmailService` class testable, we can create an `IEmailSender` interface and its implementation, `EmailSender`. The `EmailSender` class is a wrapper of the `SmtpClient` class, which is used to send emails. The following code shows the updated `EmailService` class:

```
public async Task SendEmailAsync(string to, string subject, string
body)
{
    // Mock the email sending process
    // In real world, you may use a third-party email service, such as
SendGrid, MailChimp, Azure Logic Apps, etc.
    logger.LogInformation($"Sending email to {to} with subject
{subject} and body {body}");
    try
    {
        await emailSender.SendEmailAsync(to, subject, body);
        logger.LogInformation($"Email sent to {to} with subject
{subject}");
    }
    catch (SmtpException e)
    {
        logger.LogError(e, $"SmtpClient error occurs. Failed to send
email to {to} with subject {subject}.");
```

```
    }
    catch (Exception e)
    {
        logger.LogError(e, $"Failed to send email to {to} with subject
{subject}.");
    }
}
```

So, now, `EmailService` has a dependency on the `IEmailSender` interface. To test the behavior of the `SendEmailAsync()` method of the `EmailService` class, we need to mock the `IEmailSender` interface to isolate the `EmailService` class from the `EmailSender` class. Otherwise, if any error occurs in the unit test, we cannot say for sure whether the error is caused by the `EmailService` class or the `EmailSender` class.

We can use the Moq library to mock the `IEmailSender` interface. Moq is a popular mocking library for .NET. It is available as a NuGet package. To install Moq, you can use **NuGet Package Manager** in VS 2022 or run the following command:

```
dotnet add package Moq
```

Then, we can create the unit tests for the `SendEmailAsync()` method. Because the `SendEmailAsync()` method may throw an exception if the email-sending process fails, we need to write two unit tests to test the success and failure scenarios. The following code shows the unit test for the success scenario:

```
[Fact]
public async Task SendEmailAsync_Should_Send_Email()
{
    // Arrange
    var to = "user@example.com";
    var subject = "Test Email";
    var body = "Hello, this is a test email";
    var emailSenderMock = new Mock<IEmailSender>();
    emailSenderMock.Setup(m => m.SendEmailAsync(It.IsAny<string>(),
It.IsAny<string>(), It.IsAny<string>()))
        .Returns(Task.CompletedTask);
    var loggerMock = new Mock<ILogger<IEmailService>>();
    loggerMock.Setup(l => l.Log(It.IsAny<LogLevel>(),
It.IsAny<EventId>(), It.IsAny<It.IsAnyType>(),
        It.IsAny<Exception>(), (Func<It.IsAnyType, Exception?,
string>)It.IsAny<object>())).Verifiable();
    var emailService = new EmailService(loggerMock.Object,
emailSenderMock.Object);
    // Act
    await emailService.SendEmailAsync(to, subject, body);
    // Assert
```

```
    emailSenderMock.Verify(m => m.SendEmailAsync(It.IsAny<string>(),
It.IsAny<string>(), It.IsAny<string>()), Times.Once);
    loggerMock.Verify(
        l => l.Log(
            It.IsAny<LogLevel>(),
            It.IsAny<EventId>(),
            It.Is<It.IsAnyType>((v, t) => v.ToString().
Contains($"Sending email to {to} with subject {subject} and body
{body}")),
            It.IsAny<Exception>(),
            (Func<It.IsAnyType, Exception?, string>)It.IsAny<object>()
        ),
        Times.Once
    );
    loggerMock.Verify(
        l => l.Log(
            It.IsAny<LogLevel>(),
            It.IsAny<EventId>(),
            It.Is<It.IsAnyType>((v, t) => v.ToString().
Contains($"Email sent to {to} with subject {subject}")),
            It.IsAny<Exception>(),
            (Func<It.IsAnyType, Exception?, string>)It.IsAny<object>()
        ),
        Times.Once
    );
}
```

In the preceding code, we use the Mock class to create mock objects of the IEmailSender interface and the ILogger interface. We need to set up the behavior of the methods for the mock objects. If the methods used in the unit test are not set up, the unit test will fail. For example, we use the SetUp() method to mock the SendEmailAsync method of the IEmailSender interface:

```
emailSenderMock.Setup(m => m.SendEmailAsync(It.IsAny<string>(),
It.IsAny<string>(), It.IsAny<string>()))
    .Returns(Task.CompletedTask);
```

The SetUp() method takes a lambda expression as the parameter, which is used to configure the behavior of the SendEmailAsync() method. In the preceding code, we use the It.IsAny<string>() method to specify that the SendEmailAsync() method can accept any string value as the parameter. Then, we use the Returns() method to specify the return value of the SendEmailAsync() method. In this case, we use the Task.CompletedTask property to specify that the SendEmailAsync() method will return a completed task. If you need to return a specific value, you can also use the Returns() method to return a specific value. For example, if the `SendEmailAsync()` method returns a `bool` value, you can use the following code to return a `true` value:

```
emailSenderMock.Setup(m => m.SendEmailAsync(It.IsAny<string>(),
It.IsAny<string>(), It.IsAny<string>()))
    .ReturnsAsync(true);
```

> **Mocking the ILogger interface**
>
> The `EmailService` class uses the `ILogger` interface to log the information and errors. We use the `LogInformation()` method to log the information and `LogError()` method to log the errors. However, we cannot mock the `LogInformation()` or `LogError()` method directly because they are extension methods on top of the `ILogger` interface. These extension methods, such as `LogInformation()`, `LogError()`, `LogDebug()`, `LogWarning()`, `LogCritical()`, `LogTrace()`, and others, all call the `Log()` method of the `ILogger` interface. Therefore, to verify that a given log message is logged, it is necessary to mock only the `Log()` method of the `ILogger` interface.

If the `SendEmailAsync()` method throws an exception, we need to ensure the logger will log the exception when the exception occurs. To test the failure scenario, we need to mock the `SendEmailAsync()` method so that it throws an exception. We can use the `ThrowsAsync()` method to mock the `SendEmailAsync()` method to throw an exception explicitly. The following code shows how to mock the `SendEmailAsync()` method to throw an exception:

```
emailSenderMock.Setup(m => m.SendEmailAsync(It.IsAny<string>(),
It.IsAny<string>(), It.IsAny<string>()))
  .ThrowsAsync(new SmtpException("Test SmtpException"));
```

Then, we can verify whether the `LogError()` method of the `ILogger` interface is called when the `SendEmailAsync()` method throws an exception, as shown here:

```
// Act + Assert
await Assert.ThrowsAsync<SmtpException>(() => emailService.
SendEmailAsync(to, subject, body));
loggerMock.Verify(
    l => l.Log(
        It.IsAny<LogLevel>(),
        It.IsAny<EventId>(),
        It.Is<It.IsAnyType>((v, t) =>
            v.ToString().Contains($"Failed to send email to {to} with
subject {subject}")),
        It.IsAny<SmtpException>(),
        (Func<It.IsAnyType, Exception?, string>)It.IsAny<object>()
    ),
    Times.Once
);
```

In this way, we can ensure that the `SendEmailAsync()` method will log the exception when the exception occurs.

When we write unit tests, note that the test method name should be descriptive and should indicate the purpose of the test. For example, `SendEmailAsync_ShouldLogError_WhenEmailSendingFails()` is a good name because it indicates that the `SendEmailAsync` method should log an error when the email sending fails. However, `SendEmailAsyncTest()` is not a good name because it does not indicate the purpose of the test.

For more information about how to use the `Mock` library to create mock objects, see `https://github.com/moq/moq`.

Using FluentAssertions to verify the test results

`xUnit` provides a set of static assertion methods to verify the test results. For example, we can use the `Assert.Equal()` method to verify if two objects are equal. These methods cover most of the scenarios, such as verifying objects, collections, exceptions, events, equality, types, and more. Here is a list of the assertion methods provided by xUnit:

xUnit Assertion Methods	Explanation
`Assert.Equal(expected, actual)`	Verifies that the `expected` value is equal to the `actual` value
`Assert.NotEqual(expected, actual)`	Verifies that the `expected` value is not equal to the `actual` value
`Assert.StrictEqual(expected, actual)`	Verifies that the `expected` value is strictly equal to the `actual` value, using the type's default comparer
`Assert.NotStrictEqual(expected, actual)`	Verifies that the `expected` value is strictly not equal to the `actual` value, using the type's default comparer
`Assert.Same(expected, actual)`	Verifies that the `expected` object is the same instance as the `actual` object
`Assert.NotSame(expected, actual)`	Verifies that the `expected` object is not the same instance as the `actual` object
`Assert.True(condition)`	Verifies that the `condition` is true
`Assert.False(condition)`	Verifies that the `condition` is false
`Assert.Null(object)`	Verifies that the `object` is null
`Assert.NotNull(object)`	Verifies that the `object` is not null

xUnit Assertion Methods	Explanation
`Assert.IsType(expectedType, object)`	Verifies that the `object` is exactly `expectedType`, and not a derived type
`Assert.IsNotType(unexpectedType, object)`	Verifies that the `object` is not exactly `unexpectedType`
`Assert.IsAssignableFrom(expectedType, object)`	Verifies that the `object` is assignable to `expectedType`, which means that `object` is the given type or a derived type
`Assert.Contains(expected, collection)`	Verifies that `collection` contains the `expected` object
`Assert.DoesNotContain(expected, collection)`	Verifies that `collection` does not contain the `expected` object
`Assert.Empty(collection)`	Verifies that `collection` is empty
`Assert.NotEmpty(collection)`	Verifies that `collection` is not empty
`Assert.Single(collection)`	Verifies that `collection` contains exactly one element of the given type
`Assert.InRange(actual, low, high)`	Verifies that the `actual` value is within the range of `low` and `high` (inclusive)
`Assert.NotInRange(actual, low, high)`	Verifies that the `actual` value is not within the range of `low` and `high` (inclusive)
`Assert.Throws<exceptionType>(action)`	Verifies that `action` throws an exception of the specified `exceptionType`, and not a derived exception type
`Assert.ThrowsAny<exceptionType>(action)`	Verifies that `action` throws an exception of the specified `exceptionType` or a derived exception type

Table 9.1 – List of assertion methods provided by xUnit

Note this list is not complete. You can find more assertion methods in xUnit's GitHub repository: `https://github.com/xunit/assert.xunit`.

Although the assertion methods provided by xUnit are enough for most scenarios, they are not very readable. A good way to make the unit tests more natural and readable is to use `FluentAssertions`, an open-source assertion library for .NET. It provides a set of extension methods that allow us to write assertions fluently.

To install `FluentAssertions`, we can use the following .NET CLI command:

```
dotnet add package FluentAssertions
```

You can also use NuGet Package Manager to install the `FluentAssertions` package if you use VS 2022.

Then, we can use the `Should()` method to verify the test results. For example, we can use the `Should().Be()` method to verify whether two objects are equal.

The following code shows how to use the `Should().Be()` method to verify whether the `GetInvoicesAsync()` method returns a list of invoices:

```
// Omitted code for brevity
returnResult.Should().NotBeNull();
returnResult.Should().HaveCount(2);
// Or use returnResult.Count.Should().Be(2);
returnResult.Should().Contain(i => i.InvoiceNumber == "INV-001");
returnResult.Should().Contain(i => i.InvoiceNumber == "INV-002");
```

The `FluentAssertions` method is more intuitive and readable than the `Assert.Equal()` method. For most scenarios, you can easily replace the assertion methods provided by xUnit with the `FluentAssertions` methods without searching the documentation.

Let's see how to verify the exception using `FluentAssertions`. In the `EmailServiceTests` class, there is a `SendEmailAsync_Should_Log_SmtpException()` method. This method verifies whether the `SendEmailAsync()` method will log the exception when the `SendEmailAsync()` method throws an exception. The following code shows how to use xUnit to verify the exception:

```
await Assert.ThrowsAsync<SmtpException>(() => emailService.
SendEmailAsync(to, subject, body));
```

We can use the `Should().ThrowAsync<>()` method of `FluentAssertions` to verify the exception, as shown in the following code:

```
var act = () => emailService.SendEmailAsync(to, subject, body);
await act.Should().ThrowAsync<SmtpException>().WithMessage("Test
SmtpException");
```

Using `FluentAssertions` is more readable and intuitive than the xUnit way. Here is a table comparing some common assertion methods provided by xUnit and `FluentAssertions`:

xUnit Assertion Method	FluentAssertions Assertion Method
`Assert.Equal(expected, actual)`	`.Should().Be(expected)`
`Assert.NotEqual(expected, actual)`	`.Should().NotBe(expected)`

xUnit Assertion Method	FluentAssertions Assertion Method
`Assert.True(condition)`	`.Should().BeTrue()`
`Assert.False(condition)`	`.Should().BeFalse()`
`Assert.Null(object)`	`.Should().BeNull()`
`Assert.NotNull(object)`	`.Should().NotBeNull()`
`Assert.Contains(expected, collection)`	`.Should().Contain(expected)`
`Assert.DoesNotContain(expected, collection)`	`.Should().NotContain(expected)`
`Assert.Empty(collection)`	`.Should().BeEmpty()`
`Assert.NotEmpty(collection)`	`.Should().NotBeEmpty()`
`Assert.Throws<TException>(action)`	`.Should().Throw<TException>()`
`Assert.DoesNotThrow(action)`	`.Should().NotThrow()`

Table 9.2 – Comparison of common assertion methods provided by xUnit and FluentAssertions

Note that the preceding table is not an exhaustive list. You can find more extension methods in the official documentation of `FluentAssertions`: `https://fluentassertions.com/introduction`.

Besides the fluent assertion methods, `FluentAssertions` also provides better error messages if the test fails. For example, if we use the `Assert.Equal()` method to verify whether `returnResult` contains two invoices, the code will look like this:

```
Assert.Equal(3, returnResult.Count);
```

If the test fails, the error message will be as follows:

```
InvoiceApp.UnitTests.InvoiceControllerTests.GetInvoices_
ShouldReturnInvoices
  Source: InvoiceControllerTests.cs line 21
  Duration: 372 ms

  Message:
Assert.Equal() Failure
Expected: 3
Actual:   2

  Stack Trace:
InvoiceControllerTests.GetInvoices_ShouldReturnInvoices() line 34
InvoiceControllerTests.GetInvoices_ShouldReturnInvoices() line 41
--- End of stack trace from previous location ---
```

If we have multiple `Assert.Equal()` methods in the test method, which is not recommended but something we have to do occasionally, we cannot immediately know which `Assert.Equal()` method fails. We need to check the line number of the error message to find the failing assertion. This is not very convenient.

If we use `FluentAssertions`, the assertion code will look like this:

```
returnResult.Count.Should().Be(3);
```

If the test fails with the same reason, the error message will be as follows:

```
InvoiceApp.UnitTests.InvoiceControllerTests.GetInvoices_
ShouldReturnInvoices
  Source: InvoiceControllerTests.cs line 21
  Duration: 408 ms

  Message:
Expected returnResult.Count to be 3, but found 2.
```

Now, the error message is more detailed and intuitive and tells us which assertion fails. This is very helpful when we have multiple assertions in the test method.

You can even enrich the error message by adding a custom message to the assertion method. For example, we can add a custom message to the `Should().Be()` method, as follows:

```
returnResult.Count.Should().Be(3, "The number of invoices should be
3");
```

Now, the error message will be as follows:

```
Expected returnResult.Count to be 3 because The number of invoices
should be 3, but found 2.
```

Therefore, it is highly recommended to use `FluentAssertions` in your tests. It makes your tests more readable and maintainable.

Testing the database access layer

In many web API applications, we need to access the database to perform CRUD operations. In this section, we will learn how to test the database access layer in unit tests.

How can we test the database access layer?

Currently, we inject `InvoiceDbContext` into controllers to access the database. This approach is easy for development, but it tightly couples the controllers with the `InvoiceDbContext` class. When we test the controllers, we need to create a real `InvoiceDbContext` object and use it to test the controllers, which means controllers are not tested in isolation. This problem can be addressed in a variety of ways:

- Use the `InMemoryDatabase` provider of EF Core to create an in-memory database as the fake database
- Use the SQLite in-memory database as the fake database
- Create a separate repository layer to encapsulate the database access code, inject the repository layer into controllers (or services that need to access databases), and then use `Mock` objects to mock the repository layer
- Use the real database for testing

Each approach has its pros and cons:

- The `InMemoryDatabase` provider was originally designed for internal testing of EF Core. However, it is not a good choice for testing other applications because it does not behave like a real database. For example, it does not support transactions and raw SQL queries. So, it is not a good choice for testing the database access code.
- SQLite also provides an in-memory database feature that can be used for testing. However, it has similar limitations to EF Core's `InMemoryDatabase` provider. If the production database is SQL Server, EF Core cannot guarantee that the database access code will work correctly on SQL Server if we use SQLite for testing.
- Creating a separate repository layer is to decouple the controllers from the `DbContext` class. In this pattern, a separate `IRepository` interface is created between the application code and `DbContext`, and the implementation of the `IRepository` interface is injected into controllers or services. In this way, we can use `Mock` objects to mock the `IRepository` interface to test controllers or services, which means controllers or services can be tested in isolation. However, this approach requires a lot of work to create the repository layer. Also, the `DbContext` class is already a repository pattern, so creating another repository layer is redundant if you do not need to change the database provider. But this pattern still has its advantages. Tests can focus on the application logic without worrying about the database access code. Also, if you need to change the database provider, you only need to change the implementation of the `IRepository` interface, and there is no need to change the controllers or services.
- Testing against a real database provides more benefits. One of the most important benefits is that it can ensure that the database access code works correctly on the database in production. Using a real database is also fast and reliable. However, one challenge is that we need to ensure the isolation for tests because some of them may change the data in the database. So, we need to make sure the data can be restored or recreated after the tests are completed.

In this section, we will use a separate local database for testing, such as a LocalDB database. If your application will run on SQL Server, you can use another SQL Server for testing rather than LocalDB since LocalDB does not behave the same as SQL Server. If your application will run in the cloud, such as Azure, you may use an Azure SQL database. You can use another Azure SQL database for testing but you will need to allocate a small amount of resources for it to save the cost. Keep in mind that the database for testing should keep the same environment as much as possible to avoid unexpected behaviors in production.

Regarding the controllers, we will use the `InvoiceDbContext` class directly for simplicity; we will learn the repository pattern in future chapters.

Creating a test fixture

When we test the CRUD methods against the database, we need to prepare the database before the tests are executed, and then clean up the database after the tests are completed so that the changes made by the tests will not affect other tests. xUnit provides the `IClassFixture<T>` interface to create a test fixture, which can be used to prepare and clean up the database for each test class.

First, we need to create a test fixture class in the `InvoiceApp.UnitTests` project, as follows:

```
public class TestDatabaseFixture
{
    private const string ConnectionString = @"Server=(localdb)\
mssqllocaldb;Database=InvoiceTestDb;Trusted_Connection=True";

}
```

In the `TestDatabaseFixture` class, we define a connection string to the local database. Using a `const` string is for simplicity only. In a real application, you may want to use the configuration system to read the connection string from other sources, such as the `appsettings.json` file.

Then, we add a method to create the database context object, as follows:

```
public InvoiceDbContext CreateDbContext()
    => new(new DbContextOptionsBuilder<InvoiceDbContext>()
            .UseSqlServer(ConnectionString)
            .Options, null);
```

We also need a method to initialize the database, as follows:

```
public void InitializeDatabase()
{
    using var context = CreateDbContext();
    context.Database.EnsureDeleted();
    context.Database.EnsureCreated();
    // Create a few Contacts
    var contacts = new List<Contact>
    {
```

```
        // Omitted the code for brevity
    };
    context.Contacts.AddRange(contacts);
    // Create a few Invoices
    var invoices = new List<Invoice>
    {
        // Omitted the code for brevity
    };
    context.Invoices.AddRange(invoices);
    context.SaveChanges();
}
```

In the `InitializeDatabase()` method, we create a new `InvoiceDbContext` object and then use the `EnsureDeleted()` method to delete the database if it exists. Then, we use the `EnsureCreated()` method to create the database. After that, we seed some data into the database. In this example, we create a few `Contact` and `Invoice` objects and add them to the database. Finally, we call the `SaveChanges()` method to save the changes to the database.

Now, we need to call the `InitializeDatabase()` method to initialize the database in the constructor of the `TestDatabaseFixture` class, as follows:

```
private static readonly object Lock = new();
private static bool _databaseInitialized;

public TestDatabaseFixture()
{
    // This code comes from Mirosoft Docs: https://github.com/
    dotnet/EntityFramework.Docs/blob/main/samples/core/Testing/
    TestingWithTheDatabase/TestDatabaseFixture.cs
    lock (Lock)
    {
        if (!_databaseInitialized!)
        {
            InitializeDatabase();
            databaseInitialized = true;
        }
    }
}
```

To avoid initializing the database multiple times, we use a static field, `_databaseInitialized`, to indicate whether the database has been initialized. We also define a static object, `Lock`, to ensure that the database is initialized only once. The `InitializeDatabase()` method is used to initialize the database. It will only be called once before the tests are executed.

There are several important things to note:

- xUnit creates a new instance of the test class for every test. So, the constructor of the test class is called for every test.

- Deleting and recreating the database for each test run may slow down the tests and may not be necessary. If you do not want to delete and recreate the database for each test run, you can comment out the `EnsureDeleted()` method to allow the database to be reused. However, if you need to frequently change the database schema in the development phase, you may need to delete and recreate the database for each test run to ensure the database schema is up to date.

- We use a lock object to ensure the `InitializeDatabase()` method is only called once for each test run. The reason is that the `TextDatabaseFixture` class can be used in multiple test classes, and xUnit can run multiple test classes in parallel. Using a lock can help us ensure the seed method is only called once. We will learn more about parallel test execution in the next section.

Now that the test fixture is ready, we can use it in the test classes.

Using the test fixture

Next, we will use the test fixture in the test classes. First, let's test the `GetAll()` method of the `InvoiceController` class. Create a new test class named `InvoiceControllerTests` in the `InvoiceApp.UnitTests` project, as follows:

```
public class InvoiceControllerTests(TestFixture fixture) :
IClassFixture<TestFixture>
{
}
```

We use dependency injection to inject the `TestDatabaseFixture` object into the test class. Then, we can use the text fixture to create the `InvoiceDbContext` object in the test methods, as follows:

```
[Fact]
public async Task GetInvoices_ShouldReturnInvoices()
{
    // Arrange
    await using var dbContext = fixture.CreateDbContext();
    var emailServiceMock = new Mock<IEmailService>();
    var controller = new InvoiceController(dbContext,
emailServiceMock.Object);
    // Act
    var actionResult = await controller.GetInvoicesAsync();
    // Assert
    var result = actionResult.Result as OkObjectResult;
    Assert.NotNull(result);
```

```
    var returnResult = Assert.IsAssignableFrom<List<Invoice>>(result.
Value);
    Assert.NotNull(returnResult);
    Assert.Equal(2, returnResult.Count);
    Assert.Contains(returnResult, i => i.InvoiceNumber =="INV-001");
    Assert.Contains(returnResult, i => i.InvoiceNumber =="INV-002");
}
```

In the GetInvoices_ShouldReturnInvoices() method, we use the fixture to create the InvoiceDbContext object, and then create the InvoiceController object with some mocked dependencies. Then, we call the GetInvoicesAsync() method to get the invoices from the database. Finally, we use the Assert class to verify the result.

The data we use to verify the controller is the data we seed into the database in the TestDatabaseFixture class. If you change the data in the TestDatabaseFixture class, you also need to change the expected data in the test class.

The GetInvoices_ShouldReturnInvoices() method is a simple Fact test method. We can also use the Theory test method to test the GetInvoicesAsync() method with different parameters. For example, we can test whether the controller can return correct invoices when we pass the status parameter. The test method is as follows:

```
[Theory]
[InlineData(InvoiceStatus.AwaitPayment)]
[InlineData(InvoiceStatus.Draft)]
public async Task GetInvoicesByStatus_
ShouldReturnInvoices(InvoiceStatus status)
{
    // Arrange
    await using var dbContext = _fixture.CreateDbContext();
    var emailServiceMock = new Mock<IEmailService>();
    var controller = new InvoiceController(dbContext,
emailServiceMock.Object);
    // Act
    var actionResult = await controller.GetInvoicesAsync(status:
status);
    // Assert
    var result = actionResult.Result as OkObjectResult;
    Assert.NotNull(result);
    var returnResult = Assert.IsAssignableFrom<List<Invoice>>(result.
Value);
    Assert.NotNull(returnResult);
    Assert.Single(returnResult);
    Assert.Equal(status, returnResult.First().Status);
}
```

In the preceding example, we use the `Theory` attribute to indicate that the test method is a `Theory` test method. A `Theory` test method can have one or more `InlineData` attributes. Each `InlineData` attribute can pass one value or multiple values to the test method. In this case, we use the `InlineData` attribute to pass the `InvoiceStatus` value to the test method. You can use multiple `InlineData` attributes to pass multiple values to the test method. The test method will be executed multiple times with different values.

The tests we introduced in this chapter are used to test read-only methods. They do not change the database, so we do not need to worry about the database state. In the next section, we will introduce how to write tests for methods that change the database.

Writing tests for methods that change the database

If a method changes the database, we need to ensure that the database is in a known state before we run the test, and also ensure that the database is restored to its original state so that the change will not affect other tests.

For example, a method may delete a record from the database. If the test method deletes a record from the database but does not restore the database after the test, the next test method may fail because the record is missing.

Let's create a test method for the `CreateInvoiceAsync()` method of the `InvoiceController` class. The `CreateInvoiceAsync()` method creates a new invoice in the database. The test method is as follows:

```
[Fact]
public async Task CreateInvoice_ShouldCreateInvoice()
{
    // Arrange
    await using var dbContext = fixture.CreateDbContext();
    var emailServiceMock = new Mock<IEmailService>();
    var controller = new InvoiceController(dbContext,
emailServiceMock.Object);
    // Act
    var contactId = dbContext.Contacts.First().Id;
    var invoice = new Invoice
    {
        DueDate = DateTimeOffset.Now.AddDays(30),
        ContactId = contactId,
        Status = InvoiceStatus.Draft,
        InvoiceDate = DateTimeOffset.Now,
        InvoiceItems = new List<InvoiceItem>
        {
            // Omitted for brevity
        }
```

```
    };
    var actionResult = await controller.CreateInvoiceAsync(invoice);
    // Assert
    var result = actionResult.Result as CreatedAtActionResult;
    Assert.NotNull(result);
    var returnResult = Assert.IsAssignableFrom<Invoice>(result.Value);
    var invoiceCreated = await dbContext.Invoices.
FindAsync(returnResult.Id);

    Assert.NotNull(invoiceCreated);
    Assert.Equal(InvoiceStatus.Draft, invoiceCreated.Status);
    Assert.Equal(500, invoiceCreated.Amount);
    Assert.Equal(3, dbContext.Invoices.Count());
    Assert.Equal(contactId, invoiceCreated.ContactId);

    // Clean up
    dbContext.Invoices.Remove(invoiceCreated);
    await dbContext.SaveChangesAsync();
}
```

In this test method, we create a new invoice and pass it to the `CreateInvoiceAsync()` method. Then, we use the `Assert` class to verify the result. Finally, we remove the invoice from the database and save the changes. Note that the result of the `CreateInvoiceAsync()` method is a `CreatedActionResult` object, which contains the created invoice. So, we should convert the result into a `CreatedAtActionResult` object, and then get the created invoice from the `Value` property. Also, in this test method, we have asserted that the `Amount` property of the created invoice is correct based on the invoice items.

When we run the test, an error may occur because the contact ID is incorrect, as shown here:

```
Assert.Equal() Failure
Expected: ae29a8ef-5e32-4707-8783-b6bc098c0ccb
Actual:   275de2a8-5e0f-420d-c68a-08db59a2942f
```

The error says that the `CreateInvoiceAsync()` method does not behave as expected. We can debug the application to find out why the contact ID is not saved correctly. The reason is that when we created `Invoice`, we only specified the `ContactId` property, not the `Contact` property. So, EF Core could not find the contact with the specified ID, and then it created a new contact with a new ID. To fix this issue, we need to specify the `Contact` property when we create the `Invoice` object. Add the following code before calling the `dbContext.Invoices.AddAsync()` method:

```
var contact = await dbContext.Contacts.FindAsync(invoice.ContactId);
if (contact == null)
{
```

```
        return BadRequest("Contact not found.");
    }
    invoice.Contact = contact;
```

Now, we can run the test again. This time, the test should pass. That is why unit tests are so important. They can help us find bugs early and fix them before we deploy the application to production.

In the preceding example, the data was created in the test method and then removed from the database after the test. There is another way to manage this scenario: using a transaction. We can use a transaction to wrap the test method, and then roll back the transaction after the test. So, the data created in the test method will not be saved to the database. In this way, we do not need to manually remove the data from the database.

Let's create a test for the UpdateInvoiceAsync() method of the InvoiceController class. The UpdateInvoiceAsync() method updates an invoice in the database. The test method is as follows:

```
[Fact]
public async Task  UpdateInvoice_ShouldUpdateInvoice()
{
    // Arrange
    await using var dbContext = fixture.CreateDbContext();
    var emailServiceMock = new Mock<IEmailService>();
    var controller = new InvoiceController(dbContext,
emailServiceMock.Object);
    // Act
    // Start a transaction to prevent the changes from being saved to
the database
    await dbContext.Database.BeginTransactionAsync();
    var invoice = dbContext.Invoices.First();
    invoice.Status = InvoiceStatus.Paid;
    invoice.Description = "Updated description";
    invoice.InvoiceItems.ForEach(x =>
    {
        x.Description = "Updated description";
        x.UnitPrice += 100;
    });
    var expectedAmount = invoice.InvoiceItems.Sum(x => x.UnitPrice *
x.Quantity);
    await controller.UpdateInvoiceAsync(invoice.Id, invoice);

    // Assert
    dbContext.ChangeTracker.Clear();
    var invoiceUpdated = await dbContext.Invoices.SingleAsync(x =>
x.Id == invoice.Id);
```

```
        Assert.Equal(InvoiceStatus.Paid, invoiceUpdated.Status);
        Assert.Equal("Updated description", invoiceUpdated.Description);
        Assert.Equal(expectedAmount, invoiceUpdated.Amount);
        Assert.Equal(2, dbContext.Invoices.Count());
}
```

In the `UpdateInvoice_ShouldUpdateInvoice()` method, before we call the `UpdateInvoiceAsync()` method, we start a transaction. After the test method is executed, we do not commit the transaction, so the transaction will roll back. The changes that are made in the test method will not be saved to the database. In this way, we do not need to manually remove the data from the database.

We also use the `ChangeTracker.Clear()` method to clear the change tracker. The change tracker is used to track the changes made to the entities. If we do not clear the change tracker, we will get the tracked entities instead of querying the database. So, we need to explicitly clear the change tracker before we query the database.

This approach is convenient when we test the methods that change the database. However, it can lead to a problem: what if the controller (or the service) method already starts a transaction? We cannot wrap a transaction in another transaction. In this case, we must explicitly clean up any changes made to the database after the test method is executed.

We can use the `IDisposable` interface to clean up the database in our tests. To do this, we can create a test class that implements the `IDisposable` interface, and then clean up the database in the `Dispose()` method. To set up the test context, let's create a class called `TransactionalTestDatabaseFixture`, as follows:

```
public class TransactionalTestDatabaseFixture
{
    private const string ConnectionString = @"Server=(localdb)\
mssqllocaldb;Database=InvoiceTransactionalTestDb;Trusted_
Connection=True";
    public TransactionalTestDatabaseFixture()
    {
        // This code comes from Microsoft Docs: https://github.
com/dotnet/EntityFramework.Docs/blob/main/samples/core/Testing/
TestingWithTheDatabase/TransactionalTestDatabaseFixture.cs
        using var context = CreateDbContext();
        context.Database.EnsureDeleted();
        context.Database.EnsureCreated();
        InitializeDatabase();
    }

    public InvoiceDbContext CreateDbContext()
        => new(new DbContextOptionsBuilder<InvoiceDbContext>()
            .UseSqlServer(ConnectionString)
```

```
                       .Options, null);

    public void InitializeDatabase()
    {
        using var context = CreateDbContext();
        // Create a few Contacts and Invoices
        // Omitted for brevity
        context.SaveChanges();
    }

    public void Cleanup()
    {
        using var context = CreateDbContext();
        context.Contacts.ExecuteDelete();
        context.Invoices.ExecuteDelete();
        context.SaveChanges();
        InitializeDatabase();
    }
}
```

In the preceding code, we create a database called `InvoiceTransactionalTestDb` and initialize it. This file is similar to the `InvoiceTestDatabaseFixture` class, except that it has a `Cleanup` method, which is used to clean up the database. In the `Cleanup` method, we delete all the contacts and invoices from the database and then initialize the database to restore the data.

In the `InvoiceController.cs` file, the `UpdateInvoiceStatusAsync` method uses a transaction to update the status of an invoice. It is not required; this is purely for demonstration purposes. Let's create a test class called `TransactionalInvoiceControllerTests` to test this method, as follows:

```
public class
TransactionalInvoiceControllerTests(TransactionalTestDatabaseFixture
fixture) : IClassFixture<TransactionalTestDatabaseFixture>,
IDisposable
{

    [Fact]
    public async Task UpdateInvoiceStatusAsync_ShouldUpdateStatus()
    {
        // Arrange
        await using var dbContext = _fixture.CreateDbContext();
        var emailServiceMock = new Mock<IEmailService>();
        var controller = new InvoiceController(dbContext,
emailServiceMock.Object);
        // Act
        var invoice = await dbContext.Invoices.FirstAsync(x =>
x.Status == InvoiceStatus.AwaitPayment);
```

```
        await controller.UpdateInvoiceStatusAsync(invoice.Id,
InvoiceStatus.Paid);
        // Assert
        dbContext.ChangeTracker.Clear();
        var updatedInvoice = await dbContext.Invoices.
FindAsync(invoice.Id);
        Assert.NotNull(updatedInvoice);
        Assert.Equal(InvoiceStatus.Paid, updatedInvoice.Status);
    }

    public void Dispose()
    {
        _fixture.Cleanup();
    }
}
```

In the preceding code, we use the `TransactionalTestDatabaseFixture` class to create the database context. This class implements the `IDisposable` interface and calls the `Cleanup()` method in the `Dispose()` method. If we have multiple test methods in one test class, xUnit will create a new instance of the test class for each test method and run them in sequence. Therefore, the `Dispose()` method will be called after each test method is executed to clean up the database, which will ensure that the changes made in the test methods will not affect other test methods.

What if we want to share `TransactionalTestDatabaseFixture` in multiple test classes? By default, xUnit will run the test classes in parallel. If other test classes also need to use this fixture to clean up the database, it may cause a concurrency issue when xUnit initializes the test context. To avoid this problem, we can use the `Collection` attribute to specify that the test classes that use this fixture belong to the same test collection so that xUnit will not run them in parallel. We'll discuss the parallelism of xUnit in the next section.

Parallelism of xUnit

By default, the latest version of xUnit (v2+) runs tests in parallel. This is because parallelism can improve the performance of the test. If we have a lot of tests, running them in parallel can save a lot of time. Also, it can leverage the multi-core CPU to run the tests. However, we need to understand how xUnit runs tests in parallel, just in case it causes problems.

xUnit uses a concept called **test collection** to represent a group of tests. By default, each test class is a unique test collection. Note that the tests in the same test class will not run in parallel.

For example, in the sample project, we can find an `InvoiceControllerTests.cs` file and a `ContactControllerTests.cs` file. So, xUnit will run these two test classes in parallel, but the tests in the same test class will not run in parallel.

We also introduced a `TestDatabaseFixture` class in the *Creating a test fixture* section. A class fixture is used to share a single test context among all the tests in the same test class. So, if we use a class fixture to create a database context, the database context will be shared among all the tests in the same test class. Currently, we have two test classes that use the `TestDatabaseFixture` class to provide the database context. Will xUnit create only one instance of the `TestDatabaseFixture` class for these two test classes?

The answer is no. We can set up a breakpoint in the constructor of the `TestDatabaseFixture` class, and then debug the tests by right-clicking `InvoiceApp.UnitTests` in the **Test Explorer** window of VS 2022 and then clicking **Debug**, as shown in *Figure 9.6*:

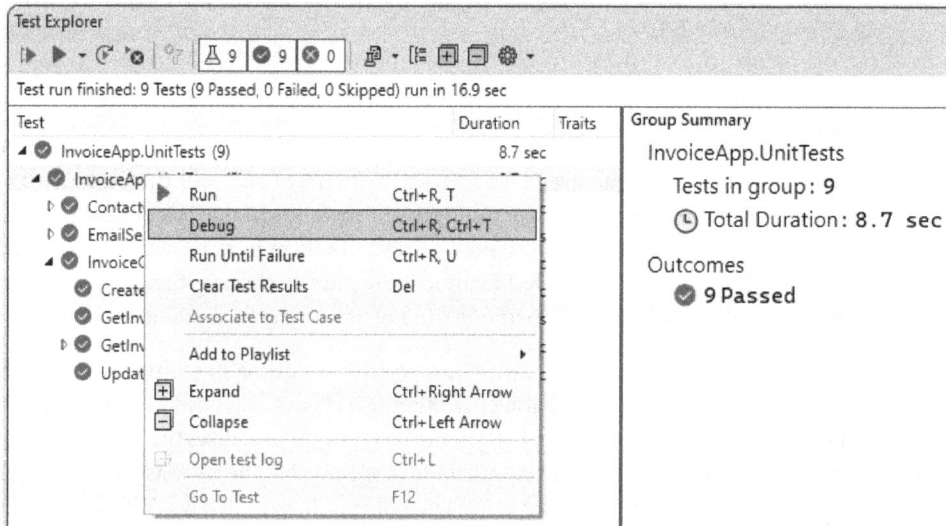

Figure 9.6 – Debug tests in VS 2022

You will find that the constructor of the `TestDatabaseFixture` class is called twice (or more, depending on how many tests use this text fixture). So, we know that xUnit will create a new instance of the `TestDatabaseFixture` class for each test class as well. That is why we use a lock to ensure that the database is only created once. If we do not use the lock, multiple test classes will try to initialize the database at the same time, which potentially causes problems.

In the *Writing tests for methods that change the database* section, we created a `Transactional-TestDatabaseFixture` class that can clean up the database. If we apply it to one test class, such as `TransactionalInvoiceControllerTests`, it will work fine. But what if we want to use it for multiple test classes? By default, xUnit will run these test classes in parallel, which means that multiple test classes will try to clean up the database at the same time. In this case, we do not want to run these test classes in parallel. To do this, we can use the `Collection` attribute to group these test classes into one collection so that xUnit will not run them in parallel. This can help us avoid the concurrency issue.

Let's see an example. In the sample project, you will find that the `ContactController` file has a method named `UpdateContactAsync()` that uses a transaction. Again, this is not required and is just for demonstration purposes. To use the collection fixture, we need to create a definition for the collection. Let's create a `TransactionalTestsCollection` class in the `InvoiceApp.UnitTests` project, as follows:

```
[CollectionDefinition("TransactionalTests")]
public class TransactionTestsCollection :
ICollectionFixture<TransactionalTestDatabaseFixture>
{
}
```

In this class, we declare that the `TransactionalTestDatabaseFixture` class is a collection fixture that uses the `CollectionDefinition` attribute. We also specify a name for this collection, which is `TransactionalTests`. Then, we use the `ICollectionFixture<T>` interface to specify that the `TransactionalTestDatabaseFixture` class is a collection fixture.

After that, we add the `Collection` attribute to the test classes, which specifies that the `TransactionalInvoiceControllerTests` and `TransactionalContactControllerTests` classes belong to the `TransactionalTests` collection, as follows:

```
[Collection("TransactionalTests")]
public class TransactionalInvoiceControllerTests : IDisposable
{
    // Omitted for brevity
}

[Collection("TransactionalTests")]
public class TransactionalContactControllerTests : IDisposable
{
    // Omitted for brevity
}
```

Now, if we debug the tests, we will find that the constructor of the `TransactionalTestDatabaseFixture` class is only called once, which means that xUnit will only create one instance of the `TransactionalTestDatabaseFixture` class for these two test classes. Also, xUnit will not run these two test classes in parallel, which means that the `Cleanup` method of the `TransactionalTestDatabaseFixture` class will not be called at the same time. So, we can use the `TransactionalTestDatabaseFixture` class to clean up the database for each test method in multiple test classes.

Let's summarize the key points of this section:

- By default, each test class is a unique test collection

- The tests in the same test class will not run in parallel

- If we want to share a single test context among all the tests within the same test class, we can use a class fixture:

  - xUnit creates a new instance of the test class for each test method

  - xUnit creates a new instance of the class fixture for each test class and shares the same instance among all the tests in the same test class

- By default, xUnit runs test classes in parallel if they are not in the same test collection

- If we do not want to run multiple test classes in parallel, we can use the `Collection` attribute to group them into one test collection

- If we want to share a single test context among several test classes and clean up the test context after each test method, we can use a collection fixture, and implement the `IDisposable` interface in each test class to clean up the test context:

  - xUnit creates a new instance of the test class for each test method

  - xUnit creates only one instance of the collection fixture for the test collection and shares the same instance among all the tests in the collection

  - xUnit does not run multiple test classes in parallel if they belong to the same test collection

xUnit provides a lot of features to customize the test execution. If you want to learn more about xUnit, you can check out the official documentation at `https://xunit.net/`.

Using the repository pattern

So far, you have learned how to use a real database to test the database access layer. There is another way to test the database access layer, which is to use a repository pattern to decouple the controllers from the `DbContext` class. In this section, we will show you how to use the repository pattern to test the database access layer.

The repository pattern is a common pattern that's used to separate the application and the database access layer. Instead of using `DbContext` directly in the controllers, we can add a separate repository layer to encapsulate the database access logic. The controllers will use the repository layer to access the database, as shown in *Figure 9.7*:

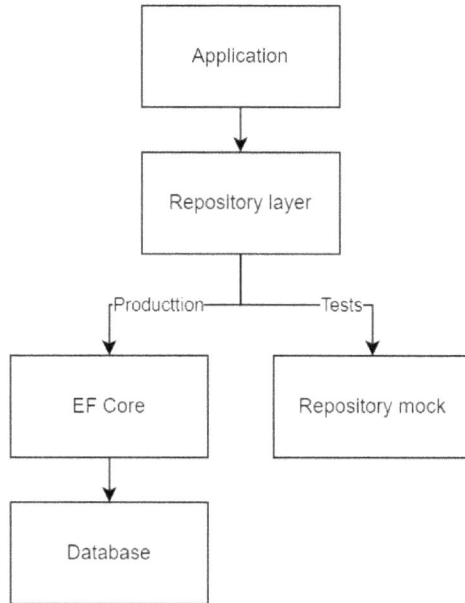

Figure 9.7– Using the repository pattern

In *Figure 9.7*, we can see that the application now has no dependency on EF Core. The application (controllers) only depends on the repository layer, and the repository layer depends on EF Core. Therefore, the repository layer can be mocked in the tests, and the controllers can be tested without a real database.

To learn how to use the repository pattern for testing, you can look at the sample project in the UnitTestsDemo\UnitTest-v2 folder. This project is based on the v1 project, and we have added a repository layer to the project.

The IInvoiceRepository interface defines the methods of the Invoice repository, as follows:

```
public interface IInvoiceRepository
{
    Task<Invoice?> GetInvoiceAsync(Guid id);
    Task<IEnumerable<Invoice>> GetInvoicesAsync(int page = 1, int
pageSize = 10, InvoiceStatus? status = null);
    Task<IEnumerable<Invoice>> GetInvoicesByContactIdAsync(Guid
contactId, int page = 1, int pageSize = 10, InvoiceStatus? status =
null);
    Task<Invoice> CreateInvoiceAsync(Invoice invoice);
    Task<Invoice?> UpdateInvoiceAsync(Invoice invoice);
    Task DeleteInvoiceAsync(Guid id);
}
```

The implementation of the `IInvoiceRepository` interface is in the `InvoiceRepository` class; it uses the DbContext class to access the database. First, we inject the `InvoiceDbContext` class into the `InvoiceRepository` class using the constructor injection, as follows:

```
public class InvoiceRepository(InvoiceDbContext dbContext) :
IInvoiceRepository
{
}
```

Next, we can implement the `IInvoiceRepository` interface in the `InvoiceRepository` class. Here is an example of the `GetInvoicesAsync` method:

```
public async Task<Invoice?> GetInvoiceAsync(Guid id)
{
    return await dbContext.Invoices.Include(i => i.Contact)
        .SingleOrDefaultAsync(i => i.Id == id);
}
```

In the `GetInvoiceAsync()` method, we use a LINQ query to get the invoice by the specified ID. Note that we use the `Include` method to include the `Contact` property in the query result. This is because we want to get the contact information of the invoice. If we do not want to include the navigation property in the query result, we can remove the `Include()` method, or add a parameter to the `GetInvoiceAsync()` method to specify whether to include the navigation property. The `Include()` method is defined in the `Microsoft.EntityFrameworkCore` namespace, so we need to add the `using Microsoft.EntityFrameworkCore;` statement to the `InvoiceRepository.cs` file.

The implementation of the `GetInvoicesAsync()` method is as follows:

```
public async Task<IEnumerable<Invoice>> GetInvoicesAsync(int page = 1,
int pageSize = 10, InvoiceStatus? status = null)
{
    return await dbContext.Invoices
        .Include(x => x.Contact)
        .Where(x => status == null || x.Status == status)
        .OrderByDescending(x => x.InvoiceDate)
        .Skip((page - 1) * pageSize)
        .Take(pageSize)
        .ToListAsync();
}
```

In the preceding `GetInvoicesAsync()` method, we use some LINQ methods, such as `Where()`, `OrderByDescending()`, `Skip()`, and `Take()`, to implement the pagination feature. Note that the `ToListAsync()` method is defined in the `Microsoft.EntityFrameworkCore` namespace, so do not forget to add the `using Microsoft.EntityFrameworkCore;` statement.

You can find the complete implementation of the `InvoiceRepository` class in the `InvoiceRepository.cs` file in the `UnitTestsDemo\UnitTest-v2` folder.

The implementation of the repository interface is just a class that uses the `DbContext` class to implement the CRUD operations. Generally, this layer does not contain any business logic. Also, we should note that the `GetInvoicesAsync()` method returns `IEnumerable<Invoice>` instead of `IQueryable<Invoice>`. This is because the `IQueryable` interface involves EF Core, but the purpose of using the repository pattern is to decouple the application from EF Core. So, we can easily mock the repository layer in the tests.

The controllers now depend on the repository layer, as shown here:

```
[Route("api/[controller]")]
[ApiController]
public class InvoiceController(IInvoiceRepository invoiceRepository,
IEmailService emailService)
    : ControllerBase

    // GET: api/Invoices
    [HttpGet]
    public async Task<ActionResult<List<Invoice>>>
GetInvoicesAsync(int page = 1, int pageSize = 10,
        InvoiceStatus? status = null)
    {
        var invoices = await invoiceRepository.GetInvoicesAsync(page,
pageSize, status);
        return Ok(invoices);
    }
    // Omitted for brevity
}
```

Now, the controller is much cleaner, and there is no dependency on EF Core. We can update the tests so that they use the repository layer instead of the `DbContext` class. Similar to that in the previous `InvoiceControllerTests`, we may need a class fixture to manage the test context, as shown here:

```
public class TestFixture
{
    public List<Invoice> Invoices { get; set; } = new();
    public List<Contact> Contacts { get; set; } = new();

    public TestFixture()
    {
        InitializeDatabase();
    }
```

```
        public void InitializeDatabase()
        {
            // Omited for brevity
        }
    }
```

In this class fixture, we have added two lists to mock the database tables. Next, we can mock the tests, like this:

```
public class InvoiceControllerTests(TestFixture fixture) :
IClassFixture<TestFixture>
{

    [Fact]
    public async Task GetInvoices_ShouldReturnInvoices()
    {
        // Arrange
        var repositoryMock = new Mock<IInvoiceRepository>();

        repositoryMock.Setup(x => x.GetInvoicesAsync(It.IsAny<int>(),
It.IsAny<int>(), It.IsAny<InvoiceStatus?>()))
            .ReturnsAsync((int page, int pageSize, InvoiceStatus?
status) =>
                fixture.Invoices.Where(x => status == null || x.Status
== status)
                    .OrderByDescending(x => x.InvoiceDate)
                    .Skip((page - 1) * pageSize)
                    .Take(pageSize)
                    .ToList());
        var emailServiceMock = new Mock<IEmailService>();
        var controller = new InvoiceController(repositoryMock.Object,
emailServiceMock.Object);
        // Act
        var actionResult = await controller.GetInvoicesAsync();
        // Assert
        var result = actionResult.Result as OkObjectResult;
        Assert.NotNull(result);
        var returnResult = Assert.
IsAssignableFrom<List<Invoice>>(result.Value);
        Assert.NotNull(returnResult);
        Assert.Equal(2, returnResult.Count);
        Assert.Contains(returnResult, i => i.InvoiceNumber == "INV-
001");
        Assert.Contains(returnResult, i => i.InvoiceNumber == "INV-
002");
```

```
    }
    // Omited for brevity
}
```

In this test method, we mock the repository layer and pass it to the controller. It follows the concept of the unit tests: focus on the unit under test and mock the dependencies. You can check the other tests in the source code and try to add more tests to cover the other scenarios, such as creating invoices, updating invoices, deleting invoices, and so on. Note that we use two instances of `List<T>` to mock the database tables. If the test methods change the data, do not forget to restore the data after the test methods are executed.

The repository pattern is a good practice for decoupling the application from the data access layer, and it also makes it possible to replace the data access layer with another one. It allows us to mock the database access layer for testing purposes. However, it increases the complexity of the application. Also, if we use the repository pattern, we may lose some features of EF Core, such as `IQueryable`. Finally, the mock behavior may be different from the real behavior. So, we should consider the trade-offs before using it.

Testing the happy path and the sad path

So far, we have written some tests to cover the happy path. However, we should also test the sad path. In testing, the terms **happy path** and **sad path** are used to describe different scenarios or test cases:

- **Happy path**: A happy path refers to the ideal or expected scenario where the code unit being tested performs as intended. The input data is valid, and the code unit produces the expected output. Happy path tests are designed to validate the typical or desired behavior of the code unit. These tests ensure that the code works as expected without any errors or exceptions when the input data is valid. For example, in the `GetInvoiceAsync(Guid id)` method, the happy path is that the invoice with the specified ID exists in the database, and the method returns the invoice.

- **Sad path**: A sad path, also known as the unhappy path, refers to the scenario where the code unit encounters an error or exception. The input data may be invalid, or the dependencies may not work as expected. Sad path tests are designed to validate whether the code unit can handle errors or exceptions gracefully. For example, in the `GetInvoiceAsync(Guid id)` method, the sad path is that the invoice with the specified ID does not exist in the database, and the method returns the `404 Not Found` error.

By combining both happy path and sad path tests, we can ensure that the code unit works as expected in different scenarios. Here is an example of a happy path for the `GetInvoiceAsync(Guid id)` method:

```
[Fact]
public async Task GetInvoice_ShouldReturnInvoice()
{
```

```
    // Arrange
    var repositoryMock = new Mock<IInvoiceRepository>();
    repositoryMock.Setup(x => x.GetInvoiceAsync(It.IsAny<Guid>()))
        .ReturnsAsync((Guid id) => fixture.Invoices.FirstOrDefault(x
=> x.Id == id));
    var emailServiceMock = new Mock<IEmailService>();
    var controller = new InvoiceController(repositoryMock.Object,
emailServiceMock.Object);
    // Act
    var invoice = fixture.Invoices.First();
    var actionResult = await controller.GetInvoiceAsync(invoice.Id);
    // Assert
    var result = actionResult.Result as OkObjectResult;
    Assert.NotNull(result);
    var returnResult = Assert.IsAssignableFrom<Invoice>(result.Value);
    Assert.NotNull(returnResult);
    Assert.Equal(invoice.Id, returnResult.Id);
    Assert.Equal(invoice.InvoiceNumber, returnResult.InvoiceNumber);
}
```

In this test method, we pass the ID of the first invoice in the Invoices list to the GetInvoiceAsync(Guid id) method. Since the invoice with the specified ID exists in the database, the method should return the invoice.

Let's create a sad path test for the GetInvoiceAsync(Guid id) method:

```
[Fact]
public async Task GetInvoice_ShouldReturnNotFound()
{
    // Arrange
    var repositoryMock = new Mock<IInvoiceRepository>();
    repositoryMock.Setup(x => x.GetInvoiceAsync(It.IsAny<Guid>()))
        .ReturnsAsync((Guid id) => _fixture.Invoices.FirstOrDefault(x
=> x.Id == id));
    var emailServiceMock = new Mock<IEmailService>();
    var coentroller = new InvoiceController(repositoryMock.Object,
emailServiceMock.Object);
    // Act
    var actionResult = await controller.GetInvoiceAsync(Guid.
NewGuid());
    // Assert
    var result = actionResult.Result as NotFoundResult;
    Assert.NotNull(result);
}
```

In this test method, we pass a new GUID to the `GetInvoiceAsync(Guid id)` method. Since the invoice with the specified ID does not exist in the database, the method should return a `404 Not Found` error. We can also create sad path tests for other methods.

> **Tip**
>
> What is the difference between `as` and `is` in C#?
>
> The `as` operator is used to perform conversions between compatible types. If the conversion is not possible, the `as` operator returns `null` instead of raising an exception. So, in the preceding test, if `result` is not `null`, we can see that the result from the controller is `NotFoundResult`, which is the expected result.
>
> The `is` operator is used to determine whether an object is compatible with a given type. If the object is compatible, the operator will return `true`; otherwise, it will return `false`. This is a useful tool for verifying the type of an object before performing an operation on it.
>
> From C# 7, we can use `is` to check and convert the type at the same time. For example, we can use `if (result is NotFoundResult notFoundResult)` to check whether `result` is `NotFoundResult`, and convert it into `NotFoundResult` at the same time.

With that, we have learned how to write unit tests for the controller. You can check the other tests in the source code and try to add more tests to cover the other scenarios, such as creating invoices, updating invoices, deleting invoices, and so on.

Summary

In this chapter, we explored the fundamentals of unit tests for ASP.NET web API applications. We discussed the use of xUnit as the testing framework and `Moq` as the mocking framework. We learned how to configure test fixtures with xUnit, and how to manage the test data with the test fixture. We also learned how to write unit tests to test the data access layer and the controller.

Unit tests are a great way to ensure that your code unit is working as expected. These tests often use mock objects to isolate the code unit from its dependencies, but this cannot guarantee that the code unit works well with its dependencies. Therefore, we also need to write integration tests to test if the code units can work together with their dependencies. For example, can the controllers handle the requests correctly?

In the next chapter, we will learn how to write integration tests for ASP.NET web API applications.

Testing in ASP.NET Core (Part 2 – Integration Testing)

In *Chapter 9*, we learned how to write unit tests for a ASP.NET Core web API application. Unit tests are used to test code units in isolation. However, a code unit often depends on other components, such as the database, external services, and so on. To test the code thoroughly, we need to test the code units in the context of the application. In other words, we need to test how the code units interact with other parts of the application. This type of testing is called integration testing.

In this chapter, we will mainly focus on integration testing. We will cover the following topics:

- Writing integration tests
- Testing with authentication and authorization
- Understanding code coverage

By the end of this chapter, you should be able to write integration tests for a ASP.NET Core web API application.

Technical requirements

The code examples in this chapter can be found at `https://github.com/PacktPublishing/Web-API-Development-with-ASP.NET-Core-8`. You can use VS Code or VS 2022 to open the solutions.

Writing integration tests

In the unit tests, we create the instances of the controllers directly. This approach does not consider some features of ASP.NET Core, such as routing, model binding, and validation and so on. To test the application thoroughly, we need to write integration tests. In this section, we will write integration tests for the application.

Unlike unit tests, which focus on isolated units, integration tests focus on the interactions between components. These integration tests may involve different layers, such as the database, the file system, the network, the HTTP request/response pipeline and so on. Integration tests ensure that the components of the application work together as expected. So, normally, integration tests use actual dependencies instead of mocks. Also, integration tests are slower than unit tests because they involve more components. Considering the cost of integration tests, we do not need to write too many integration tests. Instead, we should focus on the critical parts of the application. Most of the time, we can use unit tests to cover the other parts.

You can find the sample code of this section in the `IntegrationTestsDemo` folder. The code is based on the `InvoiceApp` project we created in *Chapter 9*. You can use VS 2022 or VS Code to open the solution. We will use the term **System Under Test (SUT)** to refer to the ASP.NET Core web API application we are testing.

Setting up the integration test project

We can continue to use **xUnit** as the test framework for integration tests. A good practice is to create a separate integration test project from the unit test project. This approach allows us to run the unit tests and integration tests separately and also makes it easier to use different configurations for the two types of tests.

If you are using VS 2022, you can create a new xUnit project by right-clicking on the solution and selecting **Add | New Project**. Then select **xUnit Test Project** from the list of project templates. You can name the project `InvoiceApp.IntegrationTests` and click **Create** to create the project. After you create the project, add a project reference to the `InvoiceApp.WebApi` project to allow the integration test project to access the classes in the web API project (see *Chapter 9*).

If you are using **.NET CLI**, you can create a new xUnit test project by running the following commands in the terminal:

```
dotnet new xunit -n InvoiceApp.IntegrationTests
dotnet sln InvoiceApp.sln add InvoiceApp.IntegrationTests/InvoiceApp.
IntegrationTests.csproj
dotnet add InvoiceApp.IntegrationTests/InvoiceApp.IntegrationTests.
csproj reference InvoiceApp.WebApi/InvoiceApp.WebApi.csproj
```

ASP.NET Core provides a built-in test web host that we can use to host the SUT to handle the HTTP requests. The benefit of using a test web host is that we can use a different configuration for the test environment, and it also saves the network traffic because the HTTP requests are handled in the same process. So, the tests using a test web host are faster than those using a real web host. To use the test web host, we need to add the `Microsoft.AspNetCore.Mvc.Testing` NuGet package to the integration test project. You can add the package in VS 2022 by right-clicking on the project and selecting **Manage NuGet Packages**. Then search for `Microsoft.AspNetCore.Mvc.Testing` and install the package.

You can also use the following command to add the package:

```
dotnet add InvoiceApp.IntegrationTests/InvoiceApp.IntegrationTests.
csproj package Microsoft.AspNetCore.Mvc.Testing
```

The default `UnitTest1.cs` file can be removed.

Feel free to install **FluentAssertions** if you want to use it in the tests, as we demonstrated in the *Using FluentAssertions to verify the test results* section in *Chapter 9*.

Now we can start to write integration tests.

Writing basic integration tests with WebApplicationFactory

Let'sstart with a simple integration test to check whether the SUT can correctly handle the HTTP requests. The sample application has a `WeatherForecastController.cs` controller that is provided by the ASP.NET Core project template. It returns a list of weather forecasts. We can write an integration test to check whether the controller returns the expected result.

Create a new file named `WeatherForecastApiTests` in the `InvoiceApp.IntegrationTests` project. Then add the following code to the file:

```
public class WeatherForecastApiTests(WebApplicationFactory<Program>
factory)
    : IClassFixture<WebApplicationFactory<Program>>
{
}
```

In the test class, we use the `WebApplicationFactory<T>` type to create a test web host and use it as the class fixture. The instance of this class fixture will be shared across the tests in the class. The `WebApplicationFactory<T>` type is provided by the `Microsoft.AspNetCore.Mvc.Testing` package. It is a generic type that allows us to create a test web host for the specified application entry point. In this case, we use the `Program` class defined in the web API project as the entry point. But you will see an error that says CS0122 `'Program' is inaccessible due to its protection level`. This is because the `Program` class is defined as `internal` by default.

To solve this issue, there are two ways:

- Open the `InvoiceApp.WebApi.csproj` file and add the following line to the file:

    ```
    <ItemGroup>
        <InternalsVisibleTo Include="MyTestProject" />
    </ItemGroup>
    ```

    Replace `MyTestProject` with the name of your test project, such as `InvoiceApp.IntegrationTests`. This approach allows the test project to access the internal members of the web API project.

- Alternatively, you can change the access modifier of the `Program` class to `public`. Add the following code to the end of the `Program.cs` file:

```
public partial class Program { }
```

You can use either approach to solve the issue. After that, we can write the test method, as follows:

```
[Fact]
public async Task GetWeatherForecast_
ReturnsSuccessAndCorrectContentType()
{
    // Arrange
    var client = factory.CreateClient();
    // Act
    var response = await client.GetAsync("/WeatherForecast");
    // Assert
    response.EnsureSuccessStatusCode(); // Status Code 200-299
    Assert.Equal("application/json; charset=utf-8", response.Content.
Headers.ContentType.ToString());
    // Deserialize the response
    var responseContent = await response.Content.ReadAsStringAsync();
    var weatherForecast = JsonSerializer.
Deserialize<List<WeatherForecast>>(responseContent, new
JsonSerializerOptions
    {
        PropertyNameCaseInsensitive = true
    });
    weatherForecast.Should().NotBeNull();
    weatherForecast.Should().HaveCount(5);
}
```

In the test method, we first create an instance of the `HttpClient` class using the `WebApplicationFactory<T>` instance. Then we send an HTTP GET request to the `/WeatherForecast` endpoint. The `EnsureSuccessStatusCode` method ensures that the response has a status code in the *200-299* range. Then we check whether the content type of the response is `application/json; charset=utf-8`. Finally, we deserialize the response content to a list of `WeatherForecast` objects and check whether the list contains five items.

Because this controller does not have any dependencies, the test is simple. What if the controller has dependencies, such as a database context, other services, or other external dependencies? We will see how to handle these scenarios in the following sections.

Testing with a database context

In the sample application, the `ContactController` class has dependencies, such as the `IContactRepository` interface. The `ContactRepository` class implements this interface and uses the `InvoiceContext` class to access the database. So, if we want to test whether the SUT can correctly handle the HTTP requests, we need to create a test database and configure the test web host to use the test database. Similar to the unit tests, we can use a separate database for integration tests.

The `WebApplicationFactory<T>` type provides a way to configure the test web host. We can override the `ConfigureWebHost` method to configure the test web host. For example, we can replace the default database context with a test database context. Let us create a new text fixture class named `CustomIntegrationTestsFixture` and add the following code to the class:

```
public class CustomIntegrationTestsFixture :
WebApplicationFactory<Program>
{
    private const string ConnectionString = @"Server=(localdb)\
mssqllocaldb;Database=InvoiceIntegrationTestDb;Trusted_
Connection=True";

    protected override void ConfigureWebHost(IWebHostBuilder builder)
    {
        // Set up a test database
        builder.ConfigureServices(services =>
        {
            var descriptor = services.SingleOrDefault(d =>
 d.ServiceType == typeof(DbContextOptions<InvoiceDbContext>));
            services.Remove(descriptor);
            services.AddDbContext<InvoiceDbContext>(options =>
            {
                options.UseSqlServer(ConnectionString);
            });
        });
    }
}
```

In the preceding code, we override the `ConfigureWebHost()` method to configure the test web host for the SUT. When the test web host is created, the `Program` class will execute first, which means the default database context defined in the `Program` class will be created. Then the `ConfigureWebHost()` method defined in the `CustomIntegrationTestsFixture` class will be executed. So we need to find the default database context using `services.SingleOrDefault(d => d.ServiceType == typeof(DbContextOptions<InvoiceDbContext>))` and then remove it from the service collection. Then we add a new database context that uses the test database. This approach allows us to use a separate database for integration tests. We also need to create the test database and seed some test data when we initialize the test fixture.

You can also add more customizations to the test web host in the `ConfigureWebHost()` method. For example, you can configure the test web host to use a different configuration file, as follows:

```
builder.ConfigureAppConfiguration((context, config) =>
{
    config.AddJsonFile("appsettings.IntegrationTest.json");
});
```

The default environment for the test web host of the SUT is `Development`. If you want to use a different environment, you can use the `UseEnvironment()` method, as follows:

```
builder.UseEnvironment("IntegrationTest");
```

Next, we need a way to create the test database and seed some test data. Create a static class named `Utilities` and add the following code to the class:

```
public static class Utilities
{
    public static void InitializeDatabase(InvoiceDbContext context)
    {
        context.Database.EnsureDeleted();
        context.Database.EnsureCreated();
        SeedDatabase(context);
    }

    public static void Cleanup(InvoiceDbContext context)
    {
        context.Contacts.ExecuteDelete();
        context.Invoices.ExecuteDelete();
        context.SaveChanges();
        SeedDatabase(context);
    }

    private static void SeedDatabase(InvoiceDbContext context)
    {
        // Omitted for brevity
    }
}
```

The `Utilities` class contains a few static methods that help us manage the test database. We need to initialize the test database before we run the tests and clean up the test database after we run the tests that change the data in the database.

When should we initialize the test database? We learned that the instance of the class fixture is created before the test class is initialized and is shared among all test methods in the test class. So, we can initialize the test database in the class fixture. Update the following code in the `ConfigureWebHost()` method of the `CustomIntegrationTestsFixture` class:

```
builder.ConfigureServices(services =>
{
    var descriptor = services.SingleOrDefault(d => d.ServiceType ==
typeof(DbContextOptions<InvoiceDbContext>));
    services.Remove(descriptor);
    services.AddDbContext<InvoiceDbContext>(options =>
    {
        options.UseSqlServer(ConnectionString);
    });
    using var scope = services.BuildServiceProvider().CreateScope();
    var scopeServices = scope.ServiceProvider;
    var dbContext = scopeServices.
GetRequiredService<InvoiceDbContext>();
    Utilities.InitializeDatabase(dbContext);
});
```

When the test web host of SUT is created, we replace the default database context with a test database context, and also initialize the test database, so that all the test methods in the test class can use the same test database.

Next, we can create a new test class named `InvoicesApiTests`, which is used to test the `/api/invoices` endpoint. Add the following code to the class:

```
public class InvoicesApiTests(CustomIntegrationTestsFixture factory) :
IClassFixture<CustomIntegrationTestsFixture>
{
}
```

The `InvoicesApiTests` test class will be initialized with an instance of the `CustomIntegrationTestsFixture` class. Then we can create some test methods to test the `/api/invoices` endpoint. A test method to test the GET `/api/invoices` endpoint might look as follows:

```
[Fact]
public async Task GetInvoices_ReturnsSuccessAndCorrectContentType()
{
    // Arrange
    var client = _factory.CreateClient();
    // Act
    var response = await client.GetAsync("/api/invoice");
    // Assert
```

```
    response.EnsureSuccessStatusCode(); // Status Code 200-299
    response.Content.Headers.ContentType.Should().NotBeNull();
    response.Content.Headers.ContentType!.ToString().Should().
Be("application/json; charset=utf-8");
    // Deserialize the response
    var responseContent = await response.Content.ReadAsStringAsync();
    var invoices = JsonSerializer.
Deserialize<List<Invoice>>(responseContent, new JsonSerializerOptions
    {
        PropertyNameCaseInsensitive = true
    });
    invoices.Should().NotBeNull();
    invoices.Should().HaveCount(2);
}
```

As this request to the GET /api/invoices endpoint does not change the database, the test method is straightforward.

Next, let's see how to test the POST /api/invoices endpoint, which changes the database. A test method to test this endpoint might look as follows:

```
[Fact]
public async Task PostInvoice_ReturnsSuccessAndCorrectContentType()
{
    // Arrange
    var client = factory.CreateClient();
    var invoice = new Invoice
    {
        DueDate = DateTimeOffset.Now.AddDays(30),
        ContactId = Guid.Parse("8a9de219-2dde-4f2a-9ebd-
b1f8df9fef03"),
        Status = InvoiceStatus.Draft,
        InvoiceItems = new List<InvoiceItem>
        {

            // Omitted for brevity        }
    };
    var json = JsonSerializer.Serialize(invoice);
    var data = new StringContent(json, Encoding.UTF8, "application/
json");
    // Act
    var response = await client.PostAsync("/api/invoice", data);
    // Assert
    response.EnsureSuccessStatusCode(); // Status Code 200-299
    response.Content.Headers.ContentType.Should().NotBeNull();
```

```
    response.Content.Headers.ContentType!.ToString().Should().
Be("application/json; charset=utf-8");
    // Deserialize the response
    var responseContent = await response.Content.ReadAsStringAsync();
    var invoiceResponse = JsonSerializer.
Deserialize<Invoice>(responseContent, new JsonSerializerOptions
    {
        PropertyNameCaseInsensitive = true
    });
    invoiceResponse.Should().NotBeNull();
    invoiceResponse!.Id.Should().NotBeEmpty();
    invoiceResponse.Amount.Should().Be(500);
    invoiceResponse.Status.Should().Be(invoice.Status);
    invoiceResponse.ContactId.Should().Be(invoice.ContactId);

    // Clean up the database
    var scope = factory.Services.CreateScope();
    var scopedServices = scope.ServiceProvider;
    var db = scopedServices.GetRequiredService<InvoiceDbContext>();
    Utilities.Cleanup(db);
}
```

As this request to the POST /api/invoices endpoint changes the database, we need to clean up the database after we run the test. To get the current instance of the database context, we need to create a new scope and get the database context from the scope. Then we can use the Cleanup method of the Utilities class to clean up the database.

A test for the sad path of the POST /api/invoices endpoint might look as follows:

```
[Fact]
public async Task PostInvoice_WhenContactIdDoesNotExist_
ReturnsBadRequest()
{
    // Arrange
    var client = factory.CreateClient();
    var invoice = new Invoice
    {
        DueDate = DateTimeOffset.Now.AddDays(30),
        ContactId = Guid.NewGuid(),
        Status = InvoiceStatus.Draft,
        InvoiceItems = new List<InvoiceItem>
        {
            // Omitted for brevity
        }
    };
```

```
    var json = JsonSerializer.Serialize(invoice);
    var data = new StringContent(json, Encoding.UTF8, "application/
json");
    // Act
    var response = await client.PostAsync("/api/invoice", data);
    // Assert
    response.StatusCode.Should().Be(HttpStatusCode.BadRequest);
}
```

You can try to add more integration tests per your needs. Note that we need to manage the test database carefully. If you have multiple test classes that change the database, you may need to follow the same pattern (using a lock or collection fixture) as that in the unit tests we introduced in the *Creating a test fixture* section in *Chapter 9*, to ensure the test database is clean before each test class runs and clean up the test database after each test class runs.

For example, if we have another `ContactsApiTests` class that also uses `CustomIntegrationTestsFixture`, xUnit will run `InvoicesApiTests` and `ContactsApiTests` in parallel. This may cause issues because both test classes try to initialize the test database at the same time. To avoid this issue, we can run these test classes in one collection to ensure that they run sequentially.

We can demonstrate this by creating a test collection named `CustomIntegrationTests` and adding the `CollectionDefinition` attribute to the `CustomIntegrationTestsCollection` class. This attribute will enable us to define the collection and its associated tests, as shown here:

```
[CollectionDefinition("CustomIntegrationTests")]
public class CustomIntegrationTestsCollection :
ICollectionFixture<CustomIntegrationTestsFixture>
{
}
```

Then we can add the `Collection` attribute to the `InvoicesApiTestsWithCollection` and `ContactsApiTestsWithCollection` classes. For example, the `InvoicesApiTestsWithCollection` class might look as follows:

```
[Collection("CustomIntegrationTests")]
public class
InvoicesApiTestsWithCollection(CustomIntegrationTestsFixture factory)
: IDisposable
{
    // Omitted for brevity
}
```

You can find the complete source code in the sample repo. Note that the normal integration test class, `InvoicesApiTests`, does not have the `Collection` attribute, so xUnit will run it in parallel with the `CustomIntegrationTests` collection. To avoid conflicts, we can skip the test methods in the `InvoicesApiTests` class, as shown here:

```
[Fact(Skip = "This test is skipped to avoid conflicts with the test
collection")]
public async Task GetInvoices_ReturnsSuccessAndCorrectContentType()
{
    // Omitted for brevity
}
```

When you run the tests in the sample repo, please add or comment out the `Skip` attributes per your needs.

Testing with mock services

As we explained in *Chapter 9*, unit tests should focus on isolated units of code. So, unit tests often use mock or stub services to isolate the code under test from other services. Integration tests, on the other hand, should test the integration between different components. So, technically, integration tests should use real services instead of mock services.

However, in some cases, using mock dependencies in integration tests can be helpful. For example, in the sample invoice application, we need to call a third-party service to send emails. If we use the real email service in the integration tests, it may have the following issues:

- The actual email service may not be available in the test environment. For example, the email service may be hosted in a different environment, and the test environment may not be able to access the email service.

- The email service may have rate limits, strict network policies, or other restrictions that may cause issues in the integration tests or slow down the test execution.

- The email service may cause unnecessary costs in the test environment, especially if the service has usage-based pricing or requires a paid subscription. If we run the integration tests frequently, it may incur high costs.

- The test may impact production and cause issues for real users.

In this case, using a mock email service in the integration can help us avoid these issues so that we can run the tests faster and more efficiently, avoid impacting production, and save costs.

Let's see how to use a mock email service in the integration tests. The service we use to send email is the IEmailSender interface. We can inject a mock service that implements the IEmailSender interface in the integration tests. Create a new test method in the test class:

```
[Theory]
[InlineData("7e096984-5919-492c-8d4f-ce93f25eaed5")]
[InlineData("b1ca459c-6874-4f2b-bc9d-f3a45a9120e4")]
public async Task SendInvoiceAsync_
ReturnsSuccessAndCorrectContentType(string invoiceId)
{
    // Arrange
    var mockEmailSender = new Mock<IEmailSender>();
    mockEmailSender.Setup(x => x.SendEmailAsync(It.IsAny<string>(),
It.IsAny<string>(), It.IsAny<string>()))
        .Returns(Task.CompletedTask).Verifiable();
    var client = factory.WithWebHostBuilder(builder =>
    {
        builder.ConfigureTestServices(services =>
        {
            var emailSender = services.SingleOrDefault(x =>
x.ServiceType == typeof(IEmailSender));
            services.Remove(emailSender);
            services.AddScoped<IEmailSender>(_ => mockEmailSender.
Object);
        });
    }).CreateClient();
    // Act
    var response = await client.PostAsync($"/api/invoice/{invoiceId}/
send", null);
    // Assert
    response.EnsureSuccessStatusCode(); // Status Code 200-299
    mockEmailSender.Verify(x => x.SendEmailAsync(It.IsAny<string>(),
It.IsAny<string>(), It.IsAny<string>()), Times.Once);
    var scope = factory.Services.CreateScope();
    var scopedServices = scope.ServiceProvider;
    var db = scopedServices.GetRequiredService<InvoiceDbContext>();
    var invoice = await db.Invoices.FindAsync(Guid.Parse(invoiceId));
    invoice!.Status.Should().Be(InvoiceStatus.AwaitPayment);
}
```

In the preceding code, we created a mock email sender service using the Moq library. The test web host of SUT provides a WithWebHostBuilder() method to configure the web host builder. In this method, we can configure the service collection of the web host using the ConfigureTestServices() method. Similar to the mock database context we introduced in the *Testing with a database context* section, we find the registered IEmailSender service and remove it from the service collection, then add the mock service to the service collection. Finally, we create the HTTP client and send the request to the API endpoint. If the mock service is used correctly, the test should pass.

In summary, whether to use mock services in integration tests or not should be decided on a case-by-case basis and depends on the specific requirements and objectives of the tests. Mocks can be useful in certain scenarios, but they should not be overused, otherwise, the integration tests may not be able to reflect the real-world scenarios.

If your web API project runs in a microservice architecture, you may need to call other microservices in the integration tests. In this case, you can use the same approach to mock the HTTP client and the HTTP responses. The integration tests may be more complicated. We will stop here and explore more when we discuss the microservice architecture in the next chapter.

Testing with authentication and authorization

A common scenario in web APIs is that some API endpoints require authentication and authorization. We introduced how to implement authentication and authorization in *Chapter 8*. In this section, we will discuss how to test the API endpoints that require authentication and authorization.

Preparing the sample application

To demonstrate testing with authentication and authorization, we will use the sample application we created in *Chapter 8*. You can find the source code in the `chapter10\AuthTestsDemo\start\` folder of the sample repo. This sample application uses claims-based authentication and authorization. You can recap the implementation details in *Chapter 8*.

In `WeatherForecastController`, there are several methods that require authentication and authorization. (Forgive the naming – we just use the default template of ASP.NET Core web APIs.)

Create a new integration test project as described in the *Setting up the integration test project* section. You need to ensure that the integration test project has the following NuGet packages installed:

- `Microsoft.AspNetCore.Mvc.Testing`: This is the test web host of SUT
- `xUnit`: This is the test framework
- `Moq`: This is the mocking library
- `FluentAssertions`: This is the assertion library (optional)

If you want to run the tests in VS 2022, you also need to install `xunit.runner.visualstudio`, the test runner for VS 2022.

Also add the reference to the sample application project. For simplicity, we will only focus on the integration tests for authentication and authorization. So, this demo does not involve the database context and the database.

Another thing you need to do is make the `Program` class public. Just add the following code at the end of the `Program` file:

```
public partial class Program { }
```

Next, we can start to write the integration tests for the API endpoints that require authentication and authorization.

Creating a test fixture

As we explained in the *Creating a test fixture* section in *Chapter 9*, we can create a test fixture to share the common code among the tests. Create a new class named `IntegrationTestsFixture` and add the following code:

```
public class IntegrationTestsFixture : WebApplicationFactory<Program>
{
    protected override void ConfigureWebHost(IWebHostBuilder builder)
    {
        // This is where you can set up your test server with the
services you need
        base.ConfigureWebHost(builder);
    }
}
```

This is a simple test fixture that inherits from the `WebApplicationFactory<Program>` class. Because we do not need to set up any services in this demo, there is no custom code in the `ConfigureWebHost` method. If you need to set up services, you can do it in this method.

Creating the test class

Next, we can create the test class. Create a new class named `AuthTests` and add the following code:

```
public class AuthTests(IntegrationTestsFixture fixture) :
IClassFixture<IntegrationTestsFixture>
{
}
```

This is similar to the test class we created in the *Using the test fixture* section in *Chapter 9*. It inherits from the `IClassFixture<IntegrationTestsFixture>` interface and has a constructor that accepts an `IntegrationTestsFixture` instance. So, the test class can use the `IntegrationTestsFixture` instance to access the test web host of SUT. So far, there is no special code in the test class.

Testing the anonymous API endpoints

Next, let us test the API endpoints that do not require authentication and authorization. In the `WeatherForecastController` class, copy the `Get()` method and paste it below the `Get()` method. Rename the new method `GetAnonymous()` and add the `AllowAnonymous` attribute. The new method should look like the following code:

```
[AllowAnonymous]
[HttpGet("anonymous")]
public IEnumerable<WeatherForecast> GetAnonymous()
{
    // Omitted for brevity
}
```

Now, we have a new API endpoint that does not require authentication and authorization. Create a new test method in the test class:

```
[Fact]
public async Task GetAnonymousWeatherForecast_ShouldReturnOk()
{
    // Arrange
    var client = fixture.CreateClient();

    // Act
    var response = await client.GetAsync("/weatherforecast/
anonymous");

    // Assert
    response.EnsureSuccessStatusCode(); // Status Code 200-299
    response.Content.Headers.ContentType.Should().NotBeNull();
    response.Content.Headers.ContentType!.ToString().Should().
Be("application/json; charset=utf-8");
    // Deserialize the response
    var responseContent = await response.Content.ReadAsStringAsync();
    var weatherForecasts = JsonSerializer.
Deserialize<List<WeatherForecast>>(responseContent, new
JsonSerializerOptions
    {
        PropertyNameCaseInsensitive = true
    });
    weatherForecasts.Should().NotBeNull();
    weatherForecasts.Should().HaveCount(5);
}
```

There are not many differences between this test method and the test method we created in the *Using the test fixture* section in *Chapter 9*. When we use the `CreateClient()` method, there is no special code to set up `HttpClient`. So, the test method can send the request to the API endpoint without any authentication and authorization. Because this endpoint allows anonymous access, the test should pass.

Testing the authorized API endpoints

The `WeatherForecastController` class has an `Authorize` attribute. So, the API endpoints that do not have the `AllowAnonymous` attribute require authentication and authorization. Let's test the sad path for the `Get()` method. Create a new test method in the test class:

```
[Fact]
public async Task GetWeatherForecast_ShouldReturnUnauthorized_
WhenNotAuthorized()
{
    // Arrange
    var client = fixture.CreateClient();
    // Act
    var response = await client.GetAsync("/weatherforecast");
    // Assert
    response.StatusCode.Should().Be(HttpStatusCode.Unauthorized);
}
```

This test method is similar to the previous `GetAnonymousWeatherForecast_ShouldReturnOk()` method, but we expect the status code to be `401 Unauthorized` because the API endpoint requires authentication and authorization. This test should pass as well.

Next, we need to set up the authentication and authorization in the test. There are several ways to do this:

- In the test, call the authentication endpoint to get the access token. Then add the access token to the `Authorization` header of the HTTP request. However, this approach is not recommended because it needs additional effort to maintain the credentials, such as the username, password, client ID, client secret, and so on. Also, the tests may not be able to access the authentication endpoint in the test environment. If the tests depend on the authentication endpoint, it increases the complexity of the tests.

- Create a helper method to generate the access token. Then add the access token to the `Authorization` header of the HTTP request. This approach does not need to call the authentication endpoint in the test. However, it means that we need to know how to generate the access token. If the authentication logic is provided by a third-party provider, we may not be able to make the same implementation in the test. So, it is only available if we have full control of the authentication logic.

- Use `WebApplicationFactory` to set up the authentication and authorization and create a custom `AuthenticationHandler` to simulate the authentication and authorization process. This approach is more practical because it does not need to call the authentication endpoint in the test. Also, it does not need to duplicate the authentication logic in the test project.

As we have the source code of the sample application, which includes the authentication logic, we can demonstrate how to use the second approach, and then we will show you how to use the third approach.

Generating the access token in the test

The code we use to generate the access token is from the `AccountController` class, which is the authentication endpoint. We can find a `GenerateToken` method in the `AccountController` class. This method is invoked when the user successfully logs in. Create a new method in the `IntegrationTestsFixture` class:

```
public string? GenerateToken(string userName)
{
    using var scope = Services.CreateScope();
    var configuration = scope.ServiceProvider.
GetRequiredService<IConfiguration>();
    var secret = configuration["JwtConfig:Secret"];
    var issuer = configuration["JwtConfig:ValidIssuer"];
    var audience = configuration["JwtConfig:ValidAudiences"];
    // Omitted for brevity
    var securityToken = tokenHandler.CreateToken(tokenDescriptor);
    var token = tokenHandler.WriteToken(securityToken);
    return token;
}
```

In the preceding method, we use the `IConfiguration` service to get the secret, issuer, and audience from the configuration. Then we copy the code from the `GenerateToken()` method in the `AccountController` class to generate the access token. Note that the configuration comes from the `appsettings.json` file in the main web API project. As we did not change the configuration of the test web host, the configuration is the same as that in the main web API project. But if you need to use a different configuration for the tests, please add proper code to the `ConfigureWebHost` method in the `IntegrationTestsFixture` class to apply any changes, as we introduced in the *Creating a test fixture* section.

Next, we can use the `GenerateToken` method in the `AuthTest` class. Create a new test method in the test class:

```
[Fact]
public async Task GetWeatherForecast_ShouldReturnOk_WhenAuthorized()
{
    // Arrange
```

```
        var token = fixture.GenerateToken("TestUser");
        var client = fixture.CreateClient();
        client.DefaultRequestHeaders.Authorization = new
    AuthenticationHeaderValue("Bearer", token);
        // Act
        var response = await client.GetAsync("/weatherforecast");
        // Assert
        response.EnsureSuccessStatusCode(); // Status Code 200-299
        // Omited for brevity
    }
```

In this test method, we call the `GenerateToken()` method to generate an access token, and then add the access token to the `Authorization` header of the HTTP request. Because the logic we use to generate the token is the same as the authentication endpoint, the test should pass.

Using a custom authentication handler

Another way to test the authorized API endpoints is to use a custom authentication handler. A custom authentication handler can simulate the authentication and authorization process. So, we can use it to test the authorized API endpoints without calling the authentication endpoint. This is the recommended approach to test the authorized API endpoints because it does not need any other dependencies, nor does it need to duplicate the authentication logic in the test project.

In the actual authentication process, we need to generate a JWT token that includes the claims of the authenticated user and add it to the `Authorization` header of the HTTP request. If we use a custom authentication handler, we can skip the process of generating the JWT token, but we still need to find a way to define the claims that we need and pass them to the custom authentication handler. We can simply add the claims in the request headers and then read the values in the custom authentication handler to create the `ClaimsPrincipal` object. Let us demonstrate how to do this.

To use a custom authentication handler, first create a new class named `TestAuthHandler`, which inherits from the `AuthenticationHandler` class:

```
public class TestAuthHandler :
AuthenticationHandler<AuthenticationSchemeOptions>
{
    public const string AuthenticationScheme = "TestScheme";
    public const string UserNameHeader = "UserName";
    public const string CountryHeader = "Country";
    public const string AccessNumberHeader = "AccessNumber";
    public const string DrivingLicenseNumberHeader =
"DrivingLicenseNumber";

    public
TestAuthHandler(IOptionsMonitor<AuthenticationSchemeOptions> options,
ILoggerFactory logger,
```

```
        UrlEncoder encoder, ISystemClock clock) : base(options,
logger, encoder, clock)
    {
    }


    protected override Task<AuthenticateResult>
HandleAuthenticateAsync()
    {
        return Task.FromResult(result);
    }
}
```

In the preceding code, we define the authentication scheme name as `TestScheme`, which is an alternative name to the actual scheme name, `Bearer`. You can find the definition in the `Program` class. Also, we define a few names for HTTP headers, which we will use to pass the claims to the custom authentication handler. The `HandleAuthenticateAsync()` method is the method that we need to override to implement the authentication logic. We will implement it in the following code.

The idea is that when we create the request in the test, we simply add the claims to the request headers. So that the custom authentication handler can read the values from the request headers, update the `HandleAuthenticateAsync()` method as follows:

```
protected override Task<AuthenticateResult> HandleAuthenticateAsync()
{
    var claims = new List<Claim>();

    if (Context.Request.Headers.TryGetValue(UserNameHeader, out var
userName))
    {
        claims.Add(new Claim(ClaimTypes.Name, userName[0]));
    }

    if (Context.Request.Headers.TryGetValue(CountryHeader, out var
country))
    {
        claims.Add(new Claim(ClaimTypes.Country, country[0]));
    }

    if (Context.Request.Headers.TryGetValue(AccessNumberHeader, out
var accessNumber))
    {
        claims.Add(new Claim(AppClaimTypes.AccessNumber,
accessNumber[0]));
    }
```

```
    if (Context.Request.Headers.
TryGetValue(DrivingLicenseNumberHeader, out var drivingLicenseNumber))
    {
        claims.Add(new Claim(AppClaimTypes.DrivingLicenseNumber,
drivingLicenseNumber[0]));
    }

    // You can add more claims here if you want

    var identity = new ClaimsIdentity(claims, AuthenticationScheme);
    var principal = new ClaimsPrincipal(identity);
    var ticket = new AuthenticationTicket(principal,
AuthenticationScheme);
    var result = AuthenticateResult.Success(ticket);
    return Task.FromResult(result);
}
```

Instead of getting claims from the JWT token, we get the claims from the request headers. If the values exist, we add them to the `ClaimsIdentity` object. Then we create the `ClaimsPrincipal` object and the `AuthenticationTicket` object. Finally, we return the `AuthenticateResult` object with the `Success` status. This method simulates the authentication process, which avoids the need to generate the JWT token, but it still creates the `ClaimsPrincipal` object that we need to test the authorized API endpoints.

Next, we can test the authorized API endpoints by using the custom authentication handler. In the `WeatherForecastController` class, we can find a `GetDrivingLicense` method, which is an authorized API endpoint that needs the `DrivingLicenseNumber` claim. We can create a new test method in the `AuthTest` class as follows:

```
[Fact]
public async Task GetDrivingLicense_ShouldReturnOk_
WhenAuthorizedWithTestAuthHandler()
{
    // Arrange
    var client = fixture.WithWebHostBuilder(builder =>
    {
        builder.ConfigureTestServices(services =>
        {
            services.AddAuthentication(options =>
                {
                    options.DefaultAuthenticateScheme =
TestAuthHandler.AuthenticationScheme;
                    options.DefaultChallengeScheme = TestAuthHandler.
AuthenticationScheme;
                    options.DefaultScheme = TestAuthHandler.
AuthenticationScheme;
```

```
                })
                .AddScheme<AuthenticationSchemeOptions,
    TestAuthHandler>(TestAuthHandler.AuthenticationScheme,
                    options => { });
        });
    }).CreateClient();
    client.DefaultRequestHeaders.Authorization = new
AuthenticationHeaderValue(TestAuthHandler.AuthenticationScheme);
    client.DefaultRequestHeaders.Add(TestAuthHandler.UserNameHeader,
"Test User");
    client.DefaultRequestHeaders.Add(TestAuthHandler.CountryHeader,
"New Zealand");
    client.DefaultRequestHeaders.Add(TestAuthHandler.
AccessNumberHeader, "123456");
    client.DefaultRequestHeaders.Add(TestAuthHandler.
DrivingLicenseNumberHeader, "12345678");
    // Act
    var response = await client.GetAsync("/weatherforecast/driving-
license");
    // Assert
    response.EnsureSuccessStatusCode(); // Status Code 200-299
    response.Content.Headers.ContentType.Should().NotBeNull();
    response.Content.Headers.ContentType!.ToString().Should().
Be("application/json; charset=utf-8");
}
```

In this test method, we specify the test web host of SUT with the `WithWebHostBuilder` method and then call the `AddAuthentication` method to specify the authentication scheme. Then we call the `AddScheme` method to apply the `TestAuthHandler` authentication handler to the authentication service. With this customized test web host, we can create a new HTTP client. Before we send the request using this HTTP client, we need to add the `Authorization` header that specifies the authentication scheme. We also add the claims to the request headers for simplicity, so that the custom authentication handler can read the values from the request headers and create the `ClaimsPrincipal` object.

Then we can call the `GetAsync` method to send the HTTP request to the API endpoint. Finally, we can verify the response status code and the response content type to ensure the request was successful.

The preceding test method is for a happy path. To test the unauthorized scenario, we can create a new test method that does not add the `DrivingLicenseNumberHeader` header to the request, and verify that the response status code is `401 Unauthorized`:

```
[Fact]
public async Task GetDrivingLicense_ShouldReturnForbidden_
WhenRequiredClaimsNotProvidedWithTestAuthHandler()
{
```

```
    // Arrange
    var client = fixture.WithWebHostBuilder(builder =>
    {
        builder.ConfigureTestServices(services =>
        {
            services.AddAuthentication(options =>
                {
                    options.DefaultAuthenticateScheme =
TestAuthHandler.AuthenticationScheme;
                    options.DefaultChallengeScheme = TestAuthHandler.
AuthenticationScheme;
                    options.DefaultScheme = TestAuthHandler.
AuthenticationScheme;
                })
                .AddScheme<AuthenticationSchemeOptions,
TestAuthHandler>(TestAuthHandler.AuthenticationScheme,
                    options => { });
        });
    }).CreateClient();
    client.DefaultRequestHeaders.Authorization = new
AuthenticationHeaderValue(TestAuthHandler.AuthenticationScheme);
    client.DefaultRequestHeaders.Add(TestAuthHandler.UserNameHeader,
"Test User");
    client.DefaultRequestHeaders.Add(TestAuthHandler.CountryHeader,
"New Zealand");
    client.DefaultRequestHeaders.Add(TestAuthHandler.
AccessNumberHeader, "123456");
    // Act
    var response = await client.GetAsync("/weatherforecast/driving-
license");
    // Assert
    response.StatusCode.Should().Be(HttpStatusCode.Forbidden);
}
```

In the preceding test method, we do not add the `DrivingLicenseNumberHeader` header to the request. So, the custom authentication handler cannot find the `DrivingLicenseNumber` claim, and it will return the `Forbidden` status code.

Now, we found that there is some duplicated code in the preceding test methods. If we need to set up the test web host and create the HTTP client for each test method, we can move the code to the `IntegrationTestsFixture` class. Create a method called `CreateClientWithAuth` in the `IntegrationTestsFixture` class:

```
public HttpClient CreateClientWithAuth(string userName, string
country, string accessNumber, string drivingLicenseNumber)
{
    var client = WithWebHostBuilder(builder =>
    {
```

```
        builder.ConfigureTestServices(services =>
        {
            services.AddAuthentication(options =>
                {
                    options.DefaultAuthenticateScheme =
TestAuthHandler.AuthenticationScheme;
                    options.DefaultChallengeScheme = TestAuthHandler.
AuthenticationScheme;
                    options.DefaultScheme = TestAuthHandler.
AuthenticationScheme;
                })
                .AddScheme<AuthenticationSchemeOptions,
TestAuthHandler>(TestAuthHandler.AuthenticationScheme,
                    options => { });
        });
    }).CreateClient();
    client.DefaultRequestHeaders.Authorization = new
AuthenticationHeaderValue(TestAuthHandler.AuthenticationScheme);
    client.DefaultRequestHeaders.Add(TestAuthHandler.UserNameHeader,
userName);
    client.DefaultRequestHeaders.Add(TestAuthHandler.CountryHeader,
country);
    client.DefaultRequestHeaders.Add(TestAuthHandler.
AccessNumberHeader, accessNumber);
    client.DefaultRequestHeaders.Add(TestAuthHandler.
DrivingLicenseNumberHeader, drivingLicenseNumber);
    return client;
}
```

The `CreateClientWithAuth()` method accepts the claims as parameters and then creates the `HttpClient` with the customized test web host. In this way, we can easily control the claims for each test method. Then we can update the test methods to use this method. For example, the test methods for the `GetCountry` endpoint can be updated as follows:

```
[Fact]
public async Task GetCountry_ShouldReturnOk_
WhenAuthorizedWithTestAuthHandler()
{
    // Arrange
    var client = fixture.CreateClientWithAuth("Test User", "New
Zealand", "123456", "12345678");
    // Act
    var response = await client.GetAsync("/weatherforecast/country");
    // Assert
    response.EnsureSuccessStatusCode(); // Status Code 200-299
    response.Content.Headers.ContentType.Should().NotBeNull();
```

```
    response.Content.Headers.ContentType!.ToString().Should().
Be("application/json; charset=utf-8");
}


[Fact]
public async Task GetCountry_ShouldReturnForbidden_
WhenRequiredClaimsNotProvidedWithTestAuthHandler()
{
    // Arrange
    // As we don't provide the country claim, the request will be
forbidden
    var client = fixture.CreateClientWithAuth("Test User", "",
"123456", "12345678");
    // Act
    var response = await client.GetAsync("/weatherforecast/country");
    // Assert
    response.StatusCode.Should().Be(HttpStatusCode.Forbidden);
}
```

Now we can verify that the test methods are still working as expected.

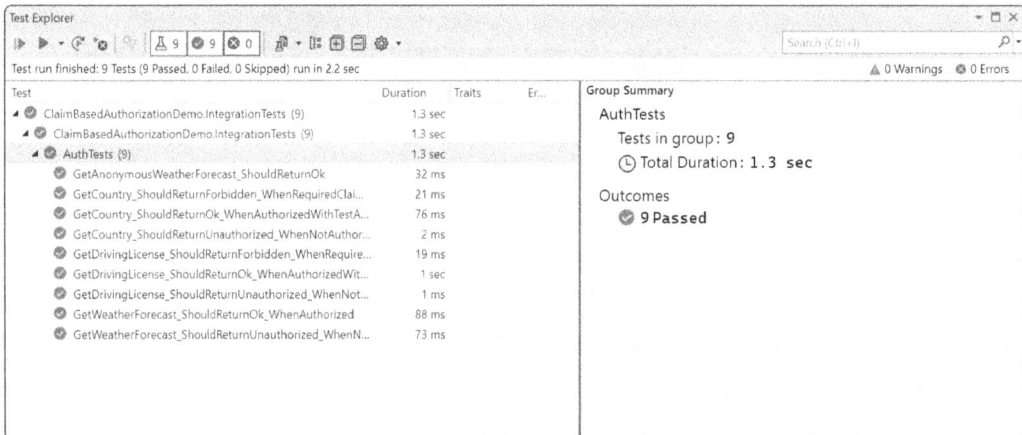

Figure 10.1 – The test methods pass as expected in VS 2022

Note that you can also customize the `AuthenticationSchemeOptions` class if you need to. For example, you can define a `TestAuthHandlerOptions` class that inherits from the `AuthenticationSchemeOptions` class, as follows:

```
public class TestAuthHandlerOptions : AuthenticationSchemeOptions
{
    public string UserName { get; set; } = string.Empty;
}
```

You can then configure `TestAuthHandlerOptions` in the `ConfigureTestServices` method:

```
var client = fixture.WithWebHostBuilder(builder =>
{
    builder.ConfigureTestServices(services =>
    {
        services.Configure<TestAuthHandlerOptions>(options =>
        {
            options.UserName = "Test User";
        });
        services.AddAuthentication(options =>
            {
                options.DefaultAuthenticateScheme = TestAuthHandler.
AuthenticationScheme;
                options.DefaultChallengeScheme = TestAuthHandler.
AuthenticationScheme;
                options.DefaultScheme = TestAuthHandler.
AuthenticationScheme;
            })
            .AddScheme<TestAuthHandlerOptions,
TestAuthHandler>(TestAuthHandler.AuthenticationScheme,
            options => { });
    });
}).CreateClient();
```

The `TestAuthHandler` class should now be updated as follows:

```
public class TestAuthHandler :
AuthenticationHandler<TestAuthHandlerOptions>
{
    public readonly string _userName;
    public TestAuthHandler(IOptionsMonitor<TestAuthHandlerOptions>
options, ILoggerFactory logger,
    UrlEncoder encoder, ISystemClock clock) : base(options, logger,
encoder, clock)
    {
        // Get the user name from the options
        _userName = options.CurrentValue.UserName;
    }
}
```

The `TestAuthHandler` class can now get the username from the `TestAuthHandlerOptions` class. You can also define other properties in the `TestAuthHandlerOptions` class, and then use them in the `TestAuthHandler` class.

If your project does not use claim-based authorization, you can also define a custom authorization handler to implement the authorization logic. Please check the official document for more information: `https://learn.microsoft.com/en-us/aspnet/core/security/authentication/`.

Code coverage

Now we have covered the unit tests and integration tests in ASP.NET Core. In this section, we will discuss **code coverage**, which is a metric that measures the extent to which the source code of the application is covered by the test suite during testing.

Code coverage is a very important metric for software quality. If the code coverage is low, it means that there are many parts of the code that are not covered by the tests. In this case, we are not confident that the code is working as expected. Also, when we make changes or refactor the code, we are not sure whether the changes will break the existing code.

Code coverage provides insights into which parts of the code are covered (or not covered) by the tests. It can help us identify areas that may require additional testing and ensure that the code is tested thoroughly.

Code coverage plays a vital role in assessing the effectiveness and reliability of the testing process. By analyzing code coverage, we can gain confidence in the quality of the code and identify potential areas of weakness or untested code. Adequate code coverage is essential to improve the quality of the code and reduce the risks of bugs and defects.

It is important to note that code coverage is not the sole indicator of code quality. While high code coverage is desirable, achieving 100% code coverage does not guarantee that the code is bug-free. Code coverage should be accompanied by other factors such as effective test design, code reviews, static analysis, manual testing, and so on. Additionally, factors such as code design, architecture, and development practices also play a role in the quality of the code. We need to find a balance between code coverage and other factors.

To analyze the code coverage, we have two steps:

- **Collecting test data**: The data collector can monitor the test execution and collect the code coverage data during test runs. It can report code coverage data in different formats, such as XML or JSON.

- **Generating the report**: The report generator can read the collected data and generate the code coverage report, often in HTML format.

Let us see how to use the data collector and report generator. We will use the `InvoiceApp` project as an example. You can find the sample project in the `chapter10\IntegrationTestsDemo\IntegrationTest-v1` folder.

Using data collectors

To use data collectors, we can use **Coverlet**. Coverlet is a cross-platform code coverage framework for .NET, with support for line, branch, and method coverage. It can be used either as a .NET Core global tool or a NuGet package. For more information, please check the Coverlet repo on GitHub: `https://github.com/coverlet-coverage/coverlet`.

The xUnit project template already includes the Coverlet package. If your test project does not include the Coverlet package, you can install it by running the following command in the Package Manager Console:

```
dotnet add package coverlet.collector
```

To get coverage data, navigate to the test project folder, and run the following command:

```
dotnet test --collect:"XPlat Code Coverage"
```

The `--collect:"XPlat Code Coverage"` option tells the `dotnet test` command to collect the code coverage data. The `XPlat Code Coverage` parameter is a friendly name for the collector. You can use any name you like but note that it is case insensitive. The code coverage data will be saved in the `TestResults` folder. You can find the code coverage data in the `coverage.cobertura.xml` file. The folder structure is `/TestResults/{GUID}/coverage.cobertura.xml`.

Here is a sample of the `coverage.cobertura.xml` file:

```xml
<?xml version="1.0" encoding="utf-8"?>
<coverage line-rate="0.1125" branch-rate="0.1875" version="1.9"
timestamp="1685100267" lines-covered="108" lines-valid="960" branches-
covered="6" branches-valid="32">
  <sources>
    <source>C:\dev\web-api-with-asp-net\example_code\chapter9\
IntegrationTestsDemo\IntegrationTest-v1\InvoiceApp\InvoiceApp.
WebApi\</source>
  </sources>
  <packages>
    <package name="InvoiceApp.WebApi" line-rate="0.1125" branch-
rate="0.1875" complexity="109">
      <classes>
        ...
        <class name="InvoiceApp.WebApi.Services.EmailService"
filename="Services\EmailService.cs" line-rate="1" branch-rate="1"
complexity="2">
          <methods>
            <method name="GenerateInvoiceEmail"
signature="(InvoiceApp.WebApi.Models.Invoice)" line-rate="1" branch-
rate="1" complexity="1">
              <lines>
```

```
                            <line number="19" hits="1" branch="False" />
                            <line number="20" hits="1" branch="False" />
                            <line number="21" hits="1" branch="False" />
                            <!-- ... -->
                            <line number="37" hits="1" branch="False" />
                            <line number="38" hits="1" branch="False" />
                        </lines>
                    </method>
                    <method name=".ctor" signature="(Microsoft.
Extensions.Logging.ILogger`1&lt;InvoiceApp.WebApi.Interfaces.
IEmailService&gt;,InvoiceApp.WebApi.Interfaces.IEmailSender)" line-
rate="1" branch-rate="1" complexity="1">
                        <lines>
                            <line number="12" hits="3" branch="False" />
                            <line number="13" hits="3" branch="False" />
                            <line number="14" hits="3" branch="False" />
                            <line number="15" hits="3" branch="False" />
                            <line number="16" hits="3" branch="False" />
                        </lines>
                    </method>
                </methods>
                <lines>
                    <line number="19" hits="1" branch="False" />
                    <line number="20" hits="1" branch="False" />
                    <line number="21" hits="1" branch="False" />
                    <!-- ... -->
                    <line number="37" hits="1" branch="False" />
                    <line number="38" hits="1" branch="False" />
                    <line number="12" hits="3" branch="False" />
                    <line number="13" hits="3" branch="False" />
                    <line number="14" hits="3" branch="False" />
                    <line number="15" hits="3" branch="False" />
                    <line number="16" hits="3" branch="False" />
                </lines>
            </class>
            ...
        </classes>
    </package>
  </packages>
</coverage>
```

In this code, we can see the following information:

- `line-rate`: This is the percentage of lines covered by tests
- `branch-rate`: This is the percentage of branches covered by tests
- `lines-covered`: This is the number of lines covered by tests
- `lines-valid`: This is the number of lines in the source code
- `branches-covered`: This is the number of branches covered by tests
- `branches-valid`: This is the number of branches in the source code

You can also use Coverlet as a .NET global tool. To do this, you can run the following command to install Coverlet as a .NET global tool:

```
dotnet tool install --global coverlet.console
```

Then you can use it as follows:

```
coverlet /path/to/InvoiceApp.UnitTests.dll --target "dotnet"
--targetargs "test /path/to/test-project --no-build"
```

Please update the paths in the preceding command to match your project structure. The `--no-build` option is used to skip building the test project, which is useful if you have already built the test project.

Now we have the code coverage data. However, the `coverage.cobertura.xml` file is not human-readable. So, we must generate a human-readable report, which we will introduce in the next section.

Generating a code coverage report

To better understand the coverage data, we can generate a code coverage report. To do this, we can use the `ReportGenerator` NuGet package. **ReportGenerator** is a tool that can convert coverage data generated by Coverlet into human-readable reports. It also supports other coverage formats, such as **OpenCover**, **dotCover**, **NCover**, and so on.

To install ReportGenerator, we can run the following command:

```
dotnet tool install -g dotnet-reportgenerator-globaltool
```

Then we can run the following command to generate a code coverage report:

```
reportgenerator "-reports:/path/to/coverage.cobertura.xml"
"-targetdir:coveragereport" "-reporttypes:Html;HtmlSummary"
```

Please update the paths in the preceding command to match your project structure.

If the command runs successfully, you will see the generated HTML report in the `coveragereport` folder. You can open the `index.html` file in the browser to view the report. The report looks like this:

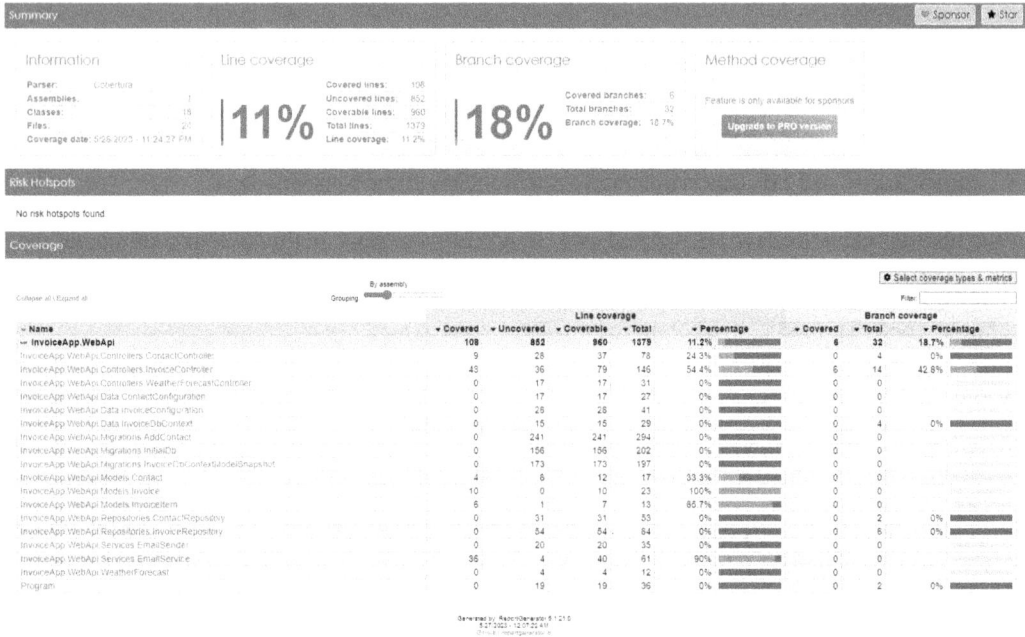

Figure 10.2 – Overview of the code coverage report

You can check each class to see the code coverage details, as shown in *Figure 10.3*:

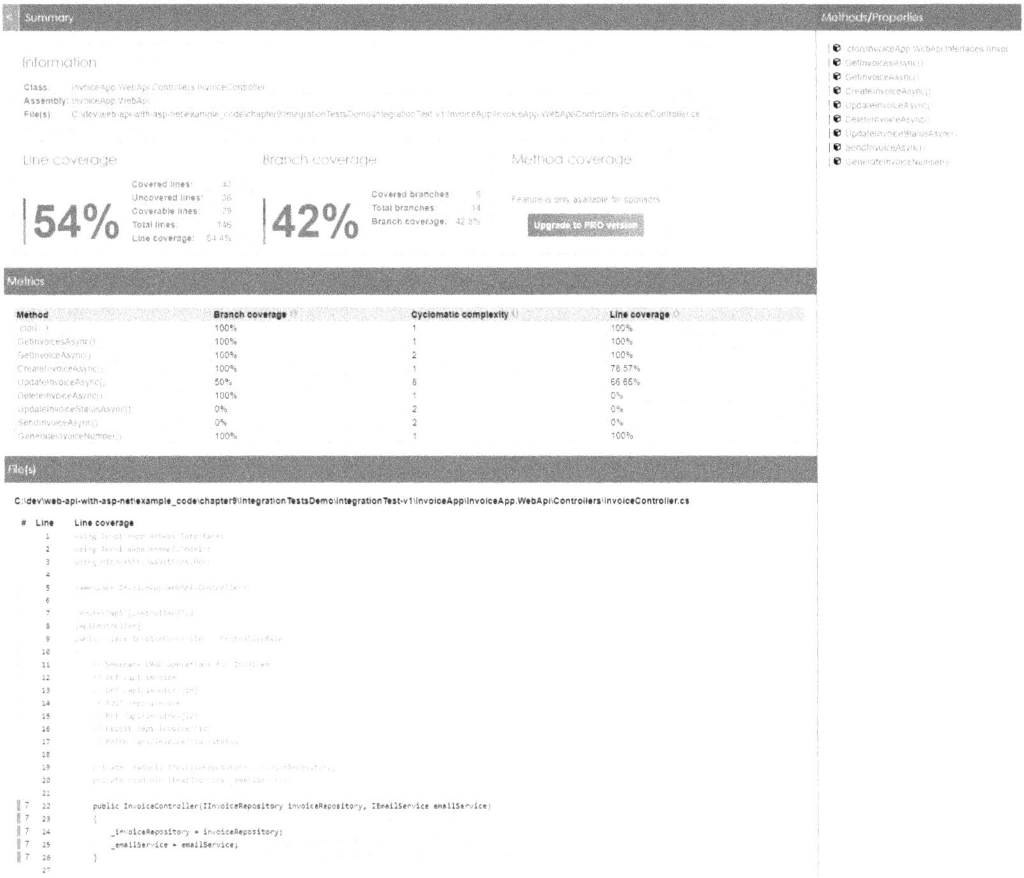

Figure 10.3 – Overview of the code coverage details

In *Figure 10.4*, we can see that some lines are not covered by tests:

```
72    // PUT: api/Invoices/5
73    [HttpPut("{id}")]
74    public async Task<IActionResult> UpdateInvoiceAsync(Guid id, Invoice invoice)
75    {
76        var existingInvoice = await _invoiceRepository.GetInvoiceAsync(id);
77        if (existingInvoice == null)
78        {
79            return NotFound();
80        }
81        // Check if the Invoice has correct Id
82        if (!string.IsNullOrWhiteSpace(invoice.Id.ToString()) && invoice.Id != Guid.Empty && invoice.Id != id)
83        {
84            return BadRequest("Invoice Id cannot be changed.");
85        }
86
87        invoice.Id = id;
88        invoice.InvoiceItems.ForEach(x => x.Amount = x.UnitPrice * x.Quantity);
89        invoice.Amount = existingInvoice.InvoiceItems.Sum(x => x.Amount);
90        await _invoiceRepository.UpdateInvoiceAsync(invoice);
91        return NoContent();
92    }
93
94    // DELETE: api/Invoices/5
95    [HttpDelete("{id}")]
96    public async Task<IActionResult> DeleteInvoiceAsync(Guid id)
97    {
98        await _invoiceRepository.DeleteInvoiceAsync(id);
99        return NoContent();
100   }
101
102   // PATCH: api/Invoices/5/status
103   [HttpPatch("{id}/status")]
104   public async Task<IActionResult> UpdateInvoiceStatusAsync(Guid id, InvoiceStatus status)
105   {
106       var invoice = await _invoiceRepository.GetInvoiceAsync(id);
107       if (invoice == null)
108       {
109           return NotFound();
110       }
111       invoice.Status = status;
112       await _invoiceRepository.UpdateInvoiceAsync(invoice);
113       return NoContent();
114   }
115
```

Figure 10.4 – Overview of the lines not covered by tests highlighted in red

Sadly, the code coverage in our sample project is awful. But luckily, it is just a sample project. In real-world projects, we should try our best to improve the code coverage!

In this section, we have learned how to use Coverlet and ReportGenerator to generate code coverage data and reports. Code coverage is an essential aspect of effective software testing. By leveraging these reports, developers and quality assurance teams can gain insights into the quality of their tests and the quality of their code, which can ultimately enhance the reliability and stability of the application and also help us confidently refactor code.

Summary

In this chapter, we discussed how to write integration tests for ASP.NET Core web API applications. We learned how to create a test fixture to set up the test web host, and how to use the test fixture in the test class. We also learned how to test authorized endpoints and generate code coverage data and reports.

As a good developer, it is important to write tests for your code. Writing tests is not only a beneficial practice but also a beneficial habit to form. You may find that you spend more time writing tests than writing features, but the effort is worth it. To ensure your ASP.NET web API applications are functioning correctly, make sure to write both unit tests and integration tests. Doing so will help to ensure your code is reliable and secure.

In the next chapter, we will explore another aspect of web APIs: gRPC, which is a high-performance, open-source, universal RPC framework.

11

Getting Started with gRPC

Besides RESTful APIs, there are other types of APIs. One of them is the **remote procedure call** (**RPC**)-based API, which we introduced in *Chapter 1*. gRPC is a high-performance RPC framework developed by Google. Now, it is an open-source project under the **Cloud Native Computing Foundation** (**CNCF**), and it is becoming more and more popular.

ASP.NET Core provides a set of gRPC tools to help us build gRPC services. In this chapter, we will introduce the fundamentals of gRPC and **Protocol Buffers** (**Protobuf**) messages. First, we will learn how to define protobuf messages and gRPC services. Then, we will learn how to implement gRPC services in ASP.NET Core, empowering us to communicate seamlessly between different applications. We will be covering the following topics in this chapter:

- Recap of gRPC

- Setting up a gRPC project

- Defining gRPC services and messages

- Implementing gRPC services and clients

- Consuming gRPC services in ASP.NET Core applications

By the end of this chapter, you should be able to understand the fundamentals of protobuf and gRPC and know how to build gRPC services in ASP.NET Core.

Technical requirements

The code examples in this chapter can be found at `https://github.com/PacktPublishing/Web-API-Development-with-ASP.NET-Core-8/tree/main/samples/chapter11`. You can use VS 2022 or VS Code to open the solutions.

Recap of gRPC

If you have read *Chapter 1*, you should be familiar with the concept of RPC – it is a protocol that allows a program to call a procedure on a remote machine. Unlike RESTful APIs, which center around resources, RPC-based APIs focus on actions. So, RPC methods support various types of actions besides CRUD operations.

gRPC is one of the most popular RPC frameworks. It provides many benefits over traditional RPC frameworks. As we mentioned in *Chapter 1*, gRPC is based on HTTP/2, which is more efficient than HTTP/1.1. gRPC uses protobuf as the default data serialization format, which is a binary format that is more compact and efficient than JSON. The tooling support for gRPC is also very good. It follows the contract-first approach, which means we can create language-neutral service definitions and generate code for different languages. It also supports streaming, which is a very useful feature for real-time communication.

With the increasing popularity of microservices, gRPC is becoming more and more popular. While gRPC offers some advantages over RESTful APIs, it is not considered a complete replacement. gRPC is an excellent choice for high-performance, low-latency communication between microservices, but RESTful APIs are more suitable for web-based applications and scenarios where simplicity, flexibility, and wide adoption are more critical. It's important to choose the right protocol based on the requirements of your specific use case. In the next section, we will learn how to set up a gRPC project in ASP.NET Core.

Setting up a gRPC project

In this section, we will build a gRPC project using the dotnet CLI. We will also create a client project to consume the gRPC service. We will be using the same project throughout this chapter.

Creating a new gRPC project

To create a new gRPC project, we can use the `dotnet new` command. The dotnet CLI provides a template for gRPC projects, which includes a basic gRPC service. We can use the following command to create a new gRPC project:

```
dotnet new grpc -o GrpcDemo
```

The `-o` option specifies the output directory. After running the command, we will see that a project named `GrpcDemo` is created.

If you prefer to use VS 2022, you can also create a new gRPC project in VS 2022 using the built-in gRPC template. You can select the **ASP.NET Core gRPC Service** template when creating a new project, as shown in *Figure 11.1*:

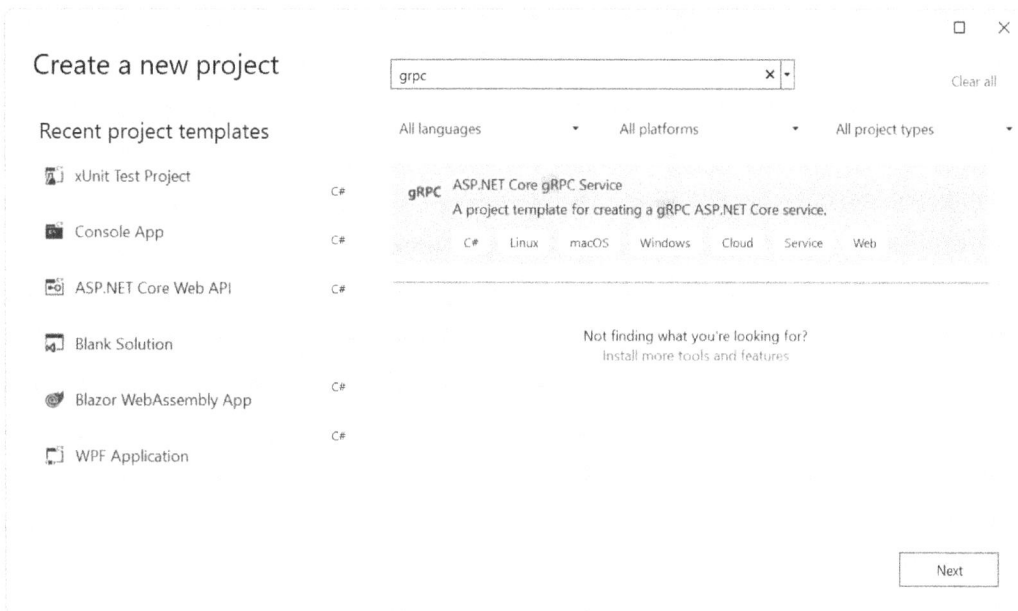

Figure 11.1 – Creating a new gRPC project in VS 2022

After creating the project, you can use VS Code or VS 2022 to open the project. Next, we'll explore the project structure.

Understanding the gRPC project structure

The project structure of a gRPC project has some differences from a RESTful API project. There is no `Controllers` folder in the gRPC project. Instead, there is a `Protos` folder, which contains the proto files. You can find a `greet.proto` file in the `Protos` folder, as follows:

```
syntax = "proto3";

option csharp_namespace = "GrpcDemo";

package greet;

// The greeting service definition.
service Greeter {
  // Sends a greeting
  rpc SayHello (HelloRequest) returns (HelloReply);
}

// The request message containing the user's name.
message HelloRequest {
```

```
      string name = 1;
}


// The response message containing the greetings.
message HelloReply {
    string message = 1;
}
```

gRPC uses protobuf as the default data serialization format. The `greet.proto` file is the proto file that defines the gRPC service and messages. If you are familiar with RESTful APIs, you can think of this file as the Swagger file (OpenAPI specification). It is the contract of the gRPC service. In the preceding proto file, we define a service named `Greeter` with a method named `SayHello()`. The `SayHello()` method takes a `HelloRequest` message as input and returns a `HelloReply` message as output. Both `HelloRequest` and `HelloReply` messages have string properties named `name` and `message`, respectively.

In a proto file, you can use `//` to add comments. To add multi-line comments, you can use `/* ... */`.

> **Important note**
>
> VS Code does not provide syntax highlighting for proto files by default. You can install some extensions, such as `vscode-proto3`, to enable syntax highlighting.

Let's check the project file. Open the `GrpcDemo.csproj` file; we will see the following content:

```
<ItemGroup>
  <Protobuf Include="Protos\greet.proto" GrpcServices="Server" />
</ItemGroup>
<ItemGroup>
  <PackageReference Include="Grpc.AspNetCore" Version="2.51.0" />
  <PackageReference Include="Google.Protobuf" Version="3.22.0-rc2" />
</ItemGroup>
```

You will see that it includes two package references:

- `Grpc.AspNetCore`: This package provides the gRPC server library for ASP.NET Core. It also references the `Grpc.Tools` package, which provides the code-generation tooling.

- `Google.Protobuf`: This package provides the `Protobuf` runtime library.

There is a `Protobuf` item group that includes the proto file. The `GrpcServices` attribute specifies the type of code generated by the proto file. It can be set to the following values:

- `None`: No code is generated

- `Client`: This option only generates client-side code

- `Server`: This option only generates server-side code

- `Both`: This option generates both client-side code and server-side code. It is the default value

In the template project, the `GrpcServices` attribute is set to `Server`, which means only server-side code is generated.

If you have multiple proto files, you can add multiple `Protobuf` items to the `ItemGroup` element.

Next, let's check the `Services` folder. You can find the `GreeterService.cs` file in the `Services` folder, which contains the implementation of the `Greeter` service:

```
public class GreeterService(ILogger<GreeterService> logger) : Greeter.
GreeterBase
{

    public override Task<HelloReply> SayHello(HelloRequest request,
ServerCallContext context)
    {
        return Task.FromResult(new HelloReply
        {
            Message = "Hello " + request.Name
        });
    }
}
```

The `GreeterService` class inherits from the `GreeterBase` class, which is generated by the proto file. It has a `SayHello()` method, which takes a `HelloRequest` object as input and returns a `HelloReply` object as output. The implementation of the `SayHello()` method is very simple – it matches the definition of the `SayHello()` method in the proto file.

If you move your mouse over the `HelloRequest` class in VS Code, you will see a pop-up message, which shows that the namespace of the `HelloRequest` class is `GrpcDemo.HelloRequest`, as shown in *Figure 11.2*:

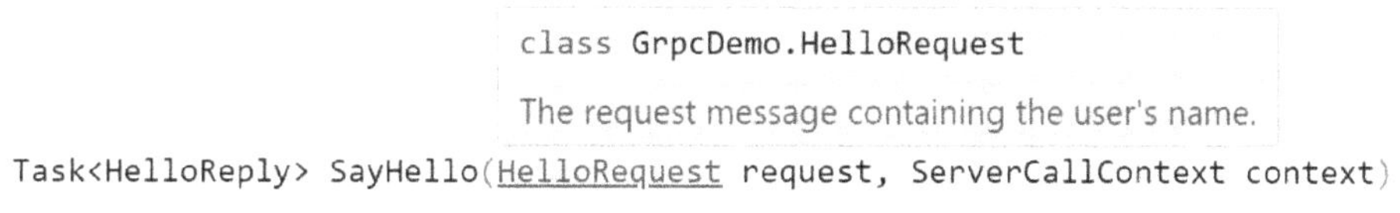

class GrpcDemo.HelloRequest

The request message containing the user's name.

Task<HelloReply> SayHello(<u>HelloRequest</u> request, ServerCallContext context)

Figure 11.2 – The namespace of the HelloRequest class

The `HelloReply` class is also similar. However, you won't be able to find the `HelloRequest` class and the `HelloReply` class in the project. Where are these classes defined?

You can press *F12* to go to the definition of the `HelloRequest` class in VS Code. You will be navigated to a `Greet.cs` file, which is located in the `obj\Debug\net8.0\Protos` folder. This file is generated by the proto file and contains the definition of the `HelloRequest` class:

```
#region Messages
/// <summary>
/// The request message containing the user's name.
/// </summary>
public sealed partial class HelloRequest :
pb::IMessage<HelloRequest>
{
    private static readonly pb::MessageParser<HelloRequest> _parser =
new pb::MessageParser<HelloRequest>(() => new HelloRequest());
    // Omitted for brevity
    public HelloRequest() {
      OnConstruction();
    }

    // Omitted for brevity

    /// <summary>Field number for the "name" field.</summary>
    public const int NameFieldNumber = 1;
    private string name_ = "";
    [global::System.Diagnostics.DebuggerNonUserCodeAttribute]
    [global::System.CodeDom.Compiler.GeneratedCode("protoc", null)]
    public string Name {
      get { return name_; }
      set {
        name_ = pb::ProtoPreconditions.CheckNotNull(value, "value");
      }
    }
    // Omitted for brevity
}
```

In the definition of the `HelloRequest` class, you can see that it implements the `IMessage<HelloRequest>` interface, which is defined in the `Google.Protobuf` package. All protobuf messages must implement this base interface. The `HelloRequest` class also has a `Name` property, which is defined in the proto file. You can find a `DebuggerNonUserCodeAttribute` attribute on the `Name` property. This attribute means that the `Name` member is not part of the user code for an application. The `Name` property also has a `GeneratedCode` attribute, which means this member is generated by the tooling. Specifically, the `Name` property is generated by the `protoc` tool, which is the protobuf compiler. Users should not modify this member.

You can also find the definition of the HelloReply class in the Greet.cs file. Next to the Greet.cs file, in the Protos folder, you can find a GreetGrpc.cs file, which defines the GreeterBase abstract class as the base class of the GreeterService class. Similarly, the GreeterBase class is also generated by the gRPC tooling. It contains the definition of the SayHello() method, as follows:

```
/// <summary>Base class for server-side implementations of Greeter</
summary>
[grpc::BindServiceMethod(typeof(Greeter), "BindService")]
public abstract partial class GreeterBase
{
  /// <summary>
  /// Sends a greeting
  /// </summary>
  /// <param name="request">The request received from the client.</
param>
  /// <param name="context">The context of the server-side call
handler being invoked.</param>
  /// <returns>The response to send back to the client (wrapped by a
task).</returns>
  [global::System.CodeDom.Compiler.GeneratedCode("grpc_csharp_plugin",
null)]
  public virtual global::System.Threading.Tasks.Task<global::GrpcDemo.
HelloReply> SayHello(global::GrpcDemo.HelloRequest request,
grpc::ServerCallContext context)
  {
    throw new grpc::RpcException(new grpc::Status(grpc::StatusCode.
Unimplemented, ""));
  }
}
```

The GreeterBase class is marked with the BindServiceMethod attribute, which means this method is the implementation of the SayHello() method defined in the proto file. The SayHello() method has an attribute called GeneratedCode that indicates that this class is generated by the gRPC C# plugin. Inside the SayHello() method, you can see that it throws an exception by default. Because this method is virtual, we need to override this method in the GreeterService class to provide the actual implementation.

Next, let's check the Program.cs file. You will find the following code in the Program.cs file:

```
var builder = WebApplication.CreateBuilder(args);
// Add services to the container.
builder.Services.AddGrpc();
var app = builder.Build();
// Configure the HTTP request pipeline.
app.MapGrpcService<GreeterService>();
```

In the preceding code block, we can see that the `AddGrpc()` method is called to add gRPC services to the service container. Then, we use the `MapGrpcService<GreeterService>()` method to map the `GreeterService` class to the gRPC service, which is similar to the `MapControllers` method in the RESTful API project.

There is another line of code in the `Program.cs` file that uses the `MapGet()` method to show a message if users access the root path of the application from a web browser:

```
app.MapGet("/", () => "Communication with gRPC endpoints must be
made through a gRPC client. To learn how to create a client, visit:
https://go.microsoft.com/fwlink/?linkid=2086909");
```

This is because gRPC services cannot be accessed by a web browser. So, we need to show a message to notify users that they need to use a gRPC client to access the gRPC service.

Let's update the proto file and see what happens. Open the `greet.proto` file and update the `HelloRequest`, as follows:

```
message HelloRequest {
  string name = 1;
  string address = 2;
}
```

Save the file and go back to the `GreeterService` class. In the `SayHello()` method, you can try to access the `Address` property of the `HelloRequest` object. You will find that the `Address` property is not available. This is because the generated code is not updated. We need to regenerate the code by using the `dotnet build` command. Alternatively, you can delete the `obj` folder and the code will be regenerated automatically.

You may find that it is not convenient to store the generated code in the `obj` folder. We can change the output directory of the generated code by using the `OutputDir` attribute in the `Protobuf` item in the `.csproj` file. For example, you can change the `Protobuf` item as follows:

```
<ItemGroup>
  <Protobuf Include="Protos\greet.proto" GrpcServices="Server"
OutputDir="Generated" />
</ItemGroup>
```

Now, the generated code will be stored in the `Generated\Protos` folder. A proto file can generate multiple files for server-side code. For example, the `greet.proto` file will generate the following files:

- `greet.cs`: This file contains the definition of the messages and the methods to serialize and deserialize the messages
- `greetGrpc.cs`: This file contains the definition of the base class of the service and the methods to bind the service to the server

Now that we understand the structure of the gRPC project, let's learn the concepts behind protobuf messages.

Creating protobuf messages

In this section, we will learn how to create protobuf messages. We will introduce the concepts of protobuf messages and how to define them in a proto file.

gRPC is a contract-first framework, meaning that the gRPC service and messages must be defined in a proto file. When we talk about messages, we are talking about the data that is sent between the client and the server. While gRPC messages may be similar to the data model in RESTful APIs, they are not the same. RESTful APIs are centered around resources, and the data model is usually a resource model that can be mapped to one or multiple database tables. In contrast, gRPC is action-based, and the message can be any other type of data model or other message sent between the client and the server. Therefore, gRPC messages may not be exactly mapped to a resource model in RESTful APIs.

For example, when creating an invoice through a RESTful API using JSON as the data format, we need to send an HTTP POST request with a JSON body to the server. The JSON body will be deserialized into a .NET object, which serves as the data model for the invoice. To retrieve an invoice, we need to send an HTTP GET request to the server and the server. The server will serialize the data model into a JSON string and send it back to the client. We may also have other actions, such as updating an invoice, deleting an invoice, and so on. All these actions are mapped to HTTP methods.

To implement the same functionality using gRPC, we need to define a gRPC service with several methods: `CreateInvoice()`, `GetInvoice()`, `UpdateInvoice()`, `DeleteInvoice()`, and others. For each of these methods, we must also define the corresponding request and response messages. For example, the `CreateInvoice()` method requires a `CreateInvoiceRequest` message containing the properties of the invoice, as well as a `CreateInvoiceResponse` message containing the ID of the created invoice. It is important to note that the request and response messages are distinct from the data model of the invoice, which is used to represent the invoice entity in the system. The request and response messages are used to send data between the client and the server.

Note that gRPC and protobuf are not the same thing. protobuf is a language-neutral, platform-neutral data serialization format. gRPC is a framework that uses protobuf as the default data serialization format. Sometimes, these two terms are used interchangeably, but we should know the difference between them.

Think about the invoice example we mentioned previously. An invoice has several properties, such as the invoice number, the invoice date, the customer's name, the total amount, and so on. A customer has a name and an address. An address has some properties, such as street, city, state, and so on.

Next, we'll define the first message that is used to create an address for the invoice service. The source code for this section can be found in the `chapter11/GrpcDemo-v2` folder. We will start with a simple message and then introduce more concepts regarding protobuf messages, including field numbers, field types, and how to use other .NET types in protobuf messages. We will also learn how to implement list and dictionary types using the `repeated` and `map` keywords.

Defining a protobuf message

Create a new `invoice.proto` file in the `Protos` folder. VS Code provides a proto file template when you create a new file, as shown in *Figure 11.3*:

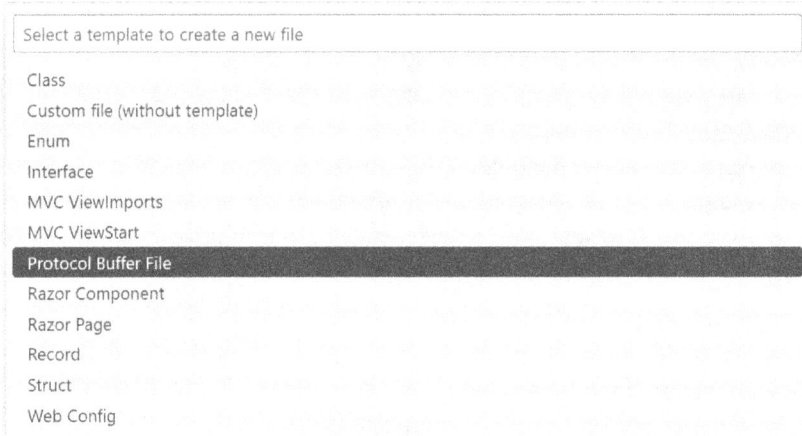

Select a template to create a new file

Class
Custom file (without template)
Enum
Interface
MVC ViewImports
MVC ViewStart
Protocol Buffer File
Razor Component
Razor Page
Record
Struct
Web Config

Figure 11.3 – The proto file template in VS Code

The proto file template creates a proto file named `Protos.proto`. Rename it `invoice.proto`. The content of the proto file is as follows:

```
syntax = "proto3";

option csharp_namespace = "MyApp.Namespace";
```

A proto file is a text file with the `.proto` extension. The first line of the proto file specifies the syntax version of the proto file. At the time of writing, the latest version of the proto file is version 3, which was released in 2016. You can find more information about the proto file syntax at `https://protobuf.dev/programming-guides/proto3/`.

The `option csharp_namespace` line specifies the namespace of the generated code in C#. You can change the namespace according to your needs. This option is used to avoid naming conflicts between different proto files. Note that even though a proto file is language-neutral, the `option csharp_namespace` attribute is only used by the C# code generator. In this sample project, we can change the namespace to `GrpcDemo` to match the namespace of the existing code.

> **Important note**
>
> Protobuf supports a `package` keyword to avoid naming conflicts, depending on the language. For example, `package com.company` is equivalent to `option csharp_namespace = "Com.Company"` in C# (the name will be converted into PascalCase), and `package com.company` is equivalent to `option java_package = "com.company"` in Java. However, `package com.company` will be ignored in Python since Python modules are organized by filesystem directories.
>
> Since we are using C#, we use the `option csharp_namespace` attribute, which can override the `package` keyword for C# applications. If you share the proto file with other applications that use other languages, you can use the `package` keyword or the language-specific option to avoid naming conflicts.

Once the proto file has been created, we need to add it to the project file. Open the `GrpcDemo.csproj` file and add the following code to an `<ItemGroup>` element:

```
<Protobuf Include="Protos\invoice.proto"
GrpcServices="Server"  OutputDir="Generated"/>
```

Now, the gRPC tooling will generate the code for the `invoice.proto` file when we build the project.

gRPC proto3 uses similar concepts as .NET classes to define messages. However, there are some differences. For example, proto3 does not support `GUID` and `decimal` types. Let's start with a simple message. We can define an `Address` message as follows:

```
message CreateAddressRequest {
  string street = 1;
  string city = 2;
  string state = 3;
  string zip_code = 4;
  string country = 5;
}
```

As we can see, it is similar to a .NET class. We use a `message` keyword to define a gRPC message. In the message body, we can use `string` to declare a string field. However, there are some questions to answer here:

- Why do we assign a number to each property? Is it the default value?

- Why does the number start with 1? Can we use 0?

- Should we use these numbers in a specific order?

Let's answer these questions before we move on.

Understanding field numbers

The numbers following the field names are called *field numbers*. Field numbers play an important role in the proto file. These field numbers are used to identify the fields in the message. What is the benefit of using field numbers instead of field names? Let's look at an example of an XML document:

```
<address>
  <street>1 Fake Street</street>
  <city>Wellington</city>
  <state>Wellington</state>
  <zip_code>6011</zip_code>
  <country>New Zealand</country>
</address>
```

In the preceding XML document, each field is wrapped in a tag. We have to open and close the tags to wrap the values of the fields. The XML syntax wastes a lot of space when transferring data. Consider the following example of a JSON document:

```
{
  "street": "1 Fake Street",
  "city": "Wellington",
  "state": "Wellington",
  "zip_code": "6011",
  "country": "New Zealand"
}
```

In the preceding JSON document, we just use each field name once. Normally, JSON format is more compact than XML format. What if we get rid of the field names? That is why we use field numbers in the proto file. By using field numbers instead of field names when encoding the message, we can make the gRPC message more compact. This is because numbers are shorter than field names. Additionally, protobuf uses a binary format, which is more compact than plain text formats such as JSON and XML. This further helps reduce the size of the message.

There are a few things to note about field numbers according to the protobuf documentation:

- The range of field numbers is from 1 to 536,870,911. So we cannot use 0 as a field number.

- The field numbers must be unique within a message.

- Field numbers 19000 to 19999 are reserved for protobuf, so you cannot use them.

- Technically, the order of the field numbers does not matter. It is recommended to use the ascending order of the field numbers. Smaller field numbers use fewer bytes to encode. For example, a field number between 1 and 15 uses only one byte to encode, but numbers from 16 to 2047 use two bytes.

- Once a field number is assigned to a field, it cannot be changed if the proto file is used in production. Changing a field number will break the backward compatibility of the proto file.

With that, we have learned what field numbers are and why we use them. Next, let's understand field types.

Understanding the field types

Similar to .NET classes, a gRPC message can have different types of fields. protobuf provides a set of native types, which are called **scalar value types**. These scalar value types have representations in most programming languages. The following table lists the mapping between protobuf scalar value types and .NET types:

Protobuf Type	.NET Type	Notes
double	double	$\pm 5.0 \times 10^{-324}$ to $\pm 1.7 \times 10^{308}$.
float	float	$\pm 1.5 \times 10^{-45}$ to $\pm 3.4 \times 10^{38}$.
int32	int	The length is variable. Use sint32 if the field has negative numbers.
int64	long	The length is variable. Use sint64 if the field has negative numbers.
uint32	uint	The length is variable. Unsigned integer. 0 to $(2^{32}-1)$.
uint64	ulong	The length is variable. Unsigned integer. 0 to $(2^{64}-1)$.
sint32	int	The length is variable. Signed integer. -2^{31} to $(2^{31}-1)$.
sint64	long	The length is variable. Signed integer. -2^{63} to $(2^{63}-1)$.
fixed32	uint	The length is always 4 bytes. This type is more efficient than uint32 for serializing or deserializing values that are greater than 2^{28}.
fixed64	ulong	The length is always 8 bytes. This type is more efficient than uint64 for serializing or deserializing values that are greater than 2^{56}.
sfixed32	int	The length is always 4 bytes.
sfixed64	long	The length is always 8 bytes.
bool	bool	

Protobuf Type	.NET Type	Notes
string	string	A string field must be encoded in UTF-8 or 7-bit ASCII. The maximum length of a string field is 2^{32}.
bytes	ByteString	This type is defined in protobuf runtime. It can be mapped to and from C#'s byte [] type.

Table 11.1 – Protobuf scalar value types and .NET types

Let's create a new message named CreateContactRequest and add some fields to it:

```
message CreateContactRequest {
  string first_name = 1;
  string last_name = 2;
  string email = 3;
  string phone = 4;
  int32 year_of_birth = 5;
  bool is_active = 6;
}
```

The CreateContactRequest message requires the first_name, last_name, email, phone, year_of_birth, and is_active fields. The types for these fields are string, int32, and bool, respectively.

Next, we can run dotnet build to generate the code. Alternatively, you can delete the existing files in the Generated folder and gRPC tooling will regenerate the code automatically based on the proto files.

The generated code files contain some complicated code. However, we can find the definition of the CreateContactRequest class, which is as follows:

```
public sealed partial class CreateContactRequest :
pb::IMessage<CreateContactRequest>
{
    private string firstName_ = "";
    public string FirstName {
      get { return firstName_; }
      set {
        firstName_ = pb::ProtoPreconditions.CheckNotNull(value,
"value");
      }
    }
```

```
    private string lastName_ = "";
    public string LastName {
      get { return lastName_; }
      set {
        lastName_ = pb::ProtoPreconditions.CheckNotNull(value,
"value");
      }
    }

    private string email_ = "";
    public string Email {
      get { return email_; }
      set {
        email_ = pb::ProtoPreconditions.CheckNotNull(value, "value");
      }
    }

    private string phone_ = "";
    public string Phone {
      get { return phone_; }
      set {
        phone_ = pb::ProtoPreconditions.CheckNotNull(value, "value");
      }
    }

    private int yearOfBirth_;
    public int YearOfBirth {
      get { return yearOfBirth_; }
      set {
        yearOfBirth_ = value;
      }
    }

    private bool isActive_;
    public bool IsActive {
      get { return isActive_; }
      set {
        isActive_ = value;
      }
    }
  }
}
```

In the preceding code block, some code has been omitted for brevity. You can see that the `CreateContactRequest` message has been converted into a .NET class, which includes the properties for each field.

> **Important note**
>
> Protobuf has a style guide for naming fields and methods. The general rules are as follows:
>
> - Use `lower_snake_case` for field names
> - Use `PascalCase` for method names
> - File names should be in `lower_snake_case`
> - Using double quotes for string literals is preferred over single quotes
> - The indentation should be two spaces in length
>
> You can find more information at `https://protobuf.dev/programming-guides/style/`.

With that, we've learned how to use protobuf scalar value types. Now, let's consider other types.

Other .NET types

The protobuf scalar data types do not support all the .NET types, such as `Guid`, `DateTime`, `decimal`, and others. There are some workarounds for these types. In this section, we will learn how to use these types in protobuf. We will also explore some other types, such as `enum` and `repeated`.

GUID values

The `GUID` type (on other platforms, it may have another name, `UUID`) is a 128-bit structure that is used to identify objects. It is a very common type in .NET applications. Normally, a `GUID` value can be represented as a string that contains 32 hexadecimal digits. For example, `31F6E4E7-7C48-4F91-8D33-7A74F6729C8B` is a GUID value.

However, protobuf does not support the `GUID` type. The best way to represent a `GUID` value in protobuf is to use a `string` field. In the .NET code, we can use `Guid.Parse()` to convert a string into a `GUID` value and use `Guid.ToString()` to convert a `GUID` value into a string.

DateTime values

.NET has several types to represent a date and time value, such as `DateTime`, `DateTimeOffset`, and `TimeSpan`. Although protobuf does not support these types directly, it provides several extensions to support them.

To use these extension types, we need to import the `google/protobuf/xxx.proto` file into the proto file. For example, here is a message that contains a timestamp and a duration:

```
syntax = "proto3";

import "google/protobuf/timestamp.proto";
import "google/protobuf/duration.proto";

message UpdateInvoiceDueDateRequest {
  string invoice_id = 1;
  google.protobuf.Timestamp due_date = 2;
  google.protobuf.Duration grace_period = 3;
}
```

Check the generated code for the `UpdateInvoiceDueDateRequest` message in the `Generated` folder. You will find that the `due_date` field is converted into a `Timestamp` type, and the `grace_period` field is converted into a `Duration` type, as follows:

```
public const int DueDateFieldNumber = 2;
private global::Google.Protobuf.WellKnownTypes.Timestamp dueDate_;
public global::Google.Protobuf.WellKnownTypes.Timestamp DueDate {
  get { return dueDate_; }
  set {
    dueDate_ = value;
  }
}

public const int GracePeriodFieldNumber = 3;
private global::Google.Protobuf.WellKnownTypes.Duration gracePeriod_;
public global::Google.Protobuf.WellKnownTypes.Duration GracePeriod {
  get { return gracePeriod_; }
  set {
    gracePeriod_ = value;
  }
}
```

The `Timestamp` type and the `Duration` type are not native .NET types. They are defined in the `Google.Protobuf.WellKnownTypes` namespace, which includes some well-known types that are not supported by protobuf. The source code for these types can be found at `https://github.com/protocolbuffers/protobuf/tree/main/csharp/src/Google.Protobuf/WellKnownTypes`.

Because these types are not native .NET types, we need to convert them into native .NET types when using them. The `Google.Protobuf.WellKnownTypes` namespace provides some methods to do the conversion. Here is an example of converting .NET types into protobuf types:

```
var updateInvoiceDueDateRequest = new UpdateInvoiceDueDateRequest
{
    InvoiceId = Guid.Parse("3193C36C-2AAB-49A7-A0B1-6BDB3B69DEA1"),
    DueDate = Timestamp.FromDateTime(DateTime.UtcNow.AddDays(30)),
    GracePeriod = Duration.FromTimeSpan(TimeSpan.FromDays(10))
};
```

We can use the `Timestamp` class to convert `DateTime` and `DateTimeOffset` values into `Timestamp` values. The `Timestamp.FromDateTime()` method is used to convert a `DateTime` value, while the `Timestamp.FromDateTimeOffset()` method is used to convert a `DateTimeOffset` value. We can also use the `Duration.FromTimeSpan()` method to convert a `TimeSpan` value into a `Duration` value. Note that if you use the `DateTimeOffset` type in your application, the offset of `DateTimeOffset` values is always 0, and the `DateTime.Kind` property is always set to `DateTimeKind.Utc`.

Similarly, we can convert protobuf types into .NET types:

```
var dueDate = updateInvoiceDueDateRequest.DueDate.ToDateTime();
var gracePeriod = updateInvoiceDueDateRequest.GracePeriod.
ToTimeSpan();
```

The `Timestamp` class provides several methods for converting its values into other types. The `ToDateTime()` method can be used to convert a `Timestamp` value into a `DateTime` value, while the `ToTimeSpan()` method can be used to convert a `Duration` value into a `TimeSpan` value. Additionally, the `ToDateTimeOffset()` method can be used to convert a `Timestamp` value into a `DateTimeOffset` value. Depending on your requirements, you can select the appropriate method for your needs.

Decimal values

At the time of writing, protobuf does not support the `decimal` type directly. There are some discussions about adding the `decimal` type to protobuf, but it hasn't been implemented yet. As a workaround, Microsoft Docs provides a `DecimalValue` type, which can be used to represent a `decimal` value in protobuf. The following code, which has been copied from Microsoft Docs, shows how to define a `decimal` value in protobuf:

```
// Example: 12345.6789 -> { units = 12345, nanos = 678900000 }
message DecimalValue {

    // Whole units part of the amount
    int64 units = 1;
```

```
    // Nano units of the amount (10^-9)
    // Must be same sign as units
    sfixed32 nanos = 2;
}
```

We will not delve into the details of the DecimalValue type in this book. You can find more information at https://learn.microsoft.com/en-us/dotnet/architecture/ grpc-for-wcf-developers/protobuf-data-types#decimals.

Enum values

The enum type is very common in .NET applications. protobuf supports the enum type. Here's an example of its usage:

```
enum InvoiceStatus {
    INVOICE_STATUS_UNKNOWN = 0;
    INVOICE_STATUS_DRAFT = 1;
    INVOICE_STATUS_AWAIT_PAYMENT = 2;
    INVOICE_STATUS_PAID = 3;
    INVOICE_STATUS_OVERDUE = 4;
    INVOICE_STATUS_CANCELLED = 5;
}
```

The preceding enum definition is similar to the enum definition in C#, but we need to define it in the proto file. In the preceding code, we define an InvoiceStatus enum type with six values. Note that every enum type must contain a 0 value, which is the default value and must be placed at the first position. The InvoiceStatus enum type is converted into a .NET enum type, as follows:

```
public enum InvoiceStatus {
    [pbr::OriginalName("INVOICE_STATUS_UNKNOWN")] Unknown = 0,
    [pbr::OriginalName("INVOICE_STATUS_DRAFT")] Draft = 1,
    [pbr::OriginalName("INVOICESTATUS_AWAIT_PAYMENT")] AwaitPayment = 2,
    [pbr::OriginalName("INVOICE_STATUS_PAID")] Paid = 3,
    [pbr::OriginalName("INVOICE_STATUS_OVERDUE")] Overdue = 4,
}
```

As you can see, the INVOICE_STATUS prefix in the original names is removed because the prefix is the same as the enum name. In the .NET code, the enum names are converted into PascalCase.

Besides the enum type, .NET also has a common type named nullable. We'll check out nullable types in the next section.

Nullable values

Protobuf scalar value types, such as `int32`, `sint32`, `fixed32`, and `bool`, cannot be `null`. But in .NET, nullable value types are very common. For example, we can use `int?` to declare an integer value that can be `null`. To support nullable value types, protobuf provides some wrapper types, which are defined in the `google/protobuf/wrappers.proto` file, to support nullable types. We can import this file into the proto file and use the wrapper types. For example, we can define a message as follows:

```
syntax = "proto3";

import "google/protobuf/wrappers.proto";

message AddInvoiceItemRequest {
  string name = 1;
  string description = 2;
  google.protobuf.DoubleValue unit_price = 3;
  google.protobuf.Int32Value quantity = 4;
  google.protobuf.BoolValue is_taxable = 5;
}
```

In the preceding code, the `google.protobuf.DoubleValue` type is used to represent a nullable double value, the `google.protobuf.Int32Value` type is used to represent a nullable `int32` value, and the `google.protobuf.BoolValue` type is used to define a nullable `bool` value. The generated code for the `AddInvoiceItemRequest` message is shown here:

```
private double? unitPrice_;
public double? UnitPrice {
  get { return unitPrice_; }
  set {
    unitPrice_ = value;
  }
}

private int? quantity_;
public int? Quantity {
  get { return quantity_; }
  set {
    quantity_ = value;
  }
}

private bool? isTaxable_;
public bool? IsTaxable {
  get { return isTaxable_; }
```

```
  set {
    isTaxable_ = value;
  }
}
```

As you can see, the unitPrice, quantity, and IsTaxable fields are converted into nullable types in .NET.

Most of .NET nullable types are supported by protobuf. Besides the google.protobuf.DoubleValue, google.protobuf.Int32Value, and google.protobuf.BoolValue types, protobuf also provides the following wrapper types:

- google.protobuf.FloatValue: This type is used to represent a float? value.
- google.protobuf.Int64Value: This type is used to represent a long? value.
- google.protobuf.UInt32Value: This type is used to represent a uint? value.
- google.protobuf.UInt64Value: This type is used to represent a ulong? value.
- google.protobuf.StringValue: This type is used to represent a string value.
- google.protobuf.BytesValue: This type is used to represent a ByteString value.

There are two special types in the preceding list: google.protobuf.StringValue and google.protobuf.BytesValue. The corresponding .NET types are string and ByteString. The ByteString type is a class that represents an immutable array of bytes, which is defined in the protobuf runtime. The default value of these two types is null.

So, if google.protobuf.StringValue is mapped to string in .NET, what is the difference between google.protobuf.StringValue and string in protobuf? The difference is the default value. We'll look at the default values of these types in the next section.

Default values

The following table lists the default values of the scalar value types:

Protobuf Type	Default Value
string	An empty string
bytes	An empty byte array
bool	false
Numeric types	0
enums	The first enum value

Table 11.2 – Default values of protobuf scalar value types

If you use `string` as the type of a field, the default value will be an empty string. However, the default value of a `google.protobuf.StringValue` field is `null`. Similarly, the default value of a `bytes` field is an empty byte array, while the default value of a `google.protobuf.BytesValue` field is `null`. All other wrapper types also have a default value of `null`.

All numeric types, including `int32`, `double`, and `float`, have a default value of 0. This applies to all numerical data types. Enum types in protobuf have a default value of the first value in the enum type, which must be 0. For instance, the `InvoiceStatus` enum type has a default value of `INVOICE_STATUS_UNKNOWN`, which is 0.

Repeated fields

Similar to .NET collections, protobuf supports repeated fields. A repeated field can contain zero or more items. The following code shows how to define a repeated field:

```
message UpdateBatchInvoicesStatusRequest {
  repeated string invoice_ids = 1;
  InvoiceStatus status = 2;
}
```

In the preceding code, we use the `repeated` keyword to define a repeated field. The generated code for the repeated `invoice_ids` field in the `UpdateInvoicesStatusRequest` message is as follows:

```
private readonly pbc::RepeatedField<string> invoiceIds_ = new
pbc::RepeatedField<string>();

public pbc::RepeatedField<string> InvoiceIds {
  get { return invoiceIds_; }
}
```

From the generated code, we can see that the repeated `string` field is converted into a `RepeatedField<string>` type. The `RepeatedField<T>` type is defined in the `Google.Protobuf.Collections` namespace, and it implements the .NET collection interfaces, as follows:

```
public sealed class RepeatedField<T> : IList<T>, ICollection<T>,
IEnumerable<T>, IEnumerable, IList, ICollection,
IDeepCloneable<RepeatedField<T>>, IEquatable<RepeatedField<T>>,
IReadOnlyList<T>, IReadOnlyCollection<T>
{
  // Omitted for brevity
}
```

The `RepeatedField<T>` type can be used as a normal .NET collection type, and any LINQ methods can be applied to it. This makes it a powerful and versatile tool for data manipulation.

You will also find that the `InvoiceIds` field is a read-only property. To add one or multiple items to the collection, the `Add()` method can be used. Here's an example:

```
var updateInvoicesStatusRequest = new
UpdateBatchInvoicesStatusRequest();
// Add one item
updateInvoicesStatusRequest.InvoiceIds.Add("3193C36C-2AAB-49A7-A0B1-
6BDB3B69DEA1");
// Add multiple items
updateInvoicesStatusRequest.InvoiceIds.Add(new[]
            { "99143291-2523-4EE8-8A4D-27B09334C980", "BB4E6CFE-6AAE-
4948-941A-26D1FBF59E8A" });
```

The default value of a repeated field is an empty collection.

Map fields

Protobuf supports map fields, which are collections of key-value pairs similar to a .NET dictionary. The following code provides an example of how to define a map field:

```
message UpdateInvoicesStatusRequest {
  map<string, InvoiceStatus> invoice_status_map = 1;
}
```

The generated code for the `invoice_status_map` field is as follows:

```
private readonly pbc::MapField<string, global::GrpcDemo.InvoiceStatus>
invoiceStatusMap_ = newpbc::MapField<string, global::GrpcDemo.
InvoiceStatus>();

public pbc::MapField<string, global::GrpcDemo.InvoiceStatus>
InvoiceStatusMap {
  get { return invoiceStatusMap_; }
}
```

The `MapField<Tkey, TValue>` type is defined in the `Google.Protobuf.Collections` namespace and it implements the `IDictionary<TKey, TValue>` interface, as follows:

```
public sealed class MapField<TKey, TValue> :
IDeepCloneable<MapField<TKey, TValue>>, IDictionary<TKey,
TValue>, ICollection<KeyValuePair<TKey, TValue>>,
IEnumerable<KeyValuePair<TKey, TValue>>, IEnumerable,
IEquatable<MapField<TKey, TValue>>, IDictionary,
ICollection, IReadOnlyDictionary<TKey, TValue>,
IReadOnlyCollection<KeyValuePair<TKey, TValue>>
{
  // Omitted for brevity
}
```

The `MapField<TKey, TValue>` type can be used as a normal .NET dictionary type. This type provides the same functionality as a standard dictionary, allowing for the storage and retrieval of key-value pairs.

Similar to the repeated field, the `InvoiceStatusMap` field is also a read-only property. We can use the `Add()` method to add one key-value pair or multiple key-value pairs to the collection, as follows:

```
var updateInvoicesStatusRequest = new UpdateInvoicesStatusRequest();
// Add one key-value pair
updateInvoicesStatusRequest.InvoiceStatusMap.Add("3193C36C-2AAB-49A7-
A0B1-6BDB3B69DEA1", InvoiceStatus.AwaitPayment);
// Add multiple key-value pairs
updateInvoicesStatusRequest.InvoiceStatusMap.Add(new
Dictionary<string, InvoiceStatus>
{
    { "99143291-2523-4EE8-8A4D-27B09334C980", InvoiceStatus.Paid },
    { "BB4E6CFE-6AAE-4948-941A-26D1FBF59E8A", InvoiceStatus.Overdue }
});
```

Note that map fields cannot be repeated. Also, the key of a map field must be a `string` or integer type. You cannot use an `enum` type as the key of a map field. The value of a map field can be any type, including a message type. But the value type cannot be another map field.

We have now acquired a comprehensive understanding of protobuf messages, including field numbers, field types, default values, repeated fields, and map fields. For further information on protobuf messages, please refer to `https://protobuf.dev/programming-guides/proto3/`.

Next, we'll examine the various protobuf services. We will explore the various types of RPC methods and how to create a gRPC client for the service. By doing so, we will gain a better understanding of how these services function and how to use them effectively.

Creating a protobuf service

Now that we have understood the definition of a protobuf message, we can move on to defining protobuf services. These services are comprised of RPC methods, each of which has a request and response message. To facilitate the implementation of these services, gRPC tooling will generate the necessary C# code, which can then be used as the base class for the service.

gRPC supports four types of RPC methods:

- **Unary RPC:** The client sends a single request message to the server and receives a single response message in return. This type of method is suitable for applications that need single request-response exchanges.

- **Server streaming RPC**: The client sends a single request message to the server and the server then responds with a stream of response messages. This type of method allows for continuous data exchange between the client and server.

- **Client streaming RPC**: The client sends a stream request message to the server and the server then responds with a response message. Similar to server streaming RPC, this type of method also allows for a continuous data exchange but the data change is initiated by the client.

- **Bidirectional streaming RPC**: The client initiates the process by sending a stream request message, to which the server responds with a stream response message. This type of method enables communication between the client and the server to be conducted in both directions.

Let's check out these RPC methods one by one. The source code for this section can be found in the `chapter11/GrpcDemo-v3` folder.

Defining a unary service

A unary service is the simplest type of RPC method. The following code shows a unary service:

```
message CreateContactRequest {
  string first_name = 1;
  string last_name = 2;
  string email = 3;
  string phone = 4;
  int32 year_of_birth = 5;
  bool is_active = 6;
}

message CreateContactResponse {
  string contact_id = 1;
}

service ContactService {
  rpc CreateContact(CreateContactRequest) returns
(CreateContactResponse);
}
```

In the preceding code, we define a `CreateContactRequest` message and a `CreateContactResponse` message, and then we define a `ContactService` service, which contains a `CreateContact()` RPC method. The `CreateContact` RPC method requires a `CreateContactRequest` request message and a `CreateContactResponse` response message.

The generated code for the `CreateContact()` RPC method is as follows:

```
public abstract partial class ContactServiceBase
{
  public virtual global::System.Threading.Tasks.Task<global::GrpcDemo.
CreateContactResponse> CreateContact(global::GrpcDemo.
CreateContactRequest request, grpc::ServerCallContext context)
  {
    throw new grpc::RpcException(new grpc::Status(grpc::StatusCode.
Unimplemented, ""));
  }
}
```

The `ContactServiceBase` class is a base class for the service implementation. It contains the `CreateContact()` method, which is a `virtual` method. By default, the `CreateContact()` method throws an exception because the method is not implemented. We need to override this method in the service implementation.

Next, create a `ContactService.cs` file in the `Service` folder. In the `ContactService.cs` file, we need to implement the `ContactService` class, which is derived from the `ContactServiceBase` class. The `ContactService` class is as follows:

```
public class ContactService(ILogger<ContactService> logger) : Contact.
ContactBase
{
    public override Task<CreateContactResponse>
CreateContact(CreateContactRequest request, ServerCallContext context)
    {
        // TODO: Save contact to database
        return Task.FromResult(new CreateContactResponse
        {
            ContactId = Guid.NewGuid().ToString()
        });
    }
}
```

In the preceding code, we override the `CreateContact()` method and implement the method. This `CreateContact()` method allows us to execute some logic we need, such as saving the contact to the database. For simplicity, we just return a new `CreateContactResponse` object with a new `ContactId` value. In reality, we may have additional logic.

Next, we need to register the `ContactService` class in the DI container. Open the `Program. cs` file and add the following code to the `ConfigureServices()` method:

```
app.MapGrpcService<ContactService>();
```

Our new unary service simplifies the process of handling HTTP requests, eliminating the need to write any code or manage different HTTP methods. All RPC calls are handled by the gRPC framework, allowing for a streamlined process.

To call a gRPC service, a gRPC client must be created as current browsers do not support this protocol. Alternatively, tools such as Postman can be used to access the service. In the following section, we will demonstrate how to create a console application to call the service.

Creating a gRPC client

A gRPC can be a console application, a web application, or any other type of application, such as a WPF application. In this section, we will create a console application as the gRPC client for the unary service we created in the previous section. You can use similar code in other types of applications. Follow these steps:

1. Use the `dotnet new` command to create a new console project:

    ```
    dotnet new console -o GrpcDemo.Client
    ```

2. Now, we have two projects. If you have not created a solution file, you can create it by running the following command:

    ```
    dotnet new sln -n GrpcDemo
    ```

3. Then, add the two projects to the solution:

    ```
    dotnet sln GrpcDemo.sln add GrpcDemo/GrpcDemo.csproj
    dotnet sln GrpcDemo.sln add GrpcDemo.Client/GrpcDemo.Client.
    csproj
    ```

4. Next, navigate to the `GrpcDemo.Client` folder and add the `Grpc.Net.Client` package to the project:

    ```
    cd GrpcDemo.Client
    dotnet add GrpcDemo.Client.csproj package Grpc.Net.Client
    ```

5. To use the gRPC tooling to generate the client code, we also need to add the following packages:

    ```
    dotnet add GrpcDemo.Client.csproj package Google.Protobuf
    dotnet add GrpcDemo.Client.csproj package Grpc.Tools
    ```

    Please change the project name and path in the preceding commands if you use a different project name or path.

    The updated `GrpcDemo.Client.csproj` file contains the following code:

    ```
    <ItemGroup>
      <PackageReference Include="Grpc.Net.Client" Version="2.55.0"
    />
    ```

```
  <PackageReference Include="Google.Protobuf"
Version="3.22.0-rc2" />
    <PackageReference Include="Grpc.Tools" Version="2.56.0">
      <IncludeAssets>runtime; build; native; contentfiles;
analyzers; buildtransitive</IncludeAssets>
      <PrivateAssets>all</PrivateAssets>
    </PackageReference>
  </ItemGroup>
```

As mentioned in the *Understanding the gRPC project structure* section, the Grpc.Tools package contains code-generation tooling for gRPC. It is a development-time dependency, which means that it is not required at runtime. So, we need to add the <PrivateAssets>all</PrivateAssets> element to the Grpc.Tools package to ensure that the package is not included in the published application.

6. Next, copy the Protos folder from the GrpcDemo project to the GrpcDemo.Client project. Then, add the following code to the GrpcDemo.Client.csproj file:

```
  <ItemGroup>
    <Protobuf Include="Protos\greet.proto" GrpcServices="Client"
OutputDir="Generated"/>
    <Protobuf Include="Protos\invoice.proto"
GrpcServices="Client"  OutputDir="Generated"/>
    <Protobuf Include="Protos\demo.proto"
GrpcServices="Client"  OutputDir="Generated"/>
  </ItemGroup>
```

Similar to the GrpcDemo project, we use the Protobuf element to specify the proto files and the output directory. The GrpcServices attribute is used to specify the type of the generated code. In this case, we use Client because we are creating a gRPC client.

When you make changes to the proto files in the GrpcDemo project, do not forget to copy the changes to the GrpcDemo.Client project to ensure that the client code is up to date.

In the Generated/Protos folder, you will find the generated code for each proto file. For example, the invoice.proto file will generate the following files:

*   Invoice.cs: This file contains the definition of the messages in the invoice.proto file

*   InvoiceGrpc.cs: This file contains the gRPC client code for the services in the invoice.proto file

7. Next, let's create an InvoiceClient.cs file in the project root folder and add the following code:

```
using Grpc.Net.Client;

namespace GrpcDemo.Client;

internal class InvoiceClient
```

```
{
    public async Task CreateContactAsync()
    {
        using var channel = GrpcChannel.ForAddress("http://
localhost:5269");
        var client = new Contact.ContactClient(channel);
        var reply = await client.CreateContactAsync(new
CreateContactRequest()
        {
            Email = "john.doe@abc.com",
            FirstName = "John",
            LastName = "Doe",
            IsActive = true,
            Phone = "1234567890",
            YearOfBirth = 1980
        });
        Console.WriteLine("Created Contact: " + reply.
ContactId);
        Console.ReadKey();
    }
}
```

In the preceding code, we use the `GrpcChannel.ForAddress()` method to create a gRPC channel, which accepts the address of the gRPC server.

8. To get the address of the gRPC server, you can use the `dotnet run` command in the `GrpcDemo` project to start the gRPC server. The following output shows the address of the gRPC server:

```
info: Microsoft.Hosting.Lifetime[14]
      Now listening on: http://localhost:5269
info: Microsoft.Hosting.Lifetime[0]
      Application started. Press Ctrl+C to shut down.
info: Microsoft.Hosting.Lifetime[0]
      Hosting environment: Development
```

9. Alternatively, you can check the `applicationUrl` property in the `Properties/launchSettings.json` file. The following code shows the `applicationUrl` property:

```
{
  "$schema": "http://json.schemastore.org/launchsettings.json",
  "profiles": {
    "http": {
      ...
      "applicationUrl": "http://localhost:5269",
      ...
    },
```

```
    "https": {
      ...
      "applicationUrl": "https://localhost:7179;http://
localhost:5269",
      ...
    }
  }
}
```

A gRPC channel is used to establish a connection to the gRPC server on the specified address and port. Once we have the gRPC channel, we can create an instance of the `ContactClient` class, which is generated from the proto file. Then, we call the `CreateContactAsync()` method to create a contact. The `CreateContactAsync()` method accepts a `CreateContactRequest` object as the parameter. The `CreateContactAsync()` method returns a `CreateContactResponse` object, which contains the `ContactId` value. At the end of the method, we print the `ContactId` value to the console.

This method is straightforward. There are a few things to note:

- Creating a gRPC channel is an expensive operation. So, it is recommended to reuse the gRPC channel. However, a gRPC client is a lightweight object, so there is no need to reuse it.

- You can create multiple gRPC clients from one gRPC channel, and you can safely use multiple gRPC clients concurrently.

10. To secure the gRPC channel using TLS, you need to run the gRPC service with HTTPS. For example, you can use the following command to run the gRPC service:

    ```
    dotnet run --urls=https://localhost:7179
    ```

11. Then, you can use the HTTPS address to create the gRPC channel:

    ```
    using var channel = GrpcChannel.ForAddress("https://
    localhost:7179");
    ```

12. In the `Program.cs` file, call the `CreateContactAsync()` method, as follows:

    ```
    var contactClient = new InvoiceClient();
    await contactClient.CreateContactAsync();
    ```

13. Run the gRPC server and the gRPC client in different terminals. By doing this, you will be able to see the following output in the gRPC client terminal:

    ```
    Created Contact: 3193c36c-2aab-49a7-a0b1-6bdb3b69dea1
    ```

This is a simple example of a gRPC client in a console application. In the next section, we will create a server streaming service and the corresponding gRPC client.

Defining a server streaming service

Similar to a unary service, a server streaming service has a request message and a response message. The difference is that the response message is a stream message. Once the server starts to send the stream response message, the client cannot send any more messages to the server, unless the server finishes sending the stream response message or the client cancels the RPC call by raising `ServerCallContext.CancellationToken`.

The server streaming service is useful when we need to send a stream of data to the client. In this case, the server can send multiple messages to the client over a single RPC call. Here are some scenarios where a server streaming service is useful:

- **Events streaming**: When the server needs to send a stream of event messages to the client so that the client can process the event messages.

- **Real-time data feeds**: When the server has a continuous stream of data to send to the client, such as stock prices, weather data, and so on.

- **File streaming**: When the server needs to send a large file to the client, the server can split the file into small chunks and send them one by one as a stream response message. This can reduce the memory usage on the server and the client because the server and the client do not need to load the entire file into memory.

The following code shows a server streaming service with the required message types:

```
message GetRandomNumbersRequest {
  int32 min = 1;
  int32 max = 2;
  int32 count = 3;
}

message GetRandomNumbersResponse {
  int32 number = 1;
}

service RandomNumbers {
  rpc GetRandomNumbers(GetRandomNumbersRequest) returns (stream
GetRandomNumbersResponse);
}
```

In the preceding proto file, we define two messages named `GetRandomNumbersRequest` and `GetRandomNumbersResponse`. Then, we define a `RandomNumbers` service, which contains a `GetRandomNumbers()` RPC method. Note that the response message of the `GetRandomNumbers` RPC method is annotated with the `stream` keyword. This means that the response message is a stream message.

The generated code for the `GetRandomNumbers()` RPC method is as follows:

```
[grpc::BindServiceMethod(typeof(RandomNumbers), "BindService")]
public abstract partial class RandomNumbersBase
{
  [global::System.CodeDom.Compiler.GeneratedCode("grpc_csharp_plugin",
null)]
  public virtual global::System.Threading.Tasks.Task
GetRandomNumbers(global::GrpcDemo.GetRandomNumbersRequest request,
grpc::IServerStreamWriter<global::GrpcDemo.GetRandomNumbersResponse>
responseStream, grpc::ServerCallContext context)
  {
    throw new grpc::RpcException(new grpc::Status(grpc::StatusCode.
Unimplemented, ""));
  }
}
```

In the generated code, we can see that the type of the response message is
`IServerStreamWriter<GetRandomNumbersResponse>`. Let's add a simple
implementation for the `RandomNumbers` service. Follow these steps:

1. Create a `RandomNumbersService.cs` file in the `Service` folder and add the following code:

```
public class RandomNumbersService(ILogger<RandomNumbersService>
logger) : RandomNumbers.RandomNumbersBase
{
    public override async Task
GetRandomNumbers(GetRandomNumbersRequest request,
        IServerStreamWriter<GetRandomNumbersResponse>
responseStream, ServerCallContext context)
    {
        var random = new Random();
        for (var i = 0; i < request.Count; i++)
        {
            await responseStream.WriteAsync(new
GetRandomNumbersResponse
            {
                Number = random.Next(request.Min, request.Max)
            });
            await Task.Delay(1000);
        }
    }
}
```

In the implementation of the `GetRandomNumbers()` method, we use a `for` loop to generate random numbers and send them to the client every second. Note that we use the `responseStream.WriteAsync()` method to send the stream response message to the client. The message finishes sending when the loop ends.

2. If we need a continuous stream response message, we can check the `ServerCallContext.CancellationToken` property of the `context` parameter. If the client cancels the RPC call, the `ServerCallContext.CancellationToken` property will be raised. The following code shows how to check the `ServerCallContext.CancellationToken` property:

```
public override async Task
GetRandomNumbers(GetRandomNumbersRequest request,
    IServerStreamWriter<GetRandomNumbersResponse>
responseStream, ServerCallContext context)
{
    var random = new Random();
    while (!context.CancellationToken.IsCancellationRequested)
    {
        await responseStream.WriteAsync(new
GetRandomNumbersResponse
        {
            Number = random.Next(request.Min, request.Max)
        });
        await Task.Delay(1000, context.CancellationToken);
    }
}
```

In the preceding code, we use a `while` loop to check the `ServerCallContext.CancellationToken` property. If the client cancels the RPC call, the `ServerCallContext.CancellationToken` property will be raised, and the `while` loop will end. If there are any other asynchronous operations in the method, we can pass the `ServerCallContext.CancellationToken` property to the asynchronous operations. This can ensure that the asynchronous operations will be canceled when the client cancels the RPC call.

3. Next, we will register the `RandomNumbersService` class in the DI container. Open the `Program.cs` file and add the following code:

```
app.MapGrpcService<RandomNumbersService>();
```

4. Next, we will create a gRPC client to call the `GetRandomNumbers()` RPC method. Create a `RandomNumbersClient.cs` file in the project root folder and add the following code:

```
internal class ServerStreamingClient
{
    public async Task GetRandomNumbers()
```

```
    {
        using var channel = GrpcChannel.ForAddress("https://
    localhost:7179");
        var client = new RandomNumbers.
    RandomNumbersClient(channel);
        var reply = client.GetRandomNumbers(new
    GetRandomNumbersRequest()
        {
            Count = 100,
            Max = 100,
            Min = 1
        });
        await foreach (var number in reply.ResponseStream.
    ReadAllAsync())
        {
            Console.WriteLine(number.Number);
        }
        Console.ReadKey();
    }
}
```

The code to create the client is similar to that of `InvoiceClient`, which we introduced in the *Creating a gRPC client* section. The only difference is in the response message, which is handled using the `await foreach` statement. The `ReadAllAsync()` method returns an `IAsyncEnumerable<T>` object, which can be iterated over using the `await foreach` statement.

5. In the `Program.cs` file of the `GrpcDemo.Client` project, call the `GetRandomNumbers()` method, as follows:

    ```
    var serverStreamingClient = new ServerStreamingClient();
    await serverStreamingClient.GetRandomNumbers();
    ```

6. Run the gRPC server and the gRPC client in different terminals. You will see that the output contains a series of random numbers.

This is an example of a server streaming service and the corresponding gRPC client. In the next section, we will create a client streaming service and the corresponding gRPC client.

Defining a client streaming service

A client streaming service allows the client to send a stream of messages to the server over a single request. The server then sends a single response message to the client when it finishes processing the stream request messages. Once the server sends the response message, the client streaming call is complete.

Here are some scenarios where a client streaming service is useful:

- **File uploading**: When the client uploads a large file to the server, the client can split the file into small chunks and send them one by one as a stream request message, which can be more efficient than sending the entire file in a single request.

- **Real-time data capture**: When the client needs to send a stream of data to the server, such as sensor data, user interactions, or any continuous stream of data, the server can process the data and respond to the batch of data.

- **Data aggregation**: When the client needs to send a batch of data to the server for aggregation or analysis.

To define a client streaming service, we need to use the `stream` keyword to annotate the request message. The following code shows a client streaming service with the required message types:

```
message SendRandomNumbersRequest {
  int32 number = 1;
}

message SendRandomNumbersResponse {
  int32 count = 1;
  int32 sum = 2;
}

service RandomNumbers {
  rpc SendRandomNumbers(stream SendRandomNumbersRequest) returns
(SendRandomNumbersResponse);
}
```

The preceding `.proto` file defines two messages: `SendRandomNumbersRequest` and `SendRandomNumbersResponse`. The client sends a stream message containing a series of numbers to the server. The server then processes the stream message and calculates the sum of the numbers. Finally, the server sends a response message to the client, which contains the count of the numbers and the sum of the numbers. It is important to note that the `SendRandomNumbers()` RPC method is annotated with the `stream` keyword, indicating that the request message is a stream message.

Similar to the server streaming service, we can implement the `SendRandomNumbers()` method, as follows:

```
public override async Task<SendRandomNumbersResponse>
SendRandomNumbers(IAsyncStreamReader<SendRandomNumbersRequest>
requestStream, ServerCallContext context)
{
    var count = 0;
    var sum = 0;
```

```
    await foreach (var request in requestStream.ReadAllAsync())
    {
        _logger.LogInformation($"Received: {request.Number}");
        count++;
        sum += request.Number;
    }
    return new SendRandomNumbersResponse
    {
        Count = count,
        Sum = sum
    };
}
```

We utilize the IAsyncStreamReader<T>.ReadAllAsync() method in the preceding code to read all the stream request messages from the client. Subsequently, we use await foreach to iterate over the stream request messages. Lastly, we compute the count and sum of the numbers and return a SendRandomNumbersResponse object.

To consume the client streaming service, we will copy the proto files from the GrpcDemo project to the GrpcDemo.Client project. Then, we will create a ClientStreamingClient class in the GrpcDemo.Client project and add the following code:

```
internal class ClientStreamingClient
{
    public async Task SendRandomNumbers()
    {
        using var channel = GrpcChannel.ForAddress("https://
localhost:7179");
        var client = new RandomNumbers.RandomNumbersClient(channel);

        // Create a streaming request
        using var clientStreamingCall = client.SendRandomNumbers();
        var random = new Random();
        for (var i = 0; i < 20; i++)
        {
            await clientStreamingCall.RequestStream.WriteAsync(new
SendRandomNumbersRequest
            {
                Number = random.Next(1, 100)
            });
            await Task.Delay(1000);
        }
        await clientStreamingCall.RequestStream.CompleteAsync();
```

```
        // Get the response
        var response = await clientStreamingCall;
        Console.WriteLine($"Count: {response.Count}, Sum: {response.
Sum}");
        Console.ReadKey();
    }
}
```

In the SendRandomNumbers() method, we create an AsyncClientStreamingCall object by calling the SendRandomNumbers() method of the RandomNumbersClient class. Note that the client streaming call starts when the SendRandomNumbers() method is called, but the client does not send any messages until the RequestStream.CompleteAsync() method is called. In a for loop, we use the RequestStream.WriteAsync() method to send the stream request message to the server. At the end of the method, we call the RequestStream.CompleteAsync() method to indicate that the stream request message is complete. The stream request message contains 20 numbers, which are generated randomly.

In the Program.cs file of the GrpcDemo.Client project, we then call the SendRandomNumbers() method, as follows:

```
var clientStreamingClient = new ClientStreamingClient();
await clientStreamingClient.SendRandomNumbers();
```

Run the gRPC server and the gRPC client in different terminals. After around 20 seconds, you will see the following output in the gRPC client terminal (the sum may be different):

```
Count: 20, Sum: 1000
```

With that, we've learned how to create a client streaming service and the corresponding gRPC client. In the next section, we will create a bidirectional streaming service and the corresponding gRPC client.

Defining a bidirectional streaming service

A bidirectional streaming service allows the client and the server to send a stream of messages to each other over a single request concurrently. Once the connection has been established, the client and the server can send messages to each other at any time in any order because the two streams are independent. For example, the server can respond to each message from the client, or the server can send a response message after receiving a series of messages from the client.

Here are some scenarios where a bidirectional streaming service is useful:

- **Chat applications**: When the client and the server need to send instant messages to each other

- **Real-time data dashboard**: When the client continuously sends data to the server and the server builds a real-time dashboard to display the data

- **Multiplayer games**: When the players need to interact with each other in real-time and the server needs to synchronize the game state between the players

Let's define a bidirectional streaming service. In this example, the client sends some sentences to the server and the server responds to each sentence with the uppercase version of the sentence. The following code shows the required message types:

```
message ChatMessage {
  string sender = 1;
  string message = 1;
}

service Chat {
  rpc SendMessage(stream ChatMessage) returns (stream ChatMessage);
}
```

In the preceding proto file, we have defined a `ChatMessage` message containing two fields: `sender` and `message`. Additionally, we have defined a `Chat` service with a `SendMessage` RPC method. It is important to note that both the request and response of this method are annotated with the `stream` keyword, indicating that they are both stream messages.

Now, we can implement the `SendMessage()` method. Follow these steps:

1.   Create a `ChatService.cs` file in the `Service` folder and add the following code:

```
public class ChatService(ILogger<ChatService> logger) : Chat.
ChatBase
{


    public override async Task
SendMessage(IAsyncStreamReader<ChatMessage> requestStream,
IServerStreamWriter<ChatMessage> responseStream,
ServerCallContext context)
    {
        await foreach (var request in requestStream.
ReadAllAsync())
        {
            logger.LogInformation($"Received: {request.
Message}");
            await responseStream.WriteAsync(new ChatMessage
            {
                Message = $"You said: {request.Message.
ToUpper()}"
            });
        }
    }
}
```

Here, we utilize the `await foreach` method to iterate over the stream request messages. For each request message, we use the `WriteAsync()` method to send a response message back to the client. This response message contains the uppercase version of the request message.

2. Next, register the `ChatService` class in the dependency injection container. Open the `Program.cs` file and add the following code:

```
app.MapGrpcService<ChatService>();
```

3. Copy the proto files from the `GrpcDemo` project to the `GrpcDemo.Client` project. Then, create a `BidirectionalStreamingClient` class in the `GrpcDemo.Client` project and add the following code:

```
internal class BidirectionalStreamingClient
{
    public async Task SendMessage()
    {
        using var channel = GrpcChannel.ForAddress("https://
localhost:7179");
        var client = new Chat.ChatClient(channel);

        // Create a streaming request
        using var streamingCall = client.SendMessage();
        Console.WriteLine("Starting a background task to receive
messages...");
        var responseReaderTask = Task.Run(async () =>
        {
            await foreach (var response in streamingCall.
ResponseStream.ReadAllAsync())
            {
                Console.WriteLine(response.Message);
            }
        });

        Console.WriteLine("Starting to send messages...");
        Console.WriteLine("Input your message then press enter
to send it.");
        while (true)
        {
            var message = Console.ReadLine();
            if (string.IsNullOrWhiteSpace(message))
            {
                break;
            }
            await streamingCall.RequestStream.WriteAsync(new
ChatMessage
```

```
            {
                Message = message
            });
        }
        Console.WriteLine("Disconnecting...");
        await streamingCall.RequestStream.CompleteAsync();
        await responseReaderTask;
    }
}
```

Because we use a console application to call the bidirectional streaming service, we need to use a background task to read the stream response messages. The `ReadAllAsync()` method returns an `IAsyncEnumerable<T>` object, which can be iterated over using the `await foreach` statement. In the background task, we use the `await foreach` statement to iterate over the stream response messages and print them to the console.

Additionally, we use a `while` loop to read the input from the console and send the stream request messages to the server in the main thread. The `while` loop ends when the user enters an empty string. At the end of the method, we call the `RequestStream.CompleteAsync()` method to indicate that the stream request message is complete so that the server can finish processing the stream request messages gracefully.

4. In the `Program.cs` file of the `GrpcDemo.Client` project, call the `SendMessage()` method, as follows:

```
var bidirectionalStreamingClient = new
BidirectionalStreamingClient();
await bidirectionalStreamingClient.SendMessage();
```

5. Run the gRPC server and the gRPC client in different terminals. You will see the following output in the gRPC client terminal:

```
Hello, World!
Starting background task to receive messages...
Starting to send messages...
Input your message then press enter to send it.
How are you?
You said: HOW ARE YOU?
What is ASP.NET Core?
You said: WHAT IS ASP.NET CORE?

Disconnecting...
```

This example is a simple demonstration of a bidirectional streaming service and the corresponding gRPC client. The bidirectional streaming service allows the client and the server to send a stream of messages to each other at any time in any order. In the preceding example, the service responds to each message from the client. However, using similar code, we can implement more complex logic per the requirements.

We have now explored four types of gRPC services: unary, server streaming, client streaming, and bidirectional streaming. We have also learned how to create a gRPC client to call each of these gRPC services. In the next section, we will learn how to use gRPC services in ASP.NET Core applications.

Consuming gRPC services in ASP.NET Core applications

In the previous section, we learned how to create console applications to consume gRPC services. In this section, we will integrate gRPC services into ASP.NET Core applications. We will reuse the gRPC services we created in the previous section, and we will create a new ASP.NET Core application to consume the gRPC services.

To get started with the steps outlined in this section, begin with the `GrpcDemo-v3` folder of the source code. The complete code for this section can be found in the `GrpcDemo-v4` folder.

In the console applications, we used the `GrpcChannel` class to create a gRPC channel, after which we used the gRPC channel to create a gRPC client, as shown in the following code:

```
using var channel = GrpcChannel.ForAddress("https://localhost:7179");
var client = new Contact.ContactClient(channel);
```

In ASP.NET Core applications, a better way to create a gRPC client is to use the `IHttpClientFactory` interface with dependency injection. Let's see how to use the DI container to create a gRPC client:

1. First, we must create a new ASP.NET Core application. In this ASP.NET Core application, we will create a REST API to consume the gRPC services we created in the previous section. Use the `dotnet new` command to create a new ASP.NET Core application:

    ```
    dotnet new webapi -o GrpcDemo.Api -controllers
    ```

2. Then, add this project to the solution:

    ```
    dotnet sln GrpcDemo.sln add GrpcDemo.Api/GrpcDemo.Api.csproj
    ```

3. Next, add the `Grpc.Net.ClientFactory` and `Grpc.Tools` packages to the project:

    ```
    cd GrpcDemo.Api
    dotnet add GrpcDemo.Api.csproj package Grpc.Net.ClientFactory
    dotnet add GrpcDemo.Api.csproj package Grpc.Tools
    ```

    The `Grpc.Net.ClientFactory` package allows developers to create a gRPC client using a dependency injection container, eliminating the need for the new keyword. Additionally, the `Grpc.Tools` package can be used to generate gRPC client code from proto files.

4. Then, copy the `Protos` folder from the `GrpcDemo` project to the `GrpcDemo.Api` project. Next, add the following code to the `GrpcDemo.Api.csproj` file:

```
<ItemGroup>
  <Protobuf Include="Protos\greet.proto" GrpcServices="Client"
OutputDir="Generated"/>
  <Protobuf Include="Protos\invoice.proto"
GrpcServices="Client"  OutputDir="Generated"/>
  <Protobuf Include="Protos\demo.proto"
GrpcServices="Client"  OutputDir="Generated"/>
</ItemGroup>
```

Similar to the `GrpcDemo.Client` project, we use the `GrpcServices="Client"` attribute to specify the type of the generated code. In this case, we use `Client` because we will create a gRPC client to consume the gRPC services in the ASP.NET Core application.

5. Next, we can register the gRPC client in the DI container. Open the `Program.cs` file and add the following code:

```
builder.Services.AddGrpcClient<Contact.ContactClient>(x =>
x.Address = new Uri("https://localhost:7179"));
```

Note that the address of the gRPC server is hardcoded in the preceding code for simplicity. In a real-world application, we should use a configuration file to store the address of the gRPC server.

6. Next, we must create a controller to consume the gRPC services. Create a `ContactController.cs` file in the `Controllers` folder and add the following code:

```
[ApiController]
[Route("[controller]")]
public class ContactController(Contact.ContactClient client,
ILogger<ContactController> logger) : ControllerBase
{

    [HttpPost]
    public async Task<IActionResult>
CreateContact(CreateContactRequest request)
    {
        var reply = await _client.CreateContactAsync(request);
        return Ok(reply);
    }
}
```

In the `ContactController` class, we use dependency injection to inject the gRPC client, `ContactClient`, which is generated from the `demo.proto` file. Then, we create a `CreateContact` action method to call the `CreateContactAsync()` method of the `ContactClient` class. The `CreateContactAsync()` method accepts a `CreateContactRequest` object as the parameter, which is also generated from the proto file. The `CreateContactAsync()` method returns a

`CreateContactResponse` object, which contains the `ContactId` value. At the end of the method, we return the `ContactId` value to the client.

7. Run the gRPC server and the ASP.NET Core application in different terminals. Note that the gRPC server address must match the address specified in the `AddGrpcClient()` method. Then, you can navigate to the Swagger UI page, such as `http://localhost:5284/ swagger/index.html`, to test the `CreateContact()` action method. For example, you can use the following JSON object as the request body:

```
{
    "firstName": "John",
    "lastName": "Doe",
    "email": "john.doe@example.com",
    "phone": "1234567890",
    "yearOfBirth": 1980,
    "isActive": true
}
```

You will see the following response (the `contactId` value may be different):

```
{
    "contactId": "8fb43c22-143f-4131-a5f5-c3700b4f3a08"
}
```

This simple example shows how to use the `AddGrpcClient()` method to register a gRPC client in the DI container in ASP.NET Core applications, and how to use the gRPC client to consume a unary gRPC service. For other types of gRPC services, you need to update the code accordingly.

Updating proto files

gRPC is a contract-first RPC framework. This means that the server and the client communicate with each other using a contract, which is defined in a proto file. Inevitably, the contract will change over time. In this section, we will learn how to update the contract and how to handle the changes in the server and the client.

Once a proto file is used in production, we need to consider backward compatibility when we update the proto file. This is because the existing clients may use the old version of the proto file, which may not be compatible with the new version of the proto file. If the new version of the contract is not backward compatible, the existing clients will break.

The following changes are backward compatible:

- **Adding new fields to a request message**: If the client does not send the new fields, the server can use the default values of the new fields.

- **Adding new fields to a response message**: If the response message contains the new fields but the client does not recognize the new fields, the client will discard the new fields in proto 3. In the future version of proto, known as 3.5, this behavior will be changed to preserve the new fields as unknown fields.

- **Adding a new RPC method to a service**: The client that uses old versions of the proto file will not be able to call the new RPC method. However, the old RPC methods will still work.

- **Adding a new service to a proto file**: Similar to adding a new RPC method, the new service will not be available to the old clients, but the old services will still work.

The following changes may cause breaking changes, which require the clients to be updated accordingly:

- Removing a field from a message

- Renaming a field in a message

- Removing or renaming a message

- Changing a data type of a field

- Changing a field number

- Removing or renaming a service

- Removing or renaming an RPC method from a service

- Renaming a package

- Changing the `csharp_namespace` option

Protobuf uses field numbers to serialize and deserialize messages. If we rename a field in a message without changing the field number and the data type, the message can still be serialized and deserialized correctly, but the field name in the .NET code will be different from the field name in the proto file. This can be confusing for developers. So, the client code needs to be updated to use the new field name.

Removing a field from a message is a breaking change as the field number cannot be reused. For example, if we remove the `year_of_birth` field from the `CreateContactRequest` message defined in the *Understanding the field types* section for the gRPC server, the server will deserialize field number 5 as an unknown field. This could lead to errors in serialization/de-serialization if a developer later decides to add a new field with field number 5 as a different data type while existing clients still send field number 5 as an integer value.

To safely remove a field, we must ensure that the removed field number is not being used in the future. To avoid any potential conflicts, we can reserve the removed field number by using the `reserved` keyword. For example, if we delete the `year_of_birth` and `is_active` fields from the `CreateContactRequest` message, we can reserve the field numbers, as follows:

```
message CreateContactRequest {
    reserved 5, 6;
    reserved "year_of_birth", "is_active";
```

```
    string first_name = 1;
    string last_name = 2;
    string email = 3;
    string phone = 4;
}
```

In the preceding code, we use the `reserved` keyword to reserve field numbers 5 and 6, and the `year_of_birth` and `is_active` field names. The reserved field numbers and field names cannot be reused in the proto file. If we try to use a reserved field number or field name, the gRPC tooling will report an error.

Note that the reserved field names should be listed, as well as the reserved field numbers. This ensures that the JSON and text formats are backward compatible. When the field names are reserved, they cannot be placed in the same `reserved` statement with the field numbers.

Summary

In this chapter, we explored the fundamentals of gRPC services and clients. We discussed the field types that are used in protobuf, including the scalar types and some other types such as `DateTime`, `enum`, repeated fields, and map fields. Then, we learned about four types of gRPC services: unary, server streaming, client streaming, and bidirectional streaming. We explored how to implement each type of gRPC service and how to create a gRPC client to consume the gRPC service. Additionally, we demonstrated how to use the `AddGrpcClient()` method to register a gRPC client in the DI container of an ASP.NET Core application and how to use the gRPC client to consume a unary gRPC service. Finally, we discussed how to update the proto files and how to handle the changes in the server and the client.

To simplify the code samples, we did not use any database access code in the gRPC services. In a real-world application, we may need to interact with a database or other external services in the gRPC services. You can follow the same approach as you do in REST API services.

gRPC is suitable for building high-performance service-to-service communication. Due to this book's content limitations, we only covered the basics of gRPC. We did not cover advanced topics such as authentication, error handling, performance tuning, and others. However, this chapter should be enough to get you started with gRPC.

In the next chapter, we will explore GraphQL, an alternative approach to web APIs. GraphQL provides clients with the ability to request only the data they need, making it easier to modify APIs over time and enabling the use of powerful developer tools.

Further reading

To learn more about gRPC on .NET Core, please refer to the following resources:

- `https://protobuf.dev/`
- `https://grpc.io/docs/languages/csharp/`
- `https://learn.microsoft.com/en-us/aspnet/core/grpc/`

12

Getting Started with GraphQL

In *Chapter 11*, we explored how to create a gRPC service in ASP.NET Core. gRPC is a high-performance RPC framework that facilitates communication between services. We discussed the field types used in protobuf messages, and how to define four types of gRPC services: unary, server streaming, client streaming, and bidirectional streaming. Additionally, we learned how to configure gRPC services in ASP.NET Core and how to call gRPC services from a client application.

Next, we will explore another shape of web APIs: GraphQL. GraphQL is a query-based API that allows clients to specify the data they need, which solves the problem of over-fetching and under-fetching data. Besides, GraphQL supports **mutations**, which allow clients to modify data. In this chapter, we will learn about some basic concepts of GraphQL and how to create a GraphQL API in ASP.NET Core. We will cover the following topics in this chapter:

- Recap of GraphQL
- Setting up a GraphQL API using HotChocolate
- Adding mutations
- Using variables in queries
- Defining a GraphQL schema
- Retrieving related objects using resolvers
- Using data loaders
- Dependency injection
- Interfaces and union types
- Filtering, sorting, and paging
- Visualizing the GraphQL schema

After reading this chapter, you will be able to understand the basic concepts of GraphQL and how to create a GraphQL API in ASP.NET Core. You will also learn how to use Apollo Federation to build a microservices-based GraphQL API.

Technical requirements

The code examples in this chapter can be found at `https://github.com/PacktPublishing/Web-API-Development-with-ASP.NET-Core-8/tree/main/samples/chapter12`. You can use VS 2022 or VS Code to open the solutions.

Recap of GraphQL

GraphQL offers a flexible way to query and mutate data. The main difference between GraphQL and REST is that GraphQL allows clients to specify the data they need, whereas REST APIs return a fixed set of data. GraphQL treats data as a graph, and it uses a query language to define the shape of the data. This addresses the issues of over-fetching and under-fetching data by enabling clients to specify their data requirements. Additionally, it supports mutations, empowering clients to modify data as needed.

While REST APIs have multiple endpoints for different resources, GraphQL is typically served over a single endpoint, usually `/graphql`, which exposes a schema that describes the data. All queries and mutations are sent to this endpoint. The schema is defined using a GraphQL Schema Definition Language, which is the contract between the client and the server. The schema defines the types of data and the operations that can be performed on the data. The client can use the schema to validate the query and mutation requests.

GraphQL can solve the problem of over-fetching and under-fetching data for clients. However, the backend development is more complex than REST APIs. GraphQL uses resolvers to fetch data from different levels of the graph. If the implementation of the resolvers is not efficient, it can lead to performance issues. GraphQL also has a steep learning curve for developers who are not familiar with it.

ASP.NET Core does not have built-in support for GraphQL. However, several third-party libraries can be used to create GraphQL APIs:

- **HotChocolate**: HotChocolate is an open-source GraphQL server for .NET. It is built on top of ASP.NET Core and supports the newest GraphQL October 2021 specification. It is supported by ChilliCream, a company that provides GraphQL tooling and consulting services. ChilliCream also provides other products, such as Banana Cake Pop, which is a GraphQL IDE to create and test GraphQL queries, and Strawberry Shake, which is a GraphQL client library for .NET. You can find more information about HotChocolate at `https://chillicream.com/docs/hotchocolate/`.

- **GraphQL.NET**: GraphQL.NET is another open-source GraphQL implementation for .NET. It provides a set of libraries that can be used to create GraphQL APIs and clients. You can find more information about GraphQL.NET at `https://graphql-dotnet.github.io/`.

In this chapter, we will use HotChocolate to create a GraphQL API in ASP.NET Core.

Setting up a GraphQL API using HotChocolate

To begin with, you can download the code example named `SchoolManagement` for this chapter from the `chapter12\start` folder. This sample project has some basic code for an `AppDbContext` class and a `Teacher` class, as well as some seed data. The `Teacher` class has the following properties:

```
public class Teacher
{
    public Guid Id { get; set; }
    public string FirstName { get; set; } = string.Empty;
    public string LastName { get; set; } = string.Empty;
    public string Email { get; set; } = string.Empty;
    public string? Phone { get; set; }
    public string? Bio { get; set; }
}
```

You can open the project in VS Code or VS 2022. We will integrate `HotChocolate` into the project to create a GraphQL API following these steps:

1. Add the `HotChocolate.AspNetCore` NuGet package to the project. This package contains the ASP.NET Core integration for HotChocolate. It also contains the GraphQL IDE, which is a GraphQL client that can be used to create and test GraphQL queries. You can use the following command to add the package to the project:

    ```
    dotnet add package HotChocolate.AspNetCore
    ```

2. Next, create a query root type, `Query`, in the `GraphQL/Queries` folder, as shown here:

    ```
    public class Query
    {
        public async Task<List<Teacher>> GetTeachers([Service]
    AppDbContext context) =>
            await context.Teachers.ToListAsync();
    }
    ```

    The `Query` class will be used to define the queries that can be executed by the client. It has one method named `GetTeachers()`, which returns a list of teachers.

3. Then, we need to register the query root type in the `Program.cs` file. Add the following code after the `AddDbContext()` method:

    ```
    builder.Services
        .AddGraphQLServer()
        .AddQueryType<Query>();
    ```

    The preceding code registers the GraphQL server and adds the `Query` type to the schema.

4. Next, we need to map the GraphQL endpoint to expose the GraphQL schema. Add the following code to the `Program.cs` file:

    ```
    app.MapGraphQL();
    ```

    The preceding code maps the GraphQL endpoint to the `/graphql` URL.

5. Run the project using `dotnet run` and open the GraphQL IDE at `https://localhost:7208/graphql/`. You should see the following screen:

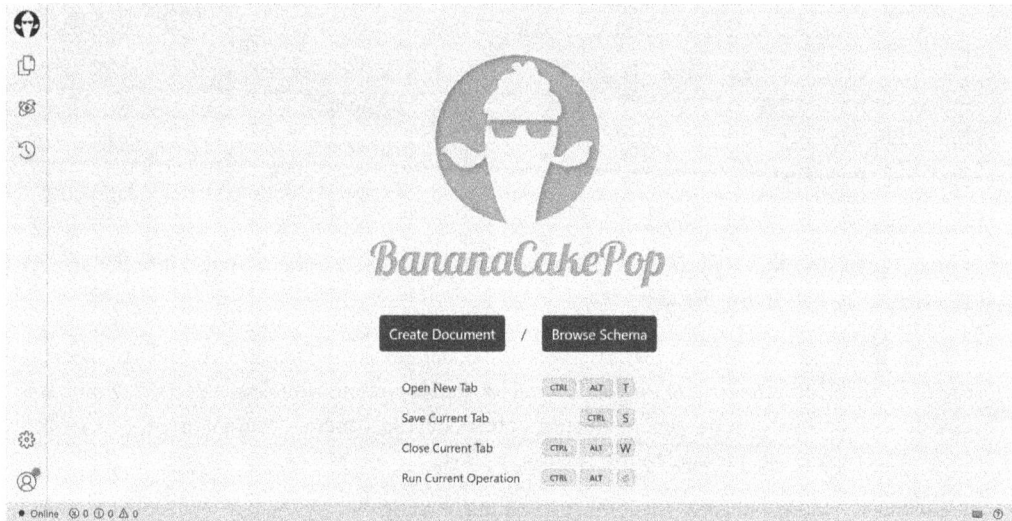

Figure 12.1 – The Banana Cake Pop GraphQL IDE

The GraphQL IDE allows you to create and test GraphQL queries.

> **Important note**
> The default launch URL is `swagger` for ASP.NET Core web API projects. You can change the launch URL in the `launchSettings.json` file to `graphql` to open the GraphQL IDE directly.

6. Click the **Browse Schema** button, then click the **Schema Definition** tab to view the GraphQL schema. You should see the following schema:

    ```
    type Query {
      teachers: [Teacher!]!
    }
    ```

```
type Teacher {
  id: UUID!
  firstName: String!
  lastName: String!
  email: String!
  phone: String
  bio: String
}
```

The preceding schema defines a query root type, Query, and a Teacher type. The Query type has one field named teachers, which returns a [Teacher!]! object. GraphQL uses ! to indicate that the field is non-nullable. By default, all fields are nullable. [Teacher!]! means that this field is a non-nullable array of non-nullable Teacher objects. When there is no data, the field will return an empty array.

The Teacher type has a few fields: id, firstName, lastName, email, phone, and bio. The id field is of the UUID type, which is a scalar type that represents a 128-bit **universally unique identifier** (**UUID**). The firstName, lastName, email, phone, and bio fields are of the String type. The client can specify which fields to be returned in the query.

7. Let us try to query the data. Click the **Create Document** button to create a new query. You can use the following query to get all teachers:

```
query {
    teachers {
        id
        firstName
        lastName
        email
        phone
        bio
    }
}
```

The preceding query will return all teachers in the database, as follows:

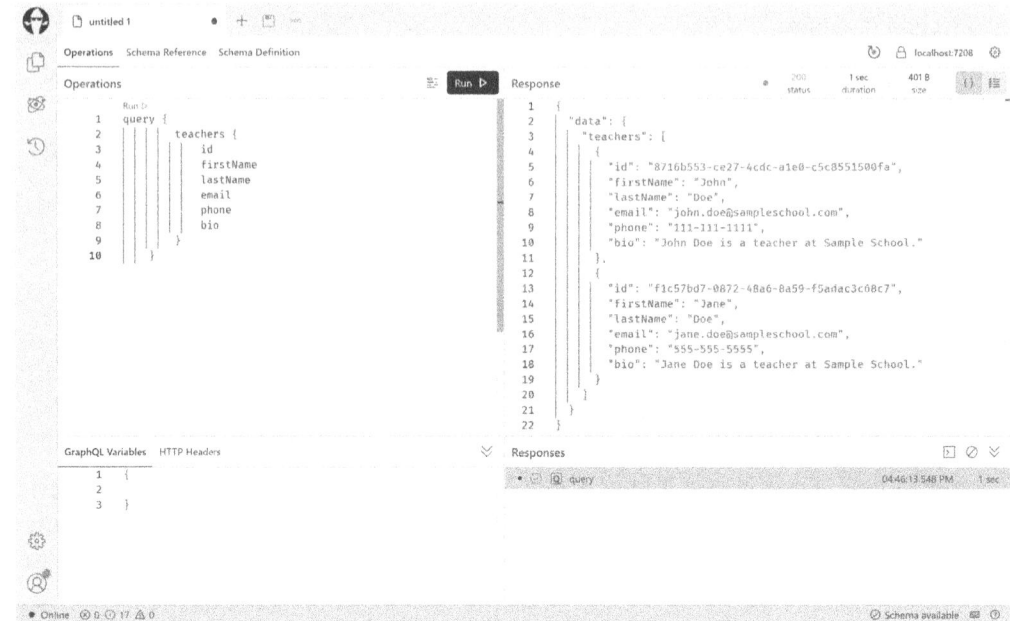

Figure 12.2 – Querying all teachers

8. You can add or remove fields in the query to specify the data to be returned. For example, to show a list of teachers on a web page, we do not need to return the id field and the bio field. We can remove the bio field from the query as follows:

```
query {
    teachers {
        firstName
        lastName
        email
        phone
    }
}
```

The preceding query will return those four fields only, which reduces the payload size.

So far, we have created a GraphQL API using HotChocolate. We have also learned how to query data using GraphQL queries. Next, we will learn how to modify data using mutations.

Adding mutations

In the previous section, we learned how to create a GraphQL API using HotChocolate. We added a query root type to query data. In this section, we will discuss how to modify data using mutations.

Mutations are used to modify data in GraphQL. A mutation consists of three parts:

- **Input**: The input is the data that will be used to modify the data. It is named with the `Input` suffix following the convention, such as `AddTeacherInput`.

- **Payload**: The payload is the data that will be returned after the mutation is executed. It is named with the `Payload` suffix following the convention, such as `AddTeacherPayload`.

- **Mutation**: The mutation is the operation that will be executed. It is named as *verb + noun* following the convention, such as `AddTeacherAsync`.

Let us add a mutation to create a new teacher. We will use the following steps:

1. Create an `AddTeacherInput` class in the `GraphQL/Mutations` folder, as shown here:

    ```
    public record AddTeacherInput(
        string FirstName,
        string LastName,
        string Email,
        string? Phone,
        string? Bio);
    ```

    The `AddTeacherInput` class is a record type that defines the input data for the `AddTeacherAsync` mutation. The `Id` property is not included in the input data because it will be generated by the code.

2. Add an `AddTeacherPayload` class in the `GraphQL/Mutations` folder, as follows:

    ```
    public class AddTeacherPayload
    {
        public Teacher Teacher { get; }

        public AddTeacherPayload(Teacher teacher)
        {
            Teacher = teacher;
        }
    }
    ```

    The `AddTeacherPayload` class defines the data that will be returned after the mutation is executed. It has a `Teacher` property of the `Teacher` type.

3. Next, we need to add the actual mutation to execute the operation. Add the `Mutation` class to the `GraphQL/Mutations` folder, as shown here:

```
public class Mutation
{
    public async Task<AddTeacherPayload> AddTeacherAsync(
        AddTeacherInput input,
        [Service] AppDbContext context)
    {
        var teacher = new Teacher
        {
            Id = Guid.NewGuid(),
            FirstName = input.FirstName,
            LastName = input.LastName,
            Email = input.Email,
            Phone = input.Phone,
            Bio = input.Bio
        };

        context.Teachers.Add(teacher);
        await context.SaveChangesAsync();

        return new AddTeacherPayload(teacher);
    }
}
```

The `Mutation` class has one method named `AddTeacherAsync`, which takes an `AddTeacherInput` object as the input data and returns an `AddTeacherPayload` object. The `AddTeacherAsync()` method creates a new `Teacher` object and adds it to the database. Then, it returns an `AddTeacherPayload` object that contains the newly created `Teacher` object.

4. Next, we need to register the mutation in the `Program.cs` file. Add the `AddMutationType` method after the `AddQueryType()` method, as follows:

```
builder.Services
    .AddGraphQLServer()
    .AddQueryType<Query>()
    .AddMutationType<Mutation>();
```

5. Run the project using `dotnet run` and open the GraphQL IDE. Check the schema definition and you should see the following mutation:

```
type Mutation {
  addTeacher(input: AddTeacherInput!): AddTeacherPayload!
```

```
  }

  input AddTeacherInput {
    firstName: String!
    lastName: String!
    email: String!
    phone: String
    bio: String
  }

  type AddTeacherPayload {
    teacher: Teacher!
  }
```

The preceding schema defines a mutation named addTeacher, which reflects the types and methods we defined in the Mutation class. Note that the AddTeacherInput type is an input type, so it uses the input keyword instead of type.

6. Click the **Create Document** button to create a new query. You can use the following mutation to create a new teacher:

```
  mutation addTeacher {
    addTeacher(
      input: {
        firstName: "John"
        lastName: "Smith"
        email: "john.smith@sampleschool.com"
        phone: "1234567890"
        bio: "John Smith is a math teacher."
      }
    ) {
      teacher {
        id
      }
    }
  }
```

The preceding mutation will create a new teacher and return the id property of the newly created teacher, as follows:

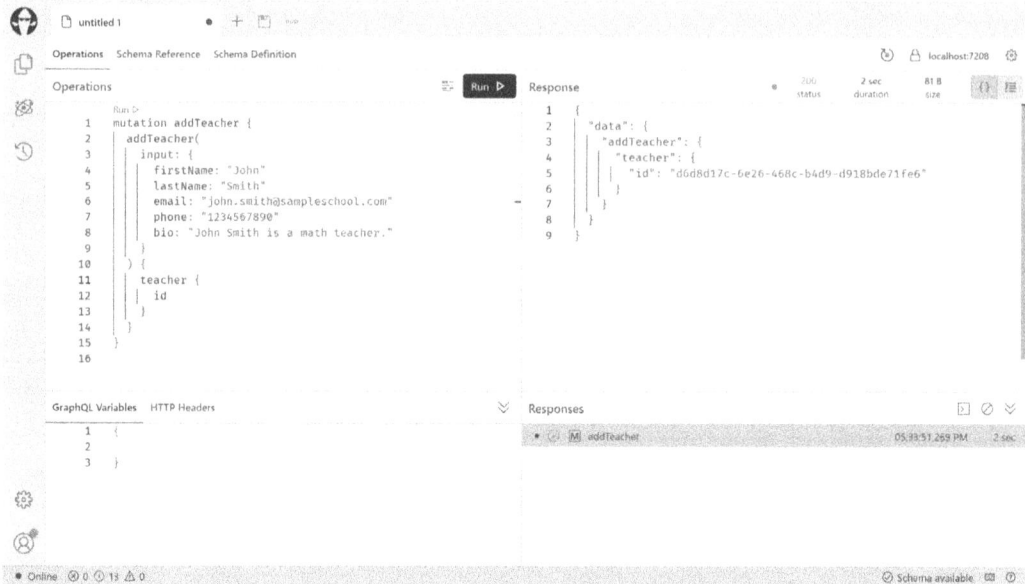

Figure 12.3 – Creating a new teacher

Then, you can query the data to verify that the new teacher has been added to the database.

Using variables in queries

In the previous section, we learned how to query data and modify data using GraphQL queries and mutations. In this section, we will discuss how to use variables in queries.

GraphQL allows you to use variables in queries. This is useful when you want to pass parameters to the query. We can create a query that accepts an id parameter and returns the teacher with the specified ID. Follow these steps to create the query:

1. Add a GetTeacher() method in the Query class, as follows:

    ```
    public async Task<Teacher?> GetTeacher(Guid id, [Service]
    AppDbContext context) =>
        await context.Teachers.FindAsync(id);
    ```

    The preceding code adds a GetTeacher() method to the Query class. It takes an id parameter and returns the teacher with the specified ID.

2. Now, you can use the $ sign to define a variable in the query. For example, you can use the following query to get a teacher by ID:

    ```
    query getTeacher($id: UUID!) {
      teacher(id: $id) {
    ```

```
        id
        firstName
        lastName
        email
        phone
    }
  }
```

The preceding query defines a variable named `id` of the `UUID!` type. The `!` sign indicates that the variable is non-nullable. The `teacher` field takes the `id` variable as the parameter and returns the teacher with the specified ID. In the **GraphQL Variables** panel, you can define the `id` variable to pass the value to the query, as follows:

```
{
  "id": "00000000-0000-0000-0000-000000000401"
}
```

You can define multiple variables in the query. Note that variables must be scalars, enums, or input object types.

Defining a GraphQL schema

Usually, a system has multiple types of data. For example, a school management system has teachers, students, departments, and courses. A department has multiple courses, and a course has multiple students. A teacher can teach multiple courses, and a course can be taught by multiple teachers as well. In this section, we will discuss how to define a GraphQL schema with multiple types of data.

Scalar types

Scalar types are the primitive types in GraphQL. The following table lists the scalar types in GraphQL:

Scalar type	Description	.NET type
Int	Signed 32-bit integer	int
Float	Signed double-precision floating-point value specified in IEEE 754	float or double
String	UTF-8 character sequence	string
Boolean	true or false	bool
ID	A unique identifier, serialized as a string	string

Table 12.1 – Scalar types in GraphQL

Besides the preceding scalar types, `HotChocolate` also supports the following scalar types:

- `Byte`: Unsigned 8-bit integer
- `ByteArray`: Byte array that is encoded as a Base64 string
- `Short`: Signed 16-bit integer
- `Long`: Signed 64-bit integer
- `Decimal`: Signed decimal value
- `Date`: ISO-8601 date
- `TimeSpan`: ISO-8601 time duration
- `DateTime`: A custom GraphQL scalar defined by the community at `https://www.graphql-scalars.com/`. It is based on RFC3339. Note that this `DateTime` scalar uses an offset to UTC instead of a time zone
- `Url`: URL
- `Uuid`: GUID
- `Any`: A special type that is used to represent any literal or output type

There are more scalar types that are not listed here. You can find more information about the scalar types at `https://chillicream.com/docs/hotchocolate/v13/defining-a-schema/scalars`.

GraphQL supports enumerations as well. Enumeration types in GraphQL are a special kind of scalar type. They are used to represent a fixed set of values. .NET supports enumeratiopn types very well so that you can use the .NET enum type directly in GraphQL. You can define an enumeration type as follows:

```
public enum CourseType
{
    Core,
    Elective,
    Lab
}
```

The preceding code defines an enumeration type named `CourseType` with three values: `Core`, `Elective`, and `Lab`. The generated GraphQL schema is as follows:

```
enum CourseType {
  CORE
  ELECTIVE
  LAB
}
```

HotChocolate automatically converts the enumeration values to uppercase according to the GraphQL specification.

Object types

The object type is the most common type in GraphQL. It can contain simple scalar types such as `Int`, `String`, and `Boolean`, as well as other object types. For example, a `Teacher` type can contain the `Department` type, as shown here:

```
public class Teacher
{
    public Guid Id { get; set; }
    public string FirstName { get; set; } = string.Empty;
    public string LastName { get; set; } = string.Empty;
    public Guid DepartmentId { get; set; }
    public Department Department { get; set; } = default!;
}

public class Department
{
    public Guid Id { get; set; }
    public string Name { get; set; } = string.Empty;
    public string? Description { get; set; }
    // other properties
}
```

The preceding code defines a `Teacher` type and a `Department` type. The `Teacher` type has a `Department` property of the `Department` type. HotChocolate will generate the schema as follows:

```
type Teacher {
  id: UUID!
  firstName: String!
  lastName: String!
  departmentId: UUID!
  department: Department!
}

type Department {
  id: UUID!
  name: String!
  description: String
}
```

As we mentioned in the previous section, all the fields in GraphQL are nullable by default. If we want to make a field non-nullable, we can use the ! sign.

The object type can contain a list of other object types. For example, a Department type can contain a list of Teacher objects, as shown here:

```
public class Department
{
    public Guid Id { get; set; }
    public string Name { get; set; } = string.Empty;
    public string? Description { get; set; }
    public List<Teacher> Teachers { get; set; } = new();
}
```

The generated schema is as follows:

```
type Department {
  id: UUID!
  name: String!
  description: String
  teachers: [Teacher!]!
}
```

The teachers field is a non-nullable array of non-nullable Teacher objects. If we want to make the teachers field nullable, we can use the ? sign as follows:

```
public class Department
{
    public Guid Id { get; set; }
    public string Name { get; set; } = string.Empty;
    public string? Description { get; set; }
    public List<Teacher>? Teachers { get; set; }
}
```

The generated schema is as follows:

```
type Department {
  id: UUID!
  name: String!
  description: String
  teachers: [Teacher!]
}
```

The preceding schema means that the teachers field is a nullable array of non-nullable Teacher objects. When there is no data, the teachers field will return null.

Let's look back at the Query type and the Mutation type we defined in the previous sections:

```
type Query {
  teachers: [Teacher!]!
  teacher(id: UUID!): Teacher
}

type Mutation {
  addTeacher(input: AddTeacherInput!): AddTeacherPayload!
}
```

These two types look like regular object types, but they have special meanings in GraphQL. The Query type and the Mutation type are two special object types in GraphQL, as they define the entry points of the GraphQL API. Every GraphQL service must have a Query type, but may or may not have a Mutation type. So the teachers query is actually a field of the Query type, just like the department field in the teacher type. Mutations work in the same way.

So far, the GraphQL types are similar to C# types. If you are familiar with object-oriented programming, you should be able to understand the GraphQL types easily. Similar to C#, GraphQL supports interfaces as well. But before we delve into interfaces, Let's discuss how to retrieve the Department object when querying the Teacher object.

Retrieving related objects using resolvers

In the previous section, we defined a Teacher type and a Department type. The Teacher type has a Department property of the Department type. When querying the Teacher object, we may also want to retrieve the Department object. How can we do that?

You may think that we can use the Include() method to retrieve the Department object, as follows:

```
public async Task<List<Teacher>> GetTeachers([Service] AppDbContext
context) =>
    await context.Teachers.Include(x => x.Department).ToListAsync();
```

Then, you can query the Department object as follows:

```
query{
  teachers{
    id
    firstName
    lastName
    department{
      id
      name
      description
```

```
        }
      }
    }
```

It does work and you will see the following result:

```
{
  "data": {
    "teachers": [
      {
        "id": "00000000-0000-0000-0000-000000000401",
        "firstName": "John",
        "lastName": "Doe",
        "department": {
          "id": "00000000-0000-0000-0000-000000000001",
          "name": "Mathematics",
          "description": "Mathematics Department"
        }
      },
      {
        "id": "00000000-0000-0000-0000-000000000402",
        "firstName": "Jane",
        "lastName": "Doe",
        "department": {
          "id": "00000000-0000-0000-0000-000000000001",
          "name": "Mathematics",
          "description": "Mathematics Department"
        }
      }
    ]
  }
}
```

But this is not the best way to do it. Remember that GraphQL allows clients to specify the data they need. If the query does not specify the department field, the Department object will still be retrieved from the database. This is not efficient. We should only retrieve the Department object when the department field is specified in the query. That leads us to the concept of resolvers.

A resolver is a function that is used to retrieve data from somewhere for a specific field. The resolver is executed when the field is requested in the query. The resolver can fetch data from a database, a web API, or any other data source. It will drill down the graph to retrieve the data for the field. For example, when the department field is requested in the teachers query, the resolver will retrieve the Department object from the database. But when the query does not specify the department field, the resolver will not be executed. This can avoid unnecessary database queries.

Field resolvers

HotChocolate supports three ways to define schemas:

- **Annotation-based**: The first way is to use the annotation-based approach, which is what we have been using so far. HotChocolate automatically converts public properties and methods to a resolver that retrieves data from the data source following conventions. If a method has a `Get` prefix or an `Async` suffix, these prefixes or suffixes will be removed from the name.

- **Code-first**: This approach allows you to define the schema using explicit types and resolvers. It uses the Fluent API to define the details of the schema. This approach is more flexible when you need to customize the schema.

- **Schema-first**: This approach allows you to define the schema using the GraphQL schema definition language. If you are familiar with GraphQL, you can use this approach to define the schema directly.

As you, our readers, are mostly .NET developers, we will use the code-first approach to define the schema in the rest of this chapter, so we can benefit from the Fluent API to fine-tune the schema.

Let us look back at the `teacher` query we defined in the previous section:

```
public async Task<Teacher?> GetTeacher(Guid id, [Service] AppDbContext
context) =>
    await context.Teachers.FindAsync(id);
```

The preceding is the annotation-based approach. HotChocolate automatically converts the `GetTeacher()` method to a resolver named `teacher`. Next, we want to retrieve the `Department` object when the `department` field is requested. Let us make some changes by following these steps:

1. First, we need to define `TeacherType` class as a GraphQL object. Create a `TeacherType` class in the `Types` folder. The code is shown here:

    ```
    public class TeacherType : ObjectType<Teacher>
    {
        protected override void
    Configure(IObjectTypeDescriptor<Teacher> descriptor)
        {
            descriptor.Field(x => x.Department)
                .Name("department")
                .Description("This is the department to which the
    teacher belongs.")
                .Resolve(async context =>
                {
                    var department = await context.
    Service<AppDbContext>().Departments.
    ```

```
FindAsync(context.    Parent<Teacher>().DepartmentId);
                return department;
            });
    }
}
```

The `TeacherType` class inherits from the `ObjectType<Teacher>` class, which has a `Configure()` method to configure the GraphQL object and specify how to resolve the fields. In the preceding code, we use the code-first approach to define the `Department` field of `TeacherType`. The `Name` method is used to specify the name of the field. If the name of the field is the same as the name of the property following the convention, we can omit the `Name` method. By convention, the `Department` field will be converted to the `department` field in the schema. Then, we use the `Description` method to define the description of the field. The description will be shown in the GraphQL IDE.

Then, we use the `Resolve()` method to define the resolver. The resolver retrieves the `Department` object from the database using the `DepartmentId` property of the `Teacher` object. Note that we use the `context.Parent<Teacher>()` method to get the `Teacher` object because the `Teacher` object is the parent object of the `Department` object.

2. As we know the `Query` type is a special object type, we will create a `QueryType` class as well. Create a new `Types` folder in the `GraphQL` folder and move the `Query.cs` file to this `Types` folder.

3. Remove the `GetTeacher()` method and add a property as follows:

```
public class Query
{
    // Omitted for brevity
    public TeacherType? Teacher { get; set; } = new();
}
```

4. Create a new class named `QueryType`, as follows:

```
public class QueryType : ObjectType<Query>
{
    protected override void
Configure(IObjectTypeDescriptor<Query> descriptor)
    {
        descriptor.Field(x => x.Teacher)
            .Name("teacher")
            .Description("This is the teacher in the school.")
            .Type<TeacherType>()
            .Argument("id", a =>
a.Type<NonNullType<UuidType>>())
            .Resolve(async context =>
```

```
          {
              var id = context.ArgumentValue<Guid>("id");
              var teacher = await context.
    Service<AppDbContext>().Teachers.FindAsync(id);
              return teacher;
          });
      }
  }
```

The preceding code defines the root query type. In this query type, we specify the type of the field to be `TeacherType`. Next, we use the `Argument()` method to define the `id` argument, which is a non-nullable `UUID` type. Then, we use the `Resolve()` method to define the resolver. The resolver takes the `id` argument and retrieves the `Teacher` object from the database. Note that `AppDbContext` is injected into the resolver from the `context` object.

5.   Next, we need to update the `Program.cs` file to register `QueryType`. Update the `Program.cs` file as follows:

```
builder.Services
    .AddGraphQLServer()
    .AddQueryType<QueryType>()
    .AddMutationType<Mutation>();
```

We use `QueryType` to replace the `Query` type we defined previously so that we can use the resolver to retrieve the `Department` object when the `department` field is requested.

6.   Now, we can test the resolvers. Run the application using `dotnet run` and send the following request to query a teacher.

This is the GraphQL request:

```
query ($id: UUID!) {
  teacher(id: $id) {
    firstName
    lastName
    email
    department {
      name
      description
    }
  }
}
```

These are the GraphQL variables:

```
{
  "id": "00000000-0000-0000-0000-000000000401"
}
```

You will see the department information in the response. Also, if you check the log, you will see that the Department object is retrieved from the database. If you remove the department field from the query, you will only see one database query in the log, which means that GraphQL does not fetch the Department object from the database.

In this example, we defined a resolver using a delegate method. We can also define a resolver in a separate class. For example, we can define a TeacherResolver class as follows:

```
public class TeacherResolvers
{
    public async Task<Department> GetDepartment([Parent] Teacher
teacher, [Service] IDbContextFactory<AppDbContext> dbContextFactory)
    {
        await using var dbContext = await dbContextFactory.
CreateDbContextAsync();
        var department = await dbContext.Departments.
FindAsync(teacher.DepartmentId);
        return department;
    }
}
```

The preceding code defines a GetDepartment() method that takes a Teacher object as the parent object and returns the Department object. Then, we can use the ResolveWith() method to define the resolver in the TeacherType class, as follows:

```
descriptor.Field(x => x.Department)
    .Description("This is the department to which the teacher
belongs.")
    .ResolveWith<TeacherResolvers>(x => x.GetDepartment(default,
default));
```

Now, the logic of the resolver is moved to a separate class. This approach is more flexible when the resolver is complex. But for simple resolvers, we can use the delegate method directly.

So far, it works well. Let us try to update the GetTeachers method using the same approach in the next section.

Resolver for a list of objects

Similarly, we can use ListType<TeacherType> to define the teachers field and then use the Resolve() method to define the resolver. The ListType class is a wrapper type for the fluent code-first API. It is used to define a list of objects. Remove the GetTeachers() method in the Query class and add a Teachers field, as shown here:

```
public class Query
{
```

```
    // Omitted for brevity
    public List<TeacherType> Teachers { get; set; } = new();
}
```

Then, configure the `Teachers` field in the `QueryType` class as follows:

```
public class QueryType : ObjectType<Query>
{
    protected override void Configure(IObjectTypeDescriptor<Query>
descriptor)
    {
        descriptor.Field(x => x.Teachers)
            .Name("teachers") // This configuration can be omitted if
the name of the field is the same as the name of the property.
            .Description("This is the list of teachers in the
school.")
            .Type<ListType<TeacherType>>()
            .Resolve(async context =>
            {
                var teachers = await context.Service<AppDbContext>().
Teachers.ToListAsync();
                return teachers;
            });
        // Omitted for brevity
    }
}
```

The preceding code defines the `Teachers` field of `QueryType`. It uses `ListType<TeacherType>` to define a list of `TeacherType`. Then, it uses the `Resolve()` method to define the resolver. The resolver retrieves all the `Teacher` objects from the database. This code is similar to the `teacher` field we defined previously. However, it retrieves a list of `TeacherType` objects instead of a single `TeacherType` object. As `TeacherType` has a resolver for the `Department` field, we can retrieve the `Department` object for each `TeacherType` object.

Now, we can test the `teachers` field using the following query:

```
query{
  teachers{
    id
    firstName
    lastName
    department{
      id
      name
      description
    }
```

```
    }
  }
```

However, you may encounter an error in the response. Some teachers can be retrieved correctly, but some may not. The error message is like this:

```
{
  "errors": [
    {
      "message": "Unexpected Execution Error",
      "locations": [
        {
          "line": 6,
          "column": 5
        }
      ],
      "path": [
        "teachers",
        7,
        "department"
      ],
      "extensions": {
        "message": "A second operation was started on this context
instance before a previous operation completed. This is usually
caused by different threads concurrently using the same instance of
DbContext. For more information on how to avoid threading issues with
DbContext, see https://go.microsoft.com/fwlink/?linkid=2097913.",

        ...
      }
    },
  ]
}
```

This is because we have multiple resolvers that execute database queries concurrently. However, AppDbContext is registered as a scoped service, and the AppDbContext class is not thread-safe. When multiple resolvers try to query the database in parallel, they will use the same AppDbContext instance, which causes the error.

To fix this issue, we need to make sure that the resolvers do not access the same AppDbContext instance concurrently. There are two ways to do that. One is to execute the resolvers sequentially, and the other is to use separate AppDbContext instances for each resolver. HotChocolate provides a RegisterDbContext<TDbContext>() method to manage DbContext for resolvers. In order to use this feature, we need to install a NuGet package named HotChocolate.Data. EntityFramework using the following command:

```
dotnet add package HotChocolate.Data.EntityFramework
```

Then, we can update the `Program.cs` file to register the `AppDbContext` class, as follows:

```
builder.Services
    .AddGraphQLServer()
    .RegisterDbContext<AppDbContext>()
    // Omitted for brevity
```

The preceding code allows HotChocolate to manage the lifetime of `AppDbContext` for resolvers.

The `RegisterDbContext<TDbContext>()` method can specify how `DbContext` should be injected. There are three options:

- `DbContextKind.Synchronized`: This is to ensure that `DbContext` is never used concurrently. `DbContext` is still injected as a scoped service.

- `DbContextKind.Resolver`: This way will resolve the scoped `DbContext` for each resolver. This option is the default configuration. From the perspective of the resolver, `DbContext` is a transient service, so HotChocolate can execute multiple resolvers concurrently without any issues.

- `DbContextKind.Pooled`: This mechanism will create a pool of `DbContext` instances. It leverages the `DbContextPool` feature of EF Core. HotChocolate will resolve `DbContext` from the pool for each resolver. When the resolver is completed, `DbContext` will be returned to the pool. In this way, `DbContext` is also like a transient service for each resolver, so HotChocolate can parallelize the resolvers as well.

To demonstrate how to benefit from the pooled `DbContext`, we will use the `DbContextKind.Pooled` option. This approach requires a couple of additional steps:

1. First, we need to register `DbContext` using the `AddPooledDbContextFactory()` method instead of the `AddDbContext()` method. Update the `Program.cs` file as follows:

    ```
    builder.Services.AddPooledDbContextFactory<AppDbContext>(options
    =>
        options.UseSqlServer(builder.Configuration.
    GetConnectionString("DefaultConnection")));

    // Register the GraphQL services
    builder.Services
        .AddGraphQLServer()
        .RegisterDbContext<AppDbContext>(DbContextKind.Pooled)
        // Omitted for brevity
    ```

    The preceding code registers `AppDbContext` as a pooled service using the `AddPooledDbContextFactory()` method. Then, we use the `RegisterDbContext()` method to register `AppDbContext` as a pooled service for HotChocolate resolvers.

2. Update the `Configure()` method in the `QueryType` file to use the pooled `DbContext`:

```
descriptor.Field(x => x.Teachers)
    // Omitted for brevity
    .Type<ListType<TeacherType>>()
    .Resolve(async context =>
    {
        var dbContextFactory = context.
Service<IDbContextFactory<AppDbContext>>();
        await using var dbContext = await dbContextFactory.
CreateDbContextAsync();
        var teachers = await dbContext.Teachers.ToListAsync();
        return teachers;
    });
```

The preceding code uses `IDbContextFactory<TDbContext>` to create a new `AppDbContext` instance for each resolver. Then, it retrieves the `Teacher` objects from the database using the new `AppDbContext` instance.

One thing to note is that we need to use the `await using` statement to dispose of the `AppDbContext` instance after the resolver is completed in order to return the `AppDbContext` instance to the pool.

3. Update the other resolvers as well. For example, the resolver of the `Teacher` type looks like this:

```
descriptor.Field(x => x.Department)
    .Description("This is the department to which the teacher
belongs.")
    .Resolve(async context =>
    {
        var dbContextFactory = context.
Service<IDbContextFactory<AppDbContext>>();
        await using var dbContext = await dbContextFactory.
CreateDbContextAsync();
        var department = await dbContext.Departments
            .FindAsync(context.Parent<Teacher>().DepartmentId);
        return department;
    });
```

Now, we can test the `teachers` field again. You will see that all the teachers with the department information can be retrieved correctly.

So, everything looks good. But wait. If you check the logs, you will find that there are many database queries for each `Department` object:

```
info: Microsoft.EntityFrameworkCore.Database.Command[20101]
      Executed DbCommand (26ms) [Parameters=[], CommandType='Text', CommandTimeout='30']
      SELECT [t].[Id], [t].[Bio], [t].[DepartmentId], [t].[Email], [t].[FirstName], [t].[LastName], [t].[Phone]
      FROM [Teachers] AS [t]
info: Microsoft.EntityFrameworkCore.Database.Command[20101]
      Executed DbCommand (12ms) [Parameters=[@__get_Item_0='?' (DbType = Guid)], CommandType='Text', CommandTimeout='30'
]
      SELECT TOP(1) [d].[Id], [d].[Description], [d].[Name]
      FROM [Departments] AS [d]
      WHERE [d].[Id] = @__get_Item_0
info: Microsoft.EntityFrameworkCore.Database.Command[20101]
      Executed DbCommand (1ms) [Parameters=[@__get_Item_0='?' (DbType = Guid)], CommandType='Text', CommandTimeout='30']
      SELECT TOP(1) [d].[Id], [d].[Description], [d].[Name]
      FROM [Departments] AS [d]
      WHERE [d].[Id] = @__get_Item_0
info: Microsoft.EntityFrameworkCore.Database.Command[20101]
      Executed DbCommand (1ms) [Parameters=[@__get_Item_0='?' (DbType = Guid)], CommandType='Text', CommandTimeout='30']
      SELECT TOP(1) [d].[Id], [d].[Description], [d].[Name]
      FROM [Departments] AS [d]
      WHERE [d].[Id] = @__get_Item_0
info: Microsoft.EntityFrameworkCore.Database.Command[20101]
      Executed DbCommand (1ms) [Parameters=[@__get_Item_0='?' (DbType = Guid)], CommandType='Text', CommandTimeout='30']
      SELECT TOP(1) [d].[Id], [d].[Description], [d].[Name]
      FROM [Departments] AS [d]
      WHERE [d].[Id] = @__get_Item_0
```

Figure 12.4 – Database queries for each Department object

What is the reason behind this? Let us find out in the next section.

Using data loaders

In the previous section, we learned how to integrate HotChocolate with EF Core. We also learned how to use the DbContextPool feature to fetch data in multiple resolvers. However, we found that there are many database queries for each Department object in the Teacher list. That is because the resolvers for each Department object are executed separately, querying the database by each DepartmentId property in the list. This is similar to the *N+1* problem we discussed in *Chapter 1*. The difference is that the *N+1* problem occurs on the client side in REST APIs, while it occurs on the server side in GraphQL. To solve this problem, we need to find a way to load the batch data efficiently.

HotChocolate provides a DataLoader mechanism to solve the *N+1* problem. The data loader fetches data in batches from the data source. Then, the resolver can retrieve the data from the data loader, rather than querying the data source directly. The data loader will cache the data for the current request. If the same data is requested again, the resolver can retrieve the data from the data loader directly. This can avoid unnecessary database queries.

Before we learn how to use the data loader to solve the *N+1* problem, let's prepare the examples. We already have a Teachers query to query the list of teachers, and each teacher has a Department object. Now, we want to add a Departments query to query the list of departments, and each department has a list of teachers. We will use the following steps to add the Departments query:

1. The Department type is defined as follows:

    ```
    public class Department
    {
        public Guid Id { get; set; }
    ```

```
        public string Name { get; set; } = string.Empty;
        public string? Description { get; set; }
        public List<Teacher> Teachers { get; set; } = new();
    }
```

2. The Department class has a list of Teacher objects. Following the convention, we can define a DepartmentType class as follows:

```
public class DepartmentType : ObjectType<Department>
{
    protected override void
Configure(IObjectTypeDescriptor<Department> descriptor)
    {
        descriptor.Field(x => x.Teachers)
            .Description("This is the list of teachers in the
department.")
            .Type<ListType<TeacherType>>()
            .ResolveWith<DepartmentResolvers>(x =>
x.GetTeachers(default, default));
    }
}


public class DepartmentResolvers
{
    public async Task<List<Teacher>> GetTeachers([Parent]
Department department,
        [Service] IDbContextFactory<AppDbContext>
dbContextFactory)
    {
        await using var dbContext = await dbContextFactory.
CreateDbContextAsync();
        var teachers = await dbContext.Teachers.Where(x =>
x.DepartmentId == department.Id).ToListAsync();
        return teachers;
    }
}
```

The preceding code is similar to TeacherType, which we defined previously. DepartmentType has a Teachers field of the ListType<TeacherType> type. Then, we use the ResolveWith() method to define the resolver. The resolver retrieves the Teacher objects from the database using the DepartmentId property of the Department object.

3. Add a new field in the Query class, as follows:

```
public class Query
{
```

```
    // Omitted for brevity
    public List<DepartmentType> Departments { get; set; } =
new();
}
```

4. Then, configure the `Departments` field in the `QueryType` as follows:

```
descriptor.Field(x => x.Departments)
    .Description("This is the list of departments in the
school.")
    .Type<ListType<DepartmentType>>()
    .Resolve(async context =>
    {
        var dbContextFactory = context.
Service<IDbContextFactory<AppDbContext>>();
        await using var dbContext = await dbContextFactory.
CreateDbContextAsync();
        var departments = await dbContext.Departments.
ToListAsync();
        return departments;
    });
```

Now we have two queries – `teachers` and `departments`:

- The `teachers` query returns a list of `Teacher` objects, and each `Teacher` object has a `Department` object

- The `departments` query returns a list of `Department` objects, and each `Department` object has a list of `Teacher` objects

You can use the following queries to test the `departments` query:

```
query{
    departments{
      id
      name
      description
      teachers{
        id
        firstName
        lastName
        bio
      }
    }
}
```

If you check the logs, you will find that the following database queries are executed multiple times:

```
info: Microsoft.EntityFrameworkCore.Database.Command[20101]
      Executed DbCommand (3ms) [Parameters=[@__
department_Id_0='?' (DbType = Guid)], CommandType='Text',
CommandTimeout='30']
      SELECT [t].[Id], [t].[Bio], [t].[DepartmentId], [t].
[Email], [t].[FirstName], [t].[LastName], [t].[Phone]
      FROM [Teachers] AS [t]
      WHERE [t].[DepartmentId] = @__department_Id_0
```

The preceding query is executed for each `Department` object. This is also an *N+1* problem, as we discussed previously. Next, we will use the data loader to solve these *N+1* problems.

Batch data loader

First, let's optimize the `teachers` query. To retrieve the `teachers` data with the department information, we want to execute two SQL queries only. One is to retrieve the `teachers` data, and the other is to retrieve the department data. Then, HotChocolate should be able to map the `department` data to the teachers in memory, instead of executing a SQL query for each teacher.

Follow these steps to use the data loader:

1. Create a folder named `DataLoaders` in the `GraphQL` folder, then create a new `DepartmentByTeacherIdBatchDataLoader` class, as follows:

```
public class DepartmentByTeacherIdBatchDataLoader(
        IDbContextFactory<AppDbContext> dbContextFactory,
        IBatchScheduler batchScheduler,
        DataLoaderOptions? options = null)
        : BatchDataLoader<Guid, Department>(batchScheduler,
options)
{

    protected override async Task<IReadOnlyDictionary<Guid,
Department>> LoadBatchAsync(IReadOnlyList<Guid>    keys,
        CancellationToken cancellationToken)
    {
        await using var dbContext = await dbContextFactory.
CreateDbContextAsync(cancellationToken);
        var departments = await dbContext.Departments.Where(x =>
keys.Contains(x.Id))
            .ToDictionaryAsync(x => x.Id, cancellationToken);
        return departments;
    }
}
```

The preceding code defines a data loader to fetch the batch data for the Department object. The parent resolver, which is the teachers query, will get a list of Teacher objects. Each Teacher object has a DepartmentId property. DepartmentByTeacherIdBatchDataLoader will fetch the Department objects for the DepartmentId values in the list. The list of the Department objects will be converted to a dictionary. The key of the dictionary is the DepartmentId property and the value is the Department object. Then, the parent resolver can map the Department object to the Teacher object in memory.

2. Update the TeacherResolvers class as follows:

```
public class TeacherResolvers
{
    public async Task<Department> GetDepartment([Parent] Teacher teacher,
        DepartmentByTeacherIdBatchDataLoader departmentByTeacherIdBatchDataLoader, CancellationToken cancellationToken)
    {
        var department = await departmentByTeacherIdBatchDataLoader.LoadAsync(teacher.DepartmentId,    cancellationToken);
        return department;
    }
}
```

Instead of querying the database directly, the resolver uses DepartmentByTeacherIdBatchDataLoader to fetch the Department object for the DepartmentId property of the Teacher object. The DepartmentByTeacherIdBatchDataLoader will be injected by HotChocolate automatically.

3. Run the application and test the teachers query again. Now, you will see only two SQL queries are executed:

```
info: Microsoft.EntityFrameworkCore.Database.Command[20101]
      Executed DbCommand (108ms) [Parameters=[],
CommandType='Text', CommandTimeout='30']
      SELECT [t].[Id], [t].[Bio], [t].[DepartmentId], [t].
[Email], [t].[FirstName], [t].[LastName], [t].   [Phone]
      FROM [Teachers] AS [t]
info: Microsoft.EntityFrameworkCore.Database.Command[20101]
      Executed DbCommand (73ms) [Parameters=[@__keys_0='?' (Size
= 4000)], CommandType='Text',    CommandTimeout='30']
      SELECT [d].[Id], [d].[Description], [d].[Name]
      FROM [Departments] AS [d]
      WHERE [d].[Id] IN (
```

```
            SELECT [k].[value]
            FROM OPENJSON(@__keys_0) WITH ([value]
uniqueidentifier '$') AS [k]
        )
```

As we see, the first query is to get the list of the teachers, and the second query is to use the IN clause to query the departments that match the `DepartmentId` values in the list. This is much more efficient than the previous approach.

As it fetches the batch data for the `Department` object, it is called a batch data loader. This data loader is often used for one-to-one relationships, such as one `Teacher` object has one `Department` object. Note that this one-to-one relationship is not the same as the one-to-one relationship in the database. In GraphQL, the one-to-one relationship means that one object has one child object.

Group data loader

Next, let's optimize the `departments` query. In this case, one `Department` object has a list of `Teacher` objects. We can use the group data loader to fetch the `Teacher` objects for each `Department` object. The group data loader is similar to the batch data loader. The difference is that the group data loader fetches a list of objects for each key. The batch data loader fetches a single object for each key.

Follow these steps to use the group data loader:

1. Create a `TeachersByDepartmentIdDataLoader` class in the `DataLoaders` folder, and add the following code:

```
public class TeachersByDepartmentIdDataLoader(
        IDbContextFactory<AppDbContext> dbContextFactory,
        IBatchScheduler batchScheduler,
        DataLoaderOptions? options = null)
        : GroupedDataLoader<Guid, Teacher>(batchScheduler,
options)
{


    protected override async Task<ILookup<Guid, Teacher>>
LoadGroupedBatchAsync(IReadOnlyList<Guid> keys,
        CancellationToken cancellationToken)
    {
        await using var dbContext = await dbContextFactory.
CreateDbContextAsync(cancellationToken);
        var teachers = await dbContext.Teachers.Where(x => keys.
Contains(x.DepartmentId))
            .ToListAsync(cancellationToken);
        return teachers.ToLookup(x => x.DepartmentId);
```

```
    }
}
```

The preceding code defines a group data loader, which returns an `ILookup<Guid, Teacher>` object in the `LoadGroupedBatchAsync()` method. The `ILookup<Guid, Teacher>` object is similar to a dictionary. The key of the dictionary is the `DepartmentId` property and the value is a list of `Teacher` objects. The parent resolver can map the `Teacher` objects to the `Department` object in memory.

2. Update the `DepartmentResolvers` class as follows:

```
public class DepartmentResolvers
{
    public async Task<List<Teacher>> GetTeachers([Parent]
Department department,
        TeachersByDepartmentIdDataLoader
teachersByDepartmentIdDataLoader, CancellationToken
cancellationToken)
    {
        var teachers = await teachersByDepartmentIdDataLoader.
LoadAsync(department.Id, cancellationToken);
        return teachers.ToList();
    }
}
```

The preceding code uses `TeachersByDepartmentIdDataLoader` to fetch the `Teacher` objects for the `Department` object. `TeachersByDepartmentIdDataLoader` will be injected by HotChocolate automatically.

3. Run the application and test the `departments` query again. Now, you will see only two SQL queries are executed:

```
info: Microsoft.EntityFrameworkCore.Database.Command[20101]
      Executed DbCommand (38ms) [Parameters=[],
CommandType='Text', CommandTimeout='30']
      SELECT [d].[Id], [d].[Description], [d].[Name]
      FROM [Departments] AS [d]
info: Microsoft.EntityFrameworkCore.Database.Command[20101]
      Executed DbCommand (36ms) [Parameters=[@__keys_0='?' (Size
= 4000)], CommandType='Text',    CommandTimeout='30']
      SELECT [t].[Id], [t].[Bio], [t].[DepartmentId], [t].
[Email], [t].[FirstName], [t].[LastName], [t].[Phone]
      FROM [Teachers] AS [t]
      WHERE [t].[DepartmentId] IN (
          SELECT [k].[value]
          FROM OPENJSON(@__keys_0) WITH ([value]
uniqueidentifier '$') AS [k]
      )
```

That is exactly what we want. The first query is to get the list of the departments, and the second query is to use the IN clause to query the teachers that match the Department Id values in the list. This approach reduces the number of database queries significantly.

In this case, each Department object has a list of Teacher objects, so this kind of data loader is called a group data loader. It is often used for one-to-many relationships, such as one Department object has a list of Teacher objects.

HotChocolate supports cache data loader as well. It also supports using multiple data loaders in a resolver. As they are not used often, we will not discuss them in this chapter. You can refer to the documentation for more details: https://github.com/PacktPublishing/Web-API-Development-with-ASP.NET-Core-8/tree/main/samples/chapter12/start.

Dependency injection

In the previous code examples, we use IDbContextFactory<AppDbContext> and AppDbContext directly in the resolvers. In order to encapsulate our data access logic, we can add a service layer to implement our business logic. HotChocolate supports dependency injection for resolvers. In this section, we will learn how to inject other services into the resolvers.

To demonstrate how to use dependency injection in HotChocolate, we will add an interface named ITeacherService and a class named TeacherService, as follows:

```
public interface ITeacherService
{
    Task<Department> GetDepartmentAsync(Guid departmentId);
    Task<List<Teacher>> GetTeachersAsync();
    Task<Teacher> GetTeacherAsync(Guid teacherId);
    // Omitted for brevity
}



public class TeacherService(IDbContextFactory<AppDbContext>
contextFactory) : ITeacherService
{

    public async Task<Department> GetDepartmentAsync(Guid
departmentId)
    {
        await using var dbContext = await contextFactory.
CreateDbContextAsync();
        var department = await dbContext.Departments.
FindAsync(departmentId);
        return department ?? throw new ArgumentException("Department
not found", nameof(departmentId));
    }
```

```
    public async Task<List<Teacher>> GetTeachersAsync()
    {
        await using var dbContext = await contextFactory.
CreateDbContextAsync();
        var teachers = await dbContext.Teachers.ToListAsync();
        return teachers;
    }

    public async Task<Teacher> GetTeacherAsync(Guid teacherId)
    {
        await using var dbContext = await contextFactory.
CreateDbContextAsync();
        var teacher = await dbContext.Teachers.FindAsync(teacherId);
        return teacher ?? throw new ArgumentException("Teacher not
found", nameof(teacherId));
    }

    // Omitted for brevity
}
```

The preceding code encapsulates the data access logic in the `TeacherService` class. Then, we need to register `ITeacherService` in the `Program.cs` file, as follows:

```
builder.Services.AddScoped<ITeacherService, TeacherService>();
```

HotChocolate uses the same approach to register the services as ASP.NET Core, but injecting the services is a little different. In ASP.NET Core, we can inject the services into the controller constructor, while HotChocolate does not recommend constructor injection. Instead, HotChocolate recommends using the method-level injection. First, the GraphQL type definitions are singleton objects. If we use constructor injection, the services will be injected as singleton objects as well. This is not what we want. Second, sometimes HotChocolate needs to synchronize the resolvers to avoid concurrency issues. If we use constructor injection, HotChocolate cannot control the lifetime of the services. Note that this applies to the HotChocolate GraphQL types and resolvers only. For other services, we can still use constructor injection.

Let us see how to use the method-level injection.

Using the Service attribute

We can use `HotChocolate.ServiceAttribute` to inject services into the resolvers. For example, we can add a `GetTeachersWithDI` method in the `Query` class as follows:

```
public async Task<List<Teacher>> GetTeachersWithDI([Service]
ITeacherService teacherService) =>
    await teacherService.GetTeachersAsync();
```

Note that the `Service` attribute is from the `HotChocolate` namespace, not the `Microsoft.AspNetCore.Mvc` namespace. With this attribute, `ITeacherService` will be injected into the `teacherService` parameter automatically.

If we have many services in the project, using the attribute for each service is tedious. HotChocolate provides a `RegisterServices()` method to simplify the injection. We can update the `Program.cs` file as follows:

```
builder.Services
    .AddGraphQLServer()
    .RegisterDbContext<AppDbContext>(DbContextKind.Pooled)
    .RegisterService<ITeacherService>()
    .AddQueryType<QueryType>()
    .AddMutationType<Mutation>();
```

Now, we can remove the `Service` attribute from the `GetTeachersWithDI()` method. HotChocolate can still inject `ITeacherService` automatically, as shown here:

```
public async Task<List<Teacher>> GetTeachersWithDI(ITeacherService
teacherService) =>
    await teacherService.GetTeachersAsync();
```

This will save us a lot of time.

Understanding the lifetime of the injected services

We have learned that, in ASP.NET Core, we can inject the services as singleton, scoped, or transient services. HotChocolate offers more options for the lifetime of the injected services. When we use the `Service` attribute or the `RegisterService()` method to inject the services, we can specify the `ServiceKind` property to control the lifetime of the services. The `ServiceKind` has the following options:

- `ServiceKind.Default`: This is the default option. The service will be injected as the same lifetime in the registered service in the DI container.

- `ServiceKind.Synchronized`: This option is similar to the synchronized `DbContext`. The resolver using the service will be executed sequentially. The synchronization only happens in the same request scope.

- `ServiceKind.Resolver`: This option is to resolve the service for each resolver scope. The service will be disposed of after the resolver is completed.

- `ServiceKind.Pooled`: This option is similar to the pooled `DbContext`. The service needs to be registered as an `ObjectPool<T>` instance in the ASP.NET Core DI container. The resolver will get the service from the pool and return it to the pool after the resolver is completed.

To specify the `ServiceKind` for the injected services, we can add a `ServiceKind` parameter in the `Service` attribute or the `RegisterService()` method. For example, we can update the `GetTeachersWithDI()` method as follows:

```
public async Task<List<Teacher>>
GetTeachersWithDI([Service(ServiceKind.Resolver)] ITeacherService
teacherService) =>
    await teacherService.GetTeachersAsync();
```

The preceding code specifies the `ServiceKind` as `ServiceKind.Resolver`. So, `ITeacherService` will be resolved for each resolver scope.

If we use the `RegisterServices()` method to register the services, we can specify the `ServiceKind` in the `RegisterServices()` method, as follows:

```
builder.Services
    .AddGraphQLServer()
    .RegisterDbContext<AppDbContext>(DbContextKind.Pooled)
    .RegisterService<ITeacherService>(ServiceKind.Resolver)
    .AddQueryType<QueryType>()
    .AddMutationType<Mutation>();
```

To inject the services in the `Resolve()` method, we can get the service from the `context` object, as follows:

```
descriptor.Field(x => x.Teachers)
    .Description("This is the list of teachers in the school.")
    .Type<ListType<TeacherType>>()
    .Resolve(async context =>
    {
        var teacherService = context.Service<ITeacherService>();
        var teachers = await teacherService.GetTeachersAsync();
        return teachers;
    });
```

The preceding code uses the `context.Service<T>()` method to get `ITeacherService` from the `context` object, which is similar to injecting `IDbContextFactory<AppDbContext>` in the previous examples.

Interface and union types

HotChocolate supports the use of interfaces and union types in GraphQL. In this section, we will explore how to incorporate these features into your GraphQL schema. Interfaces provide a way to group types that share common fields, while union types allow for the creation of a single type that can return different object types. With HotChocolate, you can easily implement these features to enhance the functionality of your GraphQL schema.

Interfaces

To prepare the examples of GraphQL interfaces, we have an `ISchoolRoom` interface and two classes that implement the interface, as follows:

```
public interface ISchoolRoom
{
    Guid Id { get; set; }
    string Name { get; set; }
    string? Description { get; set; }
    public int Capacity { get; set; }
}

public class LabRoom : ISchoolRoom
{
    public Guid Id { get; set; }
    public string Name { get; set; } = string.Empty;
    public string? Description { get; set; }
    public int Capacity { get; set; }
    public string Subject { get; set; } = string.Empty;
    public string Equipment { get; set; } = string.Empty;
    public bool HasChemicals { get; set; }
}

public class Classroom : ISchoolRoom
{
    public Guid Id { get; set; }
    public string Name { get; set; } = string.Empty;
    public string? Description { get; set; }
    public int Capacity { get; set; }
    public bool HasComputers { get; set; }
    public bool HasProjector { get; set; }
    public bool HasWhiteboard { get; set; }
}
```

The two classes both implement the `ISchoolRoom` interface, but they have some different properties. You can find the model classes and the model configurations in the sample code.

The service layer is defined in the `ISchoolRoomService` interface and the `SchoolRoomService` class, as follows:

```
public interface ISchoolRoomService
{
    Task<List<ISchoolRoom>> GetSchoolRoomsAsync();
    Task<List<LabRoom>> GetLabRoomsAsync();
```

```
    Task<List<Classroom>> GetClassroomsAsync();
    Task<LabRoom> GetLabRoomAsync(Guid labRoomId);
    Task<Classroom> GetClassroomAsync(Guid classroomId);
}


public class SchoolRoomService(IDbContextFactory<AppDbContext>
contextFactory) : ISchoolRoomService
{


    public async Task<List<ISchoolRoom>> GetSchoolRoomsAsync()
    {
        await using var dbContext = await contextFactory.
CreateDbContextAsync();
        var labRooms = await dbContext.LabRooms.ToListAsync();
        var classrooms = await dbContext.Classrooms.ToListAsync();
        var schoolRooms = new List<ISchoolRoom>();
        schoolRooms.AddRange(labRooms);
        schoolRooms.AddRange(classrooms);
        return schoolRooms;
    }


    // Omitted for brevity
}
```

The GetSchoolRoomsAsync() method retrieves a list of LabRoom objects and a list of Classroom objects from the database. Then, it combines the two lists into a single list of ISchoolRoom objects.

We also need to register ISchoolRoomService in the Program.cs file, as follows:

```
builder.Services.AddScoped<ISchoolRoomService, SchoolRoomService>();
```

To define an interface in HotChocolate, we need to use the InterfaceType<T> class. The InterfaceType<T> class is used to define an interface type in the schema. Follow these steps to define an interface type using the code-first API:

1. Create a class named SchoolRoomType in the Types folder:

    ```
    public class SchoolRoomType : InterfaceType<ISchoolRoom>
    {
        protected override void
    Configure(IInterfaceTypeDescriptor<ISchoolRoom> descriptor)
        {
    ```

```
                descriptor.Name("SchoolRoom");
        }
    }
```

The preceding code defines an interface type for the ISchoolRoom interface.

2. Create two new classes for LabRoom and Classroom, as follows:

```
public class LabRoomType : ObjectType<LabRoom>
{
    protected override void
Configure(IObjectTypeDescriptor<LabRoom> descriptor)
    {
        descriptor.Name("LabRoom");
        descriptor.Implements<SchoolRoomType>();
    }
}

public class ClassroomType : ObjectType<Classroom>
{
    protected override void
Configure(IObjectTypeDescriptor<Classroom> descriptor)
    {
        descriptor.Name("Classroom");
        descriptor.Implements<SchoolRoomType>();
    }
}
```

In the preceding code, we use the Implements() method to specify the interface implemented by the object type.

3. Add a query field in the Query class:

```
public List<SchoolRoomType> SchoolRooms { get; set; } = new();
```

4. Configure the SchoolRooms field in the QueryType class:

```
descriptor.Field(x => x.SchoolRooms)
    .Description("This is the list of school rooms in the
school.")
    .Type<ListType<SchoolRoomType>>()
    .Resolve(async context =>
    {
        var service = context.Service<ISchoolRoomService>();
        var schoolRooms = await service.GetSchoolRoomsAsync();
        return schoolRooms;
    });
```

In the preceding code, we use the `Service()` method to get `ISchoolRoomService` from the `context` object. Then, we use the `GetSchoolRoomsAsync()` method to retrieve the list of `ISchoolRoom` objects. The result includes both `LabRoom` and `Classroom` objects.

5. Next, we need to explicitly register `LabRoomType` and `ClassroomType` in `SchemaBuilder`. Update the `Program.cs` file as follows:

```
builder.Services
    .AddGraphQLServer()
    .RegisterDbContext<AppDbContext>(DbContextKind.Pooled)
    .RegisterService<ITeacherService>(ServiceKind.Resolver)
    .AddQueryType<QueryType>()
    .AddType<LabRoomType>()
    .AddType<ClassroomType>()
    .AddMutationType<Mutation>();
```

6. Run the application and check the generated schema. You will find the interface definition and its implementations, as shown here:

```
type Query {
  """
  This is the list of school rooms in the school.
  """
  schoolRooms: [SchoolRoom]
}

type LabRoom implements SchoolRoom {
  id: UUID!
  name: String!
  description: String
  capacity: Int!
  subject: String!
  equipment: String!
  hasChemicals: Boolean!
}

type Classroom implements SchoolRoom {
  id: UUID!
  name: String!
  description: String
  capacity: Int!
  hasComputers: Boolean!
  hasProjector: Boolean!
  hasWhiteboard: Boolean!
}
```

7. Next, we can use the `SchoolRoom` interface to query both the `LabRoom` and `Classroom` objects. For example, we can use the following query to retrieve the `LabRoom` objects:

```
query {
  schoolRooms {
    __typename
    id
    name
    description
    capacity
    ... on LabRoom {
      subject
      equipment
      hasChemicals
    }
    ... on Classroom {
      hasComputers
      hasProjector
      hasWhiteboard
    }
  }
}
```

The preceding query uses the `... on LabRoom` syntax to specify the `LabRoom` object. Then we can retrieve the `LabRoom` or `Classroom` properties. The `__typename` property is used to show the object type. The result is as follows:

```
{
  "data": {
    "schoolRooms": [
      {
        "__typename": "LabRoom",
        "id": "00000000-0000-0000-0000-000000000501",
        "name": "Chemistry Lab",
        "description": "Chemistry Lab",
        "capacity": 20,
        "subject": "Chemistry",
        "equipment": "Chemicals, Beakers, Bunsen Burners",
        "hasChemicals": true
      },
      {
        "__typename": "Classroom",
        "id": "00000000-0000-0000-0000-000000000601",
        "name": "Classroom 1",
        "description": "Classroom 1",
```

```
            "capacity": 20,
            "hasComputers": true,
            "hasProjector": false,
            "hasWhiteboard": true
        },
        ...
    ]
  }
}
```

In the response, you can see that the LabRoom object has the subject, equipment, and hasChemicals properties, while the Classroom object has the hasComputers, hasProjector, and hasWhiteboard properties. This can be helpful when we want to query complex objects with different properties.

Although interfaces provide flexibility for querying objects with different properties, we need to note that interfaces can be used for output types only. We cannot use interfaces for input types or arguments.

Union types

Union types are similar to interfaces. The difference is that union types do not need to define any common fields. Instead, union types can combine multiple object types into a single type.

Follow the same approach as the previous section to prepare the models for union types. You can find an Equipment class and a Furniture class in the Models folder, as follows:

```
public class Equipment
{
    public Guid Id { get; set; }
    public string Name { get; set; } = string.Empty;
    public string? Description { get; set; }
    public string Condition { get; set; } = string.Empty;
    public string Brand { get; set; } = string.Empty;
    public int Quantity { get; set; }
}

public class Furniture
{
    public Guid Id { get; set; }
    public string Name { get; set; } = string.Empty;
    public string? Description { get; set; }
    public string Color { get; set; } = string.Empty;
    public string Material { get; set; } = string.Empty;
    public int Quantity { get; set; }
}
```

The Equipment class and the Furniture class have some different properties. You can find the model configurations in the sample code. We also need to add the services for both classes. You can find the following code in the Services folder.

Here is the code for the IEquipmentService interface and the EquipmentService class:

```
public interface IEquipmentService
{
    Task<List<Equipment>> GetEquipmentListAsync();
    Task<Equipment> GetEquipmentAsync(Guid equipmentId);
}


public class EquipmentService(IDbContextFactory<AppDbContext>
contextFactory) : IEquipmentService
{

    public async Task<List<Equipment>> GetEquipmentListAsync()
    {
        await using var dbContext = await contextFactory.
CreateDbContextAsync();
        var equipment = await dbContext.Equipment.ToListAsync();
        return equipment;
    }

    public async Task<Equipment> GetEquipmentAsync(Guid equipmentId)
    {
        await using var dbContext = await contextFactory.
CreateDbContextAsync();
        var equipment = await dbContext.Equipment.
FindAsync(equipmentId);
        return equipment ?? throw new ArgumentException("Equipment not
found", nameof(equipmentId));
    }
}
```

Here is the code for the IFurnitureService interface and the FurnitureService class:

```
public interface IFurnitureService
{
    Task<List<Furniture>> GetFurnitureListAsync();
    Task<Furniture> GetFurnitureAsync(Guid furnitureId);
}
```

```
public class FurnitureService(IDbContextFactory<AppDbContext>
contextFactory) : IFurnitureService
{

    public async Task<List<Furniture>> GetFurnitureListAsync()
    {
        await using var dbContext = await contextFactory.
CreateDbContextAsync();
        var furniture = await dbContext.Furniture.ToListAsync();
        return furniture;
    }

    public async Task<Furniture> GetFurnitureAsync(Guid furnitureId)
    {
        await using var dbContext = await contextFactory.
CreateDbContextAsync();
        var furniture = await dbContext.Furniture.
FindAsync(furnitureId);
        return furniture ?? throw new ArgumentException("Furniture not
found", nameof(furnitureId));
    }
}
```

The preceding code should be straightforward. Do not forget to register the services in the `Program.cs` file, as follows:

```
builder.Services.AddScoped<IEquipmentService, EquipmentService>();
builder.Services.AddScoped<IFurnitureService, FurnitureService>();
```

Next, let's create the union types following these steps:

1. Create two classes named `EquipmentType` and `FurnitureType`, as follows:

    ```
    public class EquipmentType : ObjectType<Equipment>
    {
        protected override void
    Configure(IObjectTypeDescriptor<Equipment> descriptor)
        {
            descriptor.Name("Equipment");
        }
    }

    public class FurnitureType : ObjectType<Furniture>
    {
        protected override void
    Configure(IObjectTypeDescriptor<Furniture> descriptor)
    ```

```
    {
        descriptor.Name("Furniture");
    }
}
```

The preceding code defines the `EquipmentType` and `FurnitureType` object types. These two object types are just normal object types.

2. Create a new class named `SchoolItemType`, as follows:

```
public class SchoolItemType : UnionType
{
    protected override void Configure(IUnionTypeDescriptor
    descriptor)
    {
        descriptor.Name("SchoolItem");
        descriptor.Type<EquipmentType>();
        descriptor.Type<FurnitureType>();
    }
}
```

The preceding code defines a union type named `SchoolItem`. A union type must inherit from the `UnionType` class. Then, we use the `Type` method to specify the object types that are included in the union type. In this case, the `SchoolItem` union type includes the `EquipmentType` and `FurnitureType` object types.

As we already registered these two types in the union type, we do not need to register them again in the `Program.cs` file.

3. Add a query field in the `Query` class:

```
public List<SchoolItemType> SchoolItems { get; set; } = new();
```

4. Configure the resolver for the `SchoolItems` field in the `QueryType` class, as shown here:

```
descriptor.Field(x => x.SchoolItems)
    .Description("This is the list of school items in the
school.")
    .Type<ListType<SchoolItemType>>()
    .Resolve(async context =>
    {
        var equipmentService = context.
Service<IEquipmentService>();
        var furnitureService = context.
Service<IFurnitureService>();
        var equipmentTask = equipmentService.
GetEquipmentListAsync();
        var furnitureTask = furnitureService.
GetFurnitureListAsync();
```

```
        await Task.WhenAll(equipmentTask, furnitureTask);
        var schoolItems = new List<object>();
        schoolItems.AddRange(equipmentTask.Result);
        schoolItems.AddRange(furnitureTask.Result);
        return schoolItems;
    });
```

We retrieve a list of Equipment and Furniture objects from the database. We then combine these two lists into a single list of objects, as the object type is the base type of all types in C#. This allows us to use the object type to effectively combine the two lists.

5. Run the application and check the generated schema. You will find the union type defined as follows:

```
union SchoolItem = Equipment | Furniture

type Equipment {
  id: UUID!
  name: String!
  description: String
  condition: String!
  brand: String!
  quantity: Int!
}

type Furniture {
  id: UUID!
  name: String!
  description: String
  color: String!
  material: String!
  quantity: Int!
}
```

A union type is represented as a union of a list of object types using the | symbol. In this case, the SchoolItem union type includes the Equipment and Furniture object types.

6. Then, we can query the SchoolItem union type, as follows:

```
query {
  schoolItems {
    __typename
    ... on Equipment {
      id
      name
      description
```

```
            condition
            brand
            quantity
          }
          ... on Furniture {
            id
            name
            description
            color
            material
            quantity
          }
        }
      }
```

The preceding query uses the ... on Equipment syntax to specify the Equipment object. Then, we can retrieve the Equipment properties or Furniture properties. The result is as follows:

```
{
  "data": {
    "schoolItems": [
      {
        "__typename": "Equipment",
        "id": "00000000-0000-0000-0000-000000000701",
        "name": "Bunsen Burner",
        "description": "Bunsen Burner",
        "condition": "Good",
        "brand": "Bunsen",
        "quantity": 10
      },
      {
        "__typename": "Furniture",
        "id": "00000000-0000-0000-0000-000000000801",
        "name": "Desk",
        "description": "Desk",
        "color": "Brown",
        "material": "Wood",
        "quantity": 20
      },
      ...
    ]
  }
}
```

In the response, you can see that the `Equipment` object has the `condition` and `brand` properties, while the `Furniture` object has the `color` and `material` properties. However, even though the `Equipment` and `Furniture` objects have some of the same properties (such as `Id`, `Name`, and so on.), the query must specify the properties for each object type. For example, we cannot use the following query:

```
query {
  schoolItems {
    __typename
    id
    name
  }
}
```

The preceding query will cause an error, as follows:

```
{
  "errors": [
    {
      "message": "A union type cannot declare a field directly.
Use inline fragments or fragments instead.",
      "locations": [
        {
          "line": 2,
          "column": 15
        }
      ],
      "path": [
        "schoolItems"
      ],
      "extensions": {
        "type": "SchoolItem",
        "specifiedBy": "http://spec.graphql.org/
October2021/   #sec-Field-Selections-on-Objects-Interfaces-and-
Unions-Types"
      }
    }
  ]
}
```

Please note that the `SchoolItem` union type is not a base type of the `Equipment` and `Furniture` object types. If you want to query the common properties of the object types, you can use the interface type instead of the union type.

Filtering, sorting, and pagination

In this section, we will learn how to implement filtering, sorting, and pagination in HotChocolate. These features are very important for a real-world application. We will use the `Student` object as an example to demonstrate how to implement these features. The `Student` class is defined as follows:

```
public class Student
{
    public Guid Id { get; set; }
    public string FirstName { get; set; } = string.Empty;
    public string LastName { get; set; } = string.Empty;
    public string Email { get; set; } = string.Empty;
    public string? Phone { get; set; }
    public string Grade { get; set; } = string.Empty;
    public DateOnly? DateOfBirth { get; set; }
    public Guid GroupId { get; set; }
    public Group Group { get; set; } = default!;
    public List<Course> Courses { get; set; } = new();
    public List<StudentCourse> StudentCourses { get; set; } = new();
}
```

To use filtering, sorting, and pagination, we need to install the `HotChocolate.Data` NuGet package. If you already installed the `HotChocolate.Data.EntityFramework` package following the previous sections, you do not need to install the `HotChocolate.Data` package again. The `HotChocolate.Data` package is a dependency of the `HotChocolate.Data.EntityFramework` package. If not, you can install the `HotChocolate.Data` package using the following command:

```
dotnet add package HotChocolate.Data
```

Let's begin with filtering!

Filtering

HotChocolate supports filtering on the object type. A question is how we translate the GraphQL filter to the SQL-native queries. If the resolver exposes an `IQueryable` interface, HotChocolate can translate the GraphQL filter to SQL-native queries automatically. But we can also implement the filtering logic in the resolver manually. In this section, we will explore how to use filtering in HotChocolate.

To enable filtering on the `Student` object type, follow these steps:

1. First, we need to register the `Filtering` middleware in the `Program.cs` file, as follows:

    ```
    builder.Services
        .AddGraphQLServer()
    ```

```
    // Omitted for brevity
    .AddFiltering()
    .AddMutationType<Mutation>();
```

2. Add a query field in the Query class:

```
public List<Student> Students { get; set; } = new();
```

3. Apply the filtering in the resolver of the Students field in the QueryType:

```
descriptor.Field(x => x.Students)
    .Description("This is the list of students in the school.")
    .UseFiltering()
    .Resolve(async context =>
    {
        var dbContextFactory = context.
Service<IDbContextFactory<AppDbContext>>();
        var dbContext = await dbContextFactory.
CreateDbContextAsync();
        var students = dbContext.Students.AsQueryable();
        return students;
    });
```

The preceding code uses the UseFiltering() method to enable filtering on the Students field. Then, we use the AsQueryable() method to expose the IQueryable interface. This allows HotChocolate to translate the GraphQL filter to SQL-native queries automatically.

4. Run the application and check the generated schema. You will find the students query has a StudentFilterInput filter, as shown here:

```
students(where: StudentFilterInput): [Student!]!
```

The StudentFilterInput filter is an input type, as follows:

```
input StudentFilterInput {
  and: [StudentFilterInput!]
  or: [StudentFilterInput!]
  id: UuidOperationFilterInput
  firstName: StringOperationFilterInput
  lastName: StringOperationFilterInput
  email: StringOperationFilterInput
  phone: StringOperationFilterInput
  grade: StringOperationFilterInput
  dateOfBirth: DateOperationFilterInput
}
```

HotChocolate automatically inspects the `Student` object type and generates the filter input type. The `StudentFilterInput` filter includes all the properties of the `Student` object type by default.

5. Next, we can filter the `students` query, as follows:

```
query {
  students(where: { firstName: { eq: "John" } }) {
    id
    firstName
    lastName
    email
    phone
    grade
    dateOfBirth
  }
}
```

The preceding query uses the `where` argument to filter the `Student` objects. The `where` argument is a `StudentFilterInput` type. The `StudentFilterInput` type includes all the properties of the `Student` object type. In this case, we use the `firstName` property to filter the `Student` objects. The `firstName` property is a `StringOperationFilterInput` type. The `StringOperationFilterInput` type includes the following operators:

- `eq`: Equal to

- `neq`: Not equal to

- `in`: In the list

- `nin`: Not in the list

- `contains`: Contains

- `notContains`: Does not contain

- `startsWith`: Starts with

- `nstartsWith`: Does not start with

- `endsWith`: Ends with

- `nendsWith`: Does not end with

The result is as follows:

```
{
  "data": {
    "students": [
      {
```

```
      "id": "00000000-0000-0000-0000-000000000901",
      "firstName": "John",
      "lastName": "Doe",
      "email": "",
      "phone": null,
      "grade": "",
      "dateOfBirth": "2000-01-01"
    }
  ]
}
}
```

You can find the generated SQL query in the logs, as follows:

```
info: Microsoft.EntityFrameworkCore.Database.Command[20101]
      Executed DbCommand (36ms) [Parameters=[@__p_0='?' (Size =
32)], CommandType='Text',    CommandTimeout='30']
      SELECT [s].[Id], [s].[DateOfBirth], [s].[Email], [s].
[FirstName], [s].[Grade], [s].[GroupId], [s].    [LastName], [s].
[Phone]
      FROM [Students] AS [s]
      WHERE [s].[FirstName] = @__p_0
```

The preceding SQL query uses the WHERE clause to filter the Student objects, which means the filtering is done in the database.

6. The filtering can be defined in the variable as well. For example, we can use the following query to filter the Student objects:

```
query ($where: StudentFilterInput) {
  students(where: $where) {
    id
    firstName
    lastName
    email
    phone
    grade
    dateOfBirth
  }
}
```

The variable can be defined as follows:

```
{
  "where": {
    "firstName": {
      "eq": "John"
    }
```

```
    }
}
```

The result is the same as the previous query. You can also try other operators to filter the Student objects. For example, the following variable uses the in operator to filter the Student objects:

```
{
  "where": {
    "firstName": {
      "in": ["John", "Jane"]
    }
  }
}
```

The following variable uses the gt operator to filter the students who were born after 2001-01-01:

```
{
  "where": {
    "dateOfBirth": {
      "gt": "2001-01-01"
    }
  }
}
```

The generated filter input type contains all the properties of the object type. Sometimes, we do not need to filter all the properties. For example, we may want to allow filtering on a few properties only. In this case, we can create a custom filter input type and specify the properties we want to filter. Follow these steps to create a custom filter input type:

1. Create a Filters folder in the GraphQL folder. Then, add a new class named StudentFilterType, as follows:

    ```
    public class StudentFilterType : FilterInputType<Student>
    {
        protected override void
    Configure(IFilterInputTypeDescriptor<Student> descriptor)
        {
            descriptor.BindFieldsExplicitly();
            descriptor.Field(t => t.Id);
            descriptor.Field(t => t.GroupId);
            descriptor.Field(t => t.FirstName);
            descriptor.Field(t => t.LastName);
            descriptor.Field(t => t.DateOfBirth);
        }
    }
    ```

2. Then, we need to specify the filter input type in the resolver. Update the resolver for the `students` query, as follows:

```
descriptor.Field(x => x.Students)
    .Description("This is the list of students in the school.")
    .UseFiltering<StudentFilterType>()
    // Omitted for brevity
```

3. Check the generated schema. You will find `StudentFilterInput` only contains the fields we specified in `StudentFilterType`, as shown here:

```
input StudentFilterInput {
  and: [StudentFilterInput!]
  or: [StudentFilterInput!]
  id: UuidOperationFilterInput
  groupId: UuidOperationFilterInput
  firstName: StringOperationFilterInput
  lastName: StringOperationFilterInput
  dateOfBirth: DateOperationFilterInput
}
```

4. If the model has many properties but we only want to ignore a few properties, we can use the `Ignore()` method to ignore the properties we do not want to filter. For example, we can update `StudentFilterType` as follows:

```
override protected void
Configure(IFilterInputTypeDescriptor<Student> descriptor)
{
    descriptor.BindFieldsImplicitly();
    descriptor.Ignore(t => t.Group);
    descriptor.Ignore(t => t.Courses);
}
```

In the preceding code, all the properties of the `Student` object type will be included in the `StudentFilterInput` filter except the `Group` and `Courses` properties.

By default, `StringOperationFilterInput` includes many operations, such as eq, neq, in, nin, contains, notContains, startsWith, and endsWith. If we do not want to include all these operations, we can specify the operations by using a custom operation filter. For example, we can define a `StudentStringOperationFilterInputType` class as follows:

```
public class StudentStringOperationFilterInputType :
StringOperationFilterInputType
{
    protected override void Configure(IFilterInputTypeDescriptor
descriptor)
```

```
        {
            descriptor.Operation(DefaultFilterOperations.Equals).
    Type<StringType>();
            descriptor.Operation(DefaultFilterOperations.Contains).
    Type<StringType>();
        }
    }
```

The preceding code defines a custom `StudentStringOperationFilterInputType` filter. The `StudentStringOperationFilterInputType` filter only includes the eq and contains operations. Then, we can use the `StudentStringOperationFilterInputType` filter in `StudentFilterType`, as follows:

```
override protected void
Configure(IFilterInputTypeDescriptor<Student> descriptor)
{
    // Omitted for brevity
    descriptor.Field(t => t.FirstName).
Type<StudentStringOperationFilterInputType>();
    descriptor.Field(t => t.LastName).
Type<StudentStringOperationFilterInputType>();
}
```

Now, the `StudentFilterInput` filter only includes the eq and contains operations for the `FirstName` and `LastName` properties.

5. The filter supports and and or operations. You can find an and and or property in the `StudentFilterInput` filter. These two fields are used to combine multiple filters. The and field means the filter must match all the conditions. The or field means the filter must match at least one condition. For example, we can use the following query to filter the `Student` objects whose first name is John and who were born after 2001-01-01 using the and operation:

```
query {
  students(where: { and: [{ firstName: { eq: "John" } }, {
dateOfBirth: { gt: "2001-01-01" } }] }) {
    id
    firstName
    lastName
    email
    phone
    grade
    dateOfBirth
  }
}
```

The following query filters the Student objects whose first name is John or last name is Doe using query variables:

```
query ($where: StudentFilterInput) {
  students(where: $filter) {
    id
    firstName
    lastName
    email
    phone
    grade
    dateOfBirth
  }
}
```

The variables are defined as follows:

```
{
  "where": {
    "or": [
      {
        "firstName": {
          "eq": "John"
        }
      },
      {
        "lastName": {
          "eq": "Doe"
        }
      }
    ]
  }
}
```

In the preceding examples, we expose the IQueryable interface in the resolver, so HotChocolate can translate the GraphQL filter to SQL-native queries automatically. However, sometimes, we cannot expose the IQueryable interface in the resolver. In this case, we need to implement the filtering logic in the resolver manually. The code would be more complex. Let us see how to implement the filtering logic in the resolver manually:

1. The methods to retrieve the list of the Student type by the group ID are defined in the IStudentService interface, as follows:

    ```
    public interface IStudentService
    {
        // Omitted for brevity
    ```

```
    Task<List<Student>> GetStudentsByGroupIdAsync(Guid groupId);
    Task<List<Student>> GetStudentsByGroupIdsAsync(List<Guid>
groupIds);
}
```

We have two methods for eq and in operations. The
GetStudentsByGroupIdAsync() method retrieves the list of Student objects by
the group ID. The GetStudentsByGroupIdsAsync() method retrieves the list of
Student objects by the list of group IDs. These two methods return the list of Student
objects instead of the IQueryable interface. So, we need to implement the filtering logic
in the resolver manually.

2. Define a customized filter for the Student type as follows:

```
public class CustomStudentFilterType : FilterInputType<Student>
{
    protected override void
Configure(IFilterInputTypeDescriptor<Student> descriptor)
    {
        descriptor.BindFieldsExplicitly();
        descriptor.Name("CustomStudentFilterInput");
        descriptor.AllowAnd(false).AllowOr(false);
        descriptor.Field(t => t.GroupId).
Type<CustomStudentGuidOperationFilterInputType>();
    }
}

public class CustomStudentGuidOperationFilterInputType :
UuidOperationFilterInputType
{
    protected override void Configure(IFilterInputTypeDescriptor
descriptor)
    {
        descriptor.
Name("CustomStudentGuidOperationFilterInput");
        descriptor.Operation(DefaultFilterOperations.Equals).
Type<IdType>();
        descriptor.Operation(DefaultFilterOperations.In).
Type<ListType<IdType>>();
    }
}
```

In the preceding code, we define a custom filter input type named
CustomStudentFilterInput. The CustomStudentFilterInput
filter only includes the GroupId property. To make the filter more simple,
we disable the and and or operations. Then, we define a custom filter input

type named `CustomStudentGuidOperationFilterInput`. The `CustomStudentGuidOperationFilterInput` filter only includes the `eq` and `in` operations. Note that we need to specify the names of the filter input types. Otherwise, HotChocolate will report name conflicts because we already have a `StudentFilterInput` filter.

3. Add a new query type in the `Query` class:

    ```
    public List<Student> StudentsWithCustomFilter { get; set; } =
    new();
    ```

4. Configure the resolver and manually filter the data in the `QueryType` class, as follows:

    ```
    descriptor.Field(x => x.StudentsWithCustomFilter)
        .Description("This is the list of students in the school.")
        .UseFiltering<CustomStudentFilterType>()
        .Resolve(async context =>
        {
            var service = context.Service<IStudentService>();
            // The following code uses the custom filter.
            var filter = context.GetFilterContext()?.ToDictionary();
            if (filter != null && filter.ContainsKey("groupId"))
            {
                var groupFilter = filter["groupId"]! as
    Dictionary<string, object>;
                if (groupFilter != null && groupFilter.
    ContainsKey("eq"))
                {
                    if (!Guid.TryParse(groupFilter["eq"].ToString(),
    out var groupId))
                    {
                        throw new ArgumentException("Invalid group
    id", nameof(groupId));
                    }
                    var students = await service.
    GetStudentsByGroupIdAsync(groupId);
                    return students;
                }
                if (groupFilter != null && groupFilter.
    ContainsKey("in"))
                {
                    if (groupFilter["in"] is not IEnumerable<string>
    groupIds)
                    {
                        throw new ArgumentException("Invalid group
    ids", nameof(groupIds));
                    }
    ```

```
                        groupIds = groupIds.ToList();
                        if (groupIds.Any())
                        {
                            var students =
                                await service.
        GetStudentsByGroupIdsAsync(groupIds
                                    .Select(x => Guid.Parse(x.
        ToString())).ToList());
                            return students;
                        }
                        return new List<Student>();
                    }
                }
                var allStudents = await service.GetStudentsAsync();
                return allStudents;
            });
```

The preceding code is a bit complex. We need to get the filter from the context object. Then, we check whether the filter contains the groupId property. If the filter contains the groupId property, we check whether the eq or in operation is specified. If the eq operation is specified, we retrieve the list of Student objects by the group ID. If the in operation is specified, we retrieve the list of Student objects by the list of group IDs. If the eq or in operation is not specified, we retrieve all the Student objects.

5. Run the application and check the generated schema. You will find the studentsWithCustomFilter query has a CustomStudentFilterInput filter, as shown here:

```
input CustomStudentFilterInput {
  groupId: CustomStudentGuidOperationFilterInput
}

input CustomStudentGuidOperationFilterInput {
  and: [CustomStudentGuidOperationFilterInput!]
  or: [CustomStudentGuidOperationFilterInput!]
  eq: ID
  in: [ID]
}
```

Then, we can use the following query to filter the Student objects:

```
query ($where: CustomStudentFilterInput) {
  studentsWithCustomFilter(where: $where) {
    id
    firstName
    lastName
```

```
        email
        phone
        grade
        dateOfBirth
    }
}
```

To filter the `Student` objects by `groupId`, we can define the following variable:

```
{
    "where": {
        "groupId": {
            "eq": "00000000-0000-0000-0000-000000000201"
        }
    }
}
```

To filter the `Student` objects by the list of group IDs, we can define the following variable:

```
{
    "where": {
        "groupId": {
            "in": ["00000000-0000-0000-0000-000000000201", "00000000-
0000-0000-0000-000000000202"]
        }
    }
}
```

As the filtering variables may vary in different cases, the logic to filter the data may be different. It is recommended to use the `IQueryable` interface if possible, so that HotChocolate can translate the GraphQL filter to SQL-native queries automatically.

Sorting

In this section, we will learn how to use sorting in HotChocolate. The sorting is similar to filtering. We can use the `UseSorting()` method to enable sorting on the object type. If we use the `IQueryable` interface in the resolver, HotChocolate can translate the GraphQL sorting to SQL-native queries automatically. Otherwise, we need to implement the sorting logic in the resolver manually.

To enable sorting on the `Student` object type, the `HotChocolate.Data` package is required. Follow the step just before the *Filtering* section to install the `HotChocolate.Data` package if you have not installed it yet. Then, follow these steps to enable sorting on the `Student` object type:

1. Register the `Sorting` middleware in the `Program.cs` file, as follows:

    ```
    builder.Services
        .AddGraphQLServer()
    ```

```
            // Omitted for brevity
            .AddSorting()
            .AddMutationType<Mutation>();
```

2. Update the resolver for the `students` query:

```
descriptor.Field(x => x.Students)
        .Description("This is the list of students in the school.")
        .UseFiltering<StudentFilterType>()
        .UseSorting()
        // Omitted for brevity
```

Note that `UseSorting()` must be placed after `UseFiltering`.

3. Then, run the application and check the generated schema. You will find the `students` query has an `orderBy` argument, as shown here:

```
students(where: StudentFilterInput, order: [StudentSortInput!]):
[Student!]!
```

The `order` argument is an array of `StudentSortInput` types, as follows:

```
input StudentSortInput {
  id: SortEnumType
  firstName: SortEnumType
  lastName: SortEnumType
  email: SortEnumType
  phone: SortEnumType
  grade: SortEnumType
  dateOfBirth: SortEnumType
  groupId: SortEnumType
  group: GroupSortInput
}
```

The `StudentSortInput` type includes all the properties of the `Student` object type. The `SortEnumType` type is an enum type, as follows:

```
enum SortEnumType {
  ASC
  DESC
}
```

The `SortDirection` type includes two values: ASC and DESC. The ASC value means the sorting is ascending. The DESC value means the sorting is descending.

4. Next, we can query the `Student` type with sorting. The following query will sort the results by first name:

```
query ($order: [StudentSortInput!]) {
  students(order: $order) {
```

```
        id
        firstName
        lastName
        email
        phone
        grade
        dateOfBirth
    }
}
```

The query variable is defined as follows:

```
{
    "order": [
        {
            "firstName": "ASC"
        }
    ]
}
```

The sorting variable supports multiple properties. For example, the following query variable will sort the results by first name and last name:

```
{
    "order": [
        {
            "firstName": "ASC"
        },
        {
            "lastName": "ASC"
        }
    ]
}
```

You can check the generated SQL query in the logs as follows:

```
info: Microsoft.EntityFrameworkCore.Database.Command[20101]
      Executed DbCommand (3ms) [Parameters=[],
CommandType='Text', CommandTimeout='30']
      SELECT [s].[Id], [s].[DateOfBirth], [s].[Email], [s].
[FirstName], [s].[Grade], [s].[GroupId], [s].  [LastName], [s].
[Phone]
      FROM [Students] AS [s]
      ORDER BY [s].[FirstName], [s].[LastName]
```

The preceding SQL query uses the ORDER BY clause to sort the Student objects, which means the sorting is done in the database.

Similar to filtering, the default sorting includes all the properties of the object type. If we want to sort on specific properties only, we can create a custom sort input type and specify the properties we want to sort. Follow these steps to create a custom sort input type:

1. Create a folder named `Sorts` in the `GraphQL` folder. Add a new class named `StudentSortType`, as follows:

    ```
    public class StudentSortType : SortInputType<Student>
    {
        protected override void
    Configure(ISortInputTypeDescriptor<Student> descriptor)
        {
            descriptor.BindFieldsExplicitly();
            descriptor.Field(x => x.FirstName);
            descriptor.Field(x => x.LastName);
            descriptor.Field(x => x.DateOfBirth);
        }
    }
    ```

    The preceding code defines a custom sort input type, which only includes the `FirstName`, `LastName`, and `DateOfBirth` properties.

    Similar to the filter input type, you can explicitly specify the properties you want to sort, or you can ignore the properties you do not want to sort.

2. Update the resolver to apply the custom sort input type:

    ```
    descriptor.Field(x => x.Students)
        .UseFiltering<StudentFilterType>()
        .UseSorting<StudentSortType>()
        // Omitted for brevity
    ```

3. Run the application and check the schema. You will see that `StudentSortInput` now has three properties only:

    ```
    input StudentSortInput {
      firstName: SortEnumType
      lastName: SortEnumType
      dateOfBirth: SortEnumType
    }
    ```

The query is similar to the previous example, so we will not repeat it here.

Pagination

Pagination is a common feature in web API development. In this section, we will learn how to use pagination in HotChocolate.

Similar to filtering and sorting, we need to install the `HotChocolate.Data` package to use pagination. HotChocolate supports two types of pagination:

- **Cursor-based pagination**: This is the default pagination in HotChocolate. It uses a cursor to indicate the current position in the list. The cursor is usually an ID or a timestamp, which is opaque to the client.

- **Offset-based pagination**: This pagination uses the `skip` and `take` arguments to paginate the list.

Let's first use cursor-based pagination to paginate the `Student` objects. As we introduced before, if we use the `IQueryable` interface in the resolver, HotChocolate can translate the GraphQL pagination to SQL-native queries automatically. Follow the next steps to use cursor-based pagination for the `students` query:

1. Update the resolver for the `students` query:

```
descriptor.Field(x => x.Students)
    .Description("This is the list of students in the school.")
    .UsePaging()
    .UseFiltering<StudentFilterType>()
    .UseSorting<StudentSortType>()
    // Omitted for brevity
```

   Note that `UsePaging()` must be placed before `UseFiltering()` and `UseSorting()`.

2. Run the application and check the generated schema. You will find that the `students` query now is the `StudentsConnection` type:

```
students(
  first: Int
  after: String
  last: Int
  before: String
  where: StudentFilterInput
  order: [StudentSortInput!]
): StudentsConnection
```

   The `StudentsConnection` type is a connection type as follows:

```
type StudentsConnection {
  pageInfo: PageInfo!
  edges: [StudentsEdge!]
```

```
    nodes: [Student!]
}
```

In GraphQL, the connection type is a standard way to paginate the list. The StudentsConnection type includes three fields: pageInfo, edges, and nodes. The nodes field is a list of the Student objects. The edges and pageInfo fields are defined in the StudentsEdge and PageInfo types as follows:

```
type StudentsEdge {
  cursor: String!
  node: Student!
}

type PageInfo {
  hasNextPage: Boolean!
  hasPreviousPage: Boolean!
  startCursor: String
  endCursor: String
}
```

3. Next, we can query the paginated Student objects as follows:

```
query {
  students {
    edges {
      cursor
      node {
        id
        firstName
        dateOfBirth
      }
    }
    pageInfo {
      hasNextPage
      hasPreviousPage
    }
  }
}
```

The result is as follows:

```
{
  "data": {
    "students": {
      "edges": [
        {
```

```
          "cursor": "MA==",
          "node": {
            "id": "00000000-0000-0000-0000-000000000901",
            "firstName": "John",
            "dateOfBirth": "2000-01-01"
          }
        },
        ...
        {
          "cursor": "OQ==",
          "node": {
            "id": "00000000-0000-0000-0000-000000000910",
            "firstName": "Jack",
            "dateOfBirth": "2000-01-10"
          }
        }
      ],
      "pageInfo": {
        "hasNextPage": true,
        "hasPreviousPage": false
      }
    }
  }
}
```

The result contains a `cursor` field for each `Student` object. The `cursor` field is an opaque string, which is used to indicate the current position in the list. The `pageInfo` field indicates whether there are more pages. In this case, the `hasNextPage` field is `true`, which means there are more pages.

4. To query next page, we need to specify the `after` parameter:

```
query {
  students(after: "OQ==") {
    edges {
      cursor
      node {
        id
        firstName
        dateOfBirth
      }
    }
    pageInfo {
      hasNextPage
      hasPreviousPage
```

```
        }
      }
    }
```

You will see the next page of the `Student` objects. The generated SQL query is as follows:

```
info: Microsoft.EntityFrameworkCore.Database.Command[20101]
      Executed DbCommand (11ms) [Parameters=[@__p_0='?' (DbType
= Int32), @__p_1='?' (DbType = Int32)],    CommandType='Text',
CommandTimeout='30']
      SELECT [s].[Id], [s].[DateOfBirth], [s].[Email], [s].
[FirstName], [s].[Grade], [s].[GroupId], [s].    [LastName], [s].
[Phone]
      FROM [Students] AS [s]
      ORDER BY (SELECT 1)
      OFFSET @__p_0 ROWS FETCH NEXT @__p_1 ROWS ONLY
```

The preceding SQL query uses the `OFFSET` and `FETCH` clauses to paginate the `Student` objects, which means the pagination is handled in the database.

5. To query the previous page, we need to specify the `before` parameter, as in this example:

```
query {
  students(before: "MA==") {
    edges {
      cursor
      node {
        id
        firstName
        dateOfBirth
      }
    }
    pageInfo {
      hasNextPage
      hasPreviousPage
    }
  }
}
```

6. We can specify the options for pagination in the `UsePaging()` method. For example, we can specify the default page size and include the total count in the `UsePaging()` method, as follows:

```
descriptor.Field(x => x.Students)
    .Description("This is the list of students in the school.")
    .UsePaging(options: new PagingOptions()
    {
```

```
        MaxPageSize = 20,
        DefaultPageSize = 5,
        IncludeTotalCount = true
    })
    .UseFiltering<StudentFilterType>()
    .UseSorting<StudentSortType>()
    // Omitted for brevity
```

Now, we can include a `totalCount` field in the `pageInfo` field, as follows:

```
query {
    students {
        edges {
            cursor
            node {
                id
                firstName
                dateOfBirth
            }
        }
        totalCount
        pageInfo {
            hasNextPage
            hasPreviousPage
        }
    }
}
```

You can see the `totalCount` field in the response. The default page size is 5.

7.   We can use pagination with filtering and sorting. The following query will filter the `Student` objects by first name and sort the results by first name and then by last name:

```
query ($where: StudentFilterInput, $order: [StudentSortInput!])
{
    students(where: $where, order: $order) {
        edges {
            cursor
            node {
                id
                firstName
                dateOfBirth
            }
        }
        totalCount
        pageInfo {
```

```
      hasNextPage
      hasPreviousPage
    }
  }
}
```

The query variable is defined as follows:

```
{
    "where":{
        "dateOfBirth":{
            "gt":"2001-01-01"
        }
    },
    "order":[
        {
            "firstName":"ASC"
        },
        {
            "lastName":"ASC"
        }
    ]
}
```

After querying the first page, we can query the next page, as follows:

```
query ($where: StudentFilterInput, $order: [StudentSortInput!])
{
  students(where: $where, order: $order, after: "NA==") {
    edges {
      cursor
      node {
        id
        firstName
        dateOfBirth
      }
    }
    totalCount
    pageInfo {
      hasNextPage
      hasPreviousPage
    }
  }
}
```

You can also define the `after` parameter in the query variable, as follows:

```
{
    "where":{
        "dateOfBirth":{
            "gt":"2001-01-01"
        }
    },
    "order":[
        {
            "firstName":"ASC"
        },
        {
            "lastName":"ASC"
        }
    ],
    "after":"NA=="
}
```

The query language of GraphQL is very flexible. We cannot list all the possible queries here. You can try different queries by yourself.

HotChocolate supports offset-based pagination as well. To use offset-based pagination, we need to use the `UseOffsetPaging()` method instead of the `UsePaging()` method. Follow these steps to use offset-based pagination:

1. Update the resolver for the `students` query:

    ```
    descriptor.Field(x => x.Students)
        .Description("This is the list of students in the school.")
        .UseOffsetPaging()
        .UseFiltering<StudentFilterType>()
        .UseSorting<StudentSortType>()
        // Omitted for brevity
    ```

    The preceding code uses the `UseOffsetPaging()` method to enable offset-based pagination instead of cursor-based pagination.

2. Run the application and check the generated schema. You will find the `students` query is now the `StudentsCollectionSegment` type:

    ```
    students(
        skip: Int
        take: Int
        where: StudentFilterInput
        order: [StudentSortInput!]
    ```

```
  ): StudentsCollectionSegment

type StudentsCollectionSegment {
  pageInfo: CollectionSegmentInfo!
  items: [Student!]
}

type CollectionSegmentInfo {
  hasNextPage: Boolean!
  hasPreviousPage: Boolean!
}
```

You should be familiar with the skip and take arguments. The skip argument is used to skip the first n items. The take argument is used to take the first n items. We already used these two methods in LINQ to implement the pagination.

3. Next, we can query the paginated Student objects as follows:

```
query {
  students {
    items {
      id
      firstName
      dateOfBirth
    }
    pageInfo {
      hasNextPage
      hasPreviousPage
    }
  }
}
```

4. To query the next page, we need to specify the skip and take parameters, as follows:

```
query {
  students(skip: 5, take: 5) {
    items {
      id
      firstName
      dateOfBirth
    }
    pageInfo {
      hasNextPage
      hasPreviousPage
    }
```

```
    }
  }
```

5. You can define the `skip` and `take` parameters in the query variable as follows:

```
{
  "skip": 5,
  "take": 5
}
```

6. We can specify the pagination options in the `UseOffsetPaging` method:

```
descriptor.Field(x => x.Students)
    .Description("This is the list of students in the school.")
    .UseOffsetPaging(options: new PagingOptions()
    {
        MaxPageSize = 20,
        DefaultPageSize = 5,
        IncludeTotalCount = true
    })
    .UseFiltering<StudentFilterType>()
    .UseSorting<StudentSortType>()
    // Omitted for brevity
```

You can include the `totalCount` field in the response now.

7. We can use pagination with filtering and sorting. The following query will filter the `Student` objects by first name and sort the results by first name and then by last name, and then fetch the second page:

```
query ($where: StudentFilterInput, $order: [StudentSortInput!],
$skip: Int!, $take: Int!) {
    students(where: $where, order: $order, skip: $skip, take:
$take) {
        items {
          id
          firstName
          dateOfBirth
        }
        totalCount
        pageInfo {
          hasNextPage
          hasPreviousPage
        }
      }
    }
```

The query variable is defined as follows:

```
{
    "where":{
        "dateOfBirth":{
            "gt":"2001-01-01"
        }
    },
    "order":[
        {
            "firstName":"ASC"
        },
        {
            "lastName":"ASC"
        }
    ],
    "skip":5,
    "take":5
}
```

The generated SQL query is shown here:

```
info: Microsoft.EntityFrameworkCore.Database.Command[20101]
      Executed DbCommand (2ms) [Parameters=[@__p_0='?' (DbType
= Date), @__p_1='?' (DbType = Int32),    @__p_2='?' (DbType =
Int32)], CommandType='Text', CommandTimeout='30']
      SELECT [s].[Id], [s].[DateOfBirth], [s].[Email], [s].
[FirstName], [s].[Grade], [s].[GroupId], [s].    [LastName], [s].
[Phone]
      FROM [Students] AS [s]
      WHERE [s].[DateOfBirth] > @__p_0
      ORDER BY [s].[FirstName], [s].[LastName]
      OFFSET @__p_1 ROWS FETCH NEXT @__p_2 ROWS ONLY
```

The preceding SQL query shows that the pagination is handled in the database.

Visualizing the GraphQL schema

When the GraphQL API becomes more complex, it is difficult to understand the schema. We can use `GraphQL Voyager` to visualize the GraphQL schema. `GraphQL Voyager` is an open-source project that can visualize the GraphQL schema in an interactive graph. It is a frontend application that can be integrated with the GraphQL API. To use it in our ASP.NET Core application, we can use the `GraphQL.Server.Ui.Voyager` package. This package is part of the GraphQL.NET project.

Follow these steps to use GraphQL Voyager in our application:

1. Install the `GraphQL.Server.Ui.Voyager` package using the following command:

    ```
    dotnet add package GraphQL.Server.Ui.Voyager
    ```

2. Add the following code to the `Program.cs` file:

    ```
    app.MapGraphQLVoyager();
    ```

    The preceding code adds a middleware that maps the Voyager UI to the default URL `ui/voyager`. If you want to specify a different URL, you can pass the URL as a parameter, as in this example:

    ```
    app.MapGraphQLVoyager("/voyager");
    ```

3. Run the application and navigate to the `ui/voyager` URL. You will see the following page:

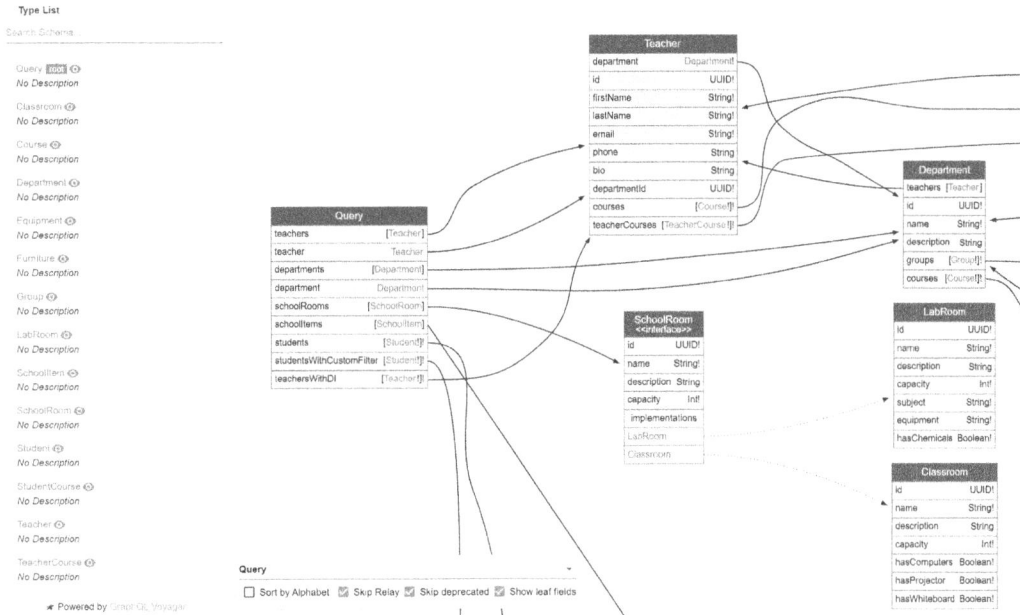

Figure 12.5 – Overview of the GraphQL Voyager UI

Visualizing the GraphQL schema can be beneficial for your team. Doing so allows you to share the schema with your team members, making it easier to collaborate and stay on the same page.

Summary

In this chapter, we explored how to use HotChocolate and Entity Framework Core to create a GraphQL API. We discussed how to define object types, queries, and mutations, as well as how to use dependency injection to inject the `DbContext` instance and services into the resolver. We also introduced the data loader, which can reduce the number of queries to the database. Additionally, we discussed interface and union types, which are useful for defining polymorphic types. Finally, we explored how to use filtering, sorting, and pagination in HotChocolate.

In the next chapter, we will discuss SignalR, which is a real-time communication library in ASP.NET Core.

Further reading

It is important to note that GraphQL is a comprehensive query language and there are many features that we were unable to cover in this chapter. For example, GraphQL supports subscriptions, which enable real-time communication with the GraphQL API. To learn more about HotChocolate and GraphQL, please refer to the following resources:

- `https://graphql.org/learn/`
- `https://chillicream.com/docs/hotchocolate/`

In a microservice architecture, we can use Apollo Federation to create a GraphQL gateway. Apollo Federation can combine multiple GraphQL APIs into a single GraphQL API. We will not cover Apollo Federation here as it is out of the scope of this book. To learn more about Apollo Federation, please refer to the following resources:

- `https://www.apollographql.com/`
- `https://github.com/apollographql`

13

Getting Started with SignalR

In *Chapter 1*, we introduced the concept of real-time web APIs. Various technologies can be used to implement real-time web APIs, such as gRPC streaming, long polling, **Server-Sent Events** (**SSE**), WebSockets, and so on. Microsoft provides an open-source library called SignalR to simplify the implementation of real-time web APIs. In this chapter, we will introduce the basics of SignalR and how to use SignalR to implement real-time web APIs.

We will cover the following topics in this chapter:

- Recap of real-time web APIs
- Setting up SignalR
- Building SignalR clients
- Using authentication and authorization in SignalR
- Managing users and groups
- Sending messages from other services
- Configuring SignalR hubs and clients

By the end of this chapter, you will be able to use SignalR to implement real-time web APIs.

Technical requirements

To follow the steps in this chapter, you can download the source code from the GitHub repository at `https://github.com/PacktPublishing/Web-API-Development-with-ASP.NET-Core-8/tree/main/samples/chapter13`. You can use VS 2022 or VS Code to open the solutions.

You also need to install the following software:

- **Node.js**: Node.js is a JavaScript runtime built on Chrome's V8 JavaScript engine. You can download the latest version of Node.js from `https://nodejs.org/en/`. We will use it to install the required packages for the TypeScript client.

Recap of real-time web APIs

We introduced a few technologies that can be used to implement real-time web APIs in *Chapter 1*. Each technology has its pros and cons. To simplify the implementation of real-time web APIs, Microsoft provides SignalR, which supports multiple transports, such as WebSockets, SSE, and long polling. SignalR will automatically choose the best transport based on the client's capabilities. On top of that, SignalR provides a simple programming model to implement real-time web APIs. Developers do not need to worry about the underlying transport details; instead, they can focus on the business logic.

SignalR was first introduced for ASP.NET in 2013. As of now, SignalR has been rewritten for ASP.NET Core and is included in the ASP.NET Core framework. So, there are two different versions of SignalR: one for ASP.NET and one for ASP.NET Core. If you are using ASP.NET Core, you do not need to install any additional packages to use SignalR. There are also some differences between the ASP.NET version and the ASP.NET Core version of SignalR. For example, ASP.NET Core SignalR does not support Microsoft Internet Explorer. However, most modern applications are targeting modern browsers, so this should not be a big issue. In this chapter, we will focus on the ASP.NET Core version of SignalR.

Unlike REST APIs, SignalR clients need to install a SignalR client library to communicate with SignalR servers. SignalR provides a couple of client libraries for different platforms:

- **JavaScript client**: This is the most used client library because it can be used in both browsers and Node.js applications

- **.NET client**: This client can be used for .NET applications, such as Xamarin, **Windows Presentation Foundation (WPF)**, and Blazor

- **Java client**: This client supports Java 8 and later

Other clients, such as the C++ client and Swift client, are not officially supported by Microsoft.

SignalR is a good choice for building real-time web APIs. For example, you can use SignalR to build a chat application, a real-time dashboard, voting applications, whiteboard applications, and so on. SignalR can push data to specific clients or groups of clients. It automatically manages connections between clients and servers. In the next section, we will introduce the basics of SignalR and build a simple chat application.

Setting up SignalR

In this section, we will build a simple chat application using SignalR. The chat application will allow users to send messages to a public chat room. The messages will be broadcast to all connected clients. This application contains four projects:

- `ChatApp.Server`: This is the ASP.NET Core web API project that provides a SignalR hub

- `ChatApp.TypeScriptClient`: This is a client application written in TypeScript

- `ChatApp.BlazorClient`: This is a client application written in Blazor, which is a web framework for building client-side applications using C#

> **Important note**
>
> The code of this sample is based on the official SignalR sample provided by Microsoft. You can find the original source code at `https://github.com/aspnet/SignalR-samples/tree/main/ChatSample`. We added the Blazor and MAUI clients to the sample.

The `ChatApp.Server` application is a simple ASP.NET Core web API application that is used to provide a SignalR hub. A SignalR hub is a class to manage connections between clients and servers. It is a high-level abstraction for SignalR real-time communication. A SignalR hub can be used to send messages to clients and receive messages from clients. A SignalR hub can also manage users and groups of clients. In ASP.NET Core SignalR, a hub is defined as a middleware component, so we can easily add it to the ASP.NET Core pipeline.

A SignalR hub has a `Clients` property to manage connections between the server and the client. When a user connects to the SignalR hub, a new connection is created. One user can have multiple connections. The `Clients` property has some methods to manage connections:

- `All` is used to call a method on all connected clients.
- `Caller` is used to call a method on the caller.
- `Others` is used to call a method on all connected clients except the caller.
- `Client` is used to call a method on a specific client.
- `Clients` is used to call a method on specific connected clients.
- `Group` is used to call a method on a group of clients.
- `Groups` is used to call a method on multiple groups of clients.
- `User` is used to call a method on a specific user. Note that one user may have multiple connections.
- `Users` is used to call a method on specified users, including all connections.
- `AllExcept` is used to call a method on all connected clients except specified clients.
- `GroupExcept` is used to call a method on a group of clients except specified clients.
- `OthersInGroup` is used to call a method on all clients in a group except the caller.

We will explore some of these methods in the following sections. You can find the complete code of the sample in the `chapter13/v1` folder of the GitHub repository.

Next, follow these steps to create a new solution and set up a `ChatApp.Server` project:

1. Create a new solution called `ChatApp` using the `dotnet new sln` command:

    ```
    dotnet new sln -n ChatApp
    ```

2. Create a new ASP.NET Core web API project called `ChatApp.Server` using the `dotnet new webapi` command and add it to the solution:

    ```
    dotnet new webapi -n ChatApp.Server
    dotnet sln add ChatApp.Server
    ```

3. Add the SignalR middleware component to the ASP.NET Core web API pipeline. Open the `Program.cs` file and add the following code:

    ```
    builder.Services.AddSignalR();
    ```

4. Create a SignalR hub class. Create a new folder named `Hubs` in the project and add a new class called `ChatHub`:

    ```
    namespace ChatApp.Server.Hubs;

    public class ChatHub : Hub
    {
        public Task SendMessage(string user, string message)
        {
            return Clients.All.SendAsync("ReceiveMessage", username,
    message);
        }
    }
    ```

    The preceding code creates a new SignalR hub class called `ChatHub` that inherits from the Hub class. The `ChatHub` class contains a method called `SendMessage()`, which is used to send a message to all connected clients. The `SendMessage()` method takes two parameters, `user` and `message`, which are used to identify the username and the message. This method uses the `Clients.All.SendAsync()` method to broadcast the message to all connected clients when the `SendMessage()` method is invoked by clients. Note the first parameter of the `SendAsync()` method (for example, `ReceiveMessage()`) is the name of the method for clients to receive the message.

5. Next, we need to map the SignalR hub to a URL. Add the following code to the `Program.cs` file:

    ```
    app.MapHub<ChatHub>("/chatHub");
    ```

    You need to add a `using ChatApp.Server.Hubs;` statement to the top of the file.

6. Check the `launchSettings.json` file in the `Properties` folder. The default `launchSettings.json` file contains `http` and `https` URLs. By default, the `dotnet run` command will use the first `http` profile. We can specify for the launch profile to use the `https` URL. Use the following command to run the application:

    ```
    dotnet run --launch-profile https
    ```

    The preceding command will use the `https` URL to run the application. Take note of the URL (for example, `https://localhost:7159`). We will use it in the next section.

The SignalR hub is now ready for use. To test it, however, a client must install a SignalR client library in order to communicate with the hub. In the following section, we will construct client applications.

Building SignalR clients

This section will demonstrate how to use SignalR in different platforms by building three SignalR clients that consume the same SignalR hub provided by the `ChatApp.Server` application. The code for SignalR is largely the same across platforms, making it easy to learn and implement in your own applications. As such, you can refer to any of these applications to gain an understanding of how to consume a SignalR service in your client applications.

Building a TypeScript client

The first client we will build is a TypeScript client. This application is just a normal HTML page that uses the SignalR JavaScript client library to communicate with the SignalR hub. TypeScript is a superset of JavaScript that provides static typing and other features to help developers write better JavaScript code. TypeScript code is compiled into JavaScript code, so it can run in any JavaScript runtime, such as browsers and Node.js. To learn more about TypeScript, you can visit the official website at `https://www.typescriptlang.org/`.

To use TypeScript in the application, we need to install it. You can do so using the following command:

```
npm install -g typescript
```

After installing TypeScript, you can use the following command to check the version:

```
tsc -v
```

If you see the version number, it means that TypeScript is installed successfully.

Next, follow these steps to create a TypeScript client:

1. Create a new folder named `ChatApp.TypeScriptClient` in the solution folder. Then, create a `src` folder in the `ChatApp.TypeScriptClient` folder. The `src` folder is used to store the source code of the TypeScript client.

2. Create a new file named `index.html` in the `src` folder and add the following code:

```html
<!DOCTYPE html>
<html lang="en">
<head>
    <meta charset="UTF-8">
    <meta name="viewport" content="width=device-width, initial-
scale=1.0">
    <title>Chat App</title>
</head>
<body>
    <div id="divChat">
        <label for="txtUsername">User Name</label>
        <input type="text" id="txtUsername" />
        <label for="txtMessage">Message</label>
        <input type="text" id="txtMessage" />
        <button id="btnSend">Send</button>
        <ul id="messages"></ul>
    </div>
</body>
</html>
```

The preceding code creates a simple HTML page that contains a textbox and a button. The `ul` element is used to display messages.

3. Create a new file named `tsconfig.json` in the `ChatApp.TypeScriptClient` folder and add the following code:

```json
{
  "compilerOptions": {
    "noEmitOnError": true,
    "noImplicitAny": true,
    "sourceMap": true,
    "target": "es6",
    "moduleResolution":"node"
  },
  "files": ["src/app.ts"],
  "compileOnSave": true
}
```

The preceding code is the configuration file for the TypeScript compiler. It specifies the target version of JavaScript, the module system, and other options. It also specifies the TypeScript files to compile, such as `app.ts`. We will create an `app.ts` file in *step 6*.

4. Next, we need to set up npm so that we can install the required packages. Use the following command to initialize npm:

```
npm init -y
```

This command creates a package.json file in the ChatApp.TypeScriptClient folder. The package.json file is used to manage the dependencies of the project. It also contains other information about the project, such as the name, version, description, and so on.

5. Next, we need to install the required packages. Use the following command to install the required packages:

```
npm install @microsoft/signalr @types/node
```

The @microsoft/signalr package is the official SignalR JavaScript client library. The @types/node package is used to provide type definitions for Node.js.

6. Create a new file named app.ts in the src folder and add the following code:

```typescript
import * as signalR from "@microsoft/signalr";

const txtUsername: HTMLInputElement = document.getElementById(
  "txtUsername"
) as HTMLInputElement;
const txtMessage: HTMLInputElement = document.getElementById(
  "txtMessage"
) as HTMLInputElement;
const btnSend: HTMLButtonElement = document.getElementById(
  "btnSend"
) as HTMLButtonElement;

btnSend.disabled = true;

const connection = new signalR.HubConnectionBuilder()
  .withUrl("https://localhost:7159/chatHub")
  .build();

connection.on("ReceiveMessage", (username: string, message:
string) => {
  const li = document.createElement("li");
  li.textContent = `${username}: ${message}`;
  const messageList = document.getElementById("messages");
  messageList.appendChild(li);
  messageList.scrollTop = messageList.scrollHeight;
});

connection
```

```
    .start()
    .then(() => (btnSend.disabled = false))
    .catch((err) => console.error(err.toString()));

txtMessage.addEventListener("keyup", (event) => {
  if (event.key === "Enter") {
    sendMessage();
  }
});

btnSend.addEventListener("click", sendMessage);

function sendMessage() {
  connection
    .invoke("SendMessage", txtUsername.value, txtMessage.value)
    .catch((err) => console.error(err.toString()))
    .then(() => (txtMessage.value = ""));
}
```

The preceding code creates a SignalR connection to the SignalR hub. The `connection` object is used to send messages to the SignalR hub and receive messages from the SignalR hub. The `withURL()` method is used to specify the URL of the SignalR hub. In this case, we use `https://localhost:7159/chatHub` as the URL. If your SignalR hub is hosted on a different URL, you need to change it accordingly.

When the page is loaded, the **Send** button is disabled, then it will be enabled when the connection is established. The `connection` object has a couple of methods used in this sample:

- **The start method**: This is used to start the connection.

- **The on method**: This method is used to receive messages from the SignalR hub. The `on()` method takes two parameters: the first parameter is the name of the method, which is `RecieveMessage()`, as we defined in the `ChatHub` class, and the second parameter is a callback function that is called when the message is received.

- **The invoke method**: The `invoke()` method is called when the user clicks the **Send** button or presses the *Enter* key. The `invoke()` method takes three parameters: the first parameter is the name of the method we want to invoke on the SignalR hub, which is `SendMessage()`, as we defined in the `ChatHub` class, the second parameter is the username, and the third parameter is the message.

Make sure to use the correct method names. Otherwise, the client will not be able to communicate with the SignalR hub.

7. Next, we need to compile the TypeScript code to JavaScript code. We will use Gulp to automate the compilation process. If you prefer to use other tools, such as Webpack, you can use them as well. Use the following command to install Gulp globally:

    ```
    npm install -g gulp
    ```

8. Install `gulp` and `gulp-typescript` in the project:

    ```
    npm install --save-dev gulp gulp-typescript browserify vinyl-
    source-stream vinyl-buffer gulp-sourcemaps tsify
    ```

    These packages are used to compile the TypeScript code to JavaScript code, generate a bundle file, and generate source map files. The bundle file is used to load the SignalR JavaScript client library so that the client can use it to communicate with the SignalR hub. The source map files are used to map the JavaScript code to the original TypeScript code. This is useful when debugging the application. We will not cover the details of frontend development in this chapter as the focus of this chapter is SignalR. If you want to learn more about these packages, you can visit the official websites of these packages.

9. Create a new file named `gulpfile.js` in the `ChatApp.TypeScriptClient` folder and add the following code:

    ```
    const gulp = require('gulp');
    const browserify = require('browserify');
    const source = require('vinyl-source-stream');
    const buffer = require('vinyl-buffer');
    const sourcemaps = require('gulp-sourcemaps');
    const tsify = require('tsify');

    // Bundle TypeScript with SignalR
    gulp.task('bundle', () => {
      return browserify({
        basedir: '.',
        debug: true,
        entries: ['src/app.ts'], // Replace with your TypeScript
    entry file
        cache: {},
        packageCache: {},
      })
        .plugin(tsify)
        .bundle()
        .pipe(source('bundle.js'))
        .pipe(buffer())
        .pipe(sourcemaps.init({ loadMaps: true }))
        .pipe(sourcemaps.write('./'))
        .pipe(gulp.dest('dist'));
    ```

```
});

// Copy HTML
gulp.task('copy-html', () => {
  return gulp.src('src/**/*.html')
    .pipe(gulp.dest('dist'));
});

// Main build task
gulp.task('default', gulp.series('bundle', 'copy-html'));
```

The `gulp` configuration file defines some tasks that are used to compile TypeScript code to JavaScript and generate a bundle file. Additionally, it copies HTML files to the `dist` folder, which is used to store the compiled JavaScript code and HTML files. If desired, the folder name can be changed. The bundle file loads the SignalR JavaScript client library and the compiled JavaScript code.

10. Add a script to the `package.json` file to run `gulp` tasks:

```
"scripts": {
  "gulp": "gulp"
}
```

The complete `package.json` file should look like this:

```
{
  "name": "chatapp.typescriptclient",
  "version": "1.0.0",
  "description": "",
  "main": "index.js",
  "scripts": {
    "gulp": "gulp"
  },
  "keywords": [],
  "author": "",
  "license": "ISC",
  "dependencies": {
    "@microsoft/signalr": "^8.0.0",
    "@types/node": "^20.9.0"
  },
  "devDependencies": {
    "@microsoft/signalr": "^8.0.0",
    "browserify": "^17.0.0",
    "gulp": "^4.0.2",
    "gulp-sourcemaps": "^3.0.0",
    "gulp-typescript": "^6.0.0-alpha.1",
```

```
        "tsify": "^5.0.4",
        "vinyl-buffer": "^1.0.1",
        "vinyl-source-stream": "^2.0.0"
    }
}
```

11. Next, update the `index.html` file in the `src` folder to load the bundle file:

```
<!-- Omitted -->
    <script src="bundle.js"></script>
</body>
</html>
```

12. Run the following command to compile the TypeScript code and copy the HTML files to the `dist` folder:

npm run gulp

This command will compile the TypeScript code and generate a bundle file in the `dist` folder. It will also copy the HTML files to the `dist` folder. If the command is executed successfully, you should see three files in the `dist` folder: `bundle.js`, `bundle.js.map`, and `index.html`. In the next sections, if you make any changes to the TypeScript code, you need to run this command again to compile the TypeScript code.

The development of the TypeScript client is now complete. To test it, we need to run a web server to host the HTML page. VS Code has some extensions that can be used to run a web server. For example, you can use the **Live Preview** extension. Once you install this extension, you can right-click the `index.html` file in the `dist` folder and select the **Show Preview** menu to run the web server. You will see VS Code opens a new tab and displays the HTML page, as shown next:

Figure 13.1 – Running the TypeScript client in VS Code

You can also try some other tools, such as `http-server`.

13. Now, start the SignalR server by running the following command:

    ```
    dotnet run --launch-profile https
    ```

    Copy the URL of the HTML page and open it in a browser. The default URL of the **Live Preview** web server is `http://127.0.0.1:3000`. However, you may find that the **Send** button is disabled. Press the *F12* key to open the developer tools and check the console. You should see an error message, as shown next:

    ```
    Access to fetch at 'https://localhost:7159/
    chatHub/negotiate?negotiateVersion=1' from origin
    'http://127.0.0.1:3000' has been blocked by CORS policy:
    Response to preflight request doesn't pass access control check:
    No 'Access-Control-Allow-Origin' header is present on the
    requested resource. If an opaque response serves your needs, set
    the request's mode to 'no-cors' to fetch the resource with CORS
    disabled.
    ```

    This error message indicates that the client cannot connect to the SignalR hub because of the **cross-origin resource sharing** (**CORS**) policy. If a web page makes requests to a different domain, this request is a cross-origin request. The browser will block cross-origin requests by default. This is a security feature called the same-origin policy. This can help prevent **cross-site scripting** (**XSS**) attacks. In this case, the URL of the client is different from the URL of the SignalR hub, so the browser will block the request by default.

14. To allow cross-origin requests, we need to add the CORS middleware component to the web API application. Add the following code to the `Program.cs` file:

    ```
    // Enable CORS
    var corsPolicy = new CorsPolicyBuilder()
        .AllowAnyHeader()
        .AllowAnyMethod()
        .AllowCredentials()
        .WithOrigins("http://127.0.0.1:3000")
        .Build();
    builder.Services.AddCors(options =>{
        options.AddPolicy("CorsPolicy", corsPolicy);
    });
    ```

    The preceding code allows cross-origin requests from `http://127.0.0.1:3000`, which is the URL of the **Live Preview** web server. You can change it to the URL of your web server if you are using a different web server. Note that this example is a very basic configuration that does not restrict any HTTP headers or HTTP methods. In a real-world application, you may need to restrict HTTP requests to improve the security of the application. For more details about CORS, you can refer to the official documentation at `https://learn.microsoft.com/en-us/aspnet/core/security/cors`.

15. Restart the SignalR server and refresh the web page. You should see the **Send** button is enabled. Enter a username and a message and click the **Send** button. You should see the message displayed in a list, as shown next:

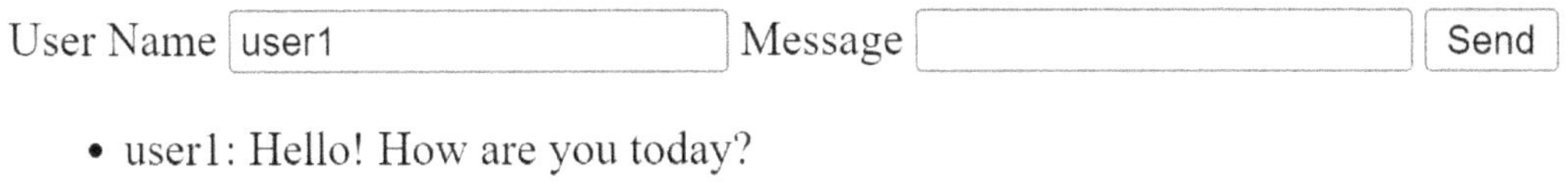

User Name `user1`   Message `[          ]` `Send`

- user1: Hello! How are you today?

Figure 13.2 – Sending a message from the TypeScript client

16. Open another browser tab and enter the same URL. Enter a different username and a message and click the **Send** button. You should see the message displayed in both browser tabs, as shown next:

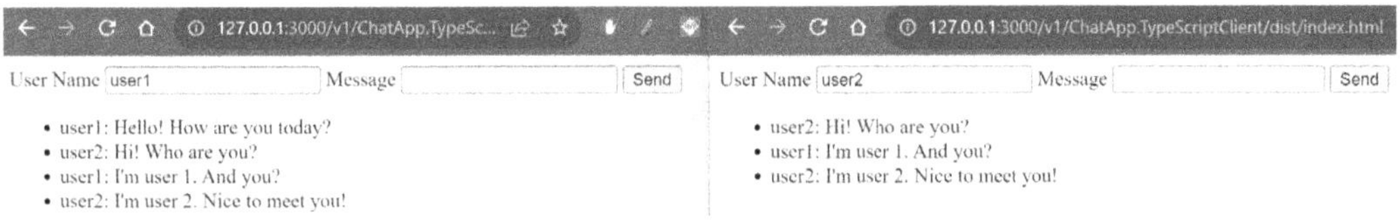

Figure 13.3 – Sending a message from another browser tab

The TypeScript client is now complete. This is a very simple client that does not use any JavaScript frameworks. The world of frontend development is changing rapidly. If you encounter any issues when testing the sample code, you can use any other JavaScript frameworks you like, such as React, Angular, or Vue.js. The code for SignalR is largely the same for different JavaScript frameworks.

Building a Blazor client

The second client we will build is a Blazor client. Blazor is a web framework for building client-side applications using C#. Blazor was first introduced as a part of ASP.NET Core 3.0 in 2018. Blazor supports different hosting models:

- **Blazor Server**: In this hosting model, the Blazor application is hosted on an ASP.NET Core server. Remote clients connect to the server using SignalR. The server is responsible for handling user interactions and updating the UI over a SignalR connection. The application can use the full power of the .NET ecosystem and all ASP.NET Core features. This hosting model also allows the client to download a small amount of code, meaning the application loads fast, but it requires a persistent connection to the server. If the SignalR connection is lost, the application will not work.

- **Blazor WebAssembly**: This hosting model runs the Blazor application on a WebAssembly .NET runtime in the browser. The Blazor application is downloaded to the client, which means that this model requires a larger download size than the Blazor Server model. When a Blazor

WebAssembly application is hosted within an ASP.NET Core application, it is called *hosted Blazor WebAssembly*. The hosted Blazor WebAssembly application can share code with the ASP. NET Core application. When a Blazor WebAssembly application is hosted in a static website without server-side code, it is called *standalone Blazor WebAssembly*. A standalone Blazor WebAssembly application acts like a pure client-side application, such as a React application, so it can be hosted on any web server or a **content delivery network** (**CDN**). Blazor WebAssembly applications can work offline, but the performance depends on the client's hardware.

- **Blazor Hybrid**: This model allows a Blazor application to run in a .NET native app framework, such as WPF, Windows Forms, and MAUI. This model combines the power of the web and native applications, and it can use the full power of the .NET platform. It is suitable for building cross-platform applications because the Blazor code can be shared across different platforms. However, it is still required to package the application for different platforms.

In this sample application, we will use standalone Blazor WebAssembly to build the client application because a web-based application is one of the most seen scenarios. But it is also possible to use similar code for other hosting models. ASP.NET Core 8 brings some improvements to Blazor. To learn more about Blazor, you can visit the official website at `https://learn.microsoft.com/en-us/aspnet/core/blazor/`.

To create a Blazor WebAssembly application, follow these steps:

1. Navigate to the root folder of the `ChatApp.sln` solution. Create a new Blazor WebAssembly application called `ChatApp.BlazorClient` and add the project to the solution using the following command:

   ```
   dotnet new blazorwasm -n ChatApp.BlazorClient
   dotnet sln add ChatApp.BlazorClient
   ```

2. Navigate to the `ChatApp.BlazorClient` folder and run the following command to install the SignalR client library:

   ```
   dotnet add package Microsoft.AspNetCore.SignalR.Client
   ```

3. Add the following code after the `page` directive in the `/Components/Pages/Home.razor` file:

   ```
   @using Microsoft.AspNetCore.SignalR.Client
   @implements IAsyncDisposable
   ```

This `using` statement imports the SignalR client library to the Home component. The `implements IAsyncDisposable` statement indicates that the Home component implements the `IAsyncDisposable` interface. The `IAsyncDisposable` interface is used to dispose of resources asynchronously. We will use it to dispose of the SignalR connection when the component is no longer in use.

4. Add the following code to the end of the Home.razor file:

```razor
@code {
    private HubConnection? _hubConnection;
    private readonly List<string> _messages = new ();
    private string? _username;
    private string? _message;
    private bool IsConnected => _hubConnection?.State ==
HubConnectionState.Connected;

    protected override async Task OnInitializedAsync()
    {
        _hubConnection = new HubConnectionBuilder()
        .WithUrl("https://localhost:7159/chatHub")
        .Build();

        _hubConnection.On<string, string>("ReceiveMessage",
(username, message) =>
        {
            var encodedMessage = $"{username}: {message}";
            _messages.Add(encodedMessage);
            StateHasChanged();
        });

        await _hubConnection.StartAsync();
    }

    private async Task SendMessage()
    {
        if (_hubConnection != null && IsConnected)
        {
            await _hubConnection!.InvokeAsync("SendMessage", _
username, _message);
            _message = string.Empty;
        }
    }

    public async ValueTask DisposeAsync()
    {
        if (_hubConnection is not null)
        {
            await _hubConnection.DisposeAsync();
        }
    }
}
```

Blazor utilizes the `@code` directive to incorporate C# code into components. In this instance, we have defined a few fields and methods for the Home component. If you compare this code to its TypeScript counterpart, you will find that the logic is very similar. The `OnInitializedAsync()` method is used to set up a SignalR connection, while the `SendMessage()` method is used to invoke the `SendMessage()` method of the SignalR hub to send a message. The `DisposeAsync()` method is used to dispose of the SignalR connection when the component is no longer in use. Additionally, the **Send** button is enabled when a SignalR connection is established. Lastly, the `StateHasChanged()` method is used to notify the component to re-render the UI.

5. Next, we need to bind these fields to the UI. Add the following code before the `@code` directive:

```
<div id="username-group">
    <label>User Name</label>
    <input type="text" @bind="_username" />
</div>
<div id="message-group">
    <label>Message</label>
    <input type="text" @bind="_message" />
</div>
<input type="button" value="Send" @onclick="SendMessage"
disabled="@(!IsConnected)" />
<ul>
    @foreach (var message in _messages)
    {
        <li>@message</li>
    }
</ul>
```

Blazor uses the `@` symbol to indicate a C# expression. The `@bind` directive is used to bind the value of the input element to the specified field. The `@onclick` directive is used to bind the click event to the specified method. The `@foreach` directive is used to iterate over the messages and display them in a list. If you are familiar with any modern JavaScript frameworks, such as React, Angular, or Vue.js, you will find some similarities between Blazor and these frameworks.

6. Next, we need to configure the CORS policy for the SignalR server so that the Blazor client can connect to the SignalR hub. Check the `launchSettings.json` file in the `Properties` folder. Similar to the SignalR server application, we can use the `http` or `https` profile to run the Blazor client application. We will use the `https` file in this case. For example, the URL of the sample code uses `https://localhost:7093` to run the Blazor client application on the HTTPS profile. We need to update the CORS policy in the SignalR server. Update the `Program.cs` file of the `ChatApp.Server` project, as shown next:

```
var corsPolicy = new CorsPolicyBuilder()
    .AllowAnyHeader()
```

```
        .AllowAnyMethod()
        .AllowCredentials()
        .WithOrigins("http://127.0.0.1:3000", "https://
localhost:7093")
        .Build();
```

Now, the SignalR server can accept cross-origin requests from the Blazor client application.

7. Run the SignalR server application and the Blazor client application in separate terminals using the `dotnet run --launch-profile https` command. You can test the Blazor client application by opening the `https://localhost:7093` URL in the browser. The Blazor client can chat with the TypeScript client, as shown next:

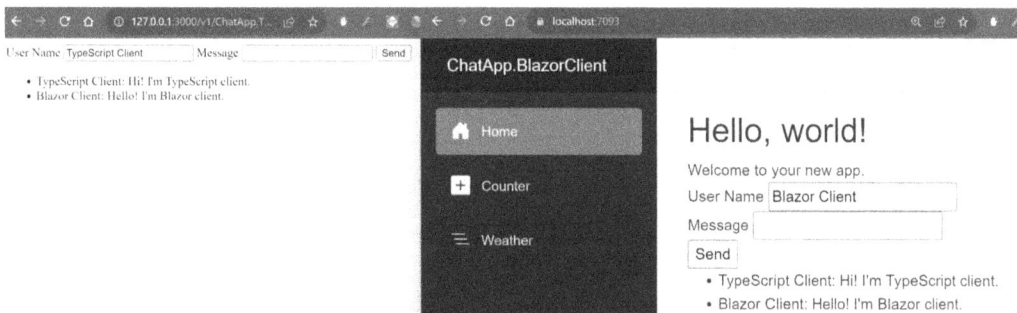

Figure 13.4 – Chatting between the Blazor client and the TypeScript client

SignalR provides the convenience of real-time communication. Developers do not need to operate the underlying transport details; instead, they can use the SignalR Hub class to send and receive messages easily. In the next section, we will explore more features of a SignalR hub.

Using authentication and authorization in SignalR

In the previous section, we used the `Hub` class to implement a simple chat app. The `Clients.All. SendAsync` method is used to send a message to all connected clients. Sometimes, we may want to send a message to a specific client or a group of clients. To manage users and groups, we need to know the identity of the user. In this section, we will explore how to use authentication and authorization in SignalR.

By default, SignalR uses a `ClaimTypes.NameIdentifier` claim to differentiate users. The `ClaimTypes.NameIdentifier` claim is used to uniquely identify a user. We introduced claim-based authorization in *Chapter 8*, so we will follow the steps from that chapter to add authentication and authorization to the SignalR server application. If you are not familiar with ASP.NET Core authentication and authorization, you can refer to *Chapter 8* for more details.

You can find the complete code of the sample in the `chapter13/v2` folder of the GitHub repository.

Adding authentication and authorization to the SignalR server

To add authentication and authorization to the SignalR server, follow these steps:

1. Install the required packages using the following command:

    ```
    dotnet add package Microsoft.AspNetCore.Identity.
    EntityFrameworkCore
    dotnet add package Microsoft.EntityFrameworkCore.SqlServer
    dotnet add package Microsoft.EntityFrameworkCore.Tools
    dotnet add package Microsoft.AspNetCore.Authentication.JwtBearer
    ```

2. Create a new folder named `Data` in the `ChatApp.Server` project. Then, create a new class called `AppbContext` in the `Data` folder. As we introduced `DbContext` in previous chapters, we will not show the code here. You can find the code in the sample application.

3. Add a connection string in the `appsettings.json` file:

    ```
    "ConnectionStrings": {
      "DefaultConnection": "Server=(localdb)\\
    mssqllocaldb;Database=ChatAppDb;Trusted_
    Connection=True;MultipleActiveResultSets=true"
    }
    ```

4. Add configurations for JWT tokens in the `appsettings.json` file:

    ```
    "JwtConfig": {
      "ValidAudiences": "http://localhost:7159",
      "ValidIssuer": "http://localhost:7159",
      "Secret": "c1708c6d-7c94-466e-aca3-e09dcd1c2042"
    }
    ```

    We will use the same SignalR server as the authentication server. So, we will use the URL of the SignalR server as the audience and issuer. If you use a different authentication server, you need to change the audience and issuer accordingly.

5. SignalR needs an `IUserIdProvider` interface to get the user ID. Create a new folder named `Services` in the `ChatApp.Server` project. Then, create a new class called `NameUserIdProvider` in the `Services` folder:

    ```
    using Microsoft.AspNetCore.SignalR;
    namespace ChatApp.Server.Services;

    public class NameUserIdProvider : IUserIdProvider
    {
        public string GetUserId(HubConnectionContext connection)
        {
    ```

```
        return connection.User?.Identity?.Name ?? string.Empty;
    }
}
```

The preceding code implements the `IUserIdProvider` interface. The `GetUserId` method returns the user ID of the current user. In this case, we use the username as the user ID. You can use any other unique value as the user ID. For example, if you want to use the email address as the user ID, you can create an `EmailBasedUserIdProvider` class as follows:

```
using System.Security.Claims;
using Microsoft.AspNetCore.SignalR;

namespace ChatApp.Server.Services;

public class EmailBasedUserIdProvider : IUserIdProvider
{
    public string GetUserId(HubConnectionContext connection)
    {
        return connection.User?.Claims.FirstOrDefault(c =>
c.Type == ClaimTypes.Email)?.Value ??    string.Empty;
    }
}
```

6. Update the `Program.cs` file to add authentication and authorization, as follows:

```
builder.Services.AddDbContext<AppDbContext>();
builder.Services.AddIdentityCore<IdentityUser>()
    .AddEntityFrameworkStores<AppDbContext>()
    .AddDefaultTokenProviders();

builder.Services.AddAuthentication(options =>
{
    options.DefaultAuthenticateScheme = JwtBearerDefaults.
AuthenticationScheme;
    options.DefaultChallengeScheme = JwtBearerDefaults.
AuthenticationScheme;
    options.DefaultScheme = JwtBearerDefaults.
AuthenticationScheme;
}).AddJwtBearer(options =>
{
    var secret = builder.Configuration["JwtConfig:Secret"];
    var issuer = builder.Configuration["JwtConfig:ValidIssuer"];
    var audience = builder.
Configuration["JwtConfig:ValidAudiences"];
    if (secret is null || issuer is null || audience is null)
    {
```

```
            throw new ApplicationException("Jwt is not set in the
    configuration");
        }
        options.SaveToken = true;
        options.RequireHttpsMetadata = false;
        options.TokenValidationParameters = new
    TokenValidationParameters()
        {
            ValidateIssuer = true,
            ValidateAudience = true,
            ValidAudience = audience,
            ValidIssuer = issuer,
            IssuerSigningKey = new SymmetricSecurityKey(Encoding.
    UTF8.GetBytes(secret))
        };

        // Hook the SignalR event to check for the token in the
    query string
        options.Events = new JwtBearerEvents
        {
            OnMessageReceived = context =>
            {
                var accessToken = context.Request.Query["access_
    token"];
                var path = context.HttpContext.Request.Path;
                if (!string.IsNullOrEmpty(accessToken) && path.
    StartsWithSegments("/chatHub"))
                {
                    context.Token = accessToken;
                }
                return Task.CompletedTask;
            }
        };
    });

    // Use the name-based user ID provider
    builder.Services.AddSingleton<IUserIdProvider,
    NameUserIdProvider>();
```

The preceding code is similar to the code in *Chapter 8*. A difference is that we configured the options.Events property of the JwtBearerOptions object. The OnMessageReceived event is used to check the token in the query string. The reason is that WebSocket APIs and SSE do not support the standard Authorization header, so it is required to attach the token to the query string. If the token is found in the query string, it will be used to authenticate the user.

We also added the `IUserIdProvider` service to the **DI** container. In this case, we use the `NameUserIdProvider` class we created earlier. If you want to use the `EmailBasedUserIdProvider` class, you need to change the code accordingly. Note that you must not use both at the same time.

7. Create a database and run migrations using the following commands:

    ```
    dotnet ef migrations add InitialDb
    dotnet ef database update
    ```

8. Next, we need to add an `Authorize` attribute to the `ChatHub` class, as shown next:

    ```
    [Authorize]
    public class ChatHub : Hub
    {
        // Omitted for brevity
    }
    ```

    The `Authorize` attribute can be applied to the Hub class or methods of the Hub class. It also supports policy-based authorization. For example, you can use the `Authorize(Policy = "Admin")` attribute to restrict access to the `ChatHub` class to administrators.

9. Run the `ChatApp.Server` application, as well as any other client applications. Unfortunately, the TypeScript and Blazor clients will not be able to connect to the SignalR hub due to the need for user authentication. To access the SignalR hub, we need to authenticate the clients.

Adding a login endpoint

To authenticate the clients, we need to provide a login endpoint. We implemented a login endpoint in *Chapter 8*. You can follow the steps in *Chapter 8* to implement the login endpoint or copy the code from the sample application. You need to create an `AccountController` class that contains register and login endpoints. You also need to add some models, such as the `LoginModel` and `AddOrUpdateUserModel` classes. With these classes, we can use the `account/register` and `account/login` endpoints to register and log in users.

One thing to note here is that when generating a JWT token, we need to add a `ClaimTypes.NameIdentifier` claim to the token. SignalR uses this claim to identify the user. The following code shows how to add a `ClaimTypes.NameIdentifier` claim to the token:

```
var tokenDescriptor = new SecurityTokenDescriptor
{
    Subject = new ClaimsIdentity(new[]
    {
        // SignalR requires the NameIdentifier claim to map the user
to the connection
        new Claim(ClaimTypes.NameIdentifier, userName),
```

```
        new Claim(ClaimTypes.Name, userName),
        // If you use the email-based user ID provider, you need to
add the email claim from the database
    }),
    Expires = DateTime.UtcNow.AddDays(1),
    Issuer = issuer,
    Audience = audience,
    SigningCredentials = new SigningCredentials(signingKey,
SecurityAlgorithms.HmacSha256Signature)
};
```

Now, we need to create some users for testing. Run the ChatApp.Server application and send a POST request to the account/register endpoint using Postman or any other HTTP client. The following code shows how to create a user using the account/register endpoint:

```
{
  "userName": "user1",
  "email": "user1@example.com",
  "password": "Passw0rd!"
}
```

Create more users such as user2, user3, and so on. We will use these users to test the Groups feature later.

Authenticating the TypeScript client

Now, we can authenticate the TypeScript client. To do so, we need to update the UI to allow the user to enter the username and password. We also need to update the TypeScript code to send the username and password to the login endpoint. Follow these steps to update the TypeScript client:

1. Update the HTML content in the <body> element as follows:

    ```html
    <body>
        <div id="divLogin">
            <label for="txtUsername">User Name</label>
            <input type="text" id="txtUsername" />
            <label for="txtPassword">Password</label>
            <input type="password" id="txtPassword" />
            <button id="btnLogin">Login</button>
        </div>
        <div id="divChat">
            <label>User Name</label>
            <label id="lblUsername" ></label>
            <label for="txtMessage">Message</label>
            <input type="text" id="txtMessage" />
    ```

```
            <button id="btnSend">Send</button>
            <ul id="messages"></ul>
        </div>
        <script type="module" src="bundle.js"></script>
    </body>
```

The preceding code adds a login form to the HTML page. The login form contains a username textbox, a password textbox, and a login button. The divChat element now has a lblUsername element to display the username. The divChat element is hidden by default. We will show it after the user is authenticated.

2. Update the app.ts file as follows:

```
import * as signalR from "@microsoft/signalr";

divChat.style.display = "none";
btnSend.disabled = true;

btnLogin.addEventListener("click", login);
let connection: signalR.HubConnection = null;
async function login() {
  const username = txtUsername.value;
  const password = txtPassword.value;

  if (username && password) {
    try {
      // Use the Fetch API to login
      const response = await fetch("https://localhost:7159/
account/login", {
          method: "POST",
          headers: { "Content-Type": "application/json" },
          body: JSON.stringify({ username, password }),
      });

      const json = await response.json();

      localStorage.setItem("token", json.token);
      localStorage.setItem("username", username);
      txtUsername.value = "";
      txtPassword.value = "";
      lblUsername.textContent = username;
      divLogin.style.display = "none";
      divChat.style.display = "block";
      txtMessage.focus();
```

```
        // Start the SignalR connection
        connection = new signalR.HubConnectionBuilder()
          .withUrl("https://localhost:7159/chatHub", {
            accessTokenFactory: () => {
              var localToken = localStorage.getItem("token");
              // You can add logic to check if the token is valid
or expired
              return localToken;
            },
          })
          .build();
        connection.on("ReceiveMessage", (username: string,
message: string) => {
          const li = document.createElement("li");
          li.textContent = `${username}: ${message}`;
          const messageList = document.getElementById("messages");
          messageList.appendChild(li);
          messageList.scrollTop = messageList.scrollHeight;
        });
        await connection.start();
        btnSend.disabled = false;
      } catch (err) {
        console.error(err.toString());
      }
    }
}

txtMessage.addEventListener("keyup", (event) => {
  if (event.key === "Enter") {
    sendMessage();
  }
});

btnSend.addEventListener("click", sendMessage);

function sendMessage() {
  connection
    .invoke("SendMessage", lblUsername.textContent, txtMessage.
value)
    .catch((err) => console.error(err.toString()))
    .then(() => (txtMessage.value = ""));
}
```

Some codes are omitted. You can find the full code from the books GitHub repository.

In the preceding code, we use the `fetch` API to send a `POST` request to the login endpoint. The login endpoint returns a JWT token if the user is authenticated. Then, we store the token in the local storage and show the username in the `divChat` element. We also adjusted the creation of the SignalR connection. The `accessTokenFactory` property is used to get the token from the local storage. You can add some logic to check whether the token is valid or expired. If the token is expired, you can redirect the user to the login page or use the **Refresh** token to get a new token.

3. Run the following command to compile the TypeScript code and copy the HTML files to the `dist` folder:

    ```
    npm run gulp
    ```

4. Use the **Live Preview** extension to run the web server. Run the SignalR server application as well. You will see a login form, as shown next:

Figure 13.5 – The login form

Use the username and password you created earlier to log in. You should see a chat form, as shown next:

Figure 13.6 – Authenticated chat

Now, the TypeScript client is authenticated. Next, we will authenticate the Blazor client.

Authenticating the Blazor client

The code to authenticate the Blazor client is very similar to the TypeScript client, so we will not list all the code here. You can find the code in the sample application. The following code shows how to log in and set a token to the SignalR connection:

```
@inject HttpClient Http

private async Task Login()
{
    if (!string.IsNullOrWhiteSpace(_username) && !string.
```

```
IsNullOrWhiteSpace(_password))
    {
        var response = await Http.PostAsJsonAsync("Account/login", new
{ Username = _username, Password = _password });
        if (response.IsSuccessStatusCode)
        {
            var jsonString = await response.Content.
ReadAsStringAsync();
            var data = System.Text.Json.JsonSerializer.
Deserialize<Dictionary<string, string>>(jsonString);
            _token = data["token"];
            if (string.IsNullOrWhiteSpace(_token))
            {
                throw new Exception("Invalid token.");
            }
            else
            {
                _showLogin = false;
                _showChat = true;
                StateHasChanged();
                // Set the token to the hub connection.
                _hubConnection = new HubConnectionBuilder()
                .WithUrl("https://localhost:7159/chatHub", options =>
                {
                    options.AccessTokenProvider = () => Task.
FromResult<string?>(_token);
                })
                .Build();
                _hubConnection.On<string, string>("ReceiveMessage",
(username, message) =>
                {
                    var encodedMessage = $"{username}: {message}";
                    _messages.Add(encodedMessage);
                    StateHasChanged();
                });
                await _hubConnection.StartAsync();
            }
        }
    }
}
```

In the preceding code, we inject HttpClient to send a POST request to the login endpoint. Then, we set a token to the SignalR connection. The AccessTokenProvider property is used to get the token from the _token field. Similar to the TypeScript client, you can add some logic to check whether the token is valid or expired.

Run the three applications. You can use different usernames to log in to the two clients and send messages. You should see messages are displayed in both clients, as shown next:

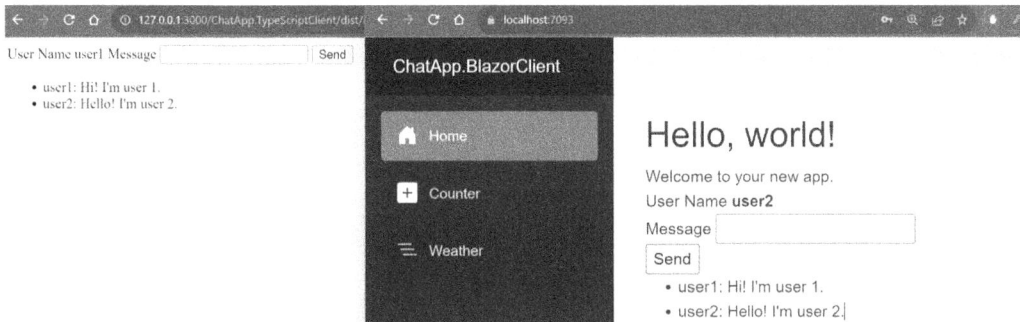

Figure 13.7 – Authenticated chat for different users

The clients now support authentication. Next, we will add more features to the chat app.

Managing users and groups

In the previous section, we implemented basic authentication and authorization for the SignalR server. We also updated clients to authenticate users. In this section, we will explore how to manage users and groups in SignalR. We want to add features to the chat app to enable the following:

- Allow users to know who is connected to the chat app

- Allow users to send a message to a specific user

- Allow users to join groups

- Allow users to send a message to a specific group

You can find the complete code of the sample in the chapter13/v3 folder of the GitHub repository. Let's start with the first feature.

Managing events in SignalR

SignalR provides events to notify clients when a user connects or disconnects. We can override the OnConnectedAsync() and OnDisconnectedAsync() methods to handle these events. The following code shows how to override the OnConnectedAsync() method:

```
public override async Task OnConnectedAsync()
{
    await Clients.All.SendAsync("UserConnected", Context.User.
Identity.Name);
    await base.OnConnectedAsync();
}
```

When a client connects to the SignalR hub, the `OnConnectedAsync()` method will be called. In this case, we use the `Clients.All.SendAsync()` method to send a message to all connected clients. The `Context.User.Identity.Name` property is used to get the username of the current user.

The following code shows how to override the `OnDisconnectAsync()` method:

```
public override async Task OnDisconnectedAsync(Exception? exception)
{
    await Clients.All.SendAsync("UserDisconnected", Context.User.
Identity.Name);
    await base.OnDisconnectedAsync(exception);
}
```

Then, we can update the TypeScript client to handle the `UserConnected` and `UserDisconnected` events. The following code shows how to handle the `UserConnected` event in the TypeScript client:

```
connection.on("UserConnected", (username: string) => {
  const li = document.createElement("li");
  li.textContent = `${username} connected`;
  const messageList = document.getElementById("messages");
  messageList.appendChild(li);
  messageList.scrollTop = messageList.scrollHeight;
});
connection.on("UserDisconnected", (username: string) => {
  const li = document.createElement("li");
  li.textContent = `${username} disconnected`;
  const messageList = document.getElementById("messages");
  messageList.appendChild(li);
  messageList.scrollTop = messageList.scrollHeight;
});
```

The code in the Blazor client is very similar:

```
_hubConnection.On<string>("UserConnected", (username) =>
{
    var encodedMessage = $"{username} connected.";
    _messages.Add(encodedMessage);
    StateHasChanged();
});
_hubConnection.On<string>("UserDisconnected", (username) =>
{
    var encodedMessage = $"{username} disconnected.";
    _messages.Add(encodedMessage);
    StateHasChanged();
});
```

Now, we can run the SignalR server and the two clients. You should see the user's connected and disconnected messages in the chat window. If you refresh the page or close the browser tab, you should see a user-disconnected message, as shown next:

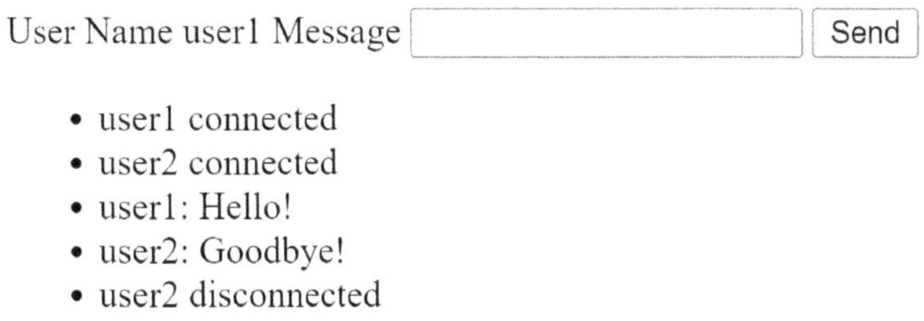

User Name user1 Message [   ] Send

- user1 connected
- user2 connected
- user1: Hello!
- user2: Goodbye!
- user2 disconnected

Figure 13.8 – User-connected and -disconnected messages

Next, we will add a feature to allow users to send a message to a specific user.

Sending a message to a specific user

The next feature we want to add is to allow users to send a message to a specific user. To do so, we need to know to whom the message is sent. SignalR uses a `ClaimTypes.NameIdentifier` claim to differentiate users. To simplify the code, we will pass the username as the target user:

```
public Task SendMessageToUser(string user, string toUser, string
message)
{
    return Clients.User(toUser).SendAsync("ReceiveMessage", user,
message);
}
```

The preceding code uses the `Clients.User(user)` method to find the connection of the specified user.

Next, update the TypeScript client to add a textbox to enter the target username. The following code shows how to update the HTML content for the `divChat` element:

```
<label for="txtToUser">To</label>
<input type="text" id="txtToUser" />
```

Then, we can invoke this method from the TypeScript client as follows:

```
function sendMessage() {
  // If the txtToUser field is not empty, send the message to the user
  if (txtToUser.value) {
    connection
      .invoke("SendMessageToUser", lblUsername.textContent, txtToUser.
value, txtMessage.value)
```

```
            .catch((err) => console.error(err.toString()))
            .then(() => (txtMessage.value = ""));
    } else {
        connection
            .invoke("SendMessage", lblUsername.textContent, txtMessage.
value)
            .catch((err) => console.error(err.toString()))
            .then(() => (txtMessage.value = ""));
    }
}
```

In the preceding code, when the txtToUser field is not empty, we use the SendMessageToUser()
method to send a message to a specified user. Otherwise, we use the SendMessage() method to
send a message to all connected users.

The code in the Blazor client is very similar:

```
private async Task SendMessage()
{
    if (_hubConnection != null && IsConnected)
    {
        if (!string.IsNullOrWhiteSpace(_toUser))
        {
            await _hubConnection.InvokeAsync("SendMessageToUser", _
username, _toUser, _message);
        }
        else
        {
            await _hubConnection.InvokeAsync("SendMessage", _username,
_message);
        }
        _message = string.Empty;
    }
}
```

Please refer to the sample application for the complete code.

Run the three applications. This time, we need to open three browser tabs for testing. Use three different
usernames to log in to the three clients. Then, we can send a message to a specific user, as shown next:

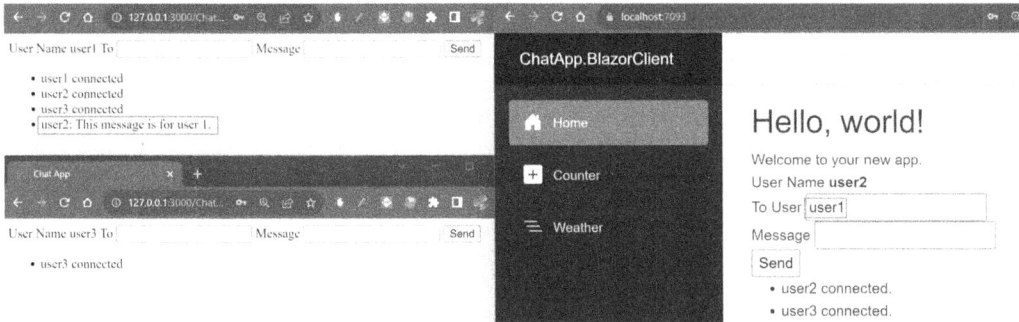

Figure 13.9 – Sending a message to a specific user

In *Figure 13.9*, we sent a message to the `user1` user from `user2`. You can see that the message is displayed in the `user1` browser tab but not in the `user3` browser tab.

You can try to log in to the same username in different browser tabs. You will find that both browser tabs will receive the message. This is because SignalR uses the `ClaimTypes.NameIdentifier` claim to differentiate users. Each browser tab has a different SignalR connection, but they use the same username. Therefore, SignalR will treat them as the same user.

Using strongly typed hubs

So far, we have added a couple of methods to the `ChatHub` class:

```
public Task SendMessage(string user, string message)
{
    await Clients.All.SendAsync("ReceiveMessage", user, message);
}
public Task SendMessageToUser(string user, string toUser, string
message)
{
    return Clients.User(toUser).SendAsync("ReceiveMessage", user,
message);
}
public override async Task OnConnectedAsync()
{
    await Clients.All.SendAsync("UserConnected", Context.User.
Identity.Name);
    await base.OnConnectedAsync();
}
public override async Task OnDisconnectedAsync(Exception? exception)
{
    await Clients.All.SendAsync("UserDisconnected", Context.User.
Identity.Name);
```

```
    await base.OnDisconnectedAsync(exception);
}
```

Each method calls the SendAsync() method with a string parameter. The string parameter is the name of the method to be invoked on the client. The SendAsync() method is a dynamic method, but it is not type-safe. If we misspell the method name, the compiler will not report any error. To improve type safety, we can use strongly typed hubs.

To use strongly typed hubs, we need to define a hub interface that contains client methods. The following code shows how to define a hub interface:

```
public interface IChatClient
{
    Task ReceiveMessage(string user, string message);
    Task UserConnected(string user);
    Task UserDisconnected(string user);
}
```

Then, we can update the ChatHub class to implement the IChatClient interface:

```
public class ChatHub : Hub<IChatClient>
{
    public Task SendMessage(string user, string message)
    {
        return Clients.All.ReceiveMessage(user, message);
    }

    public Task SendMessageToUser(string user, string toUser, string
message)
    {
        return Clients.User(toUser).ReceiveMessage(user, message);
    }

    public override async Task OnConnectedAsync()
    {
        await Clients.All.UserConnected(Context.User.Identity.Name);
        await base.OnConnectedAsync();
    }

    public override async Task OnDisconnectedAsync(Exception?
exception)
    {
        await Clients.All.UserDisconnected(Context.User.Identity.
```

```
Name);
        await base.OnDisconnectedAsync(exception);
    }
}
```

In the preceding code, the SendAsync() method is no longer used. Instead, we use the RecieveMessage(), UserConnected(), and UserDisconnected() methods defined in the IChatClient interface. The Hub class is generic, so we need to specify the IChatClient interface as the generic type argument. Now, the ChatHub class is strongly typed. Note that if you use a strongly typed hub, the SendAsync() method is no longer available.

Next, we will add a feature to allow users to join groups.

Joining groups

SignalR allows users to join groups. The Hub class has a Groups property to manage groups. The type of the Groups property is the IGroupManager interface, which provides methods such as AddToGroupAsync(), RemoveFromGroupAsync(), and so on. The following code shows how to add a user to a group and remove a user from a group:

```
public async Task AddToGroup(string user, string group)
{
    await Groups.AddToGroupAsync(Context.ConnectionId, group);
    await Clients.Group(group).ReceiveMessage(Context.User.Identity.
Name,
        $"{user} has joined the group {group}. Connection Id:
{Context.ConnectionId}");
}
public async Task RemoveFromGroup(string user, string group)
{
    await Groups.RemoveFromGroupAsync(Context.ConnectionId, group);
    await Clients.Group(group).ReceiveMessage(Context.User.Identity.
Name,
                    $"{user} has left the group {group}. Connection Id:
{Context.ConnectionId}");
}
```

In the preceding code, we use the Groups property to manage groups. The Context.ConnectionId property is used to get the connection ID of the current user. The Clients.Group method is used to send a message to all users in the specified group so that they can know who has joined or left the group.

Next, we need to update the UI to allow a user to enter the group name. Add the following code to the HTML content for the divChat element:

```
<label id="lblToGroup">Group</label>
<input type="text" id="txtToGroup" />
<button id="btnJoinGroup">Join Group</button>
<button id="btnLeaveGroup">Leave Group</button>
```

Update the TypeScript code to handle JoinGroup and LeaveGroup events. The following code shows how to handle the JoinGroup event:

```
btnJoinGroup.addEventListener("click", joinGroup);
btnLeaveGroup.addEventListener("click", leaveGroup);

function joinGroup() {
  if (txtToGroup.value) {
    connection
      .invoke("AddToGroup", lblUsername.textContent, txtToGroup.value)
      .catch((err) => console.error(err.toString()))
      .then(() => {
        btnJoinGroup.disabled = true;
        btnJoinGroup.style.display = "none";
        btnLeaveGroup.disabled = false;
        btnLeaveGroup.style.display = "inline";
        txtToGroup.readOnly = true;
      });
  }
}

function leaveGroup() {
  if (txtToGroup.value) {
    connection
      .invoke("RemoveFromGroup", lblUsername.textContent, txtToGroup.
value)
      .catch((err) => console.error(err.toString()))
      .then(() => {
        btnJoinGroup.disabled = false;
        btnJoinGroup.style.display = "inline";
        btnLeaveGroup.disabled = true;
        btnLeaveGroup.style.display = "none";
        txtToGroup.readOnly = false;
      });
  }
}
```

The preceding code shows two event handlers for the `JoinGroup` and `LeaveGroup` events, which invoke the `AddToGroup()` and `RemoveFromGroup()` methods on the SignalR hub respectively.

The code in the Blazor client is very similar. We will not list the code here. You can find the code in the sample application.

Now, the client should be able to join and leave groups. When a user joins or leaves a group, the other users in the group will receive a message, as shown next:

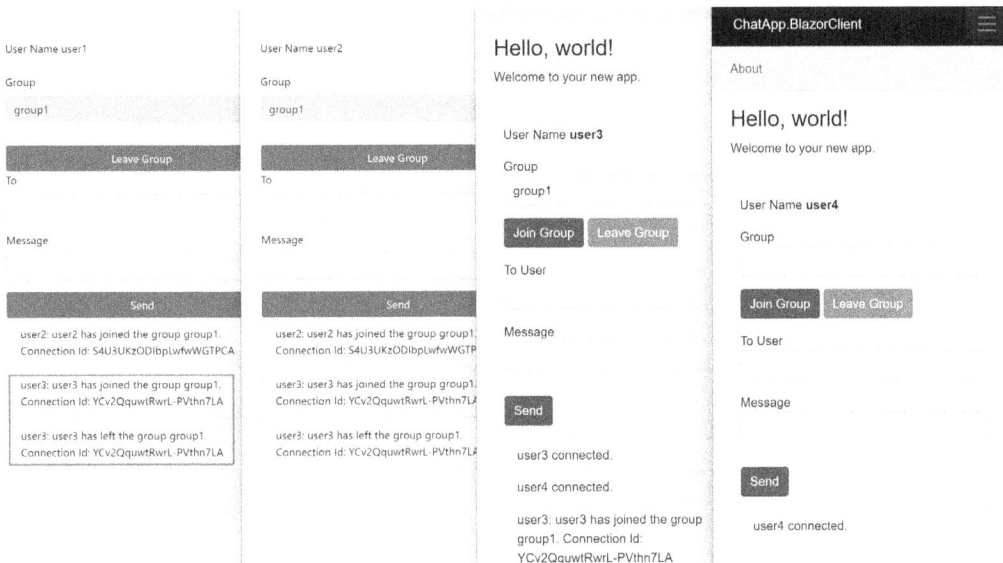

Figure 13.10 – Joining and leaving groups

In *Figure 13.10*, `user3` joined `group1` and then left `group1`. You can see that the other users in `group1` received messages.

Next, we will add a feature to allow users to send a message to a specific group.

Sending a message to a group

The code to send a message to a group is very similar to the code to send a message to a specific user. The following code shows how to send a message to a group in the `ChatHub` class:

```
public async Task SendMessageToGroup(string user, string group, string
message)
{
    await Clients.Group(group).ReceiveMessage(user, message);
}
```

The preceding code uses `Clients.Group(group)` to find connections of the users in the specified group. Then, it uses the `ReceiveMessage()` method defined in the `IChatClient` interface to send a message to users in the group.

The Blazor client can invoke this method as follows:

```
private async Task SendMessage()
{
    if (_hubConnection != null && IsConnected)
    {
        if (!string.IsNullOrWhiteSpace(_group) && _isJoinedGroup)
        {
            await _hubConnection.InvokeAsync("SendMessageToGroup",
_username, _group, _message);
        }
        // Omitted for brevity
    }
}
```

We will not list the code for the TypeScript client here. You can find the code in the sample application.

Now, the client should be able to send a message to a specific group. The following figure shows how to send a message to a group:

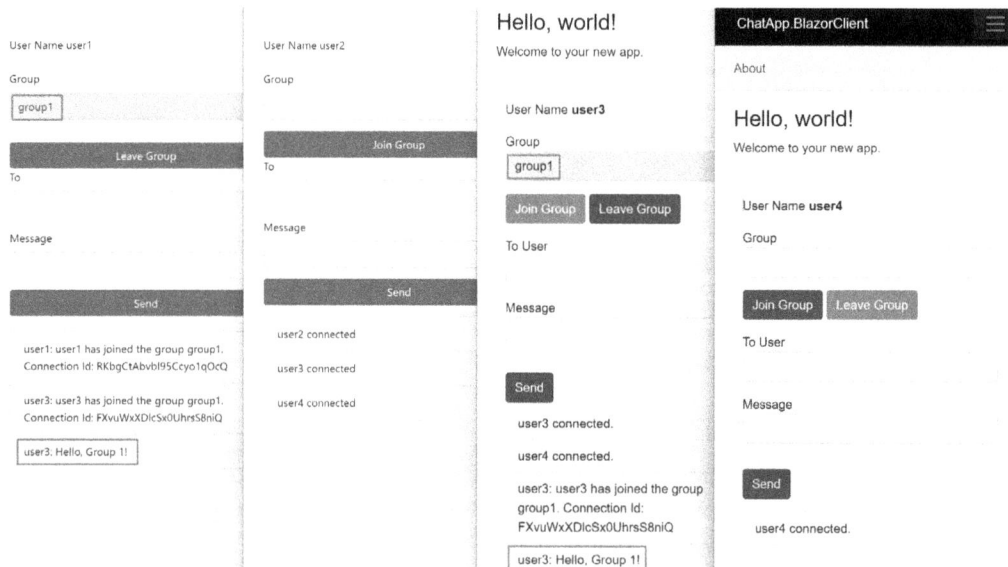

Figure 13.11 – Sending a message to a group

You will see a message displayed for `user1` and `user3` as they are in the same group. But `user2` and `user4` will not see the message because they are not in `group1`.

Sending messages from other services

So far, we have implemented a chat app that allows users to send messages to other users or groups. Sometimes, we need to send messages from other places. For example, when an event occurs, we may need to send a message to notify the users. In this section, we will explore how to send messages from other services. You can find the complete code of the sample in the `chapter13/v4` folder of the GitHub repository.

We will add a REST API endpoint to allow other systems to send messages to the SignalR hub. Follow these steps to add a REST API endpoint in the `ChatApp.Server` application:

1. Create the following models in the `Models` folder:

   ```
   public class SendToAllMessageModel
   {
       public string FromUser { get; set; } = string.Empty;
       public string Message { get; set; } = string.Empty;
   }

   public class SendToUserMessageModel
   {
       public string FromUser { get; set; } = string.Empty;
       public string ToUser { get; set; } = string.Empty;
       public string Message { get; set; } = string.Empty;
   }

   public class SendToGroupMessageModel
   {
       public string FromUser { get; set; } = string.Empty;
       public string GroupName { get; set; } = string.Empty;
       public string Message { get; set; } = string.Empty;
   }
   ```

   These models are used to send messages to the SignalR hub.

2. Create a new controller or use the existing `AccountController` class in the sample application. We will create a `ChatController` class in the `Controllers` folder.

3. Inject the `IHubContext<ChatHub, IChatClient>` service into the `ChatController` class:

   ```
   [Route("api/[controller]")]
   [ApiController]
   public class ChatController(IHubContext<ChatHub, IChatClient>
   hubContext) : ControllerBase
   {
   }
   ```

The IHubContext<ChatHub, IChatClient> service is used to send messages to clients. In this example, we use a strongly typed hub. You can also inject the IHubContext<ChatHub> service if you use a normal SignalR hub.

4. Add the following actions to send messages to all users, a specific user, and a specific group:

```
[HttpPost("/all")]
public async Task<IActionResult> SendToAllMessage([FromBody]
SendToAllMessageModel model)
{
    if (ModelState.IsValid)
    {
        await hubContext.Clients.All.ReceiveMessage(model.
FromUser, model.Message);
        return Ok();
    }
    return BadRequest(ModelState);
}


[HttpPost("/user")]
public async Task<IActionResult> SendToUserMessage([FromBody]
SendToUserMessageModel model)
{
    if (ModelState.IsValid)
    {
        await hubContext.Clients.User(model.ToUser).
ReceiveMessage(model.FromUser, model.Message);
        return Ok();
    }
    return BadRequest(ModelState);
}


[HttpPost("/group")]
public async Task<IActionResult> SendToGroupMessage([FromBody]
SendToGroupMessageModel model)
{
    if (ModelState.IsValid)
    {
        await hubContext.Clients.Group(model.GroupName).
ReceiveMessage(model.FromUser, model.Message);
        return Ok();
    }
    return BadRequest(ModelState);
}
```

The preceding code uses the `hubContext.Clients` property to send messages to clients. Note that this endpoint is not authenticated. You can add authentication and authorization to this endpoint if required.

- Run the three applications. Use different users to log in and join the group. Then, you can test the `chat/all`, `chat/user`, and `chat/group` endpoints using Postman or any other HTTP client.

This is how to send messages from external services. In the next section, we will explore how to manage SignalR connections.

Configuring SignalR hubs and clients

SignalR provides a `HubOptions` class to configure SignalR hubs. Also, SignalR clients have some configuration options. In this section, we will explore how to configure SignalR hubs and clients. You can find the complete code of the sample in the `chapter13/v5` folder of the GitHub repository.

Configuring SignalR hubs

Here are some of the configuration options for SignalR hubs:

- `KeepAliveInterval`: This property determines the interval at which a keep-alive message is sent to clients. If a client does not receive a message from the server within this period of time, it will send a `ping` message to the server in order to maintain the connection. When changing this value, it is important to also adjust the `serverTimeoutInMilliseconds` or `ServerTimeout` option in the client. For best results, it is recommended to set the `serverTimeoutInMilliseconds` or `ServerTimeout` option to a value that is double the value of the `KeepAliveInterval` property. The default value of `KeepAliveInterval` is 15 seconds.

- `ClientTimeoutInterval`: This property determines the interval at which the server will consider the client disconnected if it has not received a message from the client. It is recommended to set `ClientTimeoutInterval` to a value that is double the value of the `KeepAliveInterval` property. The default value of `ClientTimeoutInterval` is 30 seconds.

- `EnableDetailedErrors`: This property determines whether detailed error messages are sent to the client. The default value of `EnableDetailedErrors` is `false` as error messages may contain sensitive information.

- `MaximumReceiveMessageSize`: This property determines the maximum size of a message that the server will accept. The default value of `MaximumReceiveMessageSize` is 32 KB. Do not set this value to a very large value as it may cause **denial-of-service** (**DoS**) attacks and consume a lot of memory.

- **MaximumParallelInvocationsPerClient**: This property determines the maximum number of hub method invocations that can be executed in parallel per client. The default value of `MaximumParallelInvocationsPerClient` is 1.

- **StreamBufferCapacity**: This property determines the maximum number of items that can be buffered in a client upload stream. The default value of `StreamBufferCapacity` is 10. We will introduce streaming in the next section.

There are two ways to configure SignalR hubs. The first way is to provide a `HubOptions` object to all hubs. The following code shows how to configure the `ChatHub` class:

```
builder.Services.AddSignalR(options =>
{
    options.KeepAliveInterval = TimeSpan.FromSeconds(10);
    options.ClientTimeoutInterval = TimeSpan.FromSeconds(20);
    options.EnableDetailedErrors = true;
});
```

The second way is to configure the SignalR hubs for each hub. The following code shows how to configure the `ChatHub` class:

```
builder.Services.AddSignalR().AddHubOptions<ChatHub>(options =>
{
    options.KeepAliveInterval = TimeSpan.FromSeconds(10);
    options.ClientTimeoutInterval = TimeSpan.FromSeconds(20);
    options.EnableDetailedErrors = true;
});
```

The preceding code is useful if you have multiple hubs and you want to configure them differently.

Note that if you change the `KeepAliveInterval` or `ClientTimeoutInterval` property of the SignalR hub, you need to update the `serverTimeoutInMilliseconds` or `ServerTimeout` option in the clients as well. The following code shows how to configure the TypeScript client:

```
connection = new signalR.HubConnectionBuilder()
  .withUrl("https://localhost:7159/chatHub", {
    // Omitted for brevity
  })
  .build();
// The following configuration must match the configuration in the
server project
connection.keepAliveIntervalInMilliseconds = 10000;
connection.serverTimeoutInMilliseconds = 20000;
```

The HubConnection object has the keepAliveIntervalInMilliseconds property and the serverTimeoutInMilliseconds property, which can be used to match the configuration in the server project.

Similarly, you can configure the Blazor client as follows:

```
_hubConnection = new HubConnectionBuilder()
    .WithUrl("https://localhost:7159/chatHub", options =>
    {
        // Omitted for brevity
    })
    .Build();
_hubConnection.KeepAliveInterval = TimeSpan.FromSeconds(10);
_hubConnection.ServerTimeout = TimeSpan.FromSeconds(20);
You can also configure these properties on the HubConnectionBuilder
object as shown below:
_hubConnection = new HubConnectionBuilder()
    .WithUrl("https://localhost:7159/chatHub", options =>
    {
        // Omitted for brevity
    })
    .WithKeepAliveInterval(TimeSpan.FromSeconds(10))
    .WithServerTimeout(TimeSpan.FromSeconds(20))
    .Build();
```

Make sure that the values of the KeepAliveInterval and ClientTimeout/ServerTimeout properties are the same in the server and the client.

HTTP configuration options

SignalR can automatically negotiate the transport protocol with the client. The default transport protocol is WebSockets. If the client does not support WebSockets, SignalR will use SSE or long polling. You can configure HTTP options for SignalR. The following code shows how to configure HTTP options for the ChatHub class:

```
app.MapHub<ChatHub>("/chatHub", options =>
{
    options.Transports = HttpTransportType.WebSockets |
HttpTransportType.LongPolling;
    options.WebSockets.CloseTimeout = TimeSpan.FromSeconds(10);
    options.LongPolling.PollTimeout = TimeSpan.FromSeconds(120);
});
```

The preceding code configures HTTP options for the ChatHub class using a HttpConnectionDispatcherOptions object. In this sample, we configured the Transports property to use WebSockets and long polling, but not SSE. In addition, we configured the CloseTimeout property of the WebSockets property to 10 seconds, and the PollTimeout property of the LongPolling property to 120 seconds. The default value of the CloseTimeout property is 5 seconds, meaning that after the server closes, the connection will be terminated if clients cannot close the connection within 5 seconds. The default value of the PollTimeout property is 90 seconds, meaning that the server will terminate a poll request after waiting for 90 seconds and then create a new poll request.

The allowed transports can be configured in the client as well. We can configure the TypeScript client as follows:

```
connection = new signalR.HubConnectionBuilder()
  .withUrl("https://localhost:7159/chatHub", {
    transport: signalR.HttpTransportType.WebSockets | signalR.
HttpTransportType.LongPolling,
  })
  .build();
```

The following code shows how to configure the Blazor client:

```
_hubConnection = new HubConnectionBuilder()
    .WithUrl("https://localhost:7159/chatHub", options =>
    {
        options.Transports = HttpTransportType.WebSockets |
HttpTransportType.LongPolling;
    })
    .Build();
```

The HttpTransportType enum has a FlagsAttribute attribute, so you can use the bitwise OR operator to combine multiple transport protocols.

Automatically reconnecting

Sometimes, due to network issues, the SignalR connection may be disconnected. For example, if the user's device is switched from Wi-Fi to cellular, or if the user's device is in a tunnel, the SignalR connection may be disconnected. In this case, we want the client to automatically reconnect to the server:

1. SignalR allows the client to automatically reconnect to the server if the connection is dropped. The following code shows how to configure the TypeScript client to automatically reconnect to the server:

    ```
    connection = new signalR.HubConnectionBuilder()
      .withUrl("https://localhost:7159/chatHub", {
        // Omitted for brevity
    ```

```
    })
    .withAutomaticReconnect()
    .build();
```

2. Similarly, you can configure the Blazor client as follows:

```
_hubConnection = new HubConnectionBuilder()
    .WithUrl("https://localhost:7159/chatHub", options =>
    {
        // Omittted for brevity
    })
    .WithAutomaticReconnect()
    .Build();
```

3. By default, when the connection is dropped, the client will try to reconnect to the SignalR server in 0, 2, 10, and 30 seconds. You can configure the retry policy as follows:

```
connection = new signalR.HubConnectionBuilder()
    .withUrl("https://localhost:7159/chatHub", {
        // Omittted for brevity
    })
    .withAutomaticReconnect([0, 5, 20])
    .build();
```

The withAutomaticReconnect() method accepts an array of numbers to configure the delay duration in milliseconds. In the preceding code, the client will try to reconnect to the server in 0, 5, and 20 seconds.

4. In the Blazor client, you can configure the retry policy as follows:

```
_hubConnection = new HubConnectionBuilder()
    .WithUrl("https://localhost:7159/chatHub", options =>
    {
        // Omitted for brevity
    })
    .WithAutomaticReconnect(new[] { TimeSpan.FromSeconds(0),
TimeSpan.FromSeconds(5), TimeSpan.FromSeconds(20) })
    .Build();
```

The preceding code configures the same retry policy as the TypeScript client.

5. To test the automatic reconnect feature, we can add a label to show the connection status. Add the following code to the HTML content for the divChat element:

```
<div class="form-group mb-3">
    <label>Status</label>
    <label id="lblStatus"></label>
</div>
```

6. Then, update the TypeScript code to show the connection status:

```typescript
connection.onclose(() => {
  lblStatus.textContent = "Disconnected.";
});
connection.onreconnecting((error) => {
  lblStatus.textContent = `${error} Reconnecting...`;
});
connection.onreconnected((connectionId) => {
  lblStatus.textContent = `Connected. ${connectionId}`;
});
await connection.start();
lblStatus.textContent = `Connected. ${connection.connectionId}`;
```

7. We can also enable debug logging to see the connection status. The following code shows how to do this:

```typescript
connection = new signalR.HubConnectionBuilder()
  .withUrl("https://localhost:7159/chatHub", {
    // Omitted for brevity
  })
  .configureLogging(signalR.LogLevel.Debug)
  // Omitted for brevity
```

You can find the complete code in the sample application.

8. Run the SignalR server and the TypeScript client. Press *F12* to open the developer tools for the TypeScript client. Click the **Network** tab, and you can change network conditions to simulate network issues. For example, you can change the network to **Offline** to simulate network disconnection, as shown next:

Figure 13.12 – Simulating network disconnection in Chrome developer tools

9. After you change the network to **Offline**, wait for a few seconds (depending on the timeout configuration), and you should see the client automatically reconnect to the server, as shown next:

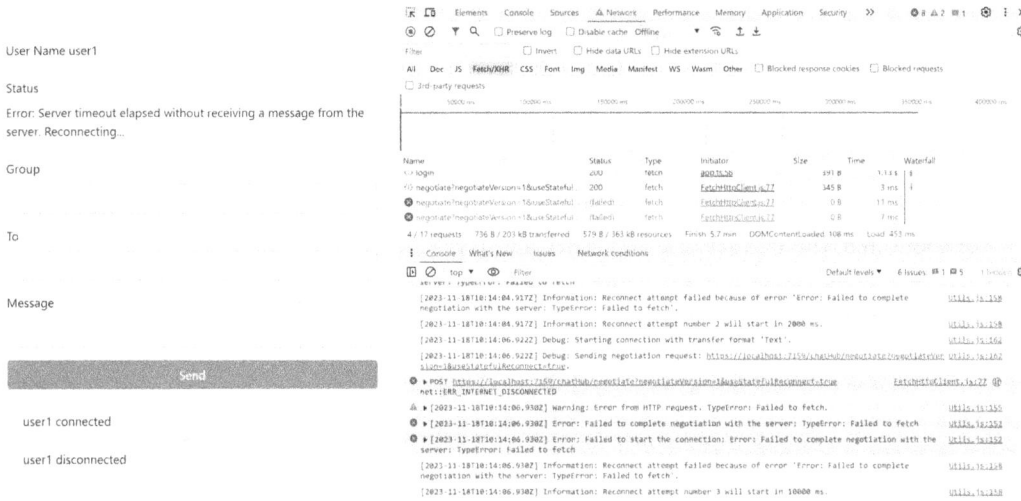

Figure 13.13 – The client automatically reconnects to the server

10. Change the network back to Online, and you should see that the client reconnects to the server, as shown next:

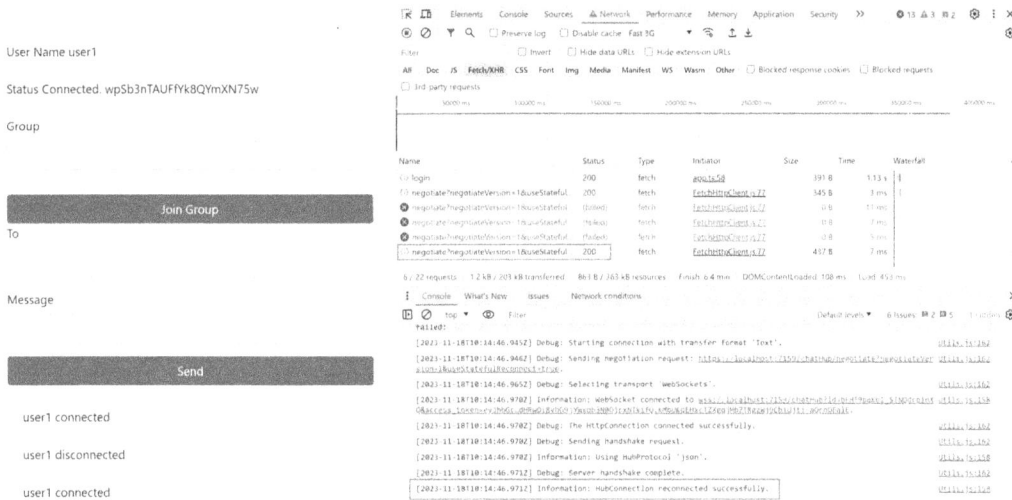

Figure 13.14 – The client reconnects to the server after the network is back online

> **Important note**
>
> If the client fails to reconnect to the server after trying four times, the `onclose` event will be triggered. You can add the event handler for the `onclose` event to handle the connection close event. For example, you can notify the user that the connection is closed and ask the user to refresh the page or manually reconnect to the server.

SignalR in ASP.NET Core 8.0 supports stateful reconnect, allowing the server to temporarily store messages when the client is disconnected. Upon reconnection, the client will use the same connection ID, and the server will replay any messages that were sent while the client was disconnected. This ensures that the client's state is maintained and that no messages are lost.

11. To enable stateful reconnect, we need to configure the `AllowStatefulReconnects` option for the SignalR hub endpoint as follows:

```
app.MapHub<ChatHub>("/chatHub", options =>
{
    // Omitted for brevity
    options.AllowStatefulReconnects = true;
});
```

12. By default, the maximum buffer size of the stateful reconnect is 100,000 bytes. You can change the buffer size as follows:

```
builder.Services.AddSignalR(options =>
{
    // Omitted for brevity
    options.StatefulReconnectBufferSize = 200000;
});
```

13. Then, we can configure the TypeScript client to use the stateful reconnect as follows:

```
connection = new signalR.HubConnectionBuilder()
  .withUrl("https://localhost:7159/chatHub", {
    // Omitted for brevity
  })
  .withAutomaticReconnect()
  .withStatefulReconnect({ bufferSize: 200000 })
  .build();
```

14. Similarly, you can configure the Blazor client as follows:

```
_hubConnection = new HubConnectionBuilder()
    .WithUrl("https://localhost:7159/chatHub", options =>
    {
        // Omitted for brevity
    })
```

```
        .WithAutomaticReconnect()
        .WithStatefulReconnect()
        .Build();
```

15. To configure the buffer size of the Blazor client, you can configure the `HubConnectionOptions` object as follows:

```
var builder = new HubConnectionBuilder()
    .WithUrl("https://localhost:7159/chatHub", options =>
    {
        // Omitted for brevity
    })
    .WithAutomaticReconnect()
    .WithStatefulReconnect();
builder.Services.Configure<HubConnectionOptions>(options =>
{
    options.StatefulReconnectBufferSize = 200000;
});
_hubConnection = builder.Build();
```

Besides the automatic reconnect feature, you can also manually reconnect to the SignalR server if the connection is dropped. You can add an event handler for the `onclose` event or `Closed` event to handle the connection close event.

Scaling SignalR

So far, we have implemented a chat app that allows users to send messages to other users or groups. We have also explored how to manage SignalR connections. You can also use a similar approach to build a real-time notification system, a real-time dashboard, and so on. However, the application can only run on a single server. If we want to scale the application, for example, using a load balancer to distribute requests to multiple servers, server *A* does not know the connections on server *B*.

SignalR requires a persistent connection between the client and the server. That means requests from the same client must be routed to the same server. This is called *sticky sessions* or *session affinity*. This is required if you have multiple SignalR servers. Besides this requirement, there are some other considerations when you scale SignalR:

- If you host the application in Azure, you can use Azure SignalR Service. Azure SignalR Service is a fully managed service that helps you scale the SignalR application without worrying about the infrastructure. With Azure SignalR Service, you do not need to use sticky sessions as all clients connect to Azure SignalR Service. This service takes on the responsibility of managing connections and freeing up resources on the SignalR servers. For more information, please refer to `https://learn.microsoft.com/en-us/azure/azure-signalr/signalr-overview`.

- If you host the application on your own infrastructure or other cloud providers, you can use Redis backplane to synchronize the connections. The Redis backplane is a Redis server that uses the pub/sub feature to forward messages to other SignalR servers. However, this approach requires sticky sessions for most cases, and the SignalR application instances require additional resources to manage connections. There are some other SignalR backplane providers, such as SQL Server, NCache, and so on.

We will not cover details of how to scale SignalR in this book. You can find more information in the official documentation.

Summary

SignalR is a powerful library that simplifies the process of building real-time web applications. In this chapter, we explored how to use SignalR to build a chat app. We introduced basic concepts of SignalR, such as hubs, clients, and connections. We created clients using TypeScript and Blazor, which demonstrated how to use both TypeScript and .NET to build SignalR clients. We also discussed how to send messages to a specific user or group and how to secure SignalR connections using JWT authentication. Additionally, we explored how to configure SignalR hubs and clients, such as configuring the keep-alive interval, configuring HTTP options, and configuring the automatic reconnect feature.

Although we have covered a lot of features, there is still more to explore, such as streaming. For more information, please refer to the official documentation: `https://learn.microsoft.com/en-us/aspnet/core/signalr/introduction`. In the next chapter, we will explore how to deploy ASP.NET Core applications.

14

CI/CD for ASP.NET Core Using Azure Pipelines and GitHub Actions

In the previous chapters, we have explored the fundamentals of building, testing, and running ASP. NET Core applications. We have also discussed how to access data from a database using **Entity Framework Core** (**EF Core**) and secure our applications using ASP.NET Core Identity. Additionally, we have discussed how to test our applications using unit tests and integration tests, as well as how to use RESTful APIs, gRPC, and GraphQL. Furthermore, we have learned how to use the `dotnet run` command to run our applications locally. Now, it is time to take the next step in our ASP.NET Core journey and learn how to deploy our applications to the cloud.

In this chapter, we will explore the concept of **continuous integration and continuous delivery/ deployment** (**CI/CD**). This chapter will focus on two popular CI/CD tools and platforms: Azure Pipelines and GitHub Actions.

We will discuss the following topics in this chapter:

- Introduction to CI/CD
- Containerizing ASP.NET Core applications using Docker
- CI/CD using Azure Pipelines
- GitHub Actions

Upon completion of this chapter, you will have a basic understanding of containerization concepts and the ability to build and deploy your ASP.NET Core applications to the cloud using either of these tools.

Technical requirements

The code examples in this chapter can be found at `https://github.com/PacktPublishing/Web-API-Development-with-ASP.NET-Core-8/tree/main/samples/chapter14/`.

Introduction to CI/CD

Developers work on code every day – they may create new features, fix bugs, or refactor existing code. In a team environment, multiple developers may be working on the same code base. A developer may create a new feature, while another developer may be fixing a bug. The code base is constantly changing, and it is important to ensure that the code changes made by different developers do not conflict with each other and do not break any existing functionalities. To avoid such issues, developers should integrate their code changes frequently.

Additionally, when the application is ready to be deployed, it is important to consider the different environments it may be deployed to, such as development, staging, or production. Different environments may have different configurations, and the deployment process may be different for each environment. To ensure that the application is deployed correctly and consistently, it is ideal to automate the deployment process. This is where CI/CD comes in.

The acronym *CI/CD* can have different interpretations depending on the context. **CI**, a development practice that allows developers to integrate code changes regularly. **CD** can refer to either **continuous delivery** or **continuous deployment**, which are often used interchangeably. It is not worth debating the exact definitions of these terms, as in most cases, *CD* means building, testing, and deploying the applications to the production environment (and, potentially, other environments) frequently and automatically.

CI/CD pipelines are key components of **DevOps**, a combination of the words **development** and **operations**. DevOps has evolved over the years and is generally defined as a set of practices, tools, and processes that enable continuous delivery of value to end users. While DevOps is a vast topic, this chapter will focus on CI/CD pipelines specifically.

A typical CI/CD process is shown in the following diagram:

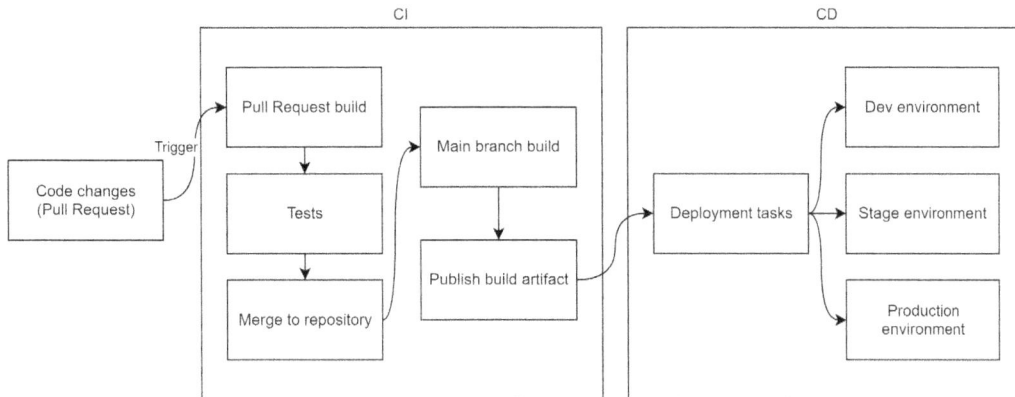

Figure 14.1 – A typical CI/CD process

The steps in *Figure 14.1* are described as follows:

1. The developer creates a new feature or fixes a bug in the code base and then commits the changes to the shared code repository. If the team is using Git as its **version control system (VCS)**, the developer will create a pull request to submit the changes.

2. The pull request will initiates the CI pipeline, which will build the application and execute tests. If the build or tests fail, the developer will be notified, allowing them to address the issue promptly.

3. If the build and tests are successful and the pull request is approved by the team, the code changes will be merged into the `main` branch. This ensures that the code is up to date and aligns with the team's standards.

4. The merge will trigger the CI pipeline to build the application and publish the artifacts (for example, binaries, configuration files, Docker images, and so on) to the artifact repository.

5. The CD pipeline can be triggered manually or automatically. The CD pipeline then deploys the application to the target environment (for example, development, staging, or production environment).

The CI/CD process can be more complex than what is shown in *Figure 14.1*. For example, the CI pipeline may require static code analysis, code test coverage, and other quality checks. The CD pipeline may need to apply different configurations for different environments or have different deployment strategies, such as blue/green deployment or canary deployment. With the increasing complexity of CI/CD pipelines, DevOps engineers are in high demand as they possess the skills to implement these pipelines using the various tools and platforms available.

Before we delve into the details of CI/CD, let us first introduce some concepts and terminologies that are commonly used in CI/CD.

CI/CD concepts and terminologies

It is essential to understand key concepts and terminologies commonly used in CI/CD. The following are some of the most common terms used in CI/CD:

- **Pipeline**: A pipeline is an automated process used to build, test, and deploy applications. They can be triggered manually or automatically and can even be set up to be triggered by other pipelines. This helps streamline the development process, ensuring that applications are built, tested, and deployed quickly and efficiently.

- **Build**: A build is a process that involves compiling the source code and creating the necessary binaries or Docker images. This ensures that the code is ready for deployment.

- **Test**: A pipeline may include automated tests, such as unit tests, integration tests, performance tests, or **end-to-end** (**E2E**) tests. These tests can be incorporated into the pipeline to ensure that the code changes do not break any existing functionalities. This helps to ensure that the software remains stable and reliable.

- **Artifact**: An artifact is a file or collection of files – normally, the output of a build process. Examples of artifacts include binary files, a Docker image, or a ZIP file containing the binary files. These artifacts can then be used as inputs for the deployment process.

- **Containerization**: Containerization is a method of packaging an application and its dependencies into a container image, which can be deployed and run in a consistent environment, regardless of the host operating system. One of the most popular containerization tools is Docker. Containerization offers numerous benefits, such as improved scalability, portability, and resource utilization.

- **VCS**: VCSs are an essential tool for software development, allowing developers to track and manage changes to source code. Git is one of the most widely used VCSs, providing developers with an effective way to manage their code base.

- **Deployment**: Deployment is the process of deploying the application to the target environment. It involves configuring the application to meet the requirements of the environment, as well as ensuring that it is secure and ready for use.

- **Trigger**: A trigger is an event that tells the pipeline when to run. A trigger can be manual or automatic. For example, a pull request can trigger the CI pipeline to validate code changes by running tests. A merge to the `main` branch can trigger the CI pipeline to build and publish artifacts. The successful CI pipeline can trigger the CD pipeline to deploy the application to non-production environments. However, the CD pipeline may need to be triggered manually to deploy the application to the production environment as a safety measure.

Gaining an understanding of the fundamental concepts and terminologies associated with CI/CD is essential for successful implementation. As many different tools and platforms can be used to implement CI/CD pipelines, we will discuss the details in the following sections.

Understanding the importance of CI/CD

CI/CD plays an important role in DevOps. It helps the team respond to changes and deliver value to end users frequently, safely, and reliably. As CI/CD pipelines are automated, they can streamline the process of delivering software and reduce the time and effort needed to deploy applications to the production environment. Additionally, CI/CD helps maintain a stable and reliable code base.

In order to successfully implement a CI/CD pipeline, the team must adhere to certain practices. Automated tests should be conducted to ensure that code changes do not break any existing functionalities. Additionally, a well-defined deployment strategy should be established, such as staging the application in a development environment before deploying it to the production environment. By following these practices, the team can reduce **time to market** (**TTM**) and deliver the application to end users faster and more frequently.

CI/CD practices help development teams in the following ways:

- **Faster feedback**: CI/CD pipelines can be triggered automatically when code changes are committed to the shared code repository. This provides developers with faster feedback on code changes, allowing them to address any issues early in the development process.

- **Reduced manual effort and risk**: CI/CD pipelines automate the deployment process, reducing manual effort and risk. This decreases the time and effort needed for production deployment, eliminating manual and error-prone processes.

- **Consistency**: Automated builds and deployments ensure consistency across different environments. This reduces the risk of deployment failures due to configuration issues or *it works on my machine* problems.

- **Enhanced quality**: Automated tests can be integrated into CI/CD pipelines, which helps to ensure that the code base remains stable and reliable. CI/CD pipelines can also run other quality checks, such as static code analysis and code test coverage, which leads to higher-quality code.

- **Rapid delivery and agility**: CI/CD pipelines enable the team to release new features and bug fixes to end users faster and more frequently. This allows businesses to respond quickly to customer needs and market changes.

With these benefits in mind, it is clear that CI/CD is a must-have for any development team. No one would want to go back to the days of manual builds and deployments anymore, as it is time-consuming and error-prone.

We have now learned some concepts of CI/CD and why it is important. In the next section, we will discuss how to containerize ASP.NET Core applications using Docker.

Containerizing ASP.NET Core applications using Docker

Many years ago, when we deployed applications to the production environment, we needed to ensure that the target environment had the correct version of the .NET Framework installed. Developers were struggling with the *it works on my machine* problem, as development environments may have had different configurations than the production environment, including software versions, operating systems, and hardware. This often led to deployment failures due to configuration issues.

The introduction of .NET Core, a cross-platform and open-source framework, has enabled us to deploy our applications on any platform, including Windows, Linux, and macOS. However, for successful deployment, we still need to ensure that the target environment has the correct runtime installed. This is where containerization comes in.

What is containerization?

Containers are lightweight, isolated, and portable environments that contain all the necessary dependencies for running an application. Unlike **virtual machines** (**VMs**), they do not require a separate guest operating system as they share the host operating system kernel. This makes them more lightweight and portable than VMs, as they can run on any platform that supports the container runtime. Containers also provide isolation, ensuring that applications are not affected by changes in the environment.

Containerization is a powerful tool that enables us to package our applications and their dependencies into a single container image. **Docker** is one of the most popular containerization solutions, offering support for Windows, Linux, and macOS for development purposes, as well as many variants of Linux, such as Ubuntu, Debian, and CentOS, for production environments. Additionally, Docker is compatible with cloud platforms, including Azure, **Amazon Web Services** (**AWS**), and **Google Cloud Platform** (**GCP**). If we use Docker as the container runtime, then the container images are called **Docker images**.

Docker images are a convenient way to package an application and its dependencies. They contain all the components necessary to run an application, such as the application code (binaries), runtime or SDK, system tools, and configurations. Docker images are immutable, meaning they cannot be changed once they are created. To store these images, they are placed in a registry, such as Docker Hub, **Azure Container Registry** (**ACR**), or AWS **Elastic Container Registry** (**ECR**). Docker Hub is a public registry that offers many pre-built images. Alternatively, a private registry can be created to store custom Docker images.

Once a Docker image has been created, it can be used to create a Docker container. A Docker container is an isolated, in-memory instance of a Docker image, with its own filesystem, network, and memory. This makes creating a container much faster than booting up a VM and also allows for fast destruction and rebuilding of a container from the same image. In addition, multiple containers can be created from the same image, which is useful for scaling out applications. If any container fails, it can be destroyed and rebuilt from the same image in a matter of seconds, making containerization a powerful tool.

The files in a Docker image are stackable. *Figure 14.2* shows an example of a container filesystem that contains an ASP.NET Core app and its dependencies:

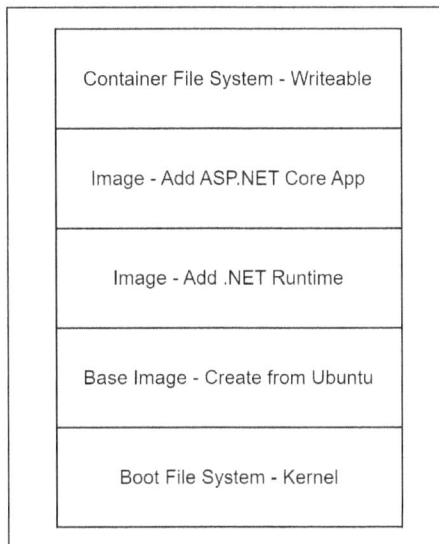

Figure 14.2 – Docker container file system

Figure 14.2 illustrates the layers of a Docker container. On top of the kernel layer is the base image layer, which is an empty container image created from Ubuntu. On top of the base image layer is the ASP.NET Core runtime layer, then the ASP.NET Core app layer. When a container is created, Docker adds a final writeable layer on top of the other layers. This writeable layer can be used to store temporary files, such as logs. However, as we mentioned earlier, Docker images are immutable, so any changes made to the writeable layer will be lost when the container is destroyed. This is why we should not store any persistent data in a container. Instead, we should store the data in a volume, which is a directory on the host machine that is mounted into the container.

This is a very simplified explanation of Docker images and containers. Next, let us install Docker and create a Docker image for our ASP.NET Core application.

Installing Docker

You can download Docker Desktop from the following links:

- Windows: `https://docs.docker.com/desktop/install/windows-install/`
- Mac: `https://docs.docker.com/desktop/install/mac-install/`
- Linux: `https://docs.docker.com/desktop/install/linux-install/`

Please follow the official documentation to install Docker on your machine.

If you use Windows, please use the **Windows Subsystem for Linux 2** (**WSL 2**) backend instead of Hyper-V. WSL 2 is a compatibility layer that allows Linux binary executables to be run natively on Windows. Using WSL 2 as the backend for Docker Desktop on Windows provides better performance than the Hyper-V backend.

To install WSL 2 on Windows, please follow the instructions at this link: `https://learn.microsoft.com/en-us/windows/wsl/install`. By default, WSL 2 uses the Ubuntu distribution. You can also install other Linux distributions, such as Debian, CentOS, or Fedora.

After installing WSL 2, you can check the version of WSL by running the following command in PowerShell:

```
wsl -l -v
```

If you see the `VERSION` field shows 2, that means WSL 2 is installed correctly. Then, you can install Docker Desktop and choose WSL 2 as the backend. If you have multiple Linux distributions installed, you can choose the default distribution to use with Docker Desktop. Go to **Settings** | **Resources** | **WSL integration**, and choose the distribution you want to use with Docker Desktop, as shown in *Figure 14.3*:

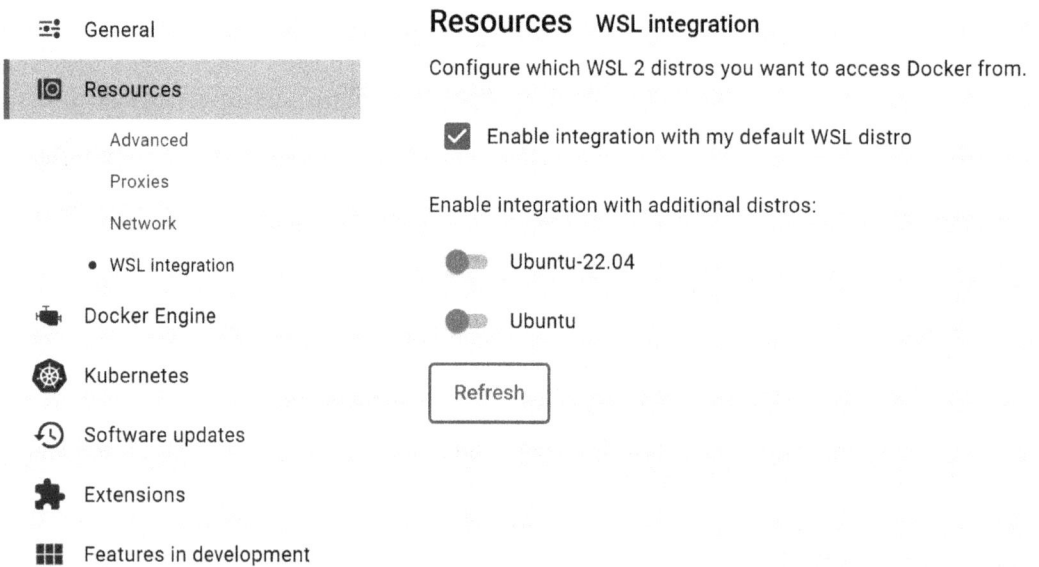

Figure 14.3 – Choosing the default Linux distribution to use with Docker Desktop

Here is the example output of the `wsl -l -v` command, which shows two Linux distros installed on this machine; the default distro is `Ubuntu-22.04`:

```
   NAME                STATE               VERSION
 * Ubuntu-22.04        Running             2
```

```
Ubuntu                    Stopped        2
docker-desktop-data       Running        2
docker-desktop            Running        2
```

Docker Desktop installs two internal Linux distros:

- `docker-desktop`: This is used to run the Docker engine
- `docker-desktop-data`: This is used to store containers and images

Note that Docker may consume a lot of resources on your machine. If you feel that Docker slows down your machine or consumes too many resources, you can configure the resources allocated to WSL 2 following the instructions in this link: `https://learn.microsoft.com/en-us/windows/wsl/wsl-config#configure-global-options-with-wslconfig`.

After installing Docker Desktop, we can now create a Docker image for our ASP.NET Core application. In the next section, we will discuss some commands that are commonly used in Docker.

Understanding Dockerfiles

To demonstrate how to build and run Docker images, we will need a sample ASP.NET Core application. You can create a new ASP.NET Core web API project using the following command:

```
dotnet new webapi -o BasicWebApiDemo -controllers
```

Alternatively, you can clone the sample code from the `/samples/chapter14/MyBasicWebApiDemo` folder in the book's GitHub repository.

Docker images can be built using a Dockerfile, a text file containing a list of instructions used to build the image. You can create a Dockerfile in the root directory of the ASP.NET Core project manually, or you can use VS 2022 to create it for you. To create a Dockerfile using VS 2022, right-click on the project in Solution Explorer, then select **Add | Docker Support**. You will see the following dialog:

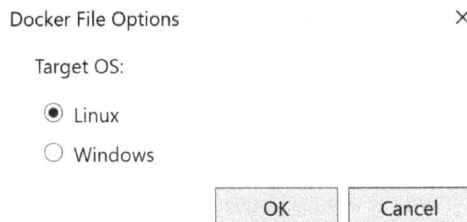

Figure 14.4 – Adding Docker support to an ASP.NET Core project in VS 2022

There are two options here: **Linux** and **Windows**. It is recommended to use Linux for development purposes, as the Linux image is smaller than the Windows image. Docker was originally designed for Linux, so it is more mature on Linux than on Windows. Many cloud platforms, such as Azure, AWS, and GCP, support Linux containers. However, not all Windows servers support Windows containers. Unless you have strong reasons to host your application on a Windows server, you should choose **Linux** here.

Once you have selected the **Linux** option, click **OK**. VS 2022 will then generate a Dockerfile for you. If you would prefer to create the Dockerfile manually, name it `Dockerfile` without any file extension. This allows us to build the Docker image using the `docker build` command without needing to specify the Dockerfile name. The default Dockerfile created by VS 2022 will resemble the following:

```
#See https://aka.ms/customizecontainer to learn how to customize your
debug container and how Visual Studio uses this Dockerfile to build
your images for faster debugging.

FROM mcr.microsoft.com/dotnet/aspnet:8.0 AS base
USER app
WORKDIR /app
EXPOSE 8080
EXPOSE 8081


FROM mcr.microsoft.com/dotnet/sdk:8.0 AS build
ARG BUILD_CONFIGURATION=Release
WORKDIR /src
COPY ["BasicWebApiDemo.csproj", "."]
RUN dotnet restore "./BasicWebApiDemo.csproj"
COPY . .
WORKDIR "/src/."
RUN dotnet build "BasicWebApiDemo.csproj" -c $BUILD_CONFIGURATION -o /
app/build


FROM build AS publish
RUN dotnet publish "BasicWebApiDemo.csproj" -c $BUILD_CONFIGURATION -o
/app/publish /p:UseAppHost=false


FROM base AS final
WORKDIR /app
COPY --from=publish /app/publish .
ENTRYPOINT ["dotnet", "BasicWebApiDemo.dll"]
```

Let us go through the Dockerfile line by line:

```
FROM mcr.microsoft.com/dotnet/aspnet:8.0 AS base
```

The FROM instruction specifies the base image to use. The FROM instruction must be the first instruction in a Dockerfile. In this case, we are using the mcr.microsoft.com/dotnet/aspnet:8.0 image, which is the ASP.NET Core runtime image. This image is provided by Microsoft. mcr.microsoft. com is the domain name of the **Microsoft Artifact Registry** (also known as **Microsoft Container Registry** or **MCR**), which is a public registry that contains many pre-built images, similar to Docker Hub. AS base means we are giving this image a name, which is base. This name can be used later in the Dockerfile to refer to this image.

```
USER app
```

Next, the USER instruction specifies the username or **unique identifier** (**UID**) to use when running the image. The app user is created by the base image. This user is not a superuser, so it is more secure than the root user.

```
WORKDIR /app
```

The WORKDIR instruction sets the working directory inside of the container for these instructions: RUN, CMD, COPY, ADD, ENTRYPOINT, and so on. This instruction is similar to the cd command in the terminal. It supports both absolute and relative paths. If the directory does not exist, it will be created. In this example, the working directory is set to /app.

```
EXPOSE 8080
EXPOSE 8081
```

The EXPOSE instruction exposes the specified port(s) to the container when it is running. Note that this instruction does not actually publish the port to the host machine. It just means the container will listen on the specified port(s). By default, the EXPOSE instruction exposes the port(s) on the TCP protocol. In this case, the container will listen on ports 8080 and 8081.

```
FROM mcr.microsoft.com/dotnet/sdk:8.0 AS build
ARG BUILD_CONFIGURATION=Release
WORKDIR /src
```

In the preceding lines, we are using a different image, which is the mcr.microsoft.com/dotnet/sdk:8.0 image, and naming it build. This image contains the .NET SDK, which is used to build the application. The ARG instruction defines a variable that can be used later in the Dockerfile. In this case, we are defining a variable called BUILD_CONFIGURATION and setting its default value to Release. The WORKDIR instruction sets the working directory to /src.

```
COPY ["BasicWebApiDemo.csproj", "."]
RUN dotnet restore "./BasicWebApiDemo.csproj"
```

The COPY instruction copies files or directories from the source (on the local machine) to the destination (the filesystem of the container). In this case, we are copying the .csproj file to the current directory. The RUN instruction executes the specified command on top of the current image and creates a new layer, then commits the results. The new layer will be used for the next step in the Dockerfile. In this case, we are running the dotnet restore command to restore the NuGet packages.

```
COPY . .
WORKDIR "/src/."
RUN dotnet build "BasicWebApiDemo.csproj" -c $BUILD_CONFIGURATION -o /
app/build
```

In the preceding lines, we are copying all the files from the local machine to the current directory in the container. Then, we are setting the working directory to /src. Finally, we are running the dotnet build command to build the application. Note that we are using the BUILD_CONFIGURATION variable defined earlier in the Dockerfile.

```
FROM build AS publish
RUN dotnet publish "BasicWebApiDemo.csproj" -c $BUILD_CONFIGURATION -o
/app/publish /p:UseAppHost=false
```

Again, the FROM instruction uses the build image we defined earlier and names it publish. Then, the RUN instruction runs the dotnet publish command to publish the application. The $BUILD_CONFIGURATION variable is used again. The published application will be placed in the /app/publish directory.

```
FROM base AS final
WORKDIR /app
COPY --from=publish /app/publish .
ENTRYPOINT ["dotnet", "BasicWebApiDemo.dll"]
```

Next, we rename the base image as final and set the working directory to /app. To run the application, we only need the runtime, so we do not need the SDK image. Then, the COPY instruction copies the published application from the app/publish directory of the publish image to the current directory. Finally, the ENTRYPOINT instruction specifies the command to run when the container starts. In this case, we are running the dotnet BasicWebApiDemo.dll command to start the ASP.NET Core application.

You can find more information about Dockerfile instructions in the official documentation provided by Docker: https://docs.docker.com/engine/reference/builder/. Next, let us move on to building a Docker image.

Building a Docker image

Docker provides a set of commands that can be used to build, run, and manage Docker images and containers. To build a Docker image, go to the root directory of the ASP.NET Core project we created earlier, then run the following command:

```
docker build -t basicwebapidemo .
```

The `-t` option is used to tag the image with a name. `.` at the end means the current directory. Docker expects to find a file named `Dockerfile` in the current directory. If you have renamed the Dockerfile or the Dockerfile is not located in the current directory, you can use the `-f` option to specify the Dockerfile name, such as `docker build -t basicwebapidemo -f MyBasicWebApiDemo/MyDockerfile MyBasicWebApiDemo`.

The output shows that Docker is building the image layer by layer. Each instruction in the Dockerfile will create a layer in the image and add more content on top of the previous layer. The layers are cached, so if a layer has not changed, it will not be rebuilt for the next build. But if a layer has changed (for example, if we update the source code), then the layer that copies the source code will be rebuilt, and all the layers after that will be affected and need to be rebuilt as well.

Now, let us review the default Dockerfile generated by VS 2022. Why does it copy all the files after running the `dotnet restore` command? It is because if we only update the source code but the NuGet packages have not changed, then the `dotnet restore` command will not be executed again as the layer is cached. This can improve the build performance. However, if we update the NuGet packages, meaning the `.csproj` file has changed, then the `dotnet restore` command will be executed again.

Here are some tips for writing Dockerfiles:

- Consider the order of layers. Layers that are less likely to change should be placed before layers that are more likely to change.

- Keep layers small as much as possible. Do not copy unnecessary files. You can configure the `.dockerignore` file to exclude files or directories from the build context. If you use VS 2022 to create the Dockerfile, it will generate a `.dockerignore` file for you. Alternatively, you can manually create a text file named `.dockerignore` and then edit it. Here is a sample `.dockerignore` file:

  ```
  # Exclude build results, Npm cache folder, and some other files
  **/bin/
  **/obj/
  **/.git
  **/.vs
  **/.vscode
  **/global.json
  **/Dockerfile
  ```

```
**/.dockerignore
**/node_modules
```

For more information about the `.dockerignore` file, please refer to the official documentation here: `https://docs.docker.com/engine/reference/builder/#dockerignore-file`.

- Keep as few layers as possible. For example, to host an ASP.NET Core application, we can reduce the number of layers by using the `mcr.microsoft.com/dotnet/aspnet` image. This image contains the ASP.NET Core runtime already, eliminating the need to install the SDK in the container. You can also combine commands into a single `RUN` instruction.

- Use multi-stage builds. Multi-stage builds allow us to use multiple `FROM` instructions in a Dockerfile. Each `FROM` instruction can be used to create a new image. The final image will only contain the layers from the last `FROM` instruction. This can reduce the size of the final image. For example, we can use the `mcr.microsoft.com/dotnet/sdk` image to build the application and then use the `mcr.microsoft.com/dotnet/aspnet` image to run the application. This way, the final image will only contain the ASP.NET Core runtime, and it will not contain the SDK, which is not needed for running the application. To learn more about multi-stage builds, please refer to the official documentation here: `https://docs.docker.com/develop/develop-images/multistage-build/`.

    For more information on optimizing Docker builds, please refer to the official documentation here: `https://docs.docker.com/build/cache/`.

We can use the following command to list all Docker images on our machine:

```
docker images
```

The output should be similar to this:

```
REPOSITORY      TAG      IMAGE ID      CREATED             SIZE
basicwebapidemo latest   b0d8d94d219c  About a minute ago  222MB
```

The `basicwebapidemo` image is the one we just built. Each image has a `TAG` value and an `IMAGE ID` value. The `TAG` value is a human-readable name for the image. By default, the tag is `latest`. We can specify a different tag when building the image. For example, we can use the following command to build the image with the `v1` tag:

```
docker build -t basicwebapidemo:v1 .
```

In the preceding command, the `-t` option is used to tag the image with a name, which is separated from the tag with a colon.

To remove an image, use the `docker rmi <container name or ID>` command followed by the image name or image ID:

```
docker rmi basicwebapidemo
```

Note that if the image is used by a container, you will need to stop the container first before removing the image.

We now have a Docker image for our ASP.NET Core application. All the necessary dependencies are included in the image. Next, let us run the Docker image.

Running a Docker container

To run a Docker image in the container, we can use the `docker run` command. The following command will run the `basicwebapidemo` image we just built:

```
docker run -d -p 80:8080 --name basicwebapidemo basicwebapidemo
```

Let's take a closer look at this:

- The `-d` option is used to run the container in detached mode, meaning the container will run in the background. You can omit this option and then the container will run in the foreground, which means if you exit the terminal, the container will stop.

- The `-p` option is used to publish a container's port(s) to the host. In this case, we are publishing port 8080 of the container to port 80 of the host.

- The `--name` option is used to specify a name for the container. The last argument is the name of the image to run.

- You can also use the `-it` option to run the container in interactive mode. This option allows you to run a command in the container.

When we wrote the Dockerfile, we explained that the `EXPOSE` instruction exposes ports 8080 and 8081 to the container only. To publish the internal container port to the host, we need to use the `-p` option. The first port number is the port of the host machine, and the second port number is the internal container port. In this example, we are exposing the container port 8080 to the host port 80. This may confuse some people. So, please check the port numbers carefully. Also, sometimes the port number of the host machine may be occupied by another process. In this case, you will need to use a different port number.

The output should return the container ID, which is a UID for the container, such as this:

```
5529b0278e5a14452a7049a7c9922797b0c1171423970f99b4481c93cfdc6a38
```

Use the `docker ps` command to list all running containers:

```
docker ps
```

The output should be similar to this:

```
CONTAINER ID    IMAGE            COMMAND                CREATE
D          STATUS          PORTS                         NAMES
403eb4952287    basicwebapidemo    "dotnet BasicWebApiD..."    5 minutes
ago    Up 5 minutes    8081/tcp, 0.0.0.0:80->8080/tcp    basicwebapidemo
```

In the output, we can see that port 8080 of the container has been mapped to port 80 of the host.

To list all containers in all states, just add a -a option:

```
docker ps -a
```

You can check the status of containers. If the container is running, the status should be Up.

We can use the following commands to manage containers:

- To pause a container: `docker pause <container name or ID>`

- To restart a container: `docker restart <container name or ID>`

- To stop a container: `docker stop <container name or ID>`

- To remove a container: `docker rm <container name or ID>`

If the container is running, you can test the endpoint by sending a request to this URL: `http://localhost/weatherforecast`. You will see the response from the ASP.NET Core application. If you change the port number of the host machine, you will need to use the correct port number in the URL. For example, if you use -p 5000:8080, then you will need to use `http://localhost:5000/weatherforecast` to access the endpoint.

We can use the `docker logs <container name or ID>` command to show logs from a container:

```
docker logs basicwebapidemo
```

You will see logs such as these:

```
info: Microsoft.Hosting.Lifetime[14]
      Now listening on: http://[::]:8080
info: Microsoft.Hosting.Lifetime[0]
      Application started. Press Ctrl+C to shut down.
info: Microsoft.Hosting.Lifetime[0]
      Hosting environment: Production
info: Microsoft.Hosting.Lifetime[0]
      Content root path: /app
```

To check the stats of a container, you can use the `docker stats <container name or ID>` command:

```
docker stats basicwebapidemo
```

You will see the stats of the container as follows:

```
CONTAINER ID    NAME                CPU %      MEM USAGE / LIMIT      MEM
%    NET I/O              BLOCK I/O    PIDS
403eb4952287    basicwebapidemo     0.01%      24.24MiB /
15.49GiB    0.15%      23.8kB / 2.95kB    0B / 0B     25
```

You can send a request to the `/weatherforecast` endpoint to get the response from the ASP.NET Core application. Note that the container is running in the production environment, so the Swagger UI is not available. This is because in the `Program.cs` file, we enabled the Swagger UI only in the development environment. To enable the Swagger UI, we can stop and delete the current container, then create a new one in development environment. Alternatively, you can create a new container with a different name. For example, you can stop the container by running the following command:

```
docker stop basicwebapidemo
```

Then remove the container by running the following command:

```
docker rm basicwebapidemo
```

Next, add an environment variable to the `docker run` command to set the environment to development:

```
docker run -d -p 80:8080 --name basicwebapidemo -e ASPNETCORE_
ENVIRONMENT=Development basicwebapidemo
```

Now, you can view the Swagger UI by navigating to the `/swagger` endpoint in the browser.

So far, we have learned how to build and run Docker images. Docker has many other commands that can be used to manage Docker images and containers. To summarize, here are some of the most commonly used Docker commands from Docker official documents:

- `docker build`: Build a Docker image

- `docker run`: Run a Docker image

- `docker images`: List all images

- `docker ps`: List all running containers

- `docker ps -a`: List all containers in all states

- `docker logs`: Show logs from a container

- `docker stats`: Show stats of a container

- `docker pause`: Pause a container
- `docker restart`: Restart a container
- `docker stop`: Stop a container
- `docker rm`: Remove a container
- `docker rmi`: Remove an image
- `docker exec`: Run a command in a running container
- `docker inspect`: Display detailed information on one or more containers or images
- `docker login`: Log in to a Docker registry
- `docker logout`: Log out from a Docker registry
- `docker pull`: Pull an image from a registry
- `docker push`: Push an image to a registry
- `docker tag`: Create a `TARGET_IMAGE` tag that refers to `SOURCE_IMAGE`
- `docker volume`: Manage volumes
- `docker network`: Manage networks
- `docker system`: Manage Docker
- `docker version`: Show the Docker version information
- `docker info`: Show Docker system-wide information
- `docker port`: List port mappings or a specific mapping for a container

You can find more information about Docker commands here: `https://docs.docker.com/engine/reference/commandline/cli/`.

We have now learned how to build a Docker image for our ASP.NET Core application and run it in a container. Even though the container is running on our local machine, there is not much difference from running it in a production environment. The container is isolated from the host machine. The portable nature of containers makes it easy to deploy the application to any environment.

We can also use Docker commands to push the image to a registry, such as Docker Hub, ACR, and so on. However, manual deployment is error-prone and time-consuming. That is why we need a CI/CD pipeline to automate the deployment process.

In the next section, we will discuss how to deploy the containerized application to the cloud using Azure DevOps and Azure Pipelines.

CI/CD using Azure DevOps and Azure Pipelines

Azure DevOps is a cloud-based service that provides a set of tools for managing the software development process. It includes the following services:

- **Azure Boards**: A service for managing work items, such as user stories, tasks, and bugs.

- **Azure Repos**: A service for hosting code repositories. It supports Git and **Team Foundation Version Control** (**TFVC**). The repositories can be public or private.

- **Azure Pipelines**: A service for building, testing, and deploying applications with any language, platform, and cloud.

- **Azure Test Plans**: A service for manual and exploratory testing tools.

- **Azure Artifacts**: A service for creating, hosting, and sharing packages, such as `Maven`, `npm`, `NuGet`, and `Python` packages.

Azure DevOps is free for open-source projects and small teams. We will not cover all the features of Azure DevOps in this book. Let us focus on Azure Pipelines. In this section, we will discuss how to use Azure Pipelines to build and deploy our ASP.NET Core application to Azure App Service.

We will need these resources before we can deploy the application to Azure:

- **Azure DevOps account**: `https://azure.microsoft.com/en-us/products/devops`.

- **Azure subscription**: `https://azure.microsoft.com/en-us/free/`. You can sign up for a free account here. You will get some free credits to use Azure services.

- **GitHub repository**: You need a GitHub repository to host the source code. The pipeline will be triggered when code changes are committed to the repository.

Preparing the source code

GitHub is one of the most popular source control solutions. It is free for public repositories. We suppose you already have a GitHub account.

Download the example code from `/chapter14/MyBasicWebApiDemo`. This is a simple ASP.NET Core web API application with its unit tests and integration tests. Create a new repository on GitHub, then push the source code to the repository. The directory structure of the repository should look like the following:

```
MyAzurePipelinesDemo
    ├── src
    │   └──MyBasicWebApiDemo
    ├── tests
    │   ├──MyBasicWebApiDemo.UnitTests
```

```
|     └──MyBasicWebApiDemo.IntegrationTests
├── MyBasicWebApiDemo.sln
├── README.md
├── LICENSE
└── .gitignore
```

In the preceding directory structure, the main ASP.NET Core web API project is placed in the `src` folder, and the unit tests and integration tests are placed in the `tests` folder. This can better organize the solution structure. But it is just a personal preference. You can also place the unit tests and integration tests in the same folder as the main project. If you use a different directory structure, please update the paths in the pipeline accordingly in the following sections.

We will use this repository for the pipeline. Next, let us create the required Azure resources.

Creating Azure resources

In today's technology landscape, cloud computing has become the backbone of modern software development. Cloud computing provides many benefits, such as scalability, **high availability** (**HA**), and cost efficiency. Among the various cloud providers, such as AWS, GCP, and Alibaba Cloud, Microsoft Azure stands out as a robust and versatile platform for hosting and orchestrating your applications. Azure provides many services for hosting applications, such as Azure App Service, **Azure Kubernetes Service** (**AKS**), **Azure Container Instances** (**ACI**), Azure VMs, and Azure Functions. In this book, we will use Azure as the cloud platform to host our applications. For other cloud platforms, the concepts are similar.

We will need the following Azure resources:

- **Azure Container Registry**: A private registry for storing Docker images. You can create a new Azure container registry in the Azure portal. Please refer to the official documentation here: `https://docs.microsoft.com/en-us/azure/container-registry/container-registry-get-started-portal`.

- **Azure Web App for Containers**: A service for hosting containerized web applications. It provides a quick and easy way to build, deploy, and scale enterprise-grade web, mobile, and API apps on any platform. You can create a new Azure Web App for Containers instance in the Azure portal. Please refer to the official documentation here: `https://learn.microsoft.com/en-us/azure/app-service/quickstart-custom-container?tabs=dotnet&pivots=container-linux-azure-portal`. Please select **Docker Container** for the **Publish** option, and select **Linux** for the **Operating System** option when creating the Web App for Containers instance because we are using Linux containers in this chapter.

Note that you can also choose Azure App Service to host your application without containerization. Azure App Service supports many programming languages and frameworks, such as .NET, .NET Core, Java, Node.js, Python, and so on. It also supports containers. You can learn more about Azure App Service here: `https://docs.microsoft.com/en-us/azure/app-service/overview`. In this section, we will explore how to deploy applications in containers, so we will use Azure Web App for Containers in the following example.

To better manage these resources, it is recommended to create a resource group to group these resources together. You can create a new resource group in the Azure portal or create a new resource group when you create a new resource. Here is key information on the resources we need to prepare for the pipelines:

Resource group

> name: `devops-lab`

Container registry

> name: `devopscrlab`

Web App for Containers instance:

> name: `azure-pipeline-demo`

You can use either the Azure portal or Azure CLI to create these resources. Defining the scripts to create the resources in code is a good practice, which is called **infrastructure as code** (**IaC**). However, we will not cover IaC here because it is out of the scope of this book. You can learn more about IaC here: `https://learn.microsoft.com/en-us/devops/deliver/what-is-infrastructure-as-code`.

Next, let us create an Azure DevOps project.

Creating an Azure DevOps project

As the official documentation provides detailed instructions on how to create an Azure DevOps project, we will not cover the details here. Please refer to the official documentation here: `https://learn.microsoft.com/en-us/azure/devops/pipelines/get-started/pipelines-sign-up?view=azure-devops`.

You need to follow the official documentation to create an Azure DevOps account. You can sign up with a Microsoft account or a GitHub account. Once you create an Azure DevOps account, you can create a new organization. An organization is a container for projects and teams. You can create multiple organizations with one Azure DevOps account.

Next, you can create a new project in Azure DevOps. Click the **New project** button on the home page, then follow the instructions to create a new project as follows:

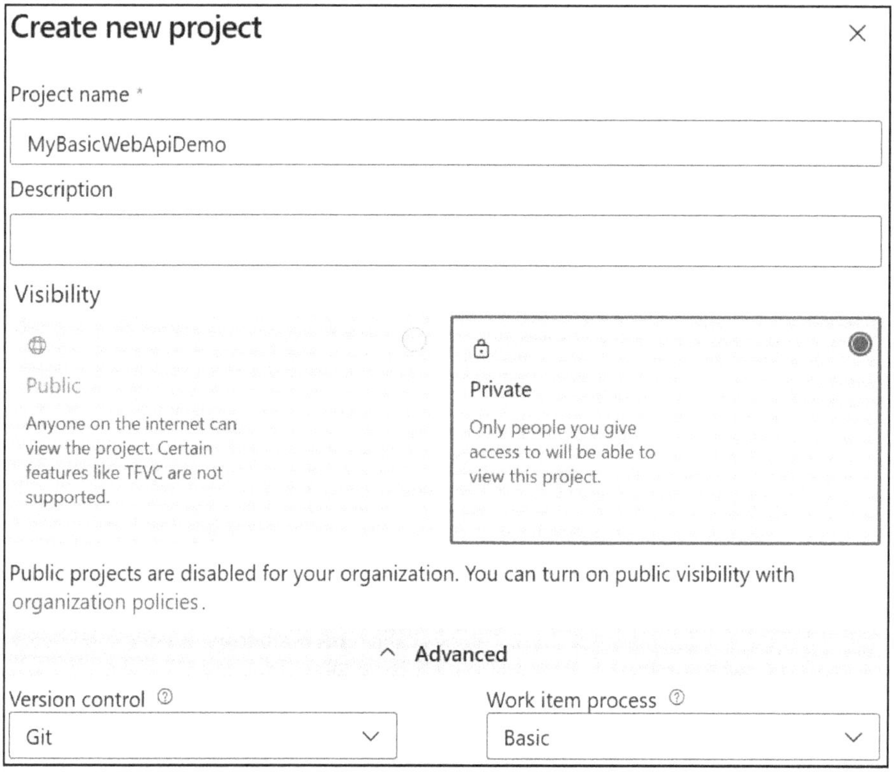

Figure 14.5 – Creating a new project in Azure DevOps

We will create these pipelines in the project in the next sections:

- **Pull request build pipeline**: A pipeline for building the application and running tests when a pull request is created. This pipeline will be triggered when a pull request is created in the GitHub repository. If the build fails or the tests fail, the pull request cannot be merged.

- **Docker build pipeline**: A pipeline for building the application into a Docker image and publishing the image to ACR. This pipeline will be triggered when a pull request is merged into the main branch.

- **Release pipeline**: A pipeline for deploying the application to Azure Container Apps. This pipeline can be triggered when a new image is published to ACR, or it can be triggered manually.

Creating a pull request pipeline

In this section, we will create a pull request build pipeline. Follow the steps:

1. Navigate to the Azure DevOps portal, click the **Pipelines** tab on the left side, and then click the **Create Pipeline** button. You will see a page like this:

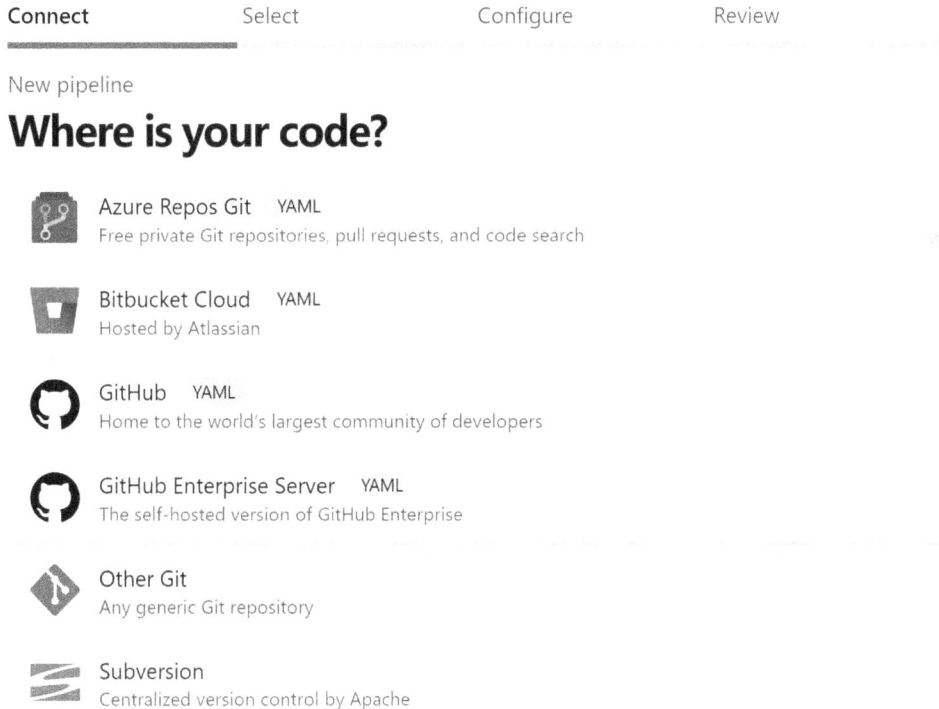

Connect   Select   Configure   Review

New pipeline

Where is your code?

Azure Repos Git   YAML
Free private Git repositories, pull requests, and code search

Bitbucket Cloud   YAML
Hosted by Atlassian

GitHub   YAML
Home to the world's largest community of developers

GitHub Enterprise Server   YAML
The self-hosted version of GitHub Enterprise

Other Git
Any generic Git repository

Subversion
Centralized version control by Apache

Figure 14.6 – Creating a new pipeline in Azure DevOps

2. Azure DevOps Pipelines supports various sources, such as Azure Repos, GitHub, Bitbucket, and Subversion. We already have a GitHub repository, so we will use GitHub as the source. Click the **GitHub** button, then follow the instructions to authorize Azure DevOps to access your GitHub account. Once you have authorized Azure DevOps to access your GitHub account, you will see a list of repositories in your GitHub account. Select the repository we created earlier; you will be navigated to GitHub and see a page where you can install Azure Pipelines to the repository. Click the **Approve and install** button to install Azure Pipelines to the repository. Then, you will be navigated back to Azure DevOps. You will see a page like this:

✓ Connect   ✓ Select   **Configure**   Review

New pipeline

Configure your pipeline

Docker
docker   Build a Docker image

Docker
docker   Build and push an image to Azure Container Registry

Deploy to Azure Kubernetes Service
Build and push image to Azure Container Registry; Deploy to Azure Kubernetes Service

ASP.NET
Build and test ASP.NET projects.

ASP.NET Core (.NET Framework)
Build and test ASP.NET Core projects targeting the full .NET Framework.

.NET Desktop
Build and run tests for .NET Desktop or Windows classic desktop solutions.

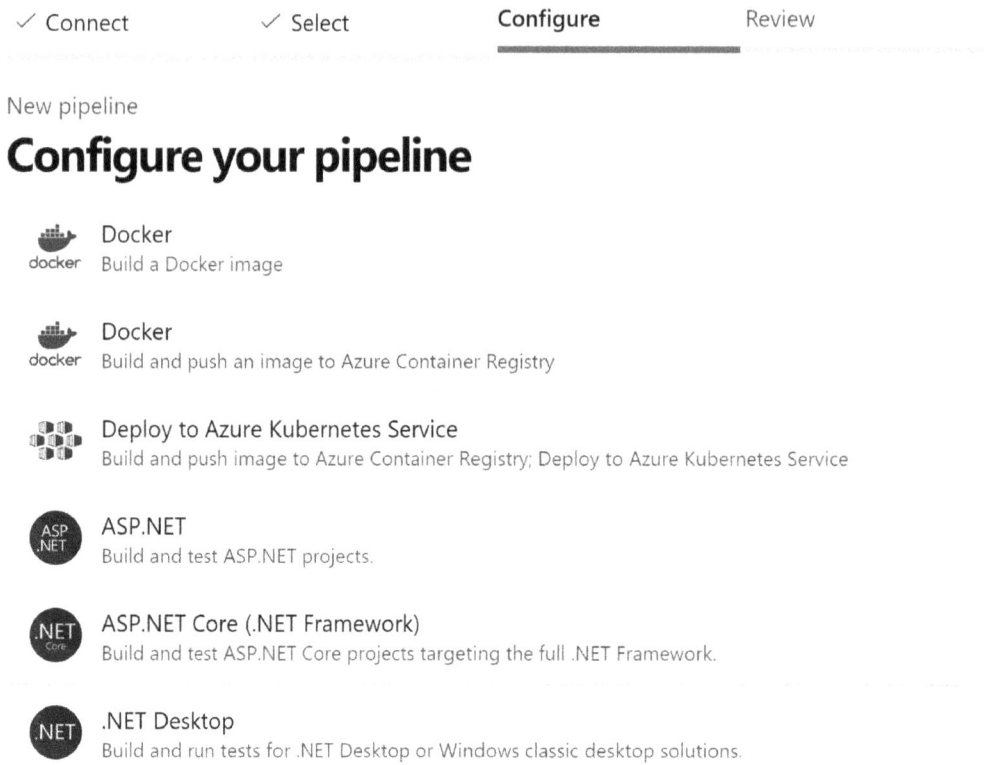

Figure 14.7 – Configuring the pipeline

3. Azure DevOps Pipelines can provide various templates for your project. In this case, you can choose the **ASP.NET** template to start with. Azure DevOps Pipelines can automatically detect the source code and generate a basic pipeline for you. The default pipeline is a YAML file, which may look like this:

```
trigger:
- main

pool:
  vmImage: 'windows-latest'

variables:
  solution: '**/*.sln'
  buildPlatform: 'Any CPU'
  buildConfiguration: 'Release'
```

```
steps:
- task: NuGetToolInstaller@1

- task: NuGetCommand@2
  inputs:
    restoreSolution: '$(solution)'

- task: VSBuild@1
  inputs:
    solution: '$(solution)'
    msbuildArgs: '/p:DeployOnBuild=true
/p:WebPublishMethod=Package /p:PackageAsSingleFile=true
/p:SkipInvalidConfigurations=true /p:PackageLocation="$(build.
artifactStagingDirectory)"'
    platform: '$(buildPlatform)'
    configuration: '$(buildConfiguration)'

- task: VSTest@2
  inputs:
    platform: '$(buildPlatform)'
    configuration: '$(buildConfiguration)'
```

4. The default pipeline is targeting Windows. We will run the application in a Linux container, so we need to make some changes to the pipeline. Delete the default content and we will start from scratch.

5. First, rename the pipeline to `pr-build-pipeline`. You can rename the pipeline by clicking the `.yml` filename:

New pipeline

Review your pipeline YAML

 yanxiaodi/MyAzurePipelinesDemo / **pr-build-pipeline.yml** *

Figure 14.8 – Renaming the pipeline

6. Then, we need to update the trigger to make the pipeline run when a pull request is opened or updated for one of the target branches. Use the `pr` keyword to indicate that the pipeline will be triggered by a pull request for the `main` branch:

```
pr:
  branches:
    include:
    - main
```

7. Next, set `pool` to use the `ubuntu-latest` image. The `ubuntu-latest` image is a Linux image, which is smaller than a Windows image. Also, the application is targeting Linux containers, so it is better to use a Linux image:

```
pool:
  vmImage: 'ubuntu-latest'
```

8. Next, create some variables for the solution path, build configuration, and so on:

```
variables:
  solution: '**/*.sln'
  buildConfiguration: 'Release'
```

9. Then, we need to add some tasks. Add a `steps:` section, then add the following tasks:

```
steps:

- task: UseDotNet@2
  displayName: 'use dotnet cli'
  inputs:
    packageType: 'sdk'
    version: '8.0.x'
    includePreviewVersions: true
```

The UseDotNet@2 task is used to install the .NET SDK so that we can use the .NET CLI to execute commands. In this case, we are installing the .NET 8.0 SDK. The `includePreviewVersions` option is used to include preview versions of the .NET SDK. We need to use the preview version because the .NET 8.0 SDK is still in preview at the time of writing this book. If you are reading this book after the .NET 8.0 SDK is released, you can remove the `includePreviewVersions` option.

The online pipeline editor provides IntelliSense for the YAML file like this:

Figure 14.9 – IntelliSense for YAML file

10. You can click the **Settings** link above the task to configure the task in a dialog box:

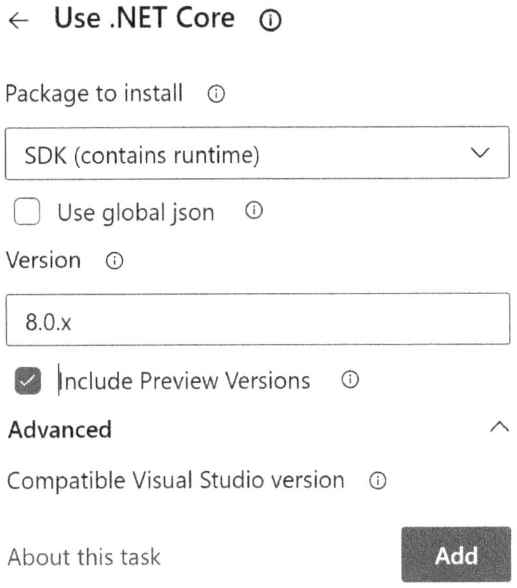

← **Use .NET Core** ⓘ

Package to install   ⓘ

SDK (contains runtime)   ∨

☐ Use global json   ⓘ

Version ⓘ

8.0.x

☑ |nclude Preview Versions   ⓘ

Advanced   ∧

Compatible Visual Studio version   ⓘ

About this task   **Add**

Figure 14.10 – Configuring the task in a dialog box

11. Next, add a DotNetCoreCLI@2 task to build the application:

```
- task: DotNetCoreCLI@2
  displayName: 'dotnet build'
  inputs:
    command: 'build'
    arguments: '--configuration $(buildConfiguration)
    projects: '$(solution)'
```

The DotNetCoreCLI@2 task is used to run dotnet CLI commands. In this case, we are running the dotnet build command to build the application. The --configuration option is used to specify the build configuration. The --runtime option is used to specify the target runtime. In this case, we are targeting the Linux runtime. The projects option is used to specify the path to the .csproj file or solution file. In this case, we are using the $(solution) variable we defined earlier.

12. Next, add tasks to run unit tests and integration tests:

```
- task: DotNetCoreCLI@2
  displayName: 'dotnet test - unit tests'
  inputs:
    command: 'test'
    arguments: '--configuration $(buildConfiguration) --no-build
--no-restore --logger trx --collect "Code coverage"'
    projects: '**/*.UnitTests.csproj'

- task: DotNetCoreCLI@2
  displayName: 'dotnet test - integration tests'
  inputs:
    command: 'test'
    arguments: '--configuration $(buildConfiguration) --no-build
--no-restore --logger trx --collect "Code coverage"'
    projects: '**/*.IntegrationTests.csproj'
```

In the preceding tasks, we are running the `dotnet test` command to run unit tests and integration tests. The `--no-build` option is used to skip building the application. The `--no-restore` option is used to skip restoring the NuGet packages because we have already restored the packages and built the application in the previous tasks. The other options are used to specify the logger and collect code coverage.

13. Click the **Save and run** button to commit the changes and run the pipeline. You will see the pipeline is running. After a while, the pipeline will be completed:

Figure 14.11 – The pipeline is successful

14. Click the **Tests** tab and you will then see the test results:

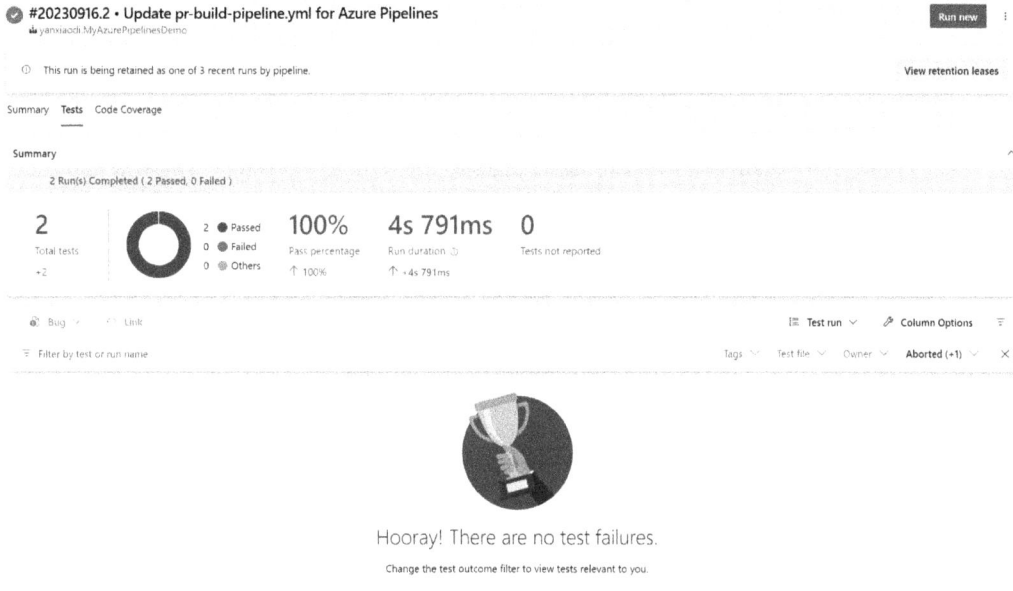

Figure 14.12 – The test results

15. You can also check the code coverage as follows:

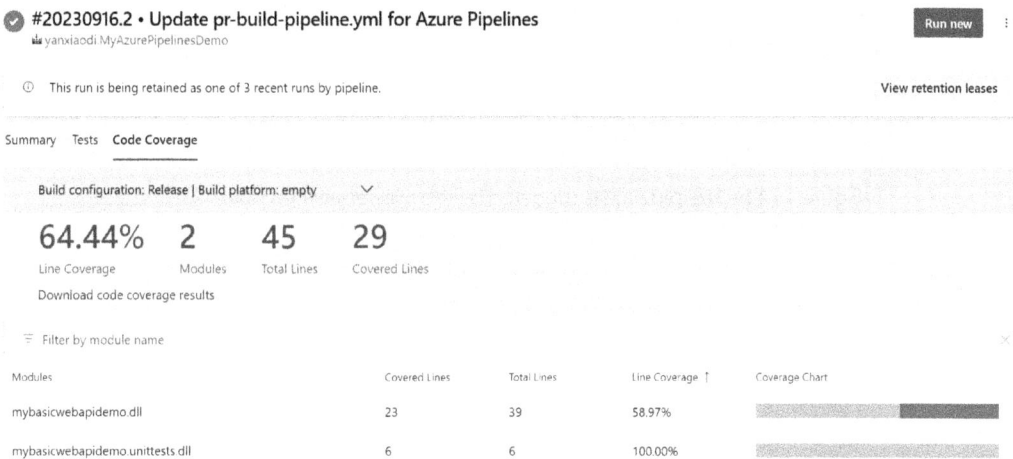

Figure 14.13 – The code coverage

16. We just manually triggered the first run of the pipeline. Next, let us create a pull request and see how the pipeline is triggered.

17. Create a new branch and make some changes to the source code. For example, we can return 6 items instead of 5 items in `WeatherForecastController`:

```
return Enumerable.Range(1, 6).Select(index => new
WeatherForecast
// Omitted for brevity
```

18. Then, commit the changes and push the changes to the remote repository. You will find the pipeline runs for the new branch as well. That is because YAML pipelines are enabled by default for all branches. You can disable this feature by using the `trigger none` option. Add the following line to the beginning of the YAML file:

    trigger: none

    Then, this pipeline will not be triggered automatically for new branches but will be triggered by pull requests.

19. Next, create a new pull request in the GitHub repository. You will see the pipeline is triggered automatically and then fails:

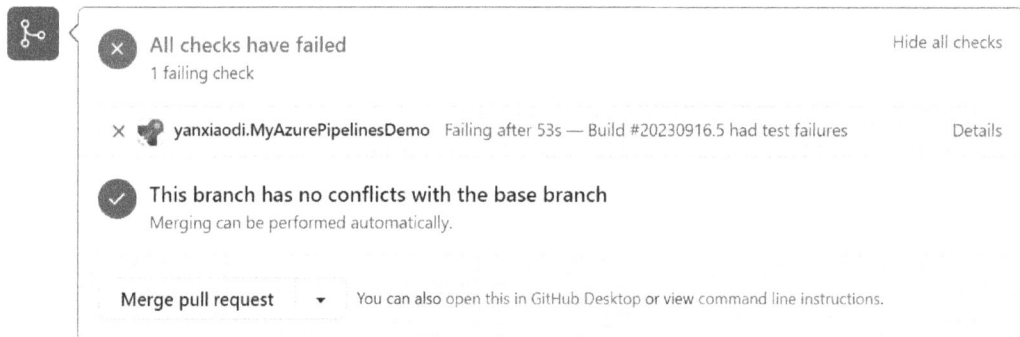

Figure 14.14 – The pipeline is triggered by the pull request but fails

20. In this case, we cannot merge the pull request because the build pipeline failed. You can check the logs to see what is wrong with the pipeline. In this case, the pipeline failed because the unit tests failed:

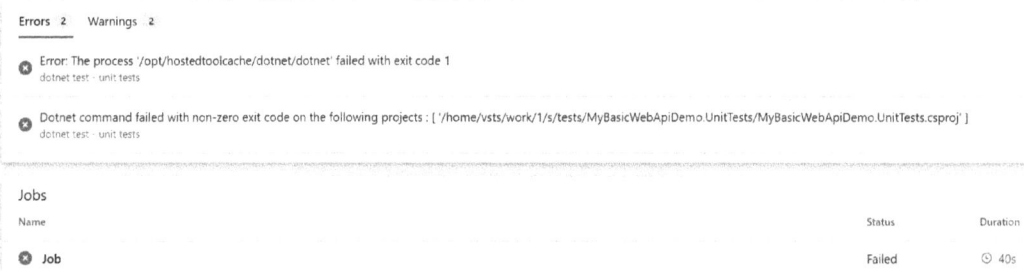

Figure 14.15 – The unit tests failed

21. Click the **Tests** tab and you will then see the test results:

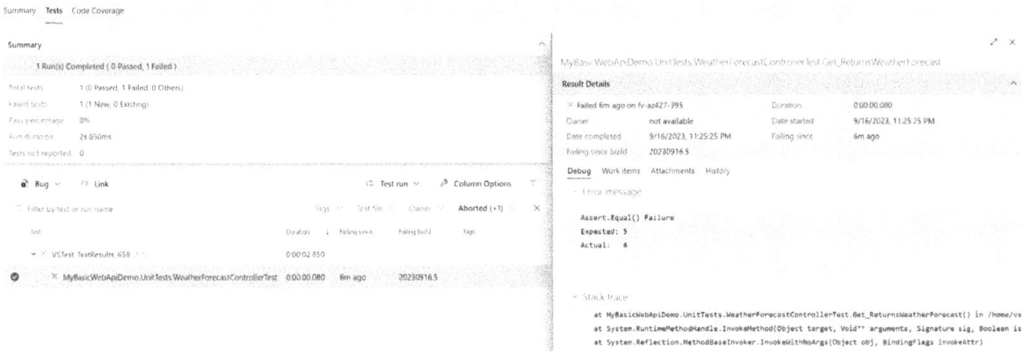

Figure 14.16 – The unit tests results

The details of the failed test case can help us to identify the problem. Next, you can fix the error and push the changes to the remote repository. Then, the pipeline will be triggered again. If the pipeline is successful, you can merge the pull request.

The pull request build pipeline is used to ensure that code changes do not break the application. It is important to run the tests before merging the pull request. Next, let us create a CI build pipeline to build the Docker image and publish it to ACR.

Publishing the Docker image to ACR

In the previous section, we created a pull request build pipeline to validate code changes in pull requests. The pipeline will be triggered when a pull request is created or updated. Once the pull request is merged to the `main` branch, we can automatically build the Docker image and publish it to ACR. To do that, we will create a CI build pipeline following these steps

1. Create a new pipeline following the same steps as we did in the previous section. In the configure step, we can choose the **Starter pipeline** template to start from scratch or choose the **Existing Azure Pipelines YAML file** template and the `pr-build-pipeline.yml` file we created in the previous section:

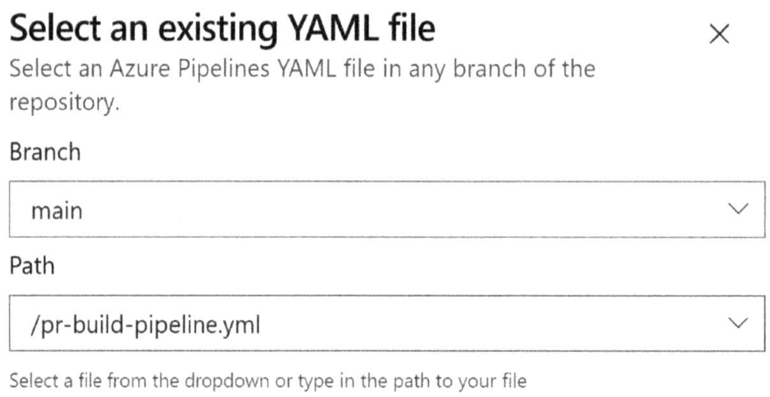

Select an existing YAML file   ✕

Select an Azure Pipelines YAML file in any branch of the repository.

Branch

```
main                                                    ⌄
```

Path

```
/pr-build-pipeline.yml                                  ⌄
```

Select a file from the dropdown or type in the path to your file

Figure 14.17 – Choosing an existing YAML file to start with

2. A better way is to choose the **Docker - Build and push an image to Azure Container Registry** template, which is exactly what we need. You will be prompted to configure your Azure subscription, ACR, and so on:

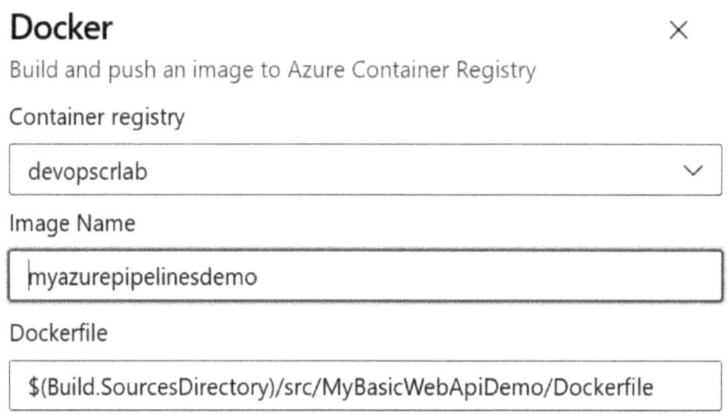

Docker   ✕

Build and push an image to Azure Container Registry

Container registry

```
devopscrlab                                             ⌄
```

Image Name

```
myazurepipelinesdemo
```

Dockerfile

```
$(Build.SourcesDirectory)/src/MyBasicWebApiDemo/Dockerfile
```

Figure 14.18 – Configuring ACR, Dockerfile, and image name

3. Azure Pipelines will automatically detect the Dockerfile and generate a pipeline for you. The default pipeline may look like this:

```
trigger:
- main

resources:
- repo: self
```

```
variables:
  # Container registry service connection established during
pipeline creation
  dockerRegistryServiceConnection: 'deb345e0-7bdd-4420-ba08-
538785d525cd'
  imageRepository: 'myazurepipelinesdemo'
  containerRegistry: 'devopscrlab.azurecr.io'
  dockerfilePath: '$(Build.SourcesDirectory)/src/
MyBasicWebApiDemo/Dockerfile'
  tag: '$(Build.BuildId)'

  # Agent VM image name
  vmImageName: 'ubuntu-latest'

stages:
- stage: Build
  displayName: Build and push stage
  jobs:
  - job: Build
    displayName: Build
    pool:
      vmImage: $(vmImageName)
    steps:
    - task: Docker@2
      displayName: Build and push an image to container registry
      inputs:
        command: buildAndPush
        repository: $(imageRepository)
        dockerfile: $(dockerfilePath)
        containerRegistry: $(dockerRegistryServiceConnection)
        tags: |
          $(tag)
          latest
```

Azure Pipelines has recognized the path of the Dockerfile in the preceding pipeline. Additionally, several variables must be configured, such as the image name and container registry. The tag variable is used to tag the image with the build ID. Note that we added a latest tag to the image. The latest tag is used to indicate the latest version of the image.

> **Azure Pipelines predefined variables**
>
> Azure Pipelines provides many predefined variables that can be used in a pipeline. For example, the `$(Build.SourcesDirectory)` variable is used to obtain the local path on the agent where the source code files have been downloaded. You can find all the predefined variables here: `https://learn.microsoft.com/en-us/azure/devops/pipelines/build/variables?view=azure-devops&tabs=yaml`.

The difference from a pull request build pipeline is that we have **stages** and **jobs** in the pipeline. *Figure 14.19* shows the structure of a pipeline:

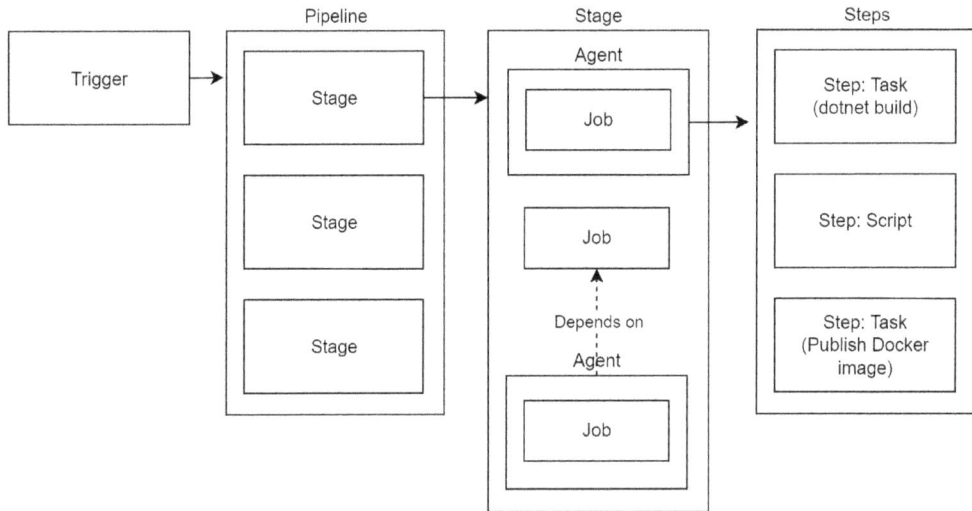

Figure 14.19 – The structure of a pipeline

The components in *Figure 14.19* are explained as follows:

- A trigger is used to specify when the pipeline should run. It can be a schedule, a pull request, a commit to a specific branch, or a manual trigger.

- A pipeline contains one or more stages. A stage is used to organize jobs. In the pull request build pipeline, the stage is omitted. You can have multiple stages in one pipeline for different purposes. For example, you can have a build stage, a test stage, and a deploy stage. The stage can be also used to set boundaries for security and approvals.

- Each stage contains one or more jobs. A job is a container of steps. A job can run on one agent or without an agent. For example, you can have a job that runs on a Windows agent, and another job that runs on a Linux agent.

- Each job contains one or more steps. A step is the smallest building block in a pipeline. A step is normally a task or a script. Azure Pipelines provides many built-in tasks, such as the `Docker@2` task we are using in this pipeline. A built-in task is a predefined packaged script that performs an action. You can also write your own custom tasks. A task can be a command-line tool, a script, or a compiled program. You can find all built-in tasks here: `https://learn.microsoft.com/en-us/azure/devops/pipelines/tasks/reference/?view=azure-pipelines`.

In the preceding pipeline, we have one stage, one job, and one step. The step is the `Docker@2` task. The `Docker@2` task is used to build and push a Docker image to a container registry. Using this task simplifies the process of building and pushing a Docker image so that we do not need to write `docker build` commands manually.

4. Rename this pipeline to `docker-build-pipeline`. Similarly to `pr-build-pipeline`, we need to exclude the `docker-build-pipeline` pipeline from running for new branches and pull requests. Add the following line to the beginning of the YAML file:

```
pr: none
```

So, when we create a new branch or a new pull request, the `docker-build-pipeline` will not be triggered automatically. We will trigger the pipeline manually or by merging a pull request.

5. Click the **Save** button to commit the changes. Next, we need to test if the pipeline can be triggered by merging a pull request. Create a new branch and make some changes to the source code. Then, create a pull request. When you merge the pull request, you will see the `docker-build-pipeline` is triggered automatically.

If everything is fine, you will see the `docker-build-pipeline` pipeline is successful and the Docker image is published to ACR:

Figure 14.20 – The Docker image is published to ACR

A CI pipeline is a good practice to ensure that code changes can be built successfully without breaking the application. However, some changes may not need running tests or rebuilding the Docker image. For example, if you only change the documentation, (for example, the README file), you do not need to run the tests or rebuild the Docker image. In this case, you have a few options:

6. You can exclude some files from the pipeline. For example, you can exclude the README.md file from the pipeline. You can use the paths option to specify paths to include or exclude, as follows:

```
trigger:
  branches:
    include:
    - main
  paths:
    exclude:
    - README.md
    - .gitignore
    - .dockerignore
    - *.yml
```

7. You can skip CI pipelines for some commits. Use [skip ci] in the commit message to skip CI pipelines. For example, you can use [skip ci] update README in the commit message to skip CI pipelines for this commit. Some variations include [ci skip], [skip azurepipelines], [skip azpipelines], [skip azp], and so on.

You may encounter some errors when creating pipelines. Here are some troubleshooting tips:

*   Check the logs carefully. The logs can help you to identify the problem.

*   Check the variables. Make sure the variables are correct.

*   Check the permissions. Make sure the service connection has the correct permissions.

*   Check the YAML syntax. Make sure the YAML file is valid. YAML files use indentation to indicate the structure. Make sure the indentation is correct.

*   Check the pipeline structure. Make sure the pipeline structure is correct. For example, make sure the steps section is under the jobs section and the jobs section is under the stages section.

*   Check the pipeline triggers. Make sure the pipeline is triggered by the correct event. You can use branches, paths, include, and exclude to specify branches and paths to trigger the pipeline.

*   Note that if you edit the pipeline YAML file in the online editor, you will commit the changes to the main branch directly. Your local feature branch will not be updated automatically. When you push a change to your feature branch, whether the pipeline should be triggered depends on the settings in the YAML file in your feature branch, not the main branch. So, make sure your feature branch keeps synchronized with the main branch.

*   If you use VS to create the Dockerfile, double-check the paths in the Dockerfile. Sometimes, VS cannot detect the correct paths if you use a custom solution structure.

We have now pushed the Docker image to ACR. Next, let's create a release pipeline to deploy the application to Azure Web App for Containers.

Deploying the application to Azure Web App for Containers

In the previous section, we created a CI build pipeline to build the Docker image and publish it to ACR. In this section, we will create a release pipeline to pull the Docker image from ACR and deploy it to Azure Web App for Containers.

Follow the steps in the previous section to create a new pipeline. When we configure the pipeline, choose the **Starter pipeline** template because we will start from scratch. The default starter pipeline has two scripts as examples. Delete the scripts, and it should look like this:

```
trigger: none

pr: none

pool:
  vmImage: ubuntu-latest
```

We can disable the triggers for the pipeline because we will manually run the pipeline when we need to deploy the application. Rename the pipeline as `release-pipeline`.

Next, add some variables to the pipeline:

```
variables:
  containerRegistry: 'devopscrlab.azurecr.io'
  imageRepository: 'myazurepipelinesdemo'
  tag: 'latest'
```

Next, we need to configure the username and password to authenticate the pipeline to pull the Docker image from ACR. You can find the username and password of your ACR instance in the Azure portal. Click the **Access keys** button on the left side, and you will then see the username and password.

As the password is a secret, we cannot use it in the YAML file directly, otherwise the password will be exposed in the GitHub repository. Azure Pipelines provides a way to store secrets in the pipeline. Click the **Variables** button in the top-right corner, then click the **New variable** button to add a new variable:

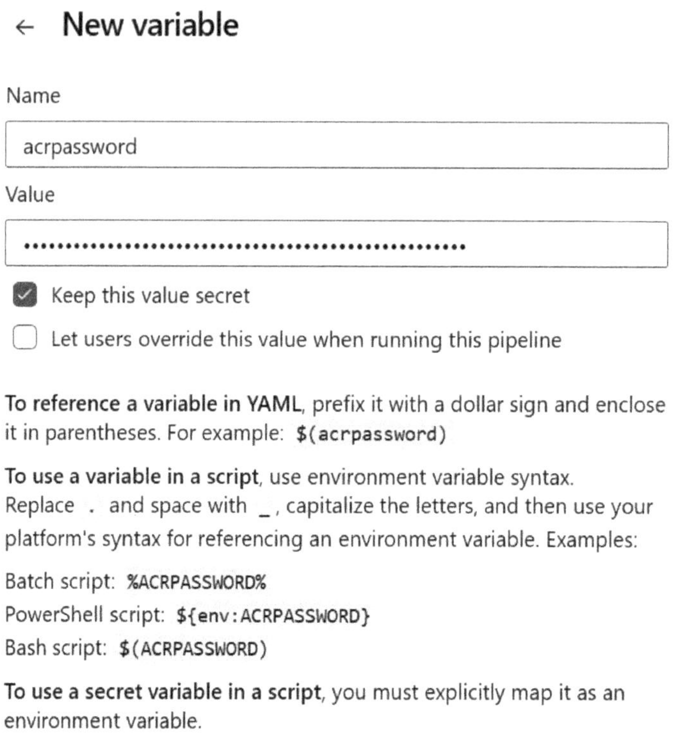

← New variable

Name

```
acrpassword
```

Value

```
••••••••••••••••••••••••••••••••••••••••••••••••••
```

☑ Keep this value secret

☐ Let users override this value when running this pipeline

To reference a variable in YAML, prefix it with a dollar sign and enclose it in parentheses. For example: `$(acrpassword)`

To use a variable in a script, use environment variable syntax. Replace `.` and space with `_`, capitalize the letters, and then use your platform's syntax for referencing an environment variable. Examples:

Batch script: `%ACRPASSWORD%`
PowerShell script: `${env:ACRPASSWORD}`
Bash script: `$(ACRPASSWORD)`

To use a secret variable in a script, you must explicitly map it as an environment variable.

Figure 14.21 – Adding a new variable to store the password

Check the **Keep this value secret** checkbox to make the variable secret. Then, click the **Add** button to add the variable. So, we can use the `$(acrpassword)` variable to refer to the password.

Next, add a task to set up the Azure App Service settings. Choose the **Azure App Service Settings** task from the task assistant. We need to add the credentials to authenticate the Azure Web App to pull the Docker image from ACR. Azure Pipelines will prompt you to configure the task. Note that the **App settings** field is a JSON string. The tasks should look like this:

```
- task: AzureAppServiceSettings@1
  displayName: Update settings
  inputs:
    azureSubscription: '<Your Azure subscription>(<guid>)'
    appName: 'azure-pipeline-demo'
    resourceGroupName: 'devops-lab'
    appSettings: |
      [
        {
          "name": "DOCKER_REGISTRY_SERVER_URL",
```

```
            "value": "$(containerRegistry)",
            "slotSetting": false
          },
          {
            "name": "DOCKER_REGISTRY_SERVER_USERNAME",
            "value": "devopscrlab",
            "slotSetting": false
          },
          {
            "name": "DOCKER_REGISTRY_SERVER_PASSWORD",
            "value": "$(acrpassword)",
            "slotSetting": false
          }
        ]
```

In the `appSettings` field, we use the `$(acrpassword)` variable to refer to the password we created earlier.

Next, click the **Show assistant** button in the top-right corner to open the assistant. The assistant helps us to use the built-in tasks easily. Choose the **Azure Web App for Containers** task that is used to deploy the Docker image from ACR to Azure Web App for Containers. Azure Pipelines will prompt you to configure the task. Configure the task as follows:

```
- task: AzureWebAppContainer@1
  displayName: Deploy to Azure Web App for Container
  inputs:
    azureSubscription: '<Your Azure subscription>'
    appName: 'azure-pipeline-demo'
    containers: '$(containerRegistry)/$(imageRepository):$(tag)'
    containerCommand: 'dotnet MyBasicWebApiDemo.dll'
```

Alternatively, you can use the **Azure App Service Deploy** task from the assistant. This task is used to deploy the application to Azure Web App that supports either native deployment or container deployment. You need to configure the Azure subscription, App Service type, and so on. When you choose the **Web App for Containers (Linux)** option for **App Service type**, you will be prompted to configure the ACR name, Docker image name, tag, and so on. The task should look like this:

```
- task: AzureRmWebAppDeployment@4
  displayName: Deploy to Web App for Container
  inputs:
    ConnectionType: 'AzureRM'
    azureSubscription: '<Your Azure subscription>'
    appType: 'webAppContainer'
    WebAppName: 'azure-pipeline-demo'
```

```
DockerNamespace: 'devopscrlab.azurecr.io'
DockerRepository: 'myazurepipelinesdemo'
DockerImageTag: 'latest'
StartupCommand: 'dotnet MyBasicWebApiDemo.dll'
```

Note that we also need to specify a `StartupCommand` value. In this case, we use `dotnet MyBasicWebApiDemo.dll` to start the application.

Now, we can manually trigger the pipeline to deploy the application. If everything is fine, you will see the deployment is successful.

Check the configuration of the Azure Web App. You will see the **Application settings** are updated:

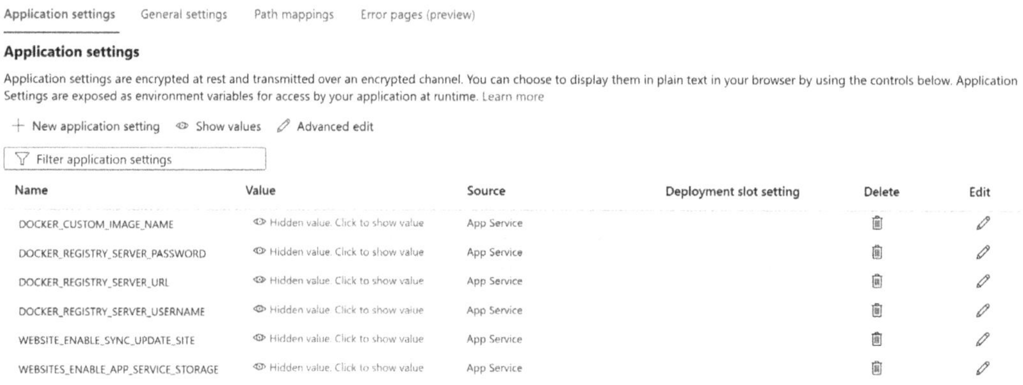

Figure 14.22 – The application settings are updated

Navigate to the Azure portal and check the details of the Azure Web App we created earlier. You can find the URL of the Azure Web App, such as `azure-pipeline-demo.azurewebsites.net`. Open the URL in the browser and check the controller endpoint, such as `https://azure-pipeline-demo.azurewebsites.net/WeatherForecast`. You will see the response from the application.

So far, we have created three pipelines:

- **Pull request build pipeline**: This pipeline is to validate the code changes in pull requests. It will be triggered when a pull request is created or updated. If the build fails or the tests fail, the pull request cannot be merged. This pipeline does not produce any artifacts.

- **Docker build pipeline**: This pipeline is to build the Docker image and publish it to ACR. It will be triggered when a pull request is merged into the `main` branch. This pipeline produces a Docker image as the artifact.

- **Release pipeline**: This pipeline is to pull the Docker image from ACR and deploy it to Azure Web App for Containers. This pipeline can be triggered manually or automatically when a new Docker image is published to ACR. This pipeline does not produce any artifacts.

Configuring settings and secrets

In the previous section, we created a release pipeline to deploy the application to Azure Web App for Containers. We may also need to deploy the application to other environments, such as staging. These different environments may have different settings, such as database connection strings, API keys, and so on. So, how can we configure the settings for different environments?

There are various ways to achieve this. A simple way is to configure variables for different environments. You can define variables in each pipeline directly. Azure Pipelines also provides a **Library** to manage variables. You can group variables into a variable group and then use the variable group in the pipeline.

Storing confidential information securely is essential to any organization. To ensure the safety of your secrets, you can use a variety of key vaults, such as Azure Key Vault and AWS Secrets Manager. Azure Pipelines offers a Key Vault task to fetch secrets from the vault. With the Azure Key Vault task, you can easily retrieve secrets from the vault and use them in your pipeline. To learn more about the Key Vault task, please visit `https://learn.microsoft.com/en-us/azure/devops/pipelines/release/key-vault-in-own-project`.

This chapter is not intended to cover all the details of Azure Pipelines, but you should now have a basic understanding of Azure Pipelines. Using CI/CD pipelines can help you to automate the build, test, and deployment process, which eliminates manual work and reduces the risk of human errors. It is more and more important to use CI/CD pipelines in modern software development. Every developer should learn how to use them.

In the next section, we will explore GitHub Actions, which is another popular CI/CD tool.

GitHub Actions

In the previous section, we explored Azure Pipelines. Next, let us explore GitHub Actions. GitHub Actions is a CI/CD tool provided by GitHub. It is quite similar to Azure Pipelines. In this section, we will use GitHub Actions to build and test the application and push the Docker image to ACR.

Preparing the project

To demonstrate how to use GitHub Actions, we will use the same source code as we used in the previous section. You can download the source code from the GitHub repository here: `https://github.com/PacktPublishing/Web-API-Development-with-ASP.NET-Core-8/tree/main/samples/chapter14/MyBasicWebApiDemo`. Create a new GitHub repository and push the source code to the repository. The directory structure of the source code is as follows:

```
MyGitHubActionsDemo
    ├── src
    │    └──MyBasicWebApiDemo
    ├── tests
    │    ├──MyBasicWebApiDemo.UnitTests
    │    └──MyBasicWebApiDemo.IntegrationTests
```

```
├── MyBasicWebApiDemo.sln
├── README.md
├── LICENSE
└── .gitignore
```

If you use a different directory structure, please update the paths in the pipeline accordingly in the following sections. We will use this repository to demonstrate how to configure GitHub Actions.

Creating GitHub Actions

We will reuse the Azure resources we created in the previous section. If you have not created the Azure resources, please refer to the *Creating Azure resources* section to create the resources.

On the GitHub repository page, click the **Actions** tab, and you can see many templates for different programming languages and frameworks:

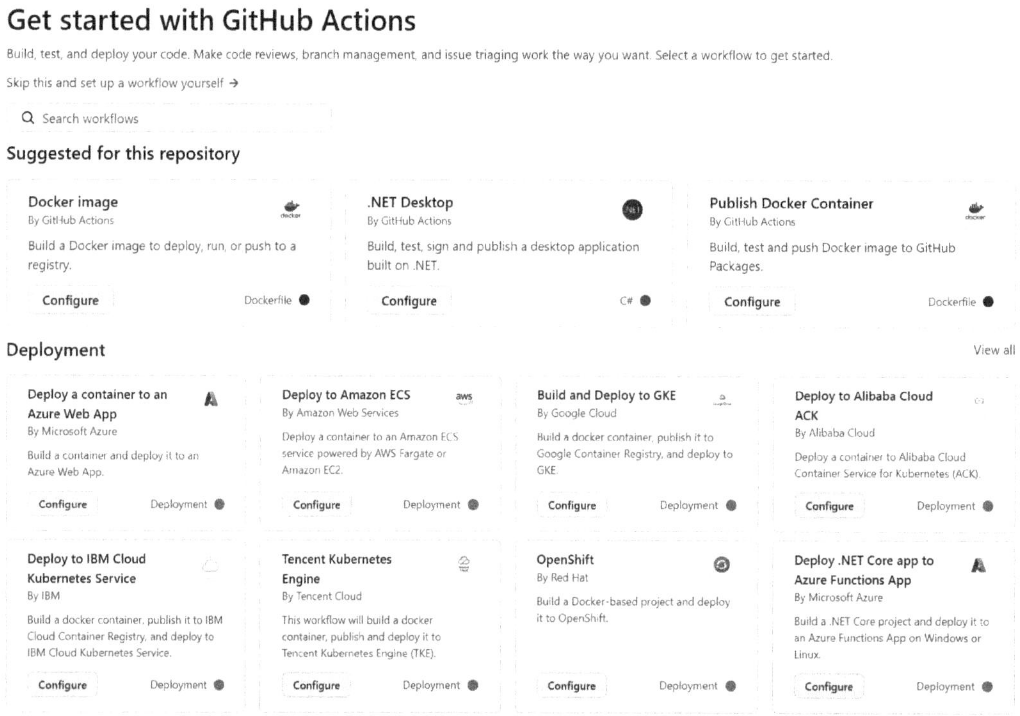

Figure 14.23 – Choosing a template for GitHub Actions

In the **Continuous integration** section, you can find the **.NET** template. Click the **Configure** button to create a new workflow. The workflow is a YAML file that defines the CI/CD pipeline. The default workflow may look like this:

```yaml
# This workflow will build a .NET project
# For more information see: https://docs.github.com/en/actions/
automating-builds-and-tests/building-and-testing-net


name: .NET

on:
  push:
    branches: [ "main" ]
  pull_request:
    branches: [ "main" ]

jobs:
  build:

    runs-on: ubuntu-latest

    steps:
    - uses: actions/checkout@v3
    - name: Setup .NET
      uses: actions/setup-dotnet@v3
      with:
        dotnet-version: 6.0.x
    - name: Restore dependencies
      run: dotnet restore
    - name: Build
      run: dotnet build --no-restore
    - name: Test
      run: dotnet test --no-build --verbosity normal
```

The syntax of the workflow is quite similar to Azure Pipelines. The workflow is triggered when a pull request is created or updated, or when a commit is pushed to the main branch.

On the right side, you can find the **Marketplace** panel. The Marketplace is similar to the assistants of Azure DevOps Pipelines and provides many built-in actions that you can use in your workflow:

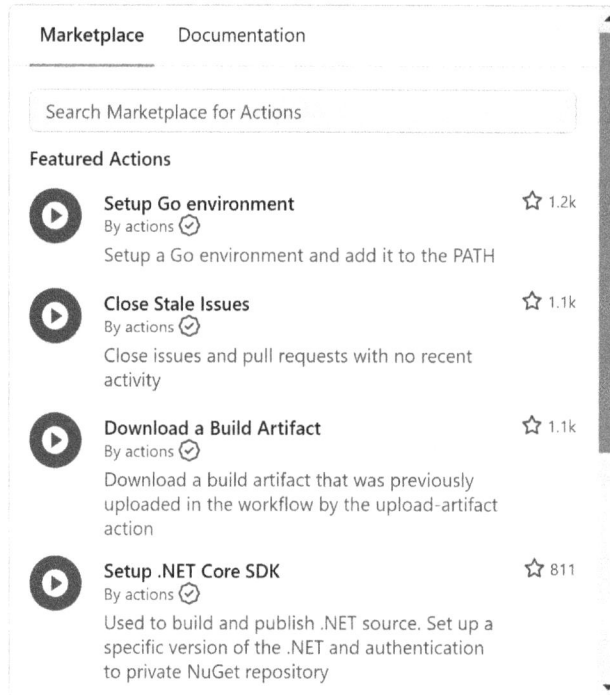

Figure 14.24 – The Marketplace panel

The sample application uses .NET 8. So, we need to update `dotnet-version` to `8.0.x`. Click the **Commit changes** button in the top-right corner. You will see the `dotnet.yml` file is committed to the `/.github/workflows` directory. Make a change to the source code or create a new branch and push the changes to the remote repository. For example, you can change the controller to return six items instead of five items:

```
public IEnumerable<WeatherForecast> Get()
{
    return Enumerable.Range(1, 6).Select(index => new WeatherForecast
    // Omitted for brevity
```

You can also create a pull request. You will see the workflow is triggered automatically and the test is failed:

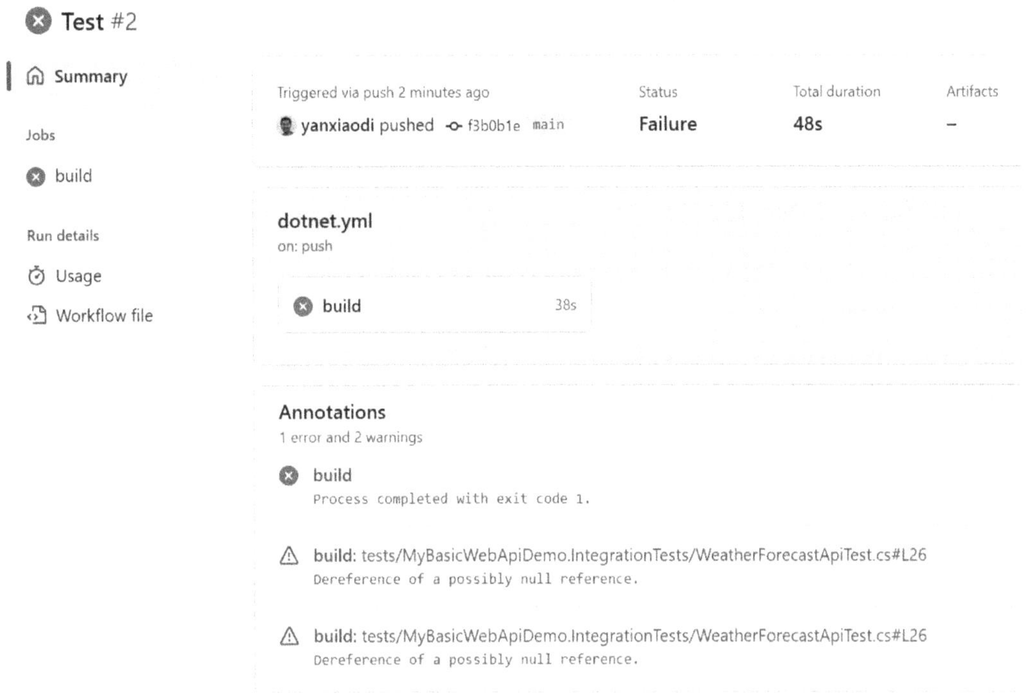

Figure 14.25 – The workflow is triggered automatically and then fails

Click the links in the logs to see details of the code that caused the test failure. It is handy to review code changes in the pull request. If everything is fine, you can merge the pull request.

To clarify that it works for pull request builds, we can rename this file to `pr-build.yml` and update the `on` section to make the workflow run for pull requests:

```
on:
  pull_request:
    branches: [ "main" ]
```

Next, let us build a Docker image and push it to ACR.

Pushing a Docker image to ACR

Create a new YAML file named `docker-build.yml` in the `.github\workflows` folder. Update the content of the file as follows:

```
name: Pull Request build

on:
  push:
    branches: [ "main" ]
```

This workflow is triggered when a commit is pushed to the `main` branch or a pull request is merged into the `main` branch.

To authenticate the action to access ACR, we need to configure the username and password of ACR in GitHub secrets. Find the username and password by clicking the **Access keys** menu of ACR in the Azure portal.

Go to **Settings** in the GitHub repository and click **Secrets and variables** in the **Security** category. Then, click **Actions**, and you will see the secrets and variables page. There are two types of secrets: environment secrets and repository secrets. You can store the username and password in repository secrets directly. To demonstrate how to use environment secrets, we will use environment secrets in this example. Click the **Manage environments** button to create a new environment named **Production**. Then, click the **Add secret** button to add a username and password:

- Name: REGISTRY_USERNAME. Value: The username of ACR.
- Name: REGISTRY_PASSWORD. Value: The password of ACR.

You can change the environment secret names, but make sure you use the same names in the following workflow.

Now, the environment secrets should look like this:

Figure 14.26 – The Actions secrets and variables page

Add the following content to the docker-build.yml file:

```yaml
jobs:
  docker_build_and_push:

    runs-on: ubuntu-latest
    environment: Production
    steps:
    - uses: actions/checkout@v3
    - name: Setup .NET
      uses: actions/setup-dotnet@v3
      with:
        dotnet-version: 8.0.x
    - name: Restore dependencies
      run: dotnet restore
    - name: Build
      run: dotnet build --no-restore
    - name: Login to Azure Container Registry
      uses: azure/docker-login@v1
      with:
        login-server: devopscrlab.azurecr.io
        username: ${{ secrets.REGISTRY_USERNAME }}
        password: ${{ secrets.REGISTRY_PASSWORD }}
    - name: Push to Azure Container Registry
      run: |
        docker build -f ${{ github.workspace }}/src/MyBasicWebApiDemo/
Dockerfile -t devopscrlab.azurecr.io/myazurepipelinesdemo:${{ github.
run_id }} -t devopscrlab.azurecr.io/myazurepipelinesdemo:latest .
        docker push devopscrlab.azurecr.io/myazurepipelinesdemo:${{
github.run_id }}
        docker push devopscrlab.azurecr.io/myazurepipelinesdemo:latest
```

The preceding is similar to the pr-build.yml file, but we made some changes:

- We added an environment section to specify the environment so that we can refer to the environment secrets later.

- We added a new azure/docker-login@v1 step to log in to ACR. In this step, we specified the login server, username, and password using the environment secrets.

- We added a new step to build the Docker image and push it to ACR. In this step, we used the github.run_id variable to tag the image with the run ID. We also tagged the image with latest.

> **GitHub Actions context**
>
> GitHub Actions supports many built-in context variables, such as `github.run_id`, `github.run_number`, `github.sha`, `github.ref`, and so on. You can find all the context variables here: `https://docs.github.com/en/actions/learn-github-actions/contexts`.

Commit the changes and push the changes to the remote repository. Next time you merge a pull request, you will see the workflow is triggered automatically and the Docker image is pushed to ACR:

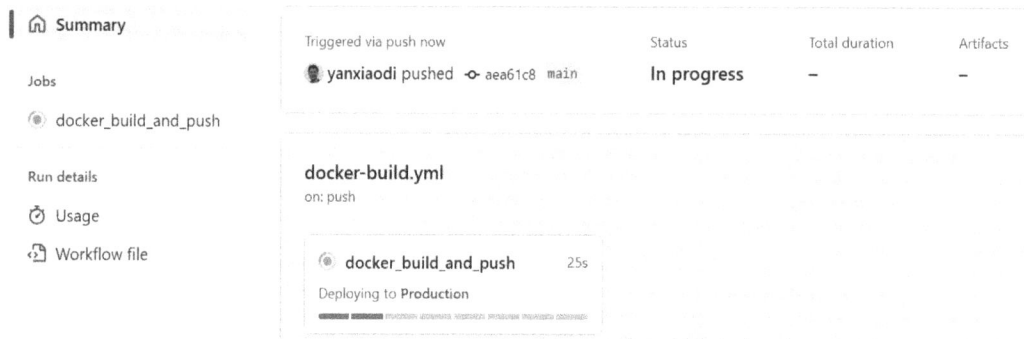

Figure 14.27 – The GitHub Actions workflow is triggered automatically

As there are many similarities between Azure Pipelines and GitHub Actions, we will not cover all the details of GitHub Actions. Maybe it is your turn to create a GitHub Actions workflow to deploy the Docker image to Azure Web App for Containers? You can find more information about GitHub Actions here: `https://docs.github.com/en/actions`.

Summary

In this chapter, we explored the fundamentals of CI/CD. We discussed how to use Docker to containerize ASP.NET web API applications, including how to create a Dockerfile, build a Docker image, and run a Docker container locally. We then looked at Azure DevOps Pipelines, a powerful CI/CD platform from Microsoft, and how to create CI/CD pipelines in YAML files. We covered configuring triggers, using built-in tasks, and using variables. We created three pipelines for builds, Docker image builds, and releases. We also briefly discussed GitHub Actions, another popular CI/CD tool, and created a GitHub Actions workflow to build and test the application, then build the Docker image and push it to ACR. After reading this chapter, you should have a basic understanding of CI/CD and be able to use CI/CD pipelines to automate the build, test, and deployment process.

In the next chapter, we will examine some common practices for building ASP.NET web APIs, including caching, `HttpClient` factory, and so on.

ASP.NET Core Web API Common Practices

We have introduced a lot of concepts in the previous chapters, including the basics of ASP.NET web API, the RESTful style, Entity Framework, unit testing and integration testing, CI/CD, and so on. You should be able to build a simple ASP.NET Core web API application by yourself. However, there are still many things that we need to learn.

You may have heard the phrase *"there is no silver bullet"* before. It means that there are often no simple, universal, or one-size-fits-all solutions to solve all problems. No matter how powerful the technologies, tools, or frameworks are, they are not omnipotent. This is true for ASP.NET Core web API development. However, there are some common practices that can help us build a better ASP. NET Core web API application.

In this chapter, we will summarize the common practices of ASP.NET Core web API development. We will cover the following topics in this chapter:

- Common practices of ASP.NET Core web API development
- Optimizing the performance by implementing caching
- Using HttpClientFactory to manage HttpClient instances

After reading this chapter, you should be able to expand your knowledge of ASP.NET Core web API development and build a better web API application.

Technical requirements

The code example in this chapter can be found at `https://github.com/PacktPublishing/Web-API-Development-with-ASP.NET-Core-8/tree/main/samples/chapter15`.

Common practices of ASP.NET web API development

In this section, we will introduce some common practices of ASP.NET web API development. Of course, we can't cover all the common practices in this book. However, we will try to cover the most important ones.

Using HTTPS instead of HTTP

In the previous chapters, we have used HTTP endpoints for simplicity; however, in the real world, HTTPS should always be used instead of HTTP. HTTPS is a secure version of HTTP, which uses TLS/SSL to encrypt the HTTP traffic, thus preventing data from being intercepted or tampered with by a third party. This ensures the safety and integrity of the data being transmitted.

HTTPS is becoming increasingly popular for websites that require secure data transmission, such as online banking and online shopping. This trend is reflected in the fact that many web browsers, such as Google Chrome, Microsoft Edge, Firefox, and so on, now mark HTTP websites as **Not Secure** to encourage users to switch to HTTPS. This is a clear indication of the growing trend toward using HTTPS for secure data transmission.

The default ASP.NET Core web API template uses HTTPS by default. You can find the following code in the `Program.cs` file:

```
app.UseHttpsRedirection();
```

This code redirects all HTTP requests to HTTPS. In the development environment, the ASP.NET Core web API application uses a self-signed certificate. When you deploy your ASP.NET Core web API application to the production environment, you need to use a valid certificate issued by a trusted **certificate authority** (**CA**), such as Let's Encrypt, DigiCert, Comodo, and so on.

Using HTTP status codes correctly

HTTP status codes are used to indicate the status of the HTTP request. There are five categories of HTTP status codes:

- 1xx: Informational
- 2xx: Success
- 3xx: Redirection
- 4xx: Client errors
- 5xx: Server errors

The following table provides a summary of some of the most commonly used HTTP status codes in RESTful web APIs:

Status code	Description
200	OK
201	Created
202	Accepted
204	No Content
301	Moved Permanently
302	Found
304	Not Modified
400	Bad Request
401	Unauthorized
403	Forbidden
404	Not Found
405	Method Not Allowed
409	Conflict
410	Gone
415	Unsupported Media Type
422	Unprocessable Entity
429	Too Many Requests
500	Internal Server Error
501	Not Implemented
503	Service Unavailable
504	Gateway Timeout

Table 15.1 – Commonly used HTTP status codes in RESTful web APIs

The following list shows the HTTP methods and their corresponding status codes for RESTful web APIs:

- GET: The GET method is used to retrieve a single resource or a collection of resources. A GET request should not modify the state of the server. It can return the following status codes:

  - 200: The resource is found and returned.

  - 404: The resource is not found. Note that if a collection exists but is empty, the GET method should return the 200 status code instead of the 404 status code.

- POST: The POST method is used to create a new single resource or a collection of resources. It can be also used to update a resource. It can return the following status codes:

  - 200: The resource is updated successfully.

  - 201: The resource is created successfully. The response should include the identifier of the newly created resource.

  - 202: The resource is accepted for processing, but the processing is not yet complete. This status code is often used for long-running operations.

  - 400: The request is invalid.

  - 409: The resource already exists.

- PUT: The PUT method is used to update a single resource or a collection of resources. It is rarely used to create resources. It can return the following status codes:

  - 200: The resource was updated successfully.

  - 204: The resource was updated successfully but there is no content to return.

  - 404: The resource was not found.

- DELETE: The DELETE method is used to delete a single resource with a specific identifier. It can be used to delete a collection of resources but it is not a common scenario. It can return the following status codes:

  - 200: The resource was deleted successfully and the response includes the deleted resource.

  - 204: The resource was deleted successfully but there is no content to return.

  - 404: The resource was not found.

It is important to note that this list is not exhaustive and only applies to RESTful web APIs. When selecting the appropriate HTTP status codes, please consider the specific scenarios. For GraphQL APIs, 200 is typically used for most responses, with the errors field indicating any errors.

Using asynchronous programming

The ASP.NET Core web API framework is designed to process requests asynchronously, so we should use asynchronous programming as much as possible. Asynchronous programming allows the application to process multiple tasks concurrently, which can improve the performance of the application. For many I/O-bound operations, such as accessing the database, sending HTTP requests, and operating files, using asynchronous programming can release the thread to process other requests while waiting for the I/O operation to complete.

In C#, you can use the `async` and `await` keywords to define and await asynchronous operations. Many methods in .NET have synchronous and asynchronous versions. For example, the `StreamReader` class has the following synchronous and asynchronous methods to read the content of the stream:

```
// Synchronous methods
public int Read();
public string ReadToEnd();
// Asynchronous methods
public Task<int> ReadAsync();
public Task<string> ReadToEndAsync();
```

In these four methods, the methods without the `Async` suffix are synchronous, which blocks the thread until the operation is completed. In contrast, the methods that have the `Async` suffix are asynchronous, which returns a `Task` object immediately and allows the thread to process other requests. When the operation is completed, the `Task` object will be completed and the thread will continue to process the request. Whenever possible, we should use asynchronous programming to improve the performance of the application.

For I/O operations, we should always use asynchronous programming. For example, when accessing `HttpRequest` and `HttpResponse` objects, we should use the asynchronous methods. Here is an example:

```
[HttpPost]
public async Task<ActionResult<Post>> PostAsync()
{
    // Read the content of the request body
    var jsonString = await new StreamReader(Request.Body).
ReadToEndAsync();
    // Do something with the content
    var result = JsonSerializer.Deserialize<Post>(jsonString);
    return Ok(result);
}
```

In the preceding code, the `ReadToEndAsync()` method is used to read the content of the request body. For this case, we should not use the synchronous `ReadToEnd()` method because it will block the thread until the operation is completed.

If there are multiple asynchronous operations that need to be executed concurrently, we can use the `Task.WhenAll()` method to wait for all the asynchronous operations to complete. Here is an example:

```
[HttpGet]
public async Task<ActionResult> GetAsync()
{
    // Simulate a long-running I/O-bound operation
    var task1 = SomeService.DoSomethingAsync();
```

```
    var task2 = SomeService.DoSomethingElseAsync();
    await Task.WhenAll(task1, task2);
    return Ok();
}
```

In the preceding code, the `Task.WhenAll()` method waits for the `task1` and `task2` tasks to complete. If you need to get the results of the tasks after they are completed, you can use the `Result` property of the `Task` object to get the results. Here is an example:

```
[HttpGet]
public async Task<ActionResult> GetAsync()
{
    // Simulate long-running I/O-bound operations
    var task1 = SomeService.DoSomethingAsync();
    var task2 = SomeService.DoSomethingElseAsync();
    await Task.WhenAll(task1, task2);
    var result1 = task1.Result;
    var result2 = task2.Result;
    // Do something with the results
    return Ok();
}
```

In the preceding code, the `Result` property of the `task1` and `task2` objects is used to get the results of the tasks. As we already used the `await` keyword to wait for the tasks to be completed, the `Result` property will return the results immediately. But if we don't use the `await` keyword to wait for the tasks to be completed, the `Result` property will block the thread until the tasks are completed. So, please be careful when using the `Result` property. Similarly, the `Wait()` method of the `Task` object will also block the thread until the task is completed. If you want to wait for a task to complete, use the `await` keyword instead of the `Wait` method.

> **Important note**
>
> Note that the `Task.WhenAll()` method is not suitable for all scenarios. For example, EF Core does not support running multiple queries in parallel on the same database context. If you need to execute multiple queries on the same database context, you should use the `await` keyword to wait for the previous query to complete before executing the next query.

When utilizing asynchronous programming, there are several important considerations to keep in mind. These include, but are not limited to, the following:

- Do not use `async void` in ASP.NET Core. The only scenario where `async void` is allowed is in event handlers. If an async method returns `void`, the exceptions thrown in the method will not be caught by the caller properly.

- Do not mix synchronous and asynchronous methods in the same method. Try to use async for the entire process if possible. This allows the entire call stack to be asynchronous.

- If you need to use the `Result` property of the `Task` object, make sure that the `Task` object is completed. Otherwise, the `Result` property will block the thread until the `Task` object is completed.

- If you have a method that only returns the result of another async method, there's no need to use the `async` keyword. Just return the `Task` object directly. For example, the following code is unnecessary:

```
public async Task<int> GetDataAsync()
{
    return await SomeService.GetDataAsync();
}
```

The following code does not use the `async/await` keywords, which is better:

```
public Task<int> GetDataAsync()
{
    return SomeService.GetDataAsync();
}
```

This is because the `async` keyword will create a state machine to manage the execution of the async method. In this case, it is unnecessary. Returning the `Task` directly does not create additional overhead.

Using pagination for large collections

It is not recommended to return a large collection of resources in a single response, as this can lead to performance issues. Such issues may include the following:

- The server may require a significant amount of time to query the database and process the response.

- The response payload may be quite large, resulting in network congestion. This can negatively impact the performance of the system, leading to increased latency and decreased throughput.

- The client may require additional time and resources to process the large response. Deserializing a large JSON object can be computationally expensive for the client. Also, rendering a large collection of items on the UI may cause the client to become unresponsive.

In order to efficiently manage large collections, it is recommended to use pagination. *Chapter 5* introduces pagination and filtering through the use of the `Skip()` and `Take()` methods of the `IQueryable` interface. Also, we mentioned that the `AsNoTracking()` method should be used to improve the performance of the read-only queries. This will result in a collection of resources being returned to the client. However, the client may not be aware of whether there are more resources available. To address this issue, we can create a custom class to represent the paginated response. An example of this is provided here:

```
public class PaginatedList<T> where T : class
{
    public int PageIndex { get; }
    public int PageSize { get; }
    public int TotalPages { get; }

    public List<T> Items { get; } = new();

    public PaginatedList(List<T> items, int count, int pageIndex = 1,
int pageSize = 10)
    {
        PageIndex = pageIndex;
        PageSize = pageSize;
        TotalPages = (int)Math.Ceiling(count / (double)pageSize);
        Items.AddRange(items);
    }

    public bool HasPreviousPage => PageIndex > 1;
    public bool HasNextPage => PageIndex < TotalPages;
}
```

In the preceding code, the PaginatedList<T> class contains a couple of properties to represent the pagination information:

- PageIndex: The current page index

- PageSize: The page size

- TotalPages: The total number of pages

- Items: The collection of items on the current page

- HasPreviousPage: Indicates whether there is a previous page

- HasNextPage: Indicates whether there is a next page

Then, we can use this class in the controller for pagination. Here is an example:

```
[HttpGet]
public async Task<ActionResult<PaginatedList<Post>>> GetPosts(int
pageIndex = 1, int pageSize = 10)
{
    var posts = _context.Posts.AsQueryable().AsNoTracking();
    var count = await posts.CountAsync();
    var items = await posts.Skip((pageIndex - 1) * pageSize).
```

```
Take(pageSize).ToListAsync();
    var result = new PaginatedList<Post>(items, count, pageIndex,
pageSize);
    return Ok(result);
}
```

In the preceding code, besides the Items property, the PaginatedList<T> class also contains the pagination information, such as PageIndex, PageSize, TotalPages, HasPreviousPage, HasNextPage, and so on. The response of the endpoint will be as follows:

```
{
  "pageIndex": 1,
  "pageSize": 10,
  "totalPages": 3,
  "items": [
    {
      "id": "3c979917-437b-406d-a784-0784170b5dd9",
      "title": "Post 26",
      "content": "Post 26 content",
      "categoryId": "ffdd0d80-3c3b-4e83-84c9-025d5650c6e5",
      "category": null
    },
    ...
  ],
  "hasPreviousPage": false,
  "hasNextPage": true
}
```

In this way, the clients can implement pagination easily. You can also include more information in the PaginatedList<T> class, such as the links to the previous page and the next page, and so on.

When implementing pagination, it is important to consider sorting and filtering. Generally, the data should be filtered first, followed by sorting and then pagination. For example, the following LINQ query can be used:

```
var posts = _context.Posts.AsQueryable().AsNoTracking();
posts = posts.Where(x => x.Title.Contains("Post")).OrderBy(x =>
x.PublishDate).Skip((pageIndex - 1) * pageSize).Take(pageSize).
ToListAsync();
```

The Where() method should be used to filter the data first in order to reduce the amount of data to be sorted. This is important, as sorting is often an expensive operation. Once the data has been filtered, the OrderBy() method can be used to sort it. Finally, the Skip() and Take() methods can be used to paginate the data.

Specifying the response types

An ASP.NET Core web API endpoint can return various types of responses, such as `ActionResult`, `ActionResult<T>`, or a specific type of the object. For example, the following code returns a `Post` object:

```
[HttpGet("{id}")]
public async Task<Post> GetAsync(Guid id)
{
    var post = await _postService.GetAsync(id);
    return post;
}
```

The preceding code works, but what if the `post` cannot be found? It is recommended to use `ActionResult<T>` instead of the specific type of the object. The `ActionResult<T>` class is a generic class that can be used to return various HTTP status codes. Here is an example:

```
public async Task<ActionResult<Post>> GetPost(Guid id)
{
    var post = await _context.Posts.FindAsync(id);

    if (post == null)
    {
        return NotFound();
    }

    return Ok(post);
}
```

In the preceding code, the `ActionResult<Post>` class is used to return a `Post` object. If the `post` cannot be found, the `NotFound` method is used to return the `404 Not Found` status code. If the `post` is found, the `Ok` method is used to return the `200 OK` status code.

We can add the `[ProducesResponseType]` attribute to specify the response types of the endpoint. Here is a complete example:

```
[HttpGet("{id}")]
[ProducesResponseType(StatusCodes.Status200OK)]
[ProducesResponseType(StatusCodes.Status404NotFound)]
public async Task<ActionResult<Post>> GetPost(Guid id)
{
    var post = await _context.Posts.FindAsync(id);

    if (post == null)
    {
```

```
        return NotFound();
    }

    return Ok(post);
}
```

In the preceding code, there are two [ProducesResponseType] attributes. The first one specifies the 200 OK status code and the second one specifies the 404 Not Found status code. The [ProducesResponseType] attribute is optional, but it is recommended to use it to specify the response types of the endpoint. The Swagger UI will use the [ProducesResponseType] attribute to generate the response types of the endpoint, as shown in *Figure 15.1*:

Responses

Code	Description
200	Success

Media type

application/json ⌄

Controls Accept header.

Example Value | Schema

```
{
  "id": "3fa85f64-5717-4562-b3fc-2c963f66afa6",
  "title": "string",
  "content": "string",
  "categoryId": "3fa85f64-5717-4562-b3fc-2c963f66afa6",
  "category": {
    "id": "3fa85f64-5717-4562-b3fc-2c963f66afa6",
    "name": "string",
    "posts": [
      "string"
    ]
  }
}
```

Code	Description
404	Not Found

Figure 15.1 – The Swagger UI uses the [ProducesResponseType]
attribute to generate the response types of the endpoint

We can see there are possible responses in the Swagger UI. This endpoint can return the 200 OK status code or the 404 Not Found status code.

To enforce the use of the [ProducesResponseType] attribute, we can use the OpenAPIAnalyzers. This analyzer can be used to report the missing [ProducesResponseType] attribute. Add the following code in the <PropertyGroup> section of the *.csproj file:

```
<IncludeOpenAPIAnalyzers>true</IncludeOpenAPIAnalyzers>
```

Then, we can see the warning in Visual Studio if the controller action does not have the [ProducesResponseType] attribute, as shown in *Figure 15.2*:

```
57    [HttpPut(template:"{id}")]
      0 references | Xiaodi Yan. 21 hours ago | 1 author. 1 change
58    public async Task<IActionResult> PutPost(Guid id, Post post)
59    {
60        if (id != post.Id)
61        {
62            return BadRequest();
63        }
64
65        _context.En
66
67        try
68        {
69            await _context.SaveChangesAsync(
70        }
71        catch (DbUpdateConcurrencyException)
72        {
73            if (!PostExists(id))
74            {
75                return NotFound();
76            }
77
78            throw;
79        }
80
81        return NoContent();
82    }
```

```
Add ProducesResponseType attributes.  ▶    ⚠ Action method returns undeclared status code '400'
Suppress or configure issues          ▶
                                           Lines 5 to 6
                                           using MyBasicWebApiDemo.Models;
                                           using Microsoft.AspNetCore.Http;

                                           Lines 57 to 58
                                               [HttpPut("{id}")]
                                               [ProducesResponseType(StatusCodes.Status204NoContent)]
                                               [ProducesResponseType(StatusCodes.Status400BadRequest)]
                                               [ProducesResponseType(StatusCodes.Status404NotFound)]
                                               [ProducesDefaultResponseType]
                                               public async Task<IActionResult> PutPost(Guid id, Post post)

                                           Preview changes
                                           Fix all occurrences in: Document | Project | Solution | Containing Member | Containing Type
```

Figure 15.2 – Visual Studio displays the warning if the controller action
does not have the [ProducesResponseType] attribute

Visual Studio will provide you with a quick fix to add these attributes. This analyzer is very useful and it is recommended to use it.

Adding comments to the endpoints

Adding XML comments to the endpoints can help other developers understand them better. These comments will be displayed in the Swagger UI, providing a comprehensive description of the endpoints. This can be a great resource for developers to use when working with the endpoints.

Adding XML comments to the endpoints is very simple. We just need to add the /// comments to them. Visual Studio will automatically generate the XML comments structure when you type ///. You need to add the description of the method, the parameters, the return value, and so on. Here is an example:

```
/// <summary>
/// Get a post by id
/// </summary>
/// <param name="id">The id of the post</param>
/// <returns>The post</returns>
[HttpGet("{id}")]
[ProducesResponseType(StatusCodes.Status200OK)]
[ProducesResponseType(StatusCodes.Status404NotFound)]
public async Task<ActionResult<Post>> GetPost(Guid id)
{
    // Omitted for brevity
}
```

You can also add comments to the model classes. Here is a simple example:

```
/// <summary>
/// The post model
/// </summary>
public class Post
{
    /// <summary>
    /// The id of the post
    /// </summary>
    public Guid Id { get; set; }

    /// <summary>
    /// The title of the post
    /// </summary>
    public string Title { get; set; }

    /// <summary>
    /// The content of the post
    /// </summary>
    public string Content { get; set; }
}
```

Then, we need to enable the XML documentation file generation in the project file. Open the `*.csproj` file and add the following code in the `<PropertyGroup>` element:

```
<GenerateDocumentationFile>true</GenerateDocumentationFile>
<NoWarn>$(NoWarn);1591</NoWarn>
```

The `GenerateDocumentationFile` property specifies whether an XML documentation file should be generated. The `NoWarn` property can be used to suppress specific warnings, such as the `1591` warning code, which is associated with missing XML comments. Suppressing this warning is beneficial, as it prevents the warning from appearing when the project is built.

Next, we need to configure the Swagger UI to use the XML documentation file. Open the `Program.cs` file and update the `builder.Services.AddSwaggerGen()` method as follows:

```
builder.Services.AddSwaggerGen(c =>
{
    // The below line is optional. It is used to describe the API.
    // c.SwaggerDoc("v1", new OpenApiInfo { Title =
"MyBasicWebApiDemo", Version = "v1" });
    c.IncludeXmlComments(Path.Combine(AppContext.BaseDirectory,
$"{Assembly.GetExecutingAssembly().GetName().Name}.xml"));
});
```

In the preceding code, the `IncludeXmlComments` method is used to specify the XML documentation file. We can use reflection `{Assembly.GetExecutingAssembly().GetName().Name}.xml`, to get the name of the XML documentation file. The `AppContext.BaseDirectory` property is used to get the base directory of the application.

To view the comments in the Swagger UI, run the application and open the Swagger UI. As shown in *Figure 15.3*, the comments will be displayed:

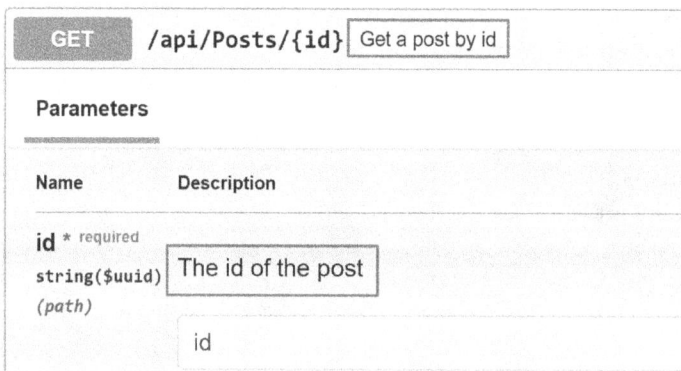

Figure 15.3 – The Swagger UI displays the comments of the endpoints

The model classes are also described in the Swagger UI, as shown in *Figure 15.4*:

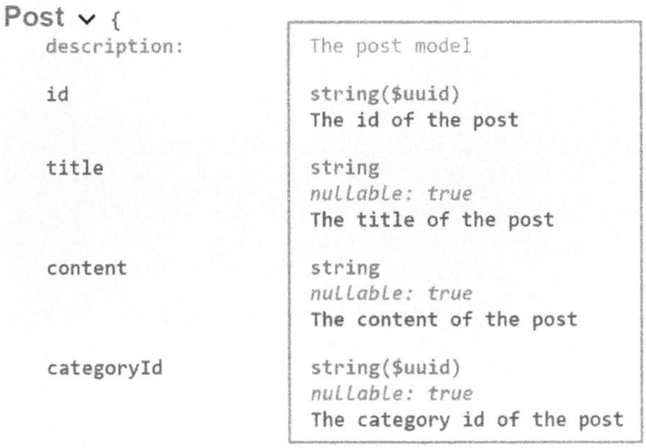

```
Post ∨ {
    description:            The post model

    id                     string($uuid)
                           The id of the post

    title                  string
                           nullable: true
                           The title of the post

    content                string
                           nullable: true
                           The content of the post

    categoryId             string($uuid)
                           nullable: true
                           The category id of the post
```

Figure 15.4 – The Swagger UI displays the comments of the model classes

Displaying the comments in the Swagger UI is a great way to provide developer-friendly API documentation. It is highly recommended to add comments to the endpoints and model classes.

Using System.Text.Json instead of Newtonsoft.Json

Newtonsoft.Json is a popular JSON library for .NET and is widely used in many projects. It was created as a personal project by James Newton-King in 2006 and has since become the number one library on NuGet, with over one billion downloads. An interesting fact is that in 2022, the downloads of Newtonsoft.Json on NuGet reached an impressive 2.1 billion, surpassing the Int32.MaxValue of 2,147,483,647. This milestone necessitated a change to NuGet to support the continued downloads of Newtonsoft.Json.

Microsoft has introduced a new JSON library, System.Text.Json, with the release of .NET Core 3.0. This library was designed to be high performing by utilizing Span<T>, which provides a type-safe and memory-safe representation of contiguous regions of arbitrary memory. Using Span<T> can reduce memory allocation and improve the performance of .NET code. System.Text.Json is included in the .NET Core SDK and is actively being developed. Although it may not have all the features of Newtonsoft.Json, it is a great choice for new projects.

The latest ASP.NET web API template uses System.Text.Json by default. It provides a simple way to serialize and deserialize JSON data. Here is an example:

```
var options = new JsonSerializerOptions
{
    PropertyNamingPolicy = JsonNamingPolicy.CamelCase,
    WriteIndented = true
};
// Serialize
```

```
var json = JsonSerializer.Serialize(post, options);
// Deserialize
var post = JsonSerializer.Deserialize<Post>(json, options);
```

If you still want to use `Newtonsoft.Json`, you can install the `Microsoft.AspNetCore.Mvc.NewtonsoftJson` NuGet package and update the `Program.cs` file as follows:

```
builder.Services.AddControllers()
    .AddNewtonsoftJson(options =>
    {
        options.SerializerSettings.ContractResolver = new
CamelCasePropertyNamesContractResolver();
        options.SerializerSettings.Formatting = Formatting.Indented;
    });
```

You can update the `options` object to configure the `Newtonsoft.Json` library. Again, it is recommended to use `System.Text.Json` instead of `Newtonsoft.Json` unless you need some specific features of `Newtonsoft.Json` because `System.Text.Json` has better performance.

Optimizing the performance by implementing caching

Caching is a common technique used to improve the performance of the application. In web API development, caching can store frequently accessed data in a temporary storage, such as memory or disk, to reduce the number of database queries and improve the responsiveness of the application. In this section, we will introduce the caching in ASP.NET Core web API development.

Caching is an effective tool when handling data that is not regularly updated but is costly to compute or obtain from the database. It is also useful when multiple clients access the same data frequently. As an example, consider an e-commerce application that displays a list of categories. The categories of the products are not often changed, yet they are frequently viewed by users. To improve the performance of the application, we can cache the categories. When a user requests the categories, the application can return the cached data directly without querying the database.

In ASP.NET Core, we have several options for implementing caching, each suited to a specific scenario:

- **In-memory caching**: This type of caching stores data in the memory of the application. It is fast and efficient and is suitable for scenarios where the data does not need to be shared across multiple instances of the application. However, the data will be lost when the application is restarted.
- **Distributed caching**: This type of caching involves storing cached data in a shared storage, such as Redis or SQL Server, which can be accessed by multiple instances of the application. It is suitable for applications that are deployed with multiple instances, such as a web farm, container orchestration, or serverless computing.
- **Response caching**: This caching technique is based on the HTTP caching mechanism.

In the following sections, we will introduce the in-memory caching and distributed caching in ASP.NET Core web API development, as well as the output caching, which is introduced in ASP.NET Core 7.0.

In-memory caching

In-memory caching is a fast and easy way to store data in an application's memory. ASP.NET Core provides the IMemoryCache interface to facilitate this process. This type of caching is highly versatile, as it can store any type of data in the form of a key-value pair.

The example project in this section can be found in the chapter15/CachingDemo folder. This is a simple ASP.NET Core web API application. It contains a /categories endpoint that returns the categories of products.

To simplify the example, we use a static list to store the categories to simulate a database. When the application queries the categories, it will print a log to indicate that the categories are queried from the database. Here is the code in the CategoryService class:

```
public async Task<IEnumerable<Category>> GetCategoriesAsync()
{
    // Simulate a database query
    _logger.LogInformation("Getting categories from the database");
    await Task.Delay(2000);
    return Categories;
}
```

In the preceding code, we use a Task.Delay() method to simulate a database query. This query takes two seconds to complete, which is slow. As the categories are not often changed, we can use in-memory caching to improve the performance of the application.

To use in-memory caching, we need to add the Microsoft.Extensions.Caching.Memory NuGet package by running the following command:

```
dotnet add package Microsoft.Extensions.Caching.Memory
```

Then, we need to register the in-memory caching in the Program class:

```
builder.Services.AddMemoryCache();
```

Next, we can use the IMemoryCache interface in other classes. Inject the IMemoryCache interface into the CategoryService class:

```
public class CategoryService(ILogger<CategoryService> logger,
IMemoryCache cache)
    : ICategoryService
{
    // Omitted for brevity
}
```

Update the `GetCategoriesAsync` method as follows:

```
public async Task<IEnumerable<Category>> GetCategoriesAsync()
{
    // Try to get the categories from the cache
    if (_cache.TryGetValue(CacheKeys.Categories, out
IEnumerable<Category>? categories))
    {
        _logger.LogInformation("Getting categories from cache");
        return categories ?? new List<Category>();
    }

    // Simulate a database query
    _logger.LogInformation("Getting categories from the database");
    await Task.Delay(2000);
    categories = Categories;
    // Cache the categories for 10 minutes
    var cacheEntryOptions = new MemoryCacheEntryOptions()
        .SetAbsoluteExpiration(TimeSpan.FromMinutes(10));
    _cache.Set(CacheKeys.Categories, categories, cacheEntryOptions);
    return Categories;
}
```

In the updated code, we first try to get the categories from the cache by the cache key. If the categories are found in the cache, we return them directly. Otherwise, we query the database and cache the categories for 10 minutes. The `SetAbsoluteExpiration()` method is used to set the absolute expiration time of the cache entry. After 10 minutes, the cache entry will be removed from the cache.

Run the application and send a request to the `/categories` endpoint. The first request will take 2 seconds to complete, and then the subsequent requests will be completed immediately. You may see the following log in the console:

```
info: CachingDemo.Services.CategoryService[0]
      Getting categories from the database
info: CachingDemo.Services.CategoryService[0]
      Getting categories from cache
```

In this way, in-memory caching can significantly improve the performance of the application.

To ensure that the cache does not become bloated with outdated entries, the cache must apply a proper expiration policy. The cache has several options for expiration, two of them are as follows:

- **Absolute expiration**: The cache entry will be removed from the cache after a specified time.

- **Sliding expiration**: The cache entry will be removed after a predetermined period of time if it is not accessed.

When using `SlidingExpiration`, the cache can be retained indefinitely if it is accessed frequently. To avoid this, we can set the `AbsoluteExpiration` property or `AbsoluteExpiration-RelativeToNow` property to limit the maximum lifetime of the cache entry. Here is an example:

```
var cacheEntryOptions = new MemoryCacheEntryOptions
{
    SlidingExpiration = TimeSpan.FromMinutes(10),
    AbsoluteExpirationRelativeToNow = TimeSpan.FromMinutes(30)
};
cache.Set(CacheKeys.Categories, categories, cacheEntryOptions);
```

In the preceding code, the `SlidingExpiration` property is set to 10 minutes, and the `AbsoluteExpirationRelativeToNow` property is set to 30 minutes. This means that the cache entry will be removed from the cache after 30 minutes, even if it is frequently accessed.

Sometimes we may need to manually update the cache entry. For example, when a new category is created, or an existing category is updated or deleted, we can remove the cache entry to force the application to query the database again and refresh the cache entry. Move the preceding code to a new method:

```
private async Task RefreshCategoriesCache()
{
    // Query the database first
    logger.LogInformation("Getting categories from the database");
    // Simulate a database query
    await Task.Delay(2000);
    var categories = Categories;
    // Then refresh the cache
    cache.Remove(CacheKeys.Categories);
    var cacheEntryOptions = new MemoryCacheEntryOptions
    {
        SlidingExpiration = TimeSpan.FromMinutes(10),
        AbsoluteExpirationRelativeToNow = TimeSpan.FromMinutes(30)
    };
    cache.Set(CacheKeys.Categories, categories, cacheEntryOptions);
}
```

Note that in the preceding code, we should query the database first, then remove the cache entry and reset it. Otherwise, the application may query the database multiple times if the cache entry is removed before the database query is completed.

Then, we can call the `RefreshCategoriesCache()` method when a new category is created or an existing category is updated or deleted. Here is an example:

```
public async Task<Category?> UpdateCategoryAsync(Category category)
{
```

```
    var existingCategory = Categories.FirstOrDefault(c => c.Id ==
category.Id);
    if (existingCategory == null)
    {
        return null;
    }

    existingCategory.Name = category.Name;
    existingCategory.Description = category.Description;
    await RefreshCategoriesCache();
    return existingCategory;
}
```

Alternatively, we can create a background task to update the cache entry periodically. A background task is a task that runs behind the scenes without user interaction. It is useful for performing tasks that are not time-sensitive, such as updating the cache entry. To create a background task, we can use the `BackgroundService` class. Create a new class named `CategoriesCacheBackgroundService` that inherits from the `BackgroundService` class:

```
public class CategoriesCacheBackgroundService(
    IServiceProvider serviceProvider,
    ILogger<CategoriesCacheBackgroundService> logger,
    IMemoryCache cache)
    : BackgroundService
{
    protected override async Task ExecuteAsync(CancellationToken
stoppingToken)
    {
        // Remove the cache every 1 hour
        while (!stoppingToken.IsCancellationRequested)
        {
            logger.LogInformation("Updating the cache in background
service");
            using var scope = serviceProvider.CreateScope();
            var categoryService = scope.ServiceProvider.
GetRequiredService<ICategoryService>();
            var categories = await categoryService.
GetCategoriesAsync();
            cache.Remove(CacheKeys.Categories);
            cache.Set(CacheKeys.Categories, categories, TimeSpan.
FromHours(1));
            await Task.Delay(TimeSpan.FromHours(1), stoppingToken);
        }
    }
}
```

In the preceding code, we use a `while` loop to reset the cache entry every one hour. Note that you cannot inject the `ICategoryService` directly because the `BackgroundService` class will be registered as a singleton service, but the `ICategoryService` is registered as a scoped service. A singleton service cannot depend on a scoped service. To solve this problem, we need to use the `IServiceProvider` interface to create a scope and get the `ICategoryService` from the scope.

Then, register the `CacheBackgroundService` class in the `Program` class:

```
builder.Services.AddHostedService<CacheBackgroundService>();
```

When the background task is executed every one hour, the cache entry will be removed from the cache. The background task should first query the database and then remove the cache entry and reset it. If the cache entry is deleted first, the application may query the database multiple times, resulting in performance issues.

When implementing caching, it is important to consider scenarios where records cannot be found in the database. Let us see how it happens. Update the `GetCategoryAsync()` method as follows:

```
public async Task<Category?> GetCategoryAsync(int id)
{
    if (cache.TryGetValue($"{CacheKeys.Categories}:{id}", out
Category? category))
    {
        logger.LogInformation($"Getting category with id {id} from
cache");
        return category;
    }
    // Simulate a database query
    logger.LogInformation($"Getting category with id {id} from the
database");
    await Task.Delay(2000);
    var result = Categories.FirstOrDefault(c => c.Id == id);
    if (result is not null)
    {
        cache.Set($"{CacheKeys.Categories}:{id}", result);
    }

    return result;
}
```

In the preceding code, if the category cannot be found in the cache, we query the database and cache the category. But there is a potential issue here. What if the category with the specified ID does not exist? In this case, the application will not set the cache and each request will query the database.

Cache is not used at all. To solve this problem, we can use the GetOrCreateAsync method of the IMemoryCache interface. Here is the updated code:

```
public async Task<Category?> GetCategoryAsync(int id)
{
    var category = await cache.GetOrCreateAsync($"{CacheKeys.
Categories}:{id}", async entry =>
    {
        // Simulate a database query
        logger.LogInformation($"Getting category with id {id} from the
database");
        await Task.Delay(2000);
        return Categories.FirstOrDefault(c => c.Id == id);
    });
    return category;
}
```

The updated code uses the GetOrCreateAsync method to retrieve the category from the cache. If the category is not present, the method will execute the specified delegate to fetch it from the database. Upon successful retrieval, the category will be cached and returned. If the category is not found, null will be returned. So, the application will not query the database every time. To avoid the issue mentioned earlier, it is recommended to use the GetOrCreateAsync method to obtain the data from the cache.

There are more important considerations when using in-memory caching:

- *Consider the expiration time of the cache entry*. If the data is not often changed, we can set a longer expiration time. Otherwise, use a shorter expiration time. Also, you can use the SlidingExpiration property and the absolute expiration time to achieve a balance between the performance and the freshness of the data.

- *The in-memory cache can cache any object, but be careful when caching large objects*. It is important to limit the size of the cache entry. We can use SetSize, Size, and SizeLimit to limit the size of the cache. Note that when using these methods, the in-memory cache must be registered as a singleton service. Please refer to the documentation for more information at https://learn.microsoft.com/en-us/aspnet/core/performance/caching/memory.

- *Define proper cache keys*. The cache keys should be unique and descriptive. Especially, when using caching for users, ensure that the cache keys are unique for each user. Otherwise, the cached data of one user may be used by another user.

- *Provide a way to fall back to the data source when the cache is not available*.

There are no hard rules for these settings. You need to consider the specific scenarios and adjust the settings accordingly.

In-memory caching is a simple and effective way to improve the performance of the application. However, it is not suitable for applications that are deployed with multiple instances. The cached data only works for the current instance. When a client requests the data from another instance, the cached data in the original instance will not be used. To solve this problem, one solution is to implement session affinity, which means the request from a user will always be routed to the same instance. This can be achieved by using a load balancer that supports session affinity, such as Nginx, Azure Application Gateway, and so on. This is out of the scope of this book. Please refer to the documentation of the load balancer for more information.

Another approach to this issue is to implement a distributed cache, as outlined in the following section.

Distributed caching

Distributed caching offloads the cache from the application to a shared storage, such as Redis or SQL Server. The data stored in the distributed cache can be accessed by multiple instances of the application. If the application restarts, the cached data will not be lost. There is no need to implement session affinity when using distributed caching.

There are several options to implement distributed caching in ASP.NET Core. The following are the most commonly used options:

- **Redis**: Redis is an open-source, in-memory data structure store. It has many features, such as caching, pub/sub, and so on.

- **SQL Server**: SQL Server can be also used as a distributed cache.

- **Azure Cache for Redis**: Azure Cache for Redis is a fully managed, open-source, in-memory data structure store. It is based on the popular open-source Redis cache. You can use a local Redis server for development and testing and use Azure Cache for Redis in production.

- **NCache**: NCache is an open-source distributed cache for .NET applications. See `https://github.com/Alachisoft/NCache`.

In this section, we will introduce the Redis cache using the same sample project as the previous section. We use a static `Dictionary<int, List<Category>>` to store the users' favorites categories, which simulates the data stored in the database. When a user requests the favorites categories, the application will use the user ID as the key to query the database. If we use in-memory caching, the caching key should include the user ID, such as `1_Favorites_Categories`. However, if this user's subsequent requests are routed to another instance, there is no way to get the cached data. That is why we need to use the distributed caching.

First, we need to prepare a Redis server. We can use the Docker to run a Redis server. Start Docker Desktop on your machine and run the following command to pull the Redis image:

```
docker pull redis
```

Then, run the Redis server:

```
docker run --name redis -p 6379:6379 -d redis
```

The Redis server will be listening on port 6379.

To access the Redis server in the terminal, we need to use the `redis-cli` command. This command is included in the Redis image. Run the following command to access the Redis server:

```
docker exec -it redis redis-cli
```

The `docker exec` command is used to execute a command in a running container. The `-it` option is used to run the command interactively. It means we want to execute the `redis-cli` command in the container. You will see the following output:

```
127.0.0.1:6379>
```

This means we have successfully accessed the Redis server. Now we can use the `redis-cli` command to access the Redis server. For example, we can use the `set` command to set the value of a key:

```
set my-key "Hello World"
```

Then, we can use the `get` command to get the value of the key:

```
get my-key
```

You will see `Hello World` in the output.

Now the Redis server is ready to use. To use the Redis cache in ASP.NET Core, we need to install the `Microsoft.Extensions.Caching.StackExchangeRedis` NuGet package:

```
dotnet add package Microsoft.Extensions.Caching.StackExchangeRedis
```

Then, we need to register the Redis cache in the `Program` class:

```
builder.Services.AddStackExchangeRedisCache(options =>    options.
Configuration = "localhost:6379";
    options.InstanceName = "CachingDemo";
});
```

In the preceding code, the `AddStackExchangeRedisCache` extension method is used to register the Redis cache. We specify the Redis server address and an optional instance name, which is used to create a logical partition for the cache. Note that these configurations can be defined in the `appsettings.json` file or environment variables, allowing for different Redis instances to be used for development and production purposes.

Next, we can use the `IDistributedCache` interface to operate the Redis cache. Inject the `IDistributedCache` interface into the `CategoryService` class:

```
public class CategoryService(ILogger<CategoryService> logger,
IMemoryCache cache, IDistributedCache distributedCache) :
ICategoryService
{
    // Omitted for brevity
}
// Update the GetFavoritesCategoriesAsync() method as follows:
public async Task<IEnumerable<Category>>
GetFavoritesCategoriesAsync(int userId)
{
    // Try to get the categories from the cache
    var cacheKey = $"{CacheKeys.FavoritesCategories}:{userId}";
    var bytes = await distributedCache.GetAsync(cacheKey);
    if (bytes is { Length: > 0 })
    {
        logger.LogInformation("Getting favorites categories from
distributed cache");
        var serializedFavoritesCategories = Encoding.UTF8.
GetString(bytes);
        var favoritesCategories = JsonSerializer.
Deserialize<IEnumerable<Category>>(serializedFavoritesCategories);
        return favoritesCategories ?? new List<Category>();
    }

    // Simulate a database query
    logger.LogInformation("Getting favorites categories from the
database");
    var categories = FavoritesCategories[userId];
    // Store the result in the distributed cache
    var cacheEntryOptions = new DistributedCacheEntryOptions
    {
        SlidingExpiration = TimeSpan.FromMinutes(10),
        AbsoluteExpirationRelativeToNow = TimeSpan.FromMinutes(30)
    };
    var serializedCategories = JsonSerializer.Serialize(categories);
    var serializedCategoriesBytes = Encoding.UTF8.
GetBytes(serializedCategories);
    await distributedCache.SetAsync(cacheKey,
serializedCategoriesBytes, cacheEntryOptions);
    await Task.Delay(2000);
    return FavoritesCategories[userId].AsEnumerable();
}
```

In the preceding code, we first try to get the favorites categories from the cache using the cache key. If the favorites categories are found in the distributed cache, we return the cached data directly. Otherwise, we query the database and store the result in the distributed cache.

As the Redis cache stores the data as `byte[]`, to store the cached data, we need to serialize the data into a JSON string and then convert the JSON string into a `byte[]` value using the `Encoding.UTF8.GetBytes()` method. Similarly, when getting the cached data, we need to convert the `byte[]` value to a JSON string using the `Encoding.UTF8.GetString()` method and then deserialize the JSON string into the strongly-typed object using the `JsonSerializer.Deserialize()` method.

In addition, the caching key must be a `string` value.

To make it easier to convert the data to and from `byte[]`, the `IDistributedCache` interface has a few extension methods as follows:

- `SetStringAsync` and `SetString`: These two methods can save `string` values directly
- `GetStringAsync` and `GetString`: These two methods can read `string` values directly

To remove a cache entry, we can use the `RemoveAsync()` method or `Remove()` method. As we mentioned before, using the asynchronous versions of these methods is preferred.

Run the application and send some requests to the `Categories/favorites/1` endpoint. You will see that the logs show the first response was from the database and the subsequent responses were from the distributed cache:

```
info: CachingDemo.Services.CategoryService[0]
      Getting favorites categories from the database
info: CachingDemo.Services.CategoryService[0]
      Getting favorites categories from distributed cache
```

You can use the `redis-cli` to examine the cached data. Run the following command to get the keys:

```
127.0.0.1:6379> keys *
```

The output should look as follows:

```
1) "CachingDemo_FavoritesCategories:1"
```

Then, use the `HGETALL` command to show the cached data:

```
127.0.0.1:6379> hgetall CachingDemo_FavoritesCategories:1
```

Note that you cannot use the `GET` command here because it is used to retrieve the string values only. The categories data is stored as `hash` in Redis, so we need to use the `HGETALL` command.

The output should look as follows, including all the fields of the cached entry:

```
1) "absexp"
2) "638322378838137428"
3) "sldexp"
4) "6000000000"
5) "data"
6) " [{\"Id\":1,\"Name\":\"Toys\",\"Description\":\"Soft toys,
action figures, dolls, and puzzles\"},{\"Id\":2,\"Name\":\"Electron-
ics\",\"Description\":\"Smartphones, tablets, laptops, and smart-
watches\"},{\"Id\":3,\"Name\":\"Clothing\",\"Description\":\"Shirts,
pants, dresses, and shoes\"}]"
```

Using a distributed cache can help make applications more scalable by allowing cached data to be shared across multiple instances. However, this does come with the potential cost of increased latency due to the extra network I/O required. Careful consideration should be taken when deciding whether to use a distributed cache.

The IDistributedCache interface does not have the GetOrCreateAsync() method. If the cached data is not found, the application still needs to query the database. To solve this problem, we can implement our own GetOrCreateAsync() method. Create an extension method for the IDistributedCache interface:

```
public static class DistributedCacheExtension
{
    public static async Task<T?> GetOrCreateAsync<T>(this
IDistributedCache cache, string key, Func<Task<T?>> createAsync,
DistributedCacheEntryOptions? options = null)
    {
        // Get the value from the cache.
        // If the value is found, return it.
        var value = await cache.GetStringAsync(key);
        if (!string.IsNullOrWhiteSpace(value))
        {
            return JsonSerializer.Deserialize<T>(value);
        }

        // If the value is not cached, then create it using the
provided function.
        var result = await createAsync();
        var json = JsonSerializer.Serialize(result);
        await cache.SetStringAsync(key, json, options ?? new
DistributedCacheEntryOptions());
        return result;
    }
}
```

Now the `GetFavoritesCategoriesAsync` method can be updated as follows:

```
public async Task<IEnumerable<Category>?>
GetFavoritesCategoriesAsync(int userId)
{
    var cacheKey = $"{CacheKeys.FavoritesCategories}:{userId}";
    var cacheEntryOptions = new DistributedCacheEntryOptions
    {
        SlidingExpiration = TimeSpan.FromMinutes(10),
        AbsoluteExpirationRelativeToNow = TimeSpan.FromMinutes(30)
    };
    var favoritesCategories = await distributedCache.
GetOrCreateAsync(cacheKey, async () =>
    {
        // Simulate a database query
        logger.LogInformation("Getting favorites categories from the
database");
        var categories = FavoritesCategories[userId];
        await Task.Delay(2000);
        return categories;
    }, cacheEntryOptions);
    return favoritesCategories?.AsEnumerable();
}
```

If the category is not found in the database, the `GetOrCreateAsync()` method will return `null` and cache the `null` value for future requests. In this way, the application will not query the database again and again.

The following table shows the differences between in-memory caching and distributed caching:

In-memory caching	Distributed caching
Cache data in the memory of the application	Cache data in a shared storage
Suitable for applications that are deployed with a single instance	Suitable for applications that are deployed with multiple instances
The cached data is lost when the application restarts	The cached data is not lost when the application restarts
The caching keys can be any `object`	The caching keys must be `string`
The cached data value can be any strongly-typed object	The cached data is persisted as `byte[]` and may need serialization and deserialization.

Table 15.2 – The differences between in-memory caching and distributed caching

If you would like to use other distributed cache, you can install other packages such as the following:

- **SqlServer**: `dotnet add package Microsoft.Extensions.Caching.SqlServer`
- **NCache**: `dotnet add package NCache.Microsoft.Extensions.Caching.OpenSource`

Please refer to their official documentation for more details.

Response caching

Response caching is defined in the RFC 9111 specification (`https://www.rfc-editor.org/rfc/rfc9111`). It uses the HTTP header `cache-control` to specify the caching behavior. The clients (such as browsers) and immediate proxies (such as CDNs and gateways), can use the `cache-control` header to determine whether to cache the response and how long to cache it for.

The `cache-control` header has several directives as follows:

- `public`: The response can be cached by the clients and the intermediate proxies.
- `private`: The response can be cached by the clients only. A shared cache, such as CDN, must not cache the response.
- `no-cache`: For requests, the clients must send the request to the server for validation before using a cached copy of the response. For responses, the clients must not use a cached copy of the response without successful validation on the server.
- `no-store`: For requests, the clients must not store any part of the request. For responses, the clients must not store any part of the response.
- `max-age`: This is the maximum age of the response in seconds. The clients can use the cached copy of the response if it is not expired. For example, `max-age=3600` means the response can be cached for one hour.

We can use the `ResponseCache` attribute to specify the caching behavior of the endpoint. Here is an example:

```
[HttpGet]
[ResponseCache(Duration = 60)]
public async Task<ActionResult<IEnumerable<Category>>> Get()
{
    var result = await categoryService.GetCategoriesAsync();
    return Ok(result);
}
```

In the preceding code, we use the `ResponseCache` attribute on the controller to specify the caching behavior of the endpoint. `Duration = 60` means the response can be cached for 60 seconds.

Run the application and test the /Categories endpoint in the Swagger UI. You will see the cache-control header in the response, as shown here:

```
cache-control: public,max-age=60
content-type: application/json; charset=utf-8
date: Sat,07 Oct 2023 03:56:06 GMT
server: Kestrel
```

If you resubmit the request, the browser will use the cached version of the response without sending the request to the server. This is managed by the max-age directive in the cache-control header. After 60 seconds have elapsed, the browser will send the request to the server for validation if the request is resubmitted.

The HTTP-based response caching takes effect on the client side. If multiple clients send requests to the same endpoint, each request will cause the server to handle the request and generate the response. ASP.NET Core provides a server-side response caching middleware to cache the response on the server side. However, this middleware has a few limitations.

- It only supports the GET and HEAD requests and it does not support requests that contain the Authorization, Set-Cookie headers, and so on.

- You cannot invalidate the client-side cached response on the server side when the data is changed.

- Additionally, most browsers, such as Chrome and Edge, automatically send requests with the cache-control: max-age=0 header, which disables response caching on the client side. As a result, the server will also respect this header and disable server-side response caching.

This book does not cover the middleware mentioned; for more information, please refer to the documentation at https://learn.microsoft.com/en-us/aspnet/core/performance/caching/middleware. We will, however, introduce output caching, which is available in ASP. NET Core 7.0 and later versions. This middleware resolves some of the limitations of the server-side response caching middleware.

Output caching

In ASP.NET Core 7.0, Microsoft introduced the output caching middleware. This middleware works in a similar way to the server-side response caching middleware, but it has a few advantages:

- It configures the caching behavior on the server side, so the client HTTP caching configuration does not affect the output caching configuration.

- It has the capability to invalidate the cached response on the server side when the data is changed.

- It can use external cache stores, such as Redis, to store the cached response.

- It can return a 304 Not Modified response to the client when the cached response is not modified. This can save the network bandwidth.

However, the output caching middleware also has similar limitations to the response caching middleware:

- It only supports GET and HEAD requests with the 200 OK status code
- It does not support Authorization and Set-Cookie headers

To enable output caching, we need to register the output caching middleware in the Program class:

```
builder.Services.AddOutputCache(options =>
{
    options.AddBasePolicy(x => x.Cache());
});
```

Then, we need to add the middleware to the HTTP request pipeline:

```
app.UseOutputCache();
```

Next, apply the OutputCache attribute to the endpoints that need to be cached. For example, we can apply the OutputCache attribute to the /categories/{id} endpoint:

```
[HttpGet("{id}")]
[OutputCache]
public async Task<ActionResult<Category?>> Get(int id)
{
    var result = await categoryService.GetCategoryAsync(id);
    if (result is null)
    {
        return NotFound();
    }
    return Ok(result);
}
```

The GetOrCreateAsync() method is shown as follows:

```
public async Task<Category?> GetCategoryAsync(int id)
{
    // Simulate a database query
    logger.LogInformation($"Getting category with id {id} from the
database");
    await Task.Delay(2000);
    return Categories.FirstOrDefault(c => c.Id == id);
}
```

Similarly, we use a `Task.Delay()` method to simulate the database query. Run the application and test the `/categories/1` endpoint in the Swagger UI. You will see that the console log shows that the first response was from the database. The headers of the response look as follows:

```
content-type: application/json; charset=utf-8
date: Sat,07 Oct 2023 06:43:02 GMT
server: Kestrel
```

Send the request again. You will not see the database query log in the console. The headers of the response look as follows:

```
age: 5
content-length: 87
content-type: application/json; charset=utf-8
date: Sat,07 Oct 2023 06:44:39 GMT
server: Kestrel
```

You can find that the headers of the response contain the `age` header, which indicates that the response is cached. The `age` header is the number of seconds since the response was generated.

By default, the expiration time of the cached response is 60 seconds. After 60 seconds have elapsed, the next request will query the database again.

We can define different caching policies for different endpoints. Update the `AddOutputCache()` method as follows:

```
builder.Services.AddOutputCache(options =>
{
    options.AddBasePolicy(x => x.Cache());
    options.AddPolicy("Expire600", x => x.Expire(TimeSpan.
FromSeconds(600)));
    options.AddPolicy("Expire3600", x => x.Expire(TimeSpan.
FromSeconds(3600)));
});
```

In the preceding code, we added two caching policies. The first `Expire600` policy will expire the cached response after 10 minutes, and the second one will expire the cached response after 1 hour. Then, we can apply the `OutputCache` attribute to the endpoints as follows:

```
[HttpGet("{id}")]
[OutputCache(PolicyName = "Expire600")]
public async Task<ActionResult<Category?>> Get(int id)
{
    // Omitted for brevity
}
```

Now, the cached response will expire in 10 minutes.

What caching strategy should I use?

Caching is a useful tool for improving the performance of applications. In this section, we introduced a couple of caching techniques, including in-memory caching, distributed caching, response caching, and output caching. Each caching technique has its suitable scenarios. We need to choose the proper caching technique based on the specific scenarios.

Response caching is relatively straightforward to implement; however, it is dependent on the client-side HTTP caching configuration. If the client-side HTTP caching is disabled, response caching will not work as intended. Output caching is more flexible and can be used independently of the client-side HTTP caching configuration. It does not need much effort to implement, but it has a few limitations.

In-memory caching is a fast and easy way to cache data in a single instance of the application. However, it needs session affinity to work properly if there are multiple instances of the application. Distributed caching supports multiple instances, but it needs extra network I/O to access the cache. So, we need to consider the trade-off between the performance and the scalability. If retrieving the data from the database is complex or needs expensive computation and the data is not often changed, we can use distributed caching to reduce the load on the database or the computation. Additionally, we can use in-memory caching and distributed caching together to leverage the advantages of both caching techniques. For example, we can query the data from the in-memory cache first, and if the data is not found, we can then query the distributed cache. Also, consider the expiration time of the cache entry. You may need various expiration policies for different data.

This section only introduces the basic concepts of caching in ASP.NET Core. To learn more about caching, please refer to the documentation at `https://docs.microsoft.com/en-us/aspnet/core/performance/caching/`.

Using HttpClientFactory to manage HttpClient instances

.NET provides the `HttpClient` class for sending HTTP requests. However, there is some confusion when using it. In the past, many developers would misuse the `using` statement to create a `HttpClient` instance, as it implements the `IDisposal` interface. This is not recommended, as the `HttpClient` class is designed to be reused for multiple requests. Creating a new instance for each request can exhaust the local socket ports.

To solve this problem, Microsoft introduced the `IHttpClientFactory` interface in ASP.NET Core 2.1. This interface simplifies the management of `HttpClient` instances. It allows us to use dependency injection to inject `HttpClient` instances into the application without worrying about the life cycle of the `HttpClient` instances. In this section, we will introduce how to use the `IHttpClientFactory` interface to manage `HttpClient` instances.

You can find the sample application for this section in the `samples/chapter15/HttpClientDemo` folder.

To demonstrate how to use the `IHttpClientFactory` interface, we need to have a web API application as the backend service. You can use any sample applications we have created in the previous chapters. In this section, we will use a fake API service: `https://jsonplaceholder.typicode.com/`. This is a free online REST API service that can be used for testing and prototyping. It provides a set of endpoints, such as `/posts`, `/comments`, `/albums`, `/photos`, `/todos`, and `/users`.

> **Tip**
>
> When you create C# models from JSON data, you can use the **Paste JSON as Classes** feature in Visual Studio. You can find this feature in the **Edit** | **Paste Special** menu.

Creating a basic HttpClient instance

The `IHttpClientFactory` interface provides an `AddHttpClient()` extension method to register the `HttpClient` instances. Add the following code in the `Program.cs` file:

```
builder.Services.AddHttpClient();
```

Then, we can inject the `IHttpClientFactory` interface into the controller and use it to create a `HttpClient` instance:

```
[ApiController]
[Route("[controller]")]
public class PostsController(IHttpClientFactory httpClientFactory) :
ControllerBase
{


    [HttpGet]
    public async Task<IActionResult> Get()
    {
        var httpClient = httpClientFactory.CreateClient();
        var httpRequestMessage = new HttpRequestMessage
        {
            Method = HttpMethod.Get,
            RequestUri = new Uri("https://jsonplaceholder.typicode.
com/posts")
        };
        var response = await httpClient.SendAsync(httpRequestMessage);
        response.EnsureSuccessStatusCode();
        var content = await response.Content.ReadAsStringAsync();
        var posts = JsonSerializerHelper.
DeserializeWithCamelCase<List<Post>>(content);
        return Ok(posts);
```

```
    }

    // Omitted for brevity
}
```

In the preceding code, we use the `CreateClient()` method to create a `HttpClient` instance. Then, we create an `HttpRequestMessage` instance and use the `SendAsync()` method to send the HTTP request. The `EnsureSuccessStatusCode()` method is used to ensure the response is successful. If the response fails, an exception will be thrown. The `ReadAsStringAsync()` method is used to read the response content as a string. Finally, we use the `JsonSerializerHelper` class to deserialize the JSON string into a list of `Post` objects.

The `JsonSerializerHelper` class is defined as follows:

```
public static class JsonSerializerHelper
{
    public static string SerializeWithCamelCase<T>(T value)
    {
        var options = new JsonSerializerOptions
        {
            PropertyNamingPolicy = JsonNamingPolicy.CamelCase,
            DictionaryKeyPolicy = JsonNamingPolicy.CamelCase,
        };
        return JsonSerializer.Serialize(value, options);
    }

    public static T? DeserializeWithCamelCase<T>(string json)
    {
        var options = new JsonSerializerOptions
        {
            PropertyNamingPolicy = JsonNamingPolicy.CamelCase,
            DictionaryKeyPolicy = JsonNamingPolicy.CamelCase,
        };
        return JsonSerializer.Deserialize<T>(json, options);
    }
}
```

This is because the JSON data returned by the API uses the camel case naming convention. We need to use the `JsonNamingPolicy.CamelCase` property to deserialize the JSON string into the strongly-typed object. We can pass a `JsonSerializerOptions` instance to the `JsonSerializer.Serialize()` and `JsonSerializer.Deserialize()` methods to specify the serialization and deserialization options. Using a helper method can simplify the code.

The `HttpRequestMessage` class is a low-level class that represents an HTTP request message. In most cases, we can use the `GetStringAsync()` method to send a GET request and get the response content as a string, as follows:

```
var content = await httpClient.GetStringAsync("https://
jsonplaceholder.typicode.com/posts");
var posts = JsonSerializerHelper.
DeserializeWithCamelCase<List<Post>>(content);
return Ok(posts);
```

The code to send a POST request is similar:

```
[HttpPost]
public async Task<IActionResult> Post(Post post)
{
    var httpClient = httpClientFactory.CreateClient();
    var json = JsonSerializer.Serialize(post);
    var data = new StringContent(json, Encoding.UTF8, "application/
json");
    var response = await httpClient.PostAsync("https://
jsonplaceholder.typicode.com/posts", data);
    var content = await response.Content.ReadAsStringAsync();
    var newPost = JsonSerializer.Deserialize<Post>(content);
    return Ok(newPost);
}
```

To send a POST request, we need to serialize the Post object to a JSON string and then convert the JSON string into a `StringContent` instance. Then, we can use the `PostAsync()` method to send the request.

The `StringContent` class is a concrete implementation of the `HttpContent` class. The `HttpContent` class is an abstract class that represents the content of an HTTP message. It has the following concrete implementations:

- `ByteArrayContent`: Represents an `HttpContent` instance based on a byte array
- `FormUrlEncodedContent`: Represents a collection of name/value pairs encoded using `application/x-www-form-urlencoded` MIME type
- `MultipartContent`: Represents a collection of `HttpContent` instances serialized using `multipart/*` MIME type
- `StreamContent`: Represents an `HttpContent` instance based on a stream
- `StringContent`: Represents an `HttpContent` instance based on a string

The `HttpClient` class has a few methods and extension methods to send HTTP requests. The following table shows the commonly used methods:

Method name	Description
`SendAsync()`	Sends an HTTP request to the specified URI. This method can send any HTTP request.
`GetAsync()`	Sends a GET request to the specified URI.
`GetStringAsync()`	Sends a GET request to the specified URI. This method returns the response body as a string.
`GetByteArrayAsync()`	Sends a GET request to the specified URI. This method returns the response body as a byte array.
`GetStreamAsync()`	Sends a GET request to the specified URI. This method returns the response body as a stream.
`GetFromJsonAsync<T>()`	Sends a GET request to the specified URI. This method returns the response body as a strongly-typed object.
`GetFromJsonAsAsyncEnumerable<T>()`	Sends a GET request to the specified URI. This method returns the response body as an `IAsyncEnumerable<T>` instance.
`PostAsync()`	Sends a POST request to the specified URI.
`PostAsJsonAsync()`	Sends a POST request to the specified URI. The request body is serialized as JSON.
`PutAsync()`	Sends a PUT request to the specified URI.
`PutAsJsonAsync()`	Sends a PUT request to the specified URI. The request body is serialized as JSON.
`DeleteAsync()`	Sends a DELETE request to the specified URI.
`DeleteFromJsonAsync<T>()`	Sends a DELETE request to the specified URI. This method returns the response body as a strongly-typed object.
`PatchAsync()`	Sends a PATCH request to the specified URI.

Table 15.3 – The commonly used methods of the HttpClient class

When we use the `HttpClient` instance created by the `IHttpClientFactory` interface, we need to specify the request URL. We can set the base address of the `HttpClient` instance when registering the `HttpClient` instance. Update the `AddHttpClient()` method in the `Program.cs` file:

```
builder.Services.AddHttpClient(client =>
{
```

```
    client.BaseAddress = new Uri("https://jsonplaceholder.typicode.
com/");
    // You can set more options like the default request headers,
timeout, and so on.
});
```

Then, we do not need to specify the base address when sending the HTTP request. However, what if we need to send requests to multiple endpoints with different base addresses? Let us see how to solve this problem in the next section.

Named HttpClient instances

It is tedious to specify the base address of the HttpClient instance or the request URI every time. We can specify some common settings when registering the HttpClient instance. For example, we can specify the base address of the HttpClient instance as follows:

```
builder.Services.AddHttpClient("JsonPlaceholder", client =>
{
    client.BaseAddress = new Uri("https://jsonplaceholder.typicode.
com/");
    // You can set more options like the default request headers,
timeout, etc.
    client.DefaultRequestHeaders.Add(HeaderNames.Accept, "application/
json");
    client.DefaultRequestHeaders.Add(HeaderNames.UserAgent,
"HttpClientDemo");
});
```

In the preceding code, we register the HttpClient instance with the name JsonPlaceholder and specify the base address of the HttpClient instance. We can also set the default request headers, such as the Accept and User-Agent headers. Then, we can use the JsonPlaceholder name to inject the HttpClient instance into the controller:

```
var httpClient = httpClientFactory.CreateClient("JsonPlaceholder");
```

This is called named HttpClient instances, which allows us to register multiple HttpClient instances with different names. This is useful when we need multiple HttpClient instances with different configurations. By using the name, we can easily access the desired instance.

Typed HttpClient instances

To better encapsulate the HttpClient instances, we can create a typed HttpClient instance for a specific type. For example, we can create a typed HttpClient instance for the User type:

```
public class UserService
{
```

```
    private readonly HttpClient _httpClient;
    private readonly JsonSerializerOptions _jsonSerializerOptions =
new()
    {
        PropertyNamingPolicy = JsonNamingPolicy.CamelCase,
        DictionaryKeyPolicy = JsonNamingPolicy.CamelCase
    };

    public UserService(HttpClient httpClient)
    {
        _httpClient = httpClient;
        _httpClient.BaseAddress = new Uri("https://jsonplaceholder.
typicode.com/");
        _httpClient.DefaultRequestHeaders.Add(HeaderNames.Accept,
"application/json");
        _httpClient.DefaultRequestHeaders.Add(HeaderNames.UserAgent,
"HttpClientDemo");
    }

    public Task<List<User>?> GetUsers()
    {
        return _httpClient.GetFromJsonAsync<List<User>>("users", _
jsonSerializerOptions);
    }

    public async Task<User?> GetUser(int id)
    {
        return await _httpClient.GetFromJsonAsync<User>($"users/{id}",
_jsonSerializerOptions);
    }
    // Omitted for brevity
}
```

In the preceding code, we create a UserService class to encapsulate the HttpClient instance. Register the UserService class in the Program class:

```
builder.Services.AddHttpClient<UserService>();
```

Then, we can inject the UserService class into the controller:

```
[ApiController]
[Route("[controller]")]
public class UsersController(UserService usersService) :
ControllerBase
{
```

```
[HttpGet]
public async Task<ActionResult<List<User>>> Get()
{
    var users = await usersService.GetUsers();
    return Ok(users);
}

[HttpGet("{id}")]
public async Task<ActionResult<User>> Get(int id)
{
    var user = await usersService.GetUser(id);
    if (user == null)
    {
        return NotFound();
    }
    return Ok(user);
}
// Omitted for brevity
}
```

In the preceding code, the controller does not need to know the details of the HttpClient instance. It only needs to call the methods of the UserService class. The code is much cleaner.

The IHttpClientFactory interface is the recommended way to manage HttpClient instances. It saves us from the tedious work of managing the lifetime of the HttpClient instances. It also allows us to configure the HttpClient instances in a centralized place. For more information, please refer to the documentation at https://learn.microsoft.com/en-us/aspnet/core/fundamentals/http-requests.

Summary

In this chapter, we discussed common practices in ASP.NET Core web API development, such as HTTP status codes, asynchronous programming, pagination, response types, and API documentation. We also explored several caching techniques, including in-memory caching, distributed caching, response caching, and output caching. Each technique has its own advantages and disadvantages, so it is important to consider the trade-offs and choose the appropriate caching strategy for the given scenario. Additionally, we discussed the IHttpClientFactory interface, which simplifies the management of HttpClient instances and allows us to use dependency injection to inject HttpClient instances into the application without worrying about their life cycle.

In the next chapter, we will discuss how to handle errors in ASP.NET Core web API applications and how to monitor the applications using OpenTelemetry.

Error Handling, Monitoring, and Observability

In *Chapter 4*, we introduced how to use logging in ASP.NET Core web API applications. Logging is a critical part of application development that helps developers understand what's happening in their applications. However, logging is not enough – we need more tools to monitor and observe how our application is running. In this chapter, we will explore the following topics:

- Error handling
- Health checks
- Monitoring and observability

After reading this chapter, you will be able to understand how to monitor ASP.NET Core web API applications. You will have gained knowledge of observability and **OpenTelemetry**, as well as how to use some tools, such as Prometheus and Grafana, to monitor applications.

Technical requirements

The code samples for this chapter can be found at `https://github.com/PacktPublishing/Web-API-Development-with-ASP.NET-Core-8/tree/main/samples/chapter16`. You can use VS 2022 or VS Code to open the solutions.

Error handling

When an exception occurs in an ASP.NET Core web API application, the application will throw an exception. If this exception is not handled, the application will crash and cause a 500 error. The response body will contain the stack trace of the exception. Displaying the stack trace to the client is acceptable during development. However, we should never expose the stack trace to the client in production. The stack trace contains sensitive information about the application that can be used by attackers to attack the application.

Handling exceptions

Let's look at an example. The MyWebApiDemo sample application has a controller named UsersController, which has an action to get a user by their user ID. This action looks as follows:

```
[HttpGet("{id:int}")]
public ActionResult<User> Get(int id)
{
    var user = Users.First(u => u.Id == id);
    if (user == null)
    {
        return NotFound();
    }

    return Ok(user);
}
```

It is not advisable to use First in this instance as it will result in an exception being thrown if the user is not found in the collection. To illustrate how to handle exceptions in the application, we will use this example.

Run the application and send a GET request to the https://localhost:5001/users/100 endpoint. You can test it in the Swagger UI directly. The application will return a 500 error because no user with an ID of 100 will be found. The response body will look as follows:

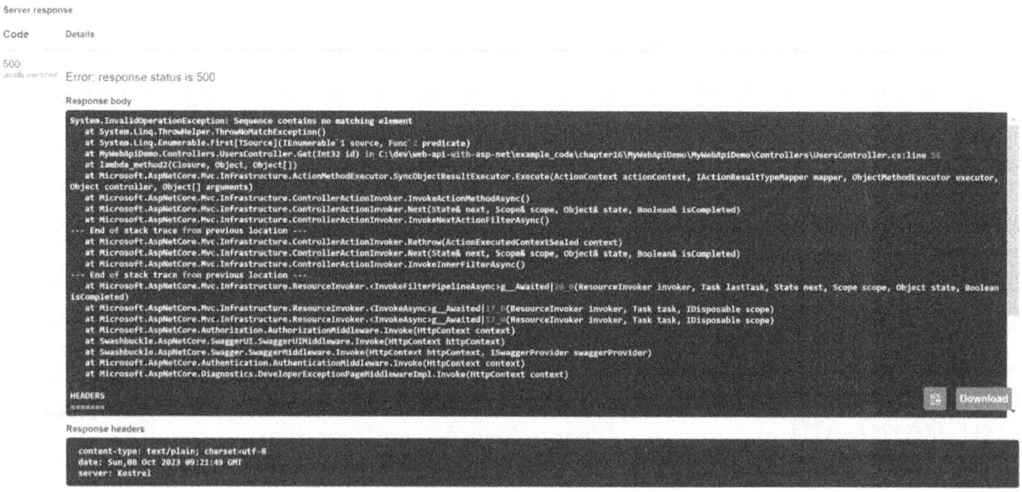

Figure 16.1 – The response body contains the stack trace

Regardless of whether the application is running in the development environment, the response body contains the stack trace. We should never show the stack trace for the production environment. Additionally, the response body is not a valid JSON payload, making it difficult for the client to parse it.

ASP.NET Core provides a built-in exception handling middleware to handle exceptions and return an error payload. The exception handling middleware can return a valid JSON payload to the client. This kind of JSON payload for error and exceptions is called **Problem Details** and is defined in RFC7807: `https://datatracker.ietf.org/doc/html/rfc7807`.

A problem details object can have the following properties:

- `type`: A URI reference that's used to identify the problem type. This reference provides helpful documentation in a human-readable format, which can assist clients in understanding the error.

- `title`: A summary that describes the problem's type in a human-readable format.

- `status`: An HTTP status code generated by the original server to indicate the status of the problem.

- `detail`: A human-readable description of the problem.

- `instance`: A URI reference that provides a specific occurrence of the problem, allowing for a more precise understanding of the issue.

The client can parse the problem details object and display a user-friendly error message. This object can be extended to include additional information about the error, though the existing properties should be sufficient for most cases.

To use the exception handling middleware, we need to create a controller to show the problem details. Create a new controller named `ErrorController` and add the following code:

```
[ApiController]
[ApiExplorerSettings(IgnoreApi = true)]
public class ErrorController(ILogger<ErrorController> logger) :
ControllerBase
{

    [Route("/error-development")]
    public IActionResult HandleErrorDevelopment(
        [FromServices] IHostEnvironment hostEnvironment)
    {
        if (!hostEnvironment.IsDevelopment())
        {
            return NotFound();
        }

        var exceptionHandlerFeature =
            HttpContext.Features.Get<IExceptionHandlerFeature>()!;
```

```
        logger.LogError(exceptionHandlerFeature.Error,
exceptionHandlerFeature.Error.Message);

        return Problem(
            detail: exceptionHandlerFeature.Error.StackTrace,
            title: exceptionHandlerFeature.Error.Message);
    }

    [Route("/error")]
    public IActionResult HandleError()
    {
        var exceptionHandlerFeature =
            HttpContext.Features.Get<IExceptionHandlerFeature>()!;
        logger.LogError(exceptionHandlerFeature.Error,
exceptionHandlerFeature.Error.Message);
        return Problem();
    }
}
```

The preceding code comes from Microsoft's official documentation: https://learn.microsoft.com/en-us/aspnet/core/web-api/handle-errors.

There are several things to note in the ErrorController class:

- The controller is marked with the [ApiExplorerSettings(IgnoreApi = true)] attribute. This attribute is used to hide this endpoint from the OpenAPI specification and Swagger UI.

- The controller has two actions. The first action is used to show a detailed error message in the development environment, so it provides the route /error-development that shows the stack trace of the exception. The second action is used to show a generic error message in the production environment, so it provides the /error route that has no additional information about the exception.

- In the actions, we use the IExceptionHandlerFeature interface to get the exception information. The IExceptionHandlerFeature interface is a feature containing the exception of the original request to be examined by an exception handler. We can log the exception information or return it to the client.

Next, we need to register the exception handling middleware in the application. Open the Program.cs file and call the UseExceptionHandler method to add the exception handling middleware:

```
if (app.Environment.IsDevelopment())
{
    app.UseSwagger();
```

```
    app.UseSwaggerUI();
    app.UseExceptionHandler("/error-development");
}
else
{
    app.UseExceptionHandler("/error");
}
```

For the development environment, we can use the `/error-development` endpoint to show the detailed error message. For the production environment, we can use the `/error` endpoint to show the generic error message. It is a good practice to hide the stack trace in the production environment.

Run the application and send a GET request to the `https://localhost:5001/users/100` endpoint. The application will return a 500 error. The response body will look as follows:

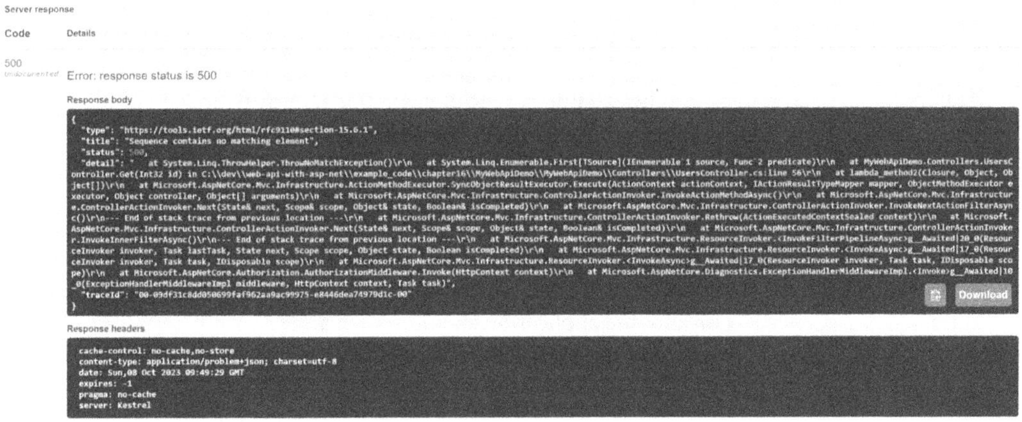

Figure 16.2 – The response body contains the problem details alongside
the stack trace in the development environment

The response body now contains a problem details JSON payload. It also contains the stack trace of the exception for troubleshooting in the development environment. The client can parse the response body and display a user-friendly error message. Meanwhile, the response headers contain the `Content-Type` header with a value of `application/problem+json`. This indicates that the response body is a problem details JSON payload.

If you run the application in the production environment, the response body will not contain the stack trace of the exception. The response body will look as follows:

Server response

Code   Details

500
Undocumented   Error: response status is 500

Response body

```
{
    "type": "https://tools.ietf.org/html/rfc9110#section-15.6.1",
    "title": "An error occurred while processing your request.",
    "status": 500,
    "traceId": "00-14376053119ff35734b6aea4fa6b1d2f-5d2eabea21eca853-00"
}
```

Response headers

```
cache-control: no-cache,no-store
content-type: application/problem+json; charset=utf-8
date: Sun,08 Oct 2023 09:52:33 GMT
expires: -1
pragma: no-cache
server: Kestrel
```

Figure 16.3 – The response body contains a general error message in the production environment

The default problem details object can be extended to include additional information about the error. We will discuss how to customize the problem details in the next section.

Model validation

When a client sends a request to the application, the application needs to validate the request. For example, when a user updates their profile, the Email property must be a valid email address. If the value of the Email property is invalid, the application should return an HTTP 400 response with a problem details object that contains the validation error message.

ASP.NET Core offers a built-in model validation feature to validate the request model. This feature is enabled using validation attributes, which are defined in the System.ComponentModel.DataAnnotations namespace. The following table outlines some available validation attributes:

Attribute name	Description
Required	Specifies that a data field is required
Range	Specifies that a numeric field must be in a specified range
StringLength	Specifies the minimum and maximum length of a string field
EmailAddress	Specifies that a data field must be a valid email address
RegularExpression	Specifies that a data field must match the specified regular expression
Url	Specifies that a data field must be a valid URL

Table 16.1 – Common model validation attributes

We can apply these validation attributes as follows:

```
public class User
{
    public int Id { get; set; }

    [Required]
    [StringLength(50, MinimumLength = 3, ErrorMessage = "The length of
FirstName must be between 3 and 50.")]
    public string FirstName { get; set; } = string.Empty;

    [Required]
    [StringLength(50, MinimumLength = 3, ErrorMessage = "The length of
LastName must be between 3 and 50.")]
    public string LastName { get; set; } = string.Empty;

    [Required]
    [Range(1, 120, ErrorMessage = "The value of Age must be between 1
and 120.")]
    public int Age { get; set; }

    [Required]
    [EmailAddress]
    public string Email { get; set; } = string.Empty;

    [Required]
    [Phone]
    public string PhoneNumber { get; set; } = string.Empty;
}
```

Run the application and send a POST request to the /users endpoint with an invalid request body, like this:

```
{
  "firstName": "ab",
  "lastName": "xy",
  "age": 20,
  "email": "user-example.com",
  "phoneNumber": "abcxyz"
}
```

The application will return an HTTP 400 response with a problem details object, as follows:

```json
{
  "type": "https://tools.ietf.org/html/rfc9110#section-15.5.1",
  "title": "One or more validation errors occurred.",
  "status": 400,
  "errors": {
    "Email": [
      "The Email field is not a valid e-mail address."
    ],
    "LastName": [
      "The length of LastName must be between 3 and 50."
    ],
    "FirstName": [
      "The length of FirstName must be between 3 and 50."
    ],
    "PhoneNumber": [
      "The PhoneNumber field is not a valid phone number."
    ]
  },
  "traceId": "00-8bafbe8952051318d15ddb570d2872b0-369effbb9978122b-00"
}
```

In this way, the client can parse the response body and display a user-friendly error message so that the user can correct the input.

Using FluentValidation to validate models

The previous section discussed the use of built-in validation attributes. However, these have certain limitations:

- The validation attributes are tightly coupled with the model. The models are polluted with validation attributes.

- The validation attributes cannot validate complex validation rules. If one property has dependencies on other properties, or the validation needs external services, the validation attributes cannot handle this.

To solve these problems, we can use `FluentValidation` to validate the models. `FluentValidation` is a popular open-source library for building strongly typed validation rules, allowing us to separate the validation logic from the models. It also supports complex validation rules.

To use `FluentValidation`, we need to install the `FluentValidation.AspNetCore` NuGet package. Run the following command in the terminal to install the package:

```
dotnet add package FluentValidation
```

> **Important note**
>
> Previously, `FluentValidation` provided a separate package for ASP.NET Core named `FluentValidation.AspNetCore`. However, this package is deprecated. It is recommended to use the `FluentValidation` package directly and use manual validation instead of using the ASP.NET Core validation pipeline. This is because the ASP.NET Core validation pipeline does not support asynchronous validation.

Next, we need to create a validator for the `User` model. Create a new class named `UserValidator` and add the following code:

```
public class UserValidator : AbstractValidator<User>
{
    public UserValidator()
    {
        RuleFor(u => u.FirstName)
            .NotEmpty()
            .WithMessage("The FirstName field is required.")
            .Length(3, 50)
            .WithMessage("The length of FirstName must be between 3
and 50.");

        // Omitted other rules for brevity

        // Create a custom rule to validate the Country and
PhoneNumber. If the country is New Zealand, the phone number must
start with 64.
        RuleFor(u => u)
            .Custom((user, context) =>
            {
                if (user.Country.ToLower() == "new zealand" && !user.
PhoneNumber.StartsWith("64"))
                {
                    context.AddFailure("The phone number must start
with 64 for New Zealand users.");
                }
            });
    }
}
```

In the preceding code, we use fluent syntax to specify validation rules for each property. We can also create a custom rule for dependent properties. In this example, we're creating a custom rule to validate the `Country` and `PhoneNumber` properties. If the country is New Zealand, we can create a custom rule that requires the phone number to start with 64. This is just one example of how to validate properties that depend on other properties; built-in validation attributes cannot handle this type of validation.

Next, we need to register the validator in the application. Add the following code to the `Program.cs` file:

```
builder.Services.AddScoped<IValidator<User>, UserValidator>();
```

The preceding code looks straightforward. But what if we have many validators? We can register all validators in a specific assembly. To do this, we need to install the `FluentValidation.DependencyInjectionExtensions` NuGet package. Run the following command in the terminal to install the package:

dotnet add package FluentValidation.DependencyInjectionExtensions

Then, we can register all validators, as follows:

```
builder.Services.AddValidatorsFromAssemblyContaining<UserValidator>();
```

Now, we can validate the model in the controller. Update the `Post` action, as follows:

```
[HttpPost]
public async Task<ActionResult<User>> Post(User user)
{
    var validationResult = await _validator.ValidateAsync(user);
    if (!validationResult.IsValid)
    {
        return BadRequest(new
ValidationProblemDetails(validationResult.ToDictionary()));
    }
    user.Id = Users.Max(u => u.Id) + 1;
    Users.Add(user);
    return CreatedAtRoute("", new { id = user.Id }, user);
}
```

In the preceding code, we utilize the `ValidateAsync()` method to validate the model. If the model is invalid, we return an HTTP `400` response containing a problem details object that contains the associated validation error message.

Send a `POST` request to the `/users` endpoint with the following payload:

```
{
  "firstName": "ab",
  "lastName": "xy",
  "age": 20,
  "email": "user-example.com",
  "country": "New Zealand",
  "phoneNumber": "12345678"
}
```

The application will return a 400 error with the following problem details object:

```
{
  "title": "One or more validation errors occurred.",
  "status": 400,
  "errors": {
    "FirstName": [
      "The length of LastName must be between 3 and 50."
    ],
    "LastName": [
      "The length of LastName must be between 3 and 50."
    ],
    "Email": [
      "The Email field is not a valid e-mail address."
    ],
    "": [
      "The phone number must start with 64 for New Zealand users."
    ]
  }
}
```

As we can see, the custom validation rule is executed and the error message is returned to the client.

`FluentValidation` has more features than just built-in validation attributes. If you have complex validation rules, you can consider using `FluentValidation`. For more details, please refer to the official documentation: `https://docs.fluentvalidation.net/en/latest/index.html`.

Health checks

To monitor the application, we need to know whether the application is running correctly or not. We can perform health checks to monitor the application. Normally, a health check is an endpoint that returns the health status of the application. This status can be *Healthy*, *Degraded*, or *Unhealthy*.

A health check is a critical part of the microservice architecture. In the microservice architecture, one API service may have multiple instances and also have dependencies on other services. A load balancer or orchestrator can be used to distribute the traffic to different instances. If one instance is unhealthy, the load balancer or orchestrator can stop sending traffic to the unhealthy instance. For example, Kubernetes – a popular container orchestrator – can use health checks to determine whether a container is healthy or not. If a container is not live, Kubernetes will restart the container.

We won't discuss the details of Kubernetes in this book. Instead, we will focus on how to implement health checks for Kubernetes in ASP.NET Core web API applications.

Implementing a basic health check

ASP.NET Core provides a straightforward way to configure health checks. We can use the AddHealthChecks method to add health checks to the application. Open the `Program.cs` file and add the following code:

```
builder.Services.AddHealthChecks();
var app = builder.Build();
app.MapHealthChecks("healthcheck");
```

The preceding code adds a basic health check to the application. The health check's endpoint is / healthcheck. Run the application and send a GET request to the /healthcheck endpoint. If successful, the application will return a 200 response with Healthy in plain text in the response body.

However, this health check is too simple. In the real world, a web API application may be more complex. It may have multiple dependencies, such as databases, message queues, and other services. We need to check the health status of these dependencies. If some core dependencies are unhealthy, the application should be unhealthy. Let's see how to implement a more complex health check.

Implementing a complex health check

A health check implementation class implements the IHealthCheck interface. The IHealthCheck interface is defined as follows:

```
public interface IHealthCheck
{
    Task<HealthCheckResult> CheckHealthAsync(
        HealthCheckContext context,
        CancellationToken cancellationToken = default);
}
```

We can create a custom health check implementation to ensure the proper functioning of our API. For instance, if the API depends on another service, we can create a health check implementation to verify the health status of the dependent service. If the dependent service is unhealthy, the API won't be able to function correctly. Here is an example of a health check implementation:

```
public class OtherServiceHealthCheck(IHttpClientFactory
httpClientFactory) : IHealthCheck
{

    public async Task<HealthCheckResult> CheckHealthAsync(
        HealthCheckContext context,
        CancellationToken cancellationToken = default)
    {
        var client = httpClientFactory.
CreateClient("JsonPlaceholder");
```

```
            var response = await client.GetAsync("posts",
cancellationToken);
        return response.IsSuccessStatusCode
            ? HealthCheckResult.Healthy("A healthy result.")
            : HealthCheckResult.Unhealthy("An unhealthy result.");
    }
}
```

In the preceding code, we create a health check implementation to check the health status of the `https://jsonplaceholder.typicode.com/posts` endpoint. If the endpoint returns a 200 response, the health check returns healthy. Otherwise, the health check returns unhealthy.

Next, we need to register the health check implementation in the application. Open the `Program.cs` file and add the following code:

```
builder.Services.AddHealthChecks()
    .AddCheck<OtherServiceHealthCheck>("OtherService");
// Omitted other code for brevity
app.MapHealthChecks("/other-service-health-check",
    new HealthCheckOptions() { Predicate = healthCheck => healthCheck.
Name == "OtherService" });
```

This code is similar to the previous health check. First, we use the `AddHealthChecks` method to register the strongly typed health check implementation. Then, we use the `MapHealthCheck` method to map the `/other-service-health-check` endpoint to the health check implementation. We also use the `HealthCheckOptions` object to specify the name of the health check, which is used to filter the health checks. If we do not specify the name of the health check, all health check implementations will be executed.

Run the application and send a `GET` request to the `/other-service-health-check` endpoint. If the dependent service, `https://jsonplaceholder.typicode.com/posts`, is healthy, the application will return a 200 response with `Healthy` in plain text in the response body.

Sometimes, we need to check multiple dependent services. We can register multiple health check implementations with a specific tag, at which point we can use this tag to filter the health checks. The following code shows how to register multiple health check implementations:

```
builder.Services.AddHealthChecks()
    .AddCheck<OtherServiceHealthCheck>("OtherService", tags: new[] {
"other-service" })
    .AddCheck<OtherService2HealthCheck>("OtherService2", tags: new[] {
"other-service" });
    .AddCheck<OtherService3HealthCheck>("OtherService3", tags: new[] {
"other-service" });
```

In the preceding code, we register three health check implementations with the same tag – that is, `other-service`. Now, we can use the tag to filter the health checks. The following code shows how to filter the health checks:

```
app.MapHealthChecks("/other-services-health-check",
    new HealthCheckOptions() { Predicate = healthCheck => healthCheck.
Tags.Contains("other-service") });
```

Like the `Name` property, we can use the `Tags` property to filter the health checks. When we send a `GET` request to the `/other-services-health-check` endpoint, the application will return a `200 OK` response with `Healthy` in plain text in the response body if all dependent services are healthy. But if one of the dependent services is unhealthy, the health check will return a `503 Service Unavailable` response with `Unhealthy` in plain text in the response body.

> **Important note**
>
> If the `MapHealthChecks()` method does not use the `HealthCheckOptions` parameter, the health check endpoint will run all registered health checks by default.

Implementing a database health check

In the previous section, we discussed how to implement a health check for a dependent service. As databases are a common component of web API applications, this section will focus on how to implement a database health check.

The approach to implementing a database health check is similar to what we covered in the previous section: we need to connect to the database and execute a simple query to check if the database is healthy. If you use EF Core to access the database, you can use the `Microsoft.Extensions.Diagnostics.HealthChecks.EntityFrameworkCore` package to implement a database health check. This package provides a health check implementation for EF Core, so we do not need to write the health check implementation ourselves. Run the following command in the terminal to install the package:

```
dotnet add package Microsoft.Extensions.Diagnostics.HealthChecks.
EntityFrameworkCore
```

In the sample project, we have an `InvoiceDbContext` class to access the database. The following code shows how to register the `InvoiceDbContext` class in the application:

```
builder.Services.AddDbContext<InvoiceDbContext>(options =>
    options.UseSqlServer(builder.Configuration.
GetConnectionString("DefaultConnection")));
```

Once you've done this, register the EF Core `DbContext` health check implementation, as follows:

```
builder.Services.AddHealthChecks().
AddDbContextCheck<InvoiceDbContext>("Database", tags: new[] {
"database" });
```

Similarly, assign a tag to the health check implementation so that we can filter the health checks. Then, we can map the health check endpoint to the health check's implementation, as follows:

```
app.MapHealthChecks("/database-health-checks",
    new HealthCheckOptions() { Predicate = healthCheck => healthCheck.
Tags.Contains("database") });
```

Run the application and send a `GET` request to the `/database-health-checks` endpoint. If the database is healthy, the application will return a `200 OK` response with `Healthy` in plain text in the response body. Additionally, you can register multiple health checks for different databases if necessary.

> **Important note**
>
> If you are using other ORMs to access the database, you can create a custom health check implementation following the previous section. This can be done by executing a simple query, such as `SELECT 1`, to determine whether the database is functioning properly.

Understanding readiness and liveness

In the previous sections, we discussed how to implement health checks for ASP.NET Core web API applications. In the real world, we may need to deploy the applications to a container orchestrator, such as Kubernetes. Kubernetes is a popular container orchestrator that can manage containerized applications, monitor health statuses, and scale up or down based on workload. Kubernetes uses the term **probe**, which is similar to health checks, to monitor the health status of the applications. While this book does not cover the details of Kubernetes, it will discuss how to implement Kubernetes probes in ASP.NET Core web API applications.

Kubernetes has three types of probes: *readiness*, *liveness*, and *startup*. Let's take a closer look:

- `liveness`: This probe indicates whether the application is running correctly. Kubernetes performs a `liveness` probe every few seconds. If the application does not respond to the `liveness` probe for a specified period, the container will be killed and Kubernetes will create a new one to replace it. The `liveness` probe can execute either an HTTP request, a command, or a TCP socket check. It also supports gRPC health checks.

- `readiness`: This probe is used to determine whether the application is ready to receive traffic. Some applications need to perform some initialization tasks before they can receive traffic, such as connecting to the database, loading configuration, checking the dependent services, and so on. During this period, the application cannot receive traffic, but this does not mean that the application is unhealthy. Kubernetes should not kill the application and restart it. After the initialization is complete and all dependent services are healthy, the readiness probe will inform Kubernetes that the application is ready to receive traffic.

- startup: This probe is similar to the readiness probe. However, the difference is that the startup probe is only executed once the application starts. It is used to determine whether the application has completed the initialization process. If this probe is configured, the liveness and readiness probes will not be executed until the startup probe is successful.

Configuring the probes incorrectly may cause cascading failures. For example, service A depends on service B and service B depends on service C. If the liveness probes are misconfigured incorrectly to check the dependent services when service C is unhealthy, service A and service B will be restarted, which does not solve the problem. This is a cascading failure. In this case, service A and service B should not be restarted. Instead, only service C should be restarted.

Here's an example of a liveness HTTP probe configuration for Kubernetes:

```
livenessProbe:
  httpGet:
    path: /liveness
    port: 8080
    httpHeaders:
    - name: Custom-Header
      value: X-Health-Check
  initialDelaySeconds: 3
  periodSeconds: 5
  timeoutSeconds: 1
  successThreshold: 1
  failureThreshold: 3
```

In the configuration, we define the following properties:

- path, port, and httpHeaders: These properties are used to configure the HTTP request. In the preceding example, we specify a custom HTTP header called Custom-Header with a value of X-Health-Check. The application can use this HTTP header to identify whether the request is a health check request. If the request does not have this HTTP header, the application can deny the request.

- initialDelaySeconds: This property is used to specify the number of seconds after the container has started before the first probe is executed. The default value is 0. Do not use a high value for this property. You can use the startup probe to check the initialization of the application instead.

- `periodSeconds`: This property is used to specify the number of seconds between each probe. The default value is 10. The minimum value is 1. You can adjust this value based on your scenarios. Make sure Kubernetes can discover the unhealthy container as soon as possible.

- `timeoutSeconds`: This property is used to specify the number of seconds after which the probe times out. The default value is 1 and the minimum value is also 1. Make sure the probe is fast.

- `successThreshold`: This property is used to determine the number of consecutive successful responses required for a probe to be considered successful after having previously failed. The value must be 1 for liveness probes.

- `failureThreshold`: This property is used to specify the number of consecutive failures for the probe to be considered failed after having succeeded. Do not use a high value for this property; otherwise, Kubernetes needs to wait a long time to restart the container.

Keep in mind that the `liveness` probe should not depend on other services. In other words, do not check the health status of other services in the `liveness` probe. Instead, this probe should only check whether the application can respond to the request.

An example of a `readiness` HTTP probe's configuration is as follows:

```
readinessProbe:
  httpGet:
    path: /readiness
    port: 8080
    httpHeaders:
    - name: Custom-Header
      value: X-Health-Check
  initialDelaySeconds: 5
  periodSeconds: 5
  timeoutSeconds: 1
  successThreshold: 3
  failureThreshold: 2
```

There are a few different considerations for the `readiness` probe:

- `successThreshold`: The default value is 1. However, we can increase this value to make sure the application is ready to receive traffic.

- `failureThreshold`: After at least `failureThreshold` probes have failed, Kubernetes will stop sending traffic to the container. As the application may have temporary problems, we can allow a few failures before the application is considered unhealthy. However, do not use a high value for this property.

If the application takes a long time to initialize, we can use the `startup` probe to check the initialization of the application. An example of a `startup` HTTP probe configuration is shown here:

```
startupProbe:
  httpGet:
    path: /startup
    port: 8080
    httpHeaders:
    - name: Custom-Header
      value: X-Health-Check
  periodSeconds: 5
  timeoutSeconds: 1
  successThreshold: 1
  failureThreshold: 30
```

In this configuration, the `startup` probe will be executed every 5 seconds, and the application will have a maximum of 150 seconds (5 * 30 = 150 seconds) to complete the initialization. `successThreshold` must be 1 so that once the `startup` probe is successful, the `liveness` and `readiness` probes will be executed. If the `startup` probe fails after 150 seconds (about 2 and a half minutes), Kubernetes will kill the container and start a new one. So, ensure that the `startup` probe has enough time to complete the initialization.

Configuring Kubernetes probes is not a simple task. We need to consider many factors. For example, should we check the dependent services in the `readiness` probe? If the application can partially operate without a specific dependent service, it can be considered as degraded instead of unhealthy. In this case, if the application has mechanisms to handle transient failures gracefully, it might be acceptable to omit specific dependent services in the `readiness` probe. So, please consider your scenarios carefully; you may need a compromise when configuring the probes.

This section is not intended to cover all the details of Kubernetes probes. For more details, please refer to the following official documentation:

- Kubernetes documentation: `https://kubernetes.io/docs/tasks/configure-pod-container/configure-liveness-readiness-startup-probes/`.

- Health checks in ASP.NET Core: `https://learn.microsoft.com/en-us/aspnet/core/host-and-deploy/health-checks`.

Monitoring and observability

In the real world, building an application is just the first step. We also need to monitor and observe how the application is performing. This is where the concept of *observability* comes in. In this section, we will discuss observability and how to use OpenTelemetry to monitor and observe applications.

What is observability?

In *Chapter 4*, we introduced logging in ASP.NET Core web API applications. We learned how to use the built-in logging framework to log messages to different logging providers. **Observability** is a more comprehensive concept than logging. Besides logging, observability allows us to gain a deeper understanding of how the application is performing. For instance, we can determine how many requests are processed in a given hour, what the request latency is, and how requests are handled by multiple services in a microservice architecture. All of these are part of observability.

In general, observability has three pillars:

- **Logs**: Logs are used to record what is happening within the application, such as incoming requests, outgoing responses, important business logic executions, exceptions, errors, warnings, and so on.

- **Metrics**: Metrics are used to measure the performance of the application, such as the number of requests, the request latency, error rates, resource usage, and so on. These metrics can be used to trigger alerts when the application is not performing well.

- **Traces**: Traces are used to track the flow of requests across multiple services to identify where the time is spent or where the errors occur. This is especially useful in the microservice architecture.

To summarize, observability is the practice of understanding the application's internal state and operational characteristics by analyzing its logs, metrics, and traces. There are a few different ways to implement observability in ASP.NET Core web API applications – we can update the source code to add logging, metrics, and traces or use tools to monitor and observe the application without changing the code. In this section, we will discuss the first approach by using OpenTelemetry to implement observability in ASP.NET Core web API applications. This gives us more flexibility to customize the observability aspect.

Using OpenTelemetry to collect observability data

OpenTelemetry is a popular cross-platform, open-source standard for collecting observability data. It provides a set of APIs, SDKs, and tools to instrument, generate, collect, and export telemetry data so that we can analyze the application's performance and behavior. It supports many platforms and languages, as well as popular cloud providers. You can find more details about OpenTelemetry at `https://opentelemetry.io/`.

The .NET OpenTelemetry implementation consists of the following components:

- **Core API**: The core API is a set of interfaces and classes that define the OpenTelemetry API. It is a platform-independent API that can be used to instrument the application.

- **Instrumentation**: This is a set of libraries that can be used to collect instrumentation from the application. This component includes multiple packages for different frameworks and platforms, such as ASP.NET Core, gRPC, HTTP calls, SQL database operations, and so on.

- **Exporters**: Exporters are used to export the collected telemetry data to different targets, such as console and **Application Performance Monitoring** (**APM**) systems, including Prometheus, Zipkin, and so on.

In *Chapter 4*, we introduced using *Serilog* and *Seq* to collect logs. In the next few sections, we will focus on how to use OpenTelemetry to collect metrics and traces. We will use *Prometheus* to collect metrics and *Grafana* to visualize the metrics. We will also use *Jaeger* to collect traces. All these tools are open-source. In addition, we will explore Azure Application Insights, a powerful APM system provided by Microsoft.

Integrating OpenTelemetry with ASP.NET Core web API applications

In this section, we will explore how to use metrics in ASP.NET Core web API applications. In the sample project, we have an `InvoiceController` class to manage invoices. We want to know how many requests are executed and the duration of each request. We have several steps to perform:

1. Define the metrics for these activities.

2. Generate and collect instrumentation data.

3. Visualize the data in a dashboard.

To start, we need to install some NuGet packages using the following commands:

```
dotnet add package OpenTelemetry.Instrumentation.AspNetCore
--prerelease
dotnet add package OpenTelemetry.Instrumentation.Http --prerelease
dotnet add package OpenTelemetry.Exporter.OpenTelemetryProtocol
dotnet add package OpenTelemetry.Exporter.Console
dotnet add package OpenTelemetry.Extensions.Hosting
```

These packages include the required .NET OpenTelemetry implementations. Note that at the time of writing, some packages did not have stable versions available, so we needed to use the `--prerelease` option to install the latest preview versions. If you are reading this book when the stable versions are available, you can omit the `--prerelease` option.

Next, we must define the metrics. In this example, we want to know how many requests are executed for each action for the `/api/Invoices` endpoint. Create a new class named `InvoiceMetrics` in the `\OpenTelemetry\Metrics` folder. The following code shows how to define the metrics:

```
public class InvoiceMetrics
{
    private readonly Counter<long> _invoiceCreateCounter;
    private readonly Counter<long> _invoiceReadCounter;
    private readonly Counter<long> _invoiceUpdateCounter;
    private readonly Counter<long> _invoiceDeleteCounter;
    public InvoiceMetrics(ImeterFactory meterFactory)
```

```
    {
        var meter = meterFactory.Create("MyWebApiDemo.Invoice");
        _invoiceCreateCounter = meter.
CreateCounter<long>("mywebapidemo.invoices.created");
        _invoiceReadCounter = meter.CreateCounter<long>("mywebapidemo.
invoices.read");
        _invoiceUpdateCounter = meter.
CreateCounter<long>("mywebapidemo.invoices.updated");
        _invoiceDeleteCounter = meter.
CreateCounter<long>("mywebapidemo.invoices.deleted");
    }

    public void IncrementCreate()
    {
        _invoiceCreateCounter.Add(1);
    }

    public void IncrementRead()
    {
        _invoiceReadCounter.Add(1);
    }
    // Omitted other methods for brevity
}
```

The IMeterFactory interface is registered in ASP.NET Core's DI container by default and is used to create a meter. This meter, which is called MyWebApiDemo.Invoice, is used to record the metrics. Additionally, four counters are created to record the number of requests for each action. To facilitate this, four public methods are exposed to increment the counters.

The name of each metric must be unique. When we create a metric or a counter, it is recommended to follow the OpenTelemetry naming guidelines: https://github.com/open-telemetry/semantic-conventions/blob/main/docs/general/metrics.md#general-guidelines.

Next, we need to register the metrics in the application. Add the following code to the Program.cs file:

```
builder.Services.AddOpenTelemetry()
    .ConfigureResource(config =>
    {
        config.AddService(nameof(MyWebApiDemo));
    })
    .WithMetrics(b =>
    {
        b.AddConsoleExporter();
        b.AddAspNetCoreInstrumentation();
```

```
        b.AddMeter("Microsoft.AspNetCore.Hosting",
            "Microsoft.AspNetCore.Server.Kestrel",
            "MyWebApiDemo.Invoice");
    });
builder.Services.AddSingleton<InvoiceMetrics>();
```

In the preceding code, we used the AddOpenTelemetry() method to register the OpenTelemetry services. The ConfigureResource() method registers the service name. Inside the WithMetrics() method, we use the AddConsoleExporter() method to add a console exporter. This console exporter is useful for local development and debugging. We also added three meters, including the ASP.NET Core hosting and Kestrel server, so that we can collect the metrics from the ASP.NET Core web API framework. Finally, we registered the InvoiceMetrics class in the dependency injection container as a singleton.

Next, we can use the InvoiceMetrics class to record the metrics. Open the InvoiceController class and call the IncrementCreate() method in the Post action, as follows:

```
[HttpPost]
public async Task<ActionResult<Invoice>> Post(Invoice invoice)
{
    // Omitted for brevity
    await dbContext.SaveChangesAsync();
    // Instrumentation
    _invoiceMetrics.IncrementCreate();
    return CreatedAtAction(nameof(Get), new { id = invoice.Id },
invoice);
}
```

The other actions are similar. Before we check the metrics in the console, we need to install the **dotnet-counters** tool, a command-line tool for viewing live metrics. Run the following command in the terminal to install the tool:

```
dotnet tool install --global dotnet-counters
```

Then, we can use the dotnet counters command to view the metrics. We can check the metrics from Microsoft.AspNetCore.Hosting using the following command:

```
dotnet-counters monitor -n MyWebApiDemo --counters Microsoft.
AspNetCore.Hosting
```

Run the application and send some requests to the /api/Invoices endpoint. You will see the following metrics:

```
Press p to pause, r to resume, q to quit.
    Status: Running
```

```
[Microsoft.AspNetCore.Hosting]
    http.server.active_requests ({request})
        http.request.method=GET,url.scheme=htt
ps                              0
        http.request.method=POST,url.scheme=htt
ps                              0
    http.server.request.duration (s)
        http.request.method=GET,http.response.status_
code=200,ht          0.006
        http.request.method=GET,http.response.status_
code=200,ht          0.006
        http.request.method=GET,http.response.status_
code=200,ht          0.006
        http.request.method=POST,http.response.status_
code=201,h           0.208
        http.request.method=POST,http.response.status_
code=201,h           0.208
        http.request.method=POST,http.response.status_
code=201,h           0.208
```

Here, you can see the metrics for the HTTP actions, including the active requests and the request duration. You can use this tool to observe more performance metrics, such as CPU usage or the rate of exceptions being thrown in the application. For more information about this tool, please refer to the official documentation: https://learn.microsoft.com/en-us/dotnet/core/diagnostics/dotnet-counters.

To check the custom metrics via InvoiceMetrics, you need to specify the --counters option in the command, as follows:

```
dotnet-counters monitor -n MyWebApiDemo --counters MyWebApiDemo.
Invoice
```

The output may look like this:

```
Press p to pause, r to resume, q to quit.
    Status: Running

[MyWebApiDemo.Invoice]
    mywebapidemo.invoices.created (Count / 1 sec)                0
    mywebapidemo.invoices.read (Count / 1 sec)                   0
```

Here, you can see the metrics we defined in the invoiceMetrics class. Note that you can include multiple counters in the --counters option, separated by commas. For example, you can use the following command to check the metrics for both Microsoft.AspNetCore.Hosting and MyWebApiDemo.Invoice:

```
dotnet-counters monitor -n MyWebApiDemo --counters Microsoft.
AspNetCore.Hosting,MyWebApiDemo.Invoice
```

In the `InvoiceMetrics` class, we defined four counters. There are more types of instruments in OpenTelemetry, such as `Gauge`, `Histogram`, and others. Here are some of the different types of instruments that are available:

- `Counter`: A counter is used to track a value that can only increase over time – for example, the number of requests after the application starts.

- `UpDownCounter`: An up-down counter is similar to a counter, but it can increase or decrease over time. An example of this is the number of active requests. When a request starts, the counter increases by 1. When the request ends, the counter decreases by 1. It can also be used to monitor the size of a queue.

- `Gauge`: A gauge measures a current value at a specific point in time, such as CPU usage or memory usage.

- `Histogram`: A histogram measures the statistical distribution of values using aggregations. For example, a histogram can measure how many requests are processed longer than a specific duration.

We can define more metrics to monitor the application. Define an `UpDownCounter` instrument to track how many active requests there are for the `/api/Invoices` endpoint. Update the `InvoiceMetrics` class, as follows:

```
private readonly UpDownCounter<long> _invoiceRequestUpDownCounter;

public InvoiceMetrics(IMeterFactory meterFactory)
{
    // Omitted for brevity
    _invoiceRequestUpDownCounter = meter.
CreateUpDownCounter<long>("mywebapidemo.invoices.requests");
}

public void IncrementRequest()
{
    _invoiceRequestUpDownCounter.Add(1);
}

public void DecrementRequest()
{
    _invoiceRequestUpDownCounter.Add(-1);
}
```

Then, update the `InvoiceController` class so that it increments and decrements the counter. For simplicity, we'll just call the `IncrementRequest()` and `DecrementRequest()` methods in the controller. In the real world, it is recommended to use an ASP.NET Core middleware to handle this. The following code shows how to update the `InvoiceController` class:

```
[HttpGet("{id}")]
public async Task<ActionResult<Invoice>> Get(Guid id)
{
    _invoiceMetrics.IncrementRequest();
    // Omitted for brevity
    _invoiceMetrics.DecrementRequest();
    return Ok(result);
}
```

An example of `Histogram` is shown here:

```
private readonly Histogram<double> _invoiceRequestDurationHistogram;

public InvoiceMetrics(IMeterFactory meterFactory)
{
    // Omitted for brevity
    _invoiceRequestDurationHistogram = meter.
CreateHistogram<double>("mywebapidemo.invoices.request_duration");
}

public void RecordRequestDuration(double duration)
{
    _invoiceRequestDurationHistogram.Record(duration);
}
```

Then, update the `InvoiceController` class so that it records the request's duration. Similarly, we can just use the `RecordRequestDuration()` method in the controller. The following code shows how to update the `InvoiceController` class:

```
[HttpGet("{id}")]
public async Task<ActionResult<Invoice>> Get(Guid id)
{
    var stopwatch = Stopwatch.StartNew();
    // Omitted for brevity
    // Simulate a latency
    await Task.Delay(_random.Next(0, 500));
    // Omitted for brevity
    stopwatch.Stop();
    _invoiceMetrics.RecordRequestDuration(stopwatch.Elapsed.
TotalMilliseconds);
    return Ok(result);
}
```

Here, we use a `Task.Delay()` method to simulate latency. Run the application and send some requests to the `/api/Invoices` endpoint. Then, check the metrics using the `dotnet-counters` tool. You will see the metrics, as follows:

```
Press p to pause, r to resume, q to quit.
    Status: Running

[Microsoft.AspNetCore.Hosting]
    http.server.active_requests ({request})
        http.request.method=GET,url.scheme=htt
ps                                0
        http.request.method=POST,url.scheme=htt
ps                                0
    http.server.request.duration (s)
        http.request.method=GET,http.response.status_
code=200,ht            0.075
        http.request.method=GET,http.response.status_
code=200,ht            0.24
        http.request.method=GET,http.response.status_
code=200,ht            0.24
        http.request.method=POST,http.response.status_
code=201,h            0.06
        http.request.method=POST,http.response.status_
code=201,h            0.06
        http.request.method=POST,http.response.status_
code=201,h            0.06
[MyWebApiDemo.Invoice]
    mywebapidemo.invoices.created (Count / 1
sec)                              0
    mywebapidemo.invoices.read (Count / 1 sec
)                                 0
    mywebapidemo.invoices.request_duration
        Percen-
tile=50                                                        74.25
        Percen-
tile=95                                                        239.5
        Percen-
tile=99                                                        239.5
    mywebapidemo.invoices.reques
ts                                0
```

In the preceding output, the histogram instruments are shown as `Percentile=50`, `Percentile=95`, and `Percentile=99`. This is the default configuration for the `dotnet-counters` tool. We can use other tools, such as Prometheus and Grafana, to provide more visualization options. We will discuss this in the next section.

Using Prometheus to collect and query metrics

Prometheus is a widely used open-source monitoring system. Prometheus was originally developed by SoundCloud (`https://soundcloud.com/`), then joined the Cloud Native Computing Foundation (`https://cncf.io/`) in 2016. Prometheus is capable of collecting metrics from a variety of sources, including applications, databases, operating systems, and more. It also offers a powerful query language for querying the collected metrics, as well as a dashboard to visualize them.

In this section, we will use Prometheus to collect metrics from the ASP.NET Core web API application and visualize the metrics.

To install Prometheus, navigate to the official website: `https://prometheus.io/download/`. Download the latest version of Prometheus for your operating system.

Next, we need to configure the ASP.NET web API application to export metrics for Prometheus. Install the `OpenTelemetry.Exporter.Prometheus.AspNetCore` package in the ASP.NET Core web API project using the following command:

```
dotnet add package OpenTelemetry.Exporter.Prometheus.AspNetCore
--prerelease
```

Then, register the Prometheus exporter in the `Program.cs` file, as follows:

```
builder.Services.AddOpenTelemetry()
    .ConfigureResource(config =>
    {
        config.AddService(nameof(MyWebApiDemo));
    })
    .WithMetrics(metrics =>
    {
        metrics.AddAspNetCoreInstrumentation()
            .AddMeter("Microsoft.AspNetCore.Hosting")
            .AddMeter("Microsoft.AspNetCore.Server.Kestrel")
            .AddMeter("MyWebApiDemo.Invoice")
            .AddConsoleExporter()
            .AddPrometheusExporter();
    });
// Omitted for brevity
// Add the Prometheus scraping endpoint
app.MapPrometheusScrapingEndpoint();
```

Now, we have two exporters: the console exporter and the Prometheus exporter. If you don't need the console exporter, you can remove it. We're also using the `MapPrometheusScrapingEndpoint()` method to map the `/metrics` endpoint for the Prometheus exporter. This endpoint is used by Prometheus to scrape metrics from the application.

Next, we need to configure Prometheus to collect metrics from the ASP.NET Core web API application. Find the port number of the ASP.NET Core web API application. In the sample project, we use port number 5125 for HTTP. You can find the relevant port numbers in the `launchSettings.json` file.

Open the `prometheus.yml` file in the Prometheus folder. Add a job at the end of the file, as follows:

```
- job_name: 'MyWebApiDemo'
  scrape_interval: 5s # Set the scrape interval to 5 seconds so we can
see the metrics update immediately.
  static_configs:
    - targets: ['localhost:5125']
```

The `scrape_interval` property is set to specify the interval at which metrics should be scrapped. For testing purposes, this can be set to 5 seconds so that you can view metrics immediately. However, in production scenarios, it is recommended to set this to a higher value, such as 15 seconds. Additionally, ensure that the `targets` property is set to the correct port number before saving the file.

If you use HTTPS for the ASP.NET Core web API application, you need to specify the `schema` property, as follows:

```
- job_name: 'MyWebApiDemo'
  scrape_interval: 5s # Set the scrape interval to 5 seconds so we can
see the metrics update immediately.
  scheme: https
  static_configs:
    - targets: ['localhost:7003']
```

Run the application and send some requests to the `/api/Invoices` endpoint. Navigate to the `/metrics` endpoint; you will see the relevant metrics:

```
# TYPE kestrel_active_connections gauge
# HELP kestrel_active_connections Number of connections that are currently active on the server.
kestrel_active_connections{network_transport="tcp",network_type="ipv6",server_address="::1",server_port="7003"} 2 1697320877828
kestrel_active_connections{network_transport="tcp",network_type="ipv4",server_address="127.0.0.1",server_port="7003"} 0 1697320877828

# TYPE kestrel_connection_duration_seconds histogram
# UNIT kestrel_connection_duration_seconds seconds
# HELP kestrel_connection_duration_seconds The duration of connections on the server.
kestrel_connection_duration_seconds_bucket{network_transport="tcp",network_type="ipv6",server_address="::1",server_port="7003",le="0"} 0 1697320877828
kestrel_connection_duration_seconds_bucket{network_transport="tcp",network_type="ipv6",server_address="::1",server_port="7003",le="0.005"} 0 1697320877828
kestrel_connection_duration_seconds_bucket{network_transport="tcp",network_type="ipv6",server_address="::1",server_port="7003",le="0.01"} 0 1697320877828
kestrel_connection_duration_seconds_bucket{network_transport="tcp",network_type="ipv6",server_address="::1",server_port="7003",le="0.025"} 1 1697320877828
kestrel_connection_duration_seconds_bucket{network_transport="tcp",network_type="ipv6",server_address="::1",server_port="7003",le="0.05"} 1 1697320877828
kestrel_connection_duration_seconds_bucket{network_transport="tcp",network_type="ipv6",server_address="::1",server_port="7003",le="0.075"} 1 1697320877828
```

Figure 16.4 – Metrics for Prometheus

Now, we can run Prometheus by executing the `prometheus.exe` file. In the output, you will find the following line:

```
ts=2023-10-14T10:44:55.133Z caller=web.go:566 level=info component=web
msg="Start listening for connections" address=0.0.0.0:9090
```

This means that Prometheus is running on port `9090`. Navigate to `http://localhost:9090`. You will see the Prometheus dashboard, as follows:

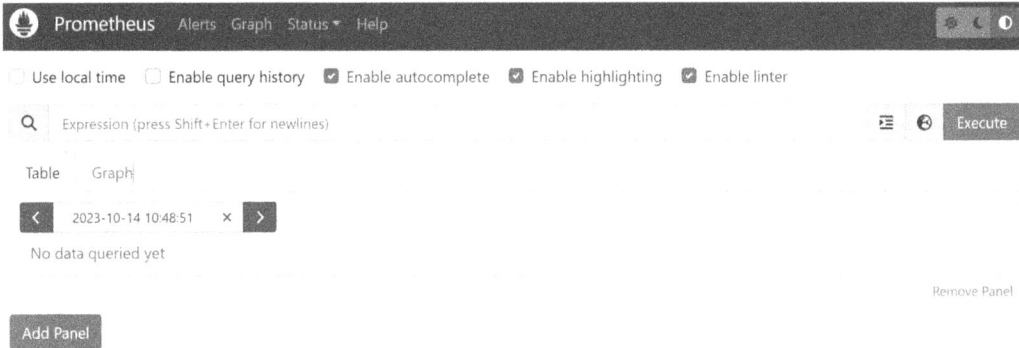

Figure 16.5 – Prometheus dashboard

Prometheus will start to scrape metrics from the ASP.NET Core web API application we configured in the `prometheus.yml` file. Click **Status** | **Targets** at the top. You will see the following page, which shows the status of the ASP.NET Core web API application:

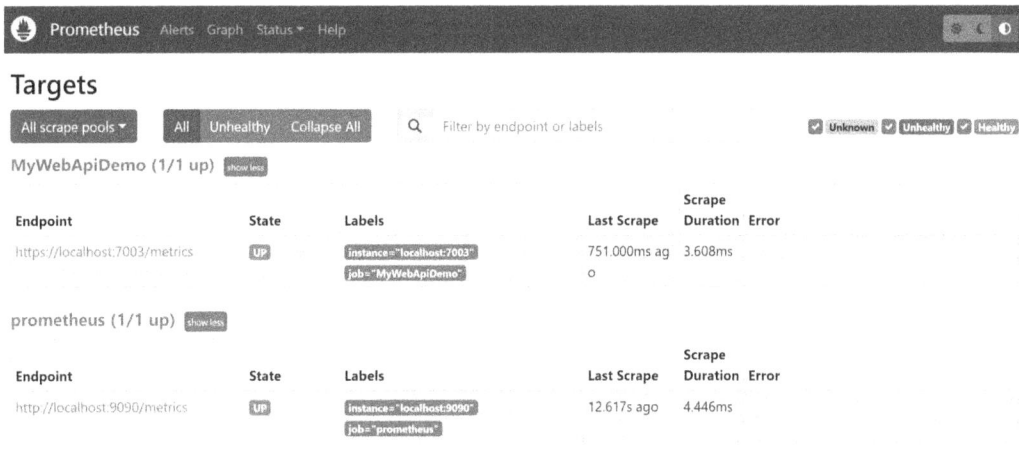

Figure 16.6 – Prometheus targets

Click **Graph** in the top menu. You will see the following page, which shows the available metrics. Click the **Open Metrics Explorer** button (highlighted in *Figure 16.7*) to open **Metrics Explorer**:

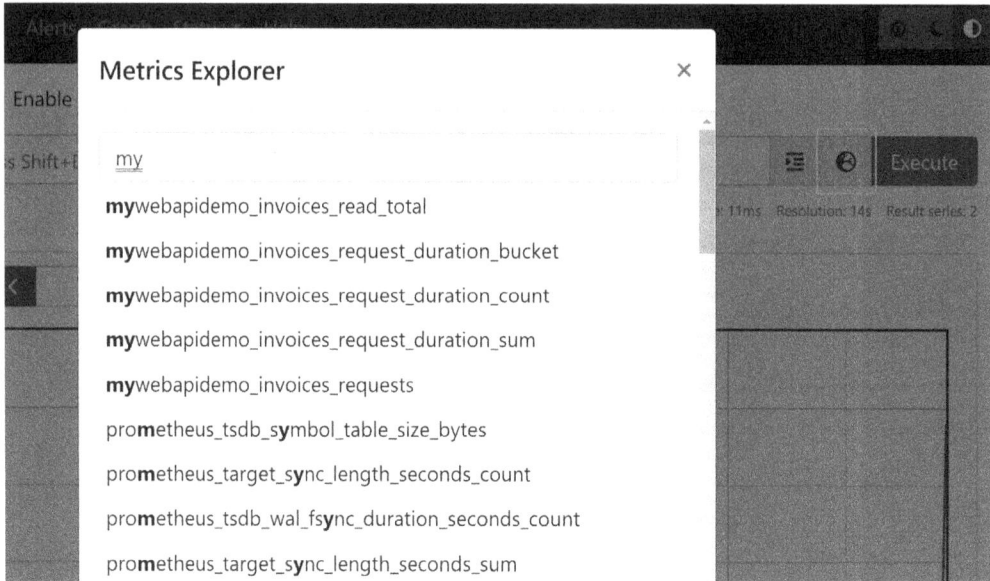

Figure 16.7 – Prometheus Metrics Explorer

Choose one metric, such as mywebapidemo_invoices_read_total, and click the **Execute** button. Then, click the **Graph** tab; you will see the following page, which shows the metric graph:

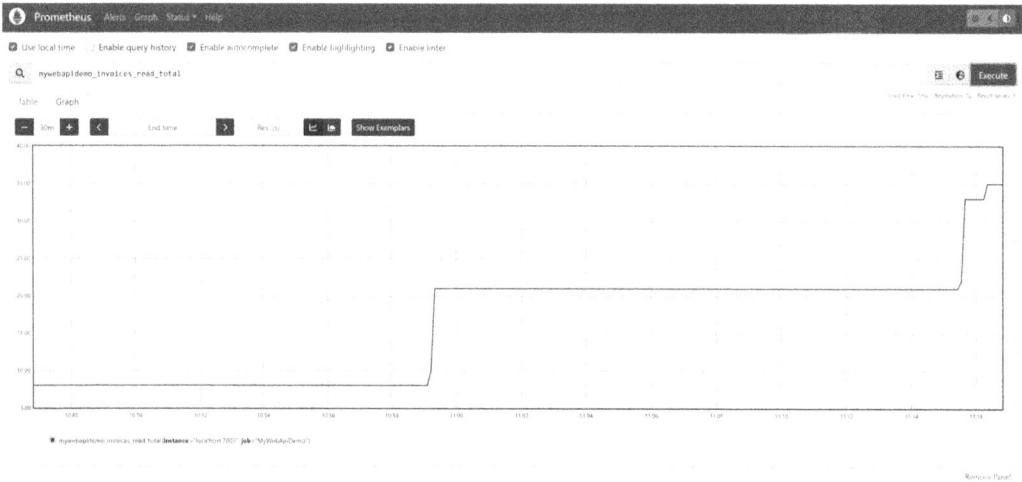

Figure 16.8 – Prometheus graph

Prometheus provides a powerful query language to query the metrics. For example, we can use the following query to get the mywebapidemo.invoices.read counter per minute:

```
rate(mywebapidemo_invoices_read_total[1m])
```

You will see the following graph:

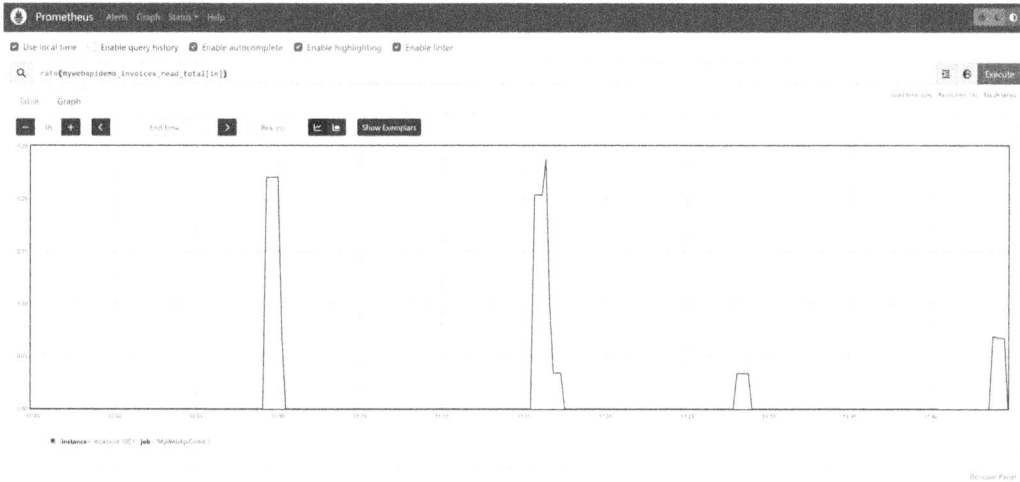

Figure 16.9 – Requests per minute

The following query can get the requests that take longer than 100 milliseconds:

```
histogram_quantile(0.95, sum(rate(mywebapidemo_invoices_request_
duration_bucket[1m])) by (le)) > 100
```

Figure 16.10 shows the result:

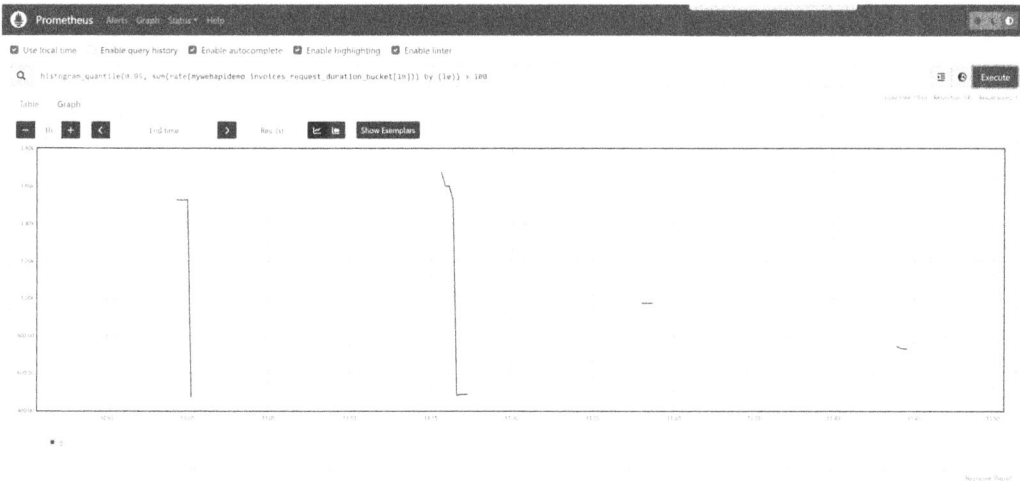

Figure 16.10 – Requests that take longer than 100 milliseconds

This section provided a brief introduction to Prometheus. For more information about the querying language syntax, please refer to the official documentation: `https://prometheus.io/docs/prometheus/latest/querying/basics/`.

Prometheus is a powerful tool for collecting and querying metrics. To gain better visualization of these metrics, Grafana can be used to create dashboards. In the following section, we will explore how to use Grafana to read metrics from Prometheus and create informative dashboards.

Using Grafana to create dashboards

Grafana is a popular opensource analytics and dashboarding tool. It can visualize metrics from multiple data sources, such as Prometheus, Elasticsearch, Azure Monitor, and others. Grafana can create beautiful dashboards to help us understand the application's performance and behavior. In this section, we will use Grafana to create dashboards for the ASP.NET Core web API application.

Grafana also provides a managed service called **Grafana Cloud**. The free tier of Grafana Cloud has a limit of 10,000 metrics, 3 users, and 50 GB of logs. You can check the pricing here: `https://grafana.com/pricing/`. In this book, we will install Grafana locally. Download the latest version of Grafana from the official website: `https://grafana.com/oss/grafana/`. Then, choose the version for your operating system.

Run Grafana by executing the `grafana-server.exe` file if you're using Windows. You may see a Windows Security Alert dialog box. Click the **Allow** button to allow Grafana to communicate on these networks. You will find the following line in the output:

```
INFO [10-15|12:31:31] Validated license
token                   logger=licensing appURL=http://localhost:3000/
source=disk status=NotFound
```

This means that Grafana is running on port `3000`. Navigate to `http://localhost:3000`. The default username and password are both `admin`. Once you've logged in, you will be prompted to change the password.

> **Important note**
> The default theme of Grafana is dark. If you prefer a light theme, you can change it by going to the **Preferences** page. We're using the light theme in this book for better readability.

Click the hamburger menu in the top-left corner, and then click **Connections**. This page shows the data sources that Grafana supports:

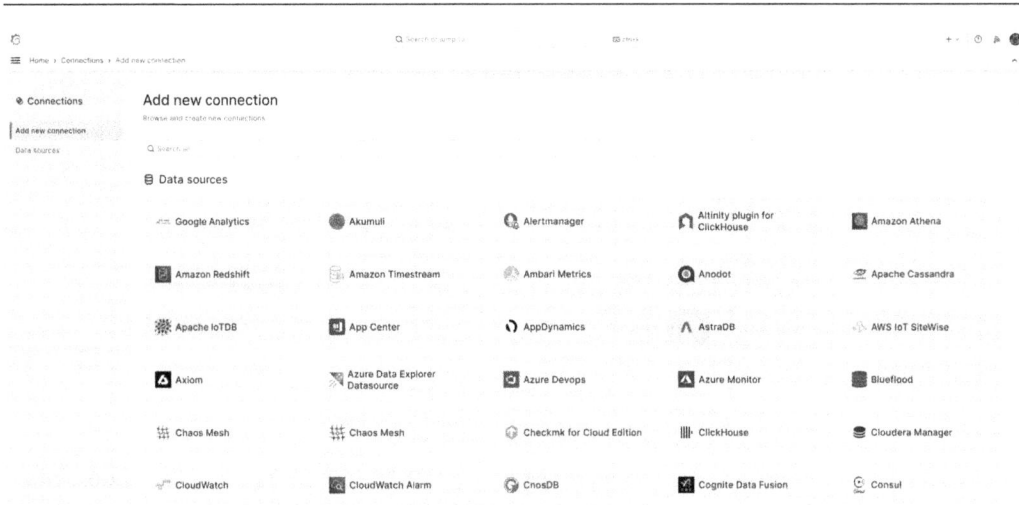

Figure 16.11 – Grafana data sources

Search for *Prometheus* and click on it. Then, click the **Create a Prometheus data source** button in the top-right corner. On the **Settings** page, we can configure the data source, as follows:

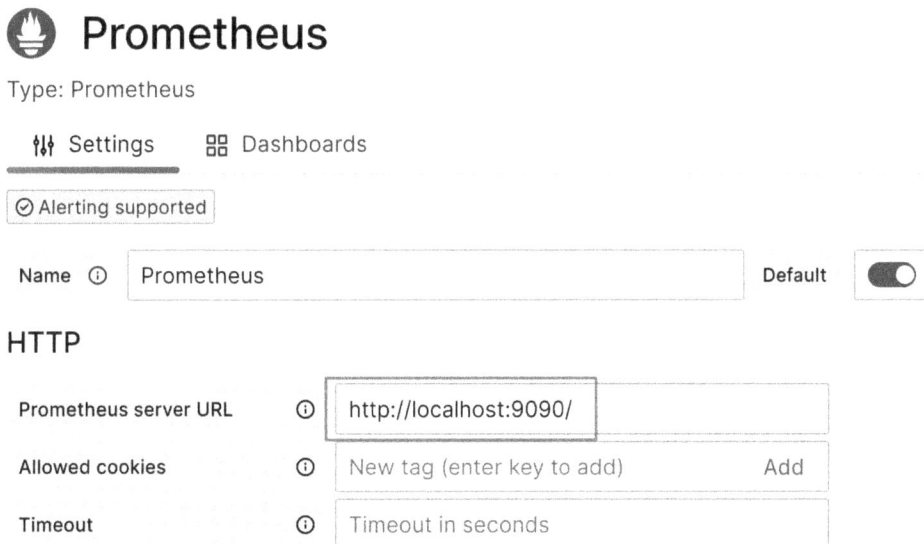

Figure 16.12 – Configuring the Prometheus data source

Use `http://localhost:9090` as the URL. Then, click the **Save & Test** button. If the data source has been configured correctly, you will see a message box that states `Successfully queried the Prometheus API`. At this point, we can create dashboards to visualize the metrics.

Navigate to the **Dashboards** page and click the **New** button. From the drop-down list, click **New Dashboard**. You will be navigated to the new dashboard page. Click the **Add visualization** button, then choose Prometheus as the data source, as follows:

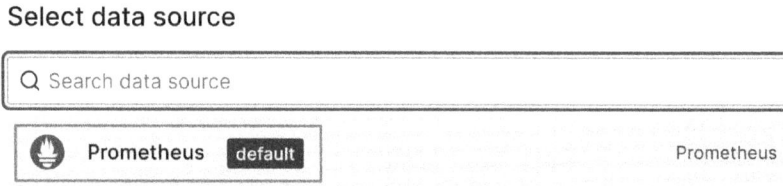

Figure 16.13 – Adding a visualization

Then, we can use the query language to query the metrics. In the **Query** tab, you can choose the relevant metrics from the drop-down list. You can also filter the metrics by job name. For example, we can use the following query to get the number of `mywebapidemo.invoices.read` requests for the `/api/Invoices` endpoint:

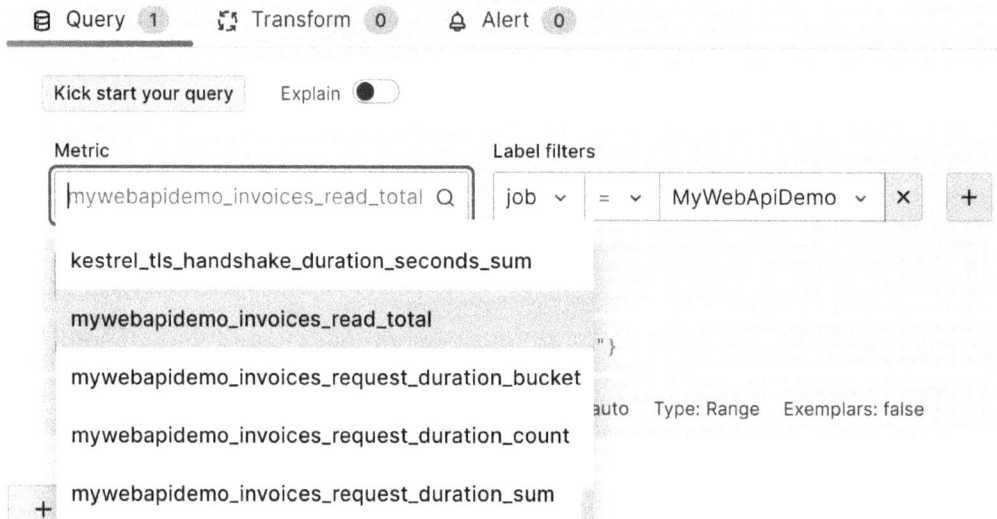

Figure 16.14 – Querying the metrics

Click the **Run queries** button; you will see the following output in the panel:

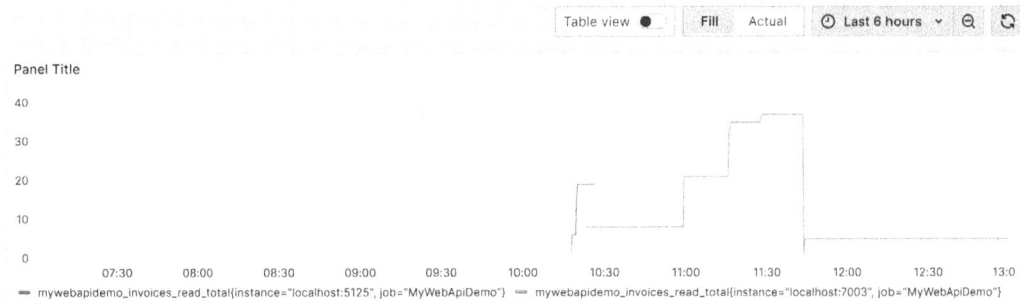

Figure 16.15 – Query result

Then, click the **Apply** button; you will see the graph in the dashboard:

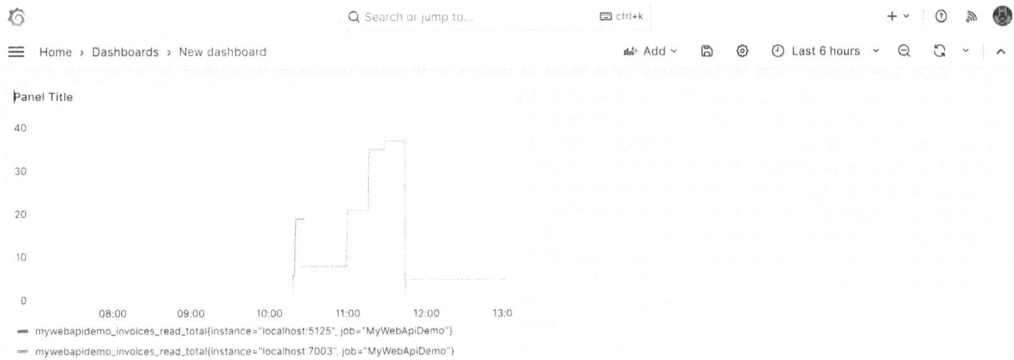

Figure 16.16 – Grafana dashboard

You can adjust the size of the dashboard as necessary. Feel free to add more dashboard panels to visualize the metrics. Before you leave the dashboard, click the **Save** button in the top-right corner to save it. You can also export the dashboard as a JSON file and import it later:

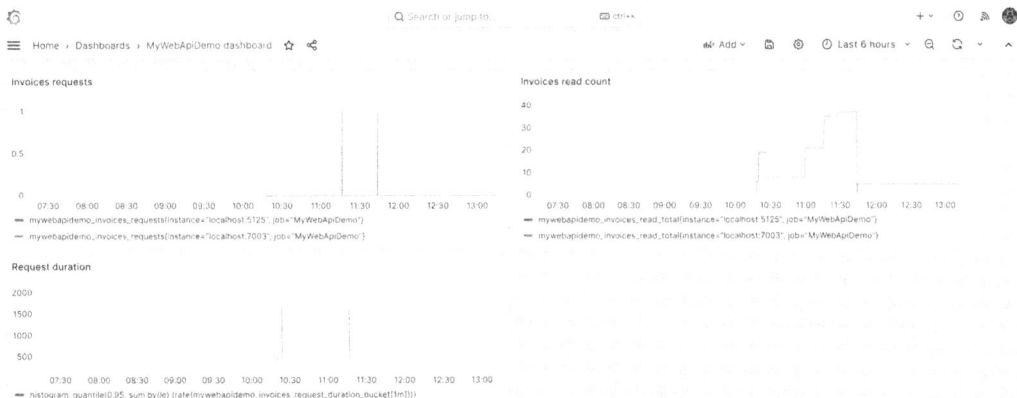

Figure 16.17 – Adding more panels

To simplify the process of creating Grafana dashboards, James Newton-King, the esteemed author of JSON.NET, has provided a Grafana dashboard template for ASP.NET Core web API applications. You can find the template here: `https://github.com/JamesNK/aspnetcore-grafana/blob/main/README.md`. There are two dashboards in this repository:

- `ASP.NET Core.json`: This dashboard shows an overview of the ASP.NET Core web API application

- `ASP.NET Core Endpoint.json`: This dashboard shows the details for specific endpoints

Create a new dashboard and click the **Import** button this time. Then, upload the `ASP.NET Core.json` file or paste the content of the file into the textbox, as follows:

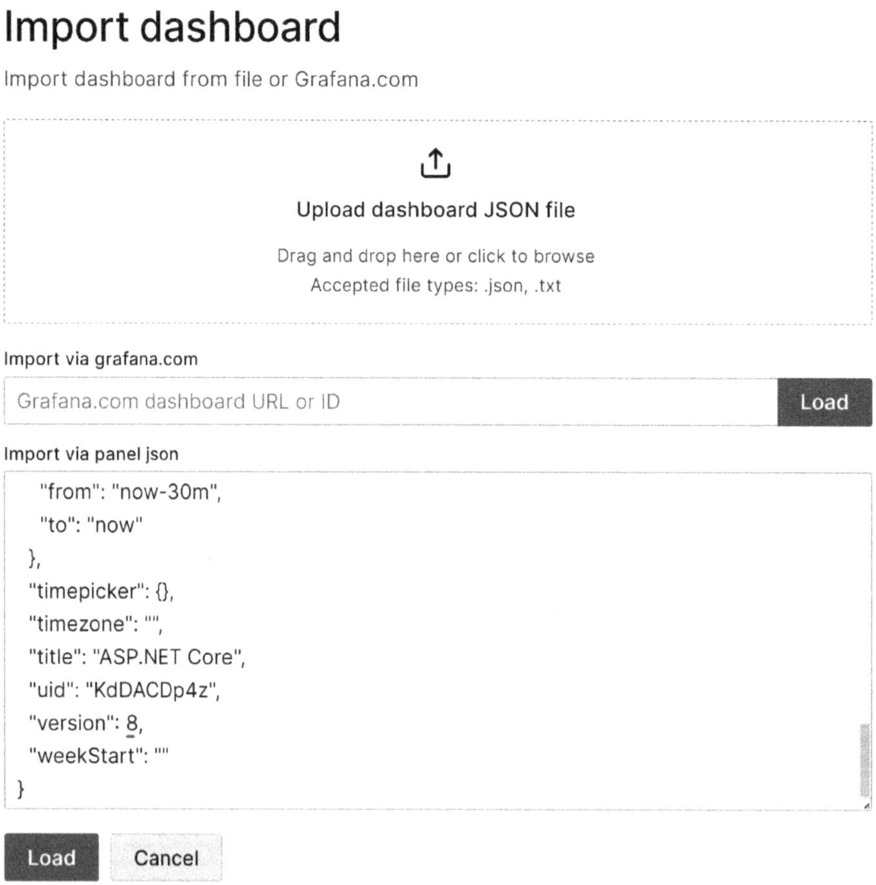

Import dashboard

Import dashboard from file or Grafana.com

⬆

Upload dashboard JSON file

Drag and drop here or click to browse
Accepted file types: .json, .txt

Import via grafana.com

| Grafana.com dashboard URL or ID | Load |

Import via panel json

```
   "from": "now-30m",
   "to": "now"
 },
 "timepicker": {},
 "timezone": "",
 "title": "ASP.NET Core",
 "uid": "KdDACDp4z",
 "version": 8,
 "weekStart": ""
}
```

Load   Cancel

Figure 16.18 – Importing the dashboard

Click the **Load** button. On the next page, choose the Prometheus data source and click the **Import** button. You will see the following dashboard:

Figure 16.19 – Overview of the ASP.NET Core dashboard provided by James Newton-King

This dashboard provides an overview of the ASP.NET Core web API application. Here, you can see the number of requests, the request's duration, the number of active requests, and so on. You can also see the error rate, which is important for monitoring the application.

Figure 16.20 shows the **ASP.NET Core Endpoint** dashboard:

Figure 16.20 – Overview of the ASP.NET Core Endpoint dashboard provided by James Newton-King

You can choose the endpoint from the drop-down list. Once you've done this, you will see the metrics for the endpoint. For example, *Figure 16.20* shows the metrics for the /api/Invoices endpoint.

Grafana offers many options to customize dashboards. On any dashboard panel, you can click the three dots in the top-right corner and then click **Edit** to edit the panel. You can change the title, the visualization type, the query, and so on. You can also use the **Builder** or **Code** editor to edit the query, as shown in *Figure 16.21* and *Figure 16.22*, respectively. Here's what the **Builder** editor looks like:

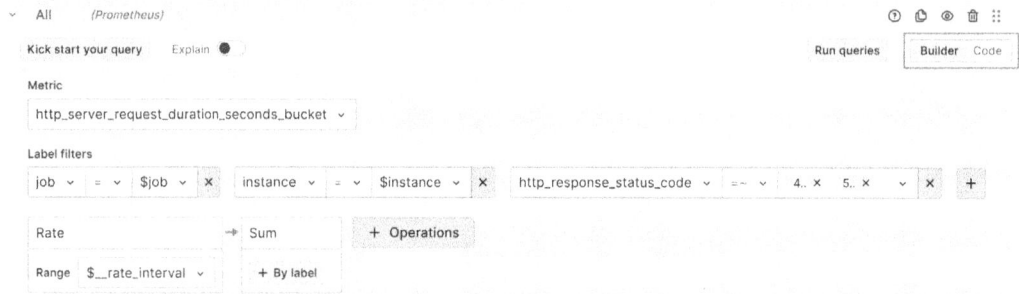

Figure 16.21 – Editing the query using the Builder editor

Here's what the **Code** editor looks like:

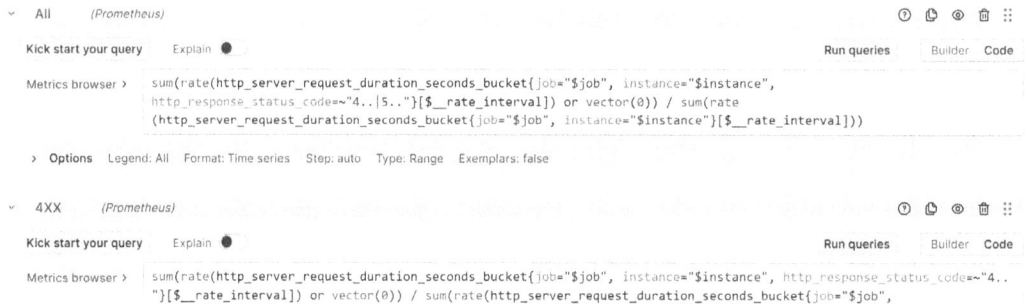

Figure 16.22 – Editing the query using the Code editor

Grafana provides a better visualization of the metrics. You can learn more about Grafana by reading the official documentation: `https://grafana.com/docs/grafana/latest/`.

In the next section, we will explore how to use OpenTelemetry and Jaeger to collect traces.

Using Jaeger to collect traces

Traces are important for understanding how requests are handled by the application. In a microservice architecture, a request will be handled by multiple services. Distributed tracing can be used to track the flow of requests across multiple services. In this section, we will learn about the basic concepts of distributed tracing and how to use OpenTelemetry and Jaeger to collect traces.

For example, in a microservice architecture, service A calls service B, and service B calls service C and service D. When a client sends a request to service A, the request will be passed through service B, service C, and service D. In this case, if any of these services fails to process the request, or the request takes too long to process, we need to know which service is responsible for the failure and which part of the request contributes to the latency or errors. Distributed tracing can give us the big picture of how the request is processed by these services.

We have not discussed microservice architecture in detail in this book. To demonstrate distributed tracing, we added two web API projects to the sample project. You can find a controller named `OrdersController` in the `MyWebApiDemo` project. In this controller, we can call the `Post` action to create an order. The `Post` action will call two external services:

- `CustomerService`: A service to check whether the customer exists
- `ProductService`: A service to check whether the product exists

To create an order, we must ensure the customer ID is valid by calling the `/api/customers/{id}` endpoint of `CustomerService`. Additionally, we must verify that the products are valid by calling the `/api/products/{id}` endpoint of `ProductService`.

Note that these services are for demonstration purposes only and should not be used for production purposes. As such, there is no real database access layer; instead, a static list is used to store the temporary data. Additionally, there is no consideration for transaction and concurrency management.

First, let's enable tracing in the `MyWebApiDemo` project. Open the `Program.cs` file and add the following code to the `MyWebApiDemo` project:

```
builder.Services.AddOpenTelemetry()
    .ConfigureResource(config =>
    {
        config.AddService(nameof(MyWebApiDemo));
    })
    .WithMetrics(metrics =>
    {
        // Omitted for brevity
    })
    .WithTracing(tracing =>
    {
        tracing.AddAspNetCoreInstrumentation()
            .AddHttpClientInstrumentation()
            .AddConsoleExporter();
    });
```

In the preceding code, we enabled tracing in our code using the `WithTracing` method. To further instrument our application, we added ASP.NET Core and HTTP client instrumentation. The HTTP client instrumentation is used to trace the HTTP calls to the external services. Finally, we added a console exporter to export the traces to the console.

Run the application and send some requests to the `api/orders` endpoint. You will see some tracing information in the terminal output:

In the console trace, you will find two important properties: `Activity.TraceId` and `Activity.SpanId`. The `Activity.TraceId` property is used to identify a trace, which is a collection of spans. A span is a unit of work in a trace. For example, if we send a `POST` request to the `api/Orders` endpoint to create an order, the application will call `ProductService` and `CustomerService`. Each call is a span. However, it is not convenient to search for a specific span in the console output. Next, we will use Jaeger to collect and visualize the traces.

Jaeger is an open-source distributed tracing platform that is used to monitor and troubleshoot distributed workflows and identify performance bottlenecks. Jaeger was originally developed by Uber Technologies (`http://uber.github.io/`) and joined the Cloud Native Computing Foundation in 2017.

Install Jaeger from the official website: `https://www.jaegertracing.io/download/`. You can choose the version for your operating system or use the Docker image. In this book, we will use the executable binaries on Windows. Navigate to the Jaeger folder in the terminal and run the following command to start Jaeger:

```
./jaeger-all-in-one --collector.otlp.enabled
```

The `jaeger-all-in-one` command is for quick local testing. It starts all the components of Jaeger, including the Jaeger UI, `jaeger-collector`, `jaeger-agent`, `jaeger-query`, and in-memory storage. The `--collector.otlp.enabled` option is used to specify that `jaeger-collector` should accept traces in OTLP format. In the output, you can find the following line, which indicates that Jaeger is receiving data from OTLP:

```
{"level":"info","ts":1697354213.8840668,"caller":"otlpreceiver@
v0.86.0/otlp.go:83","msg":"Starting GRPC
server","endpoint":"0.0.0.0:4317"}
```

`jaeger-collector` utilizes port `4317` to receive data via the gRPC protocol and port `4318` via the HTTP protocol. This allows for efficient communication between `jaeger-collector` and other services.

Next, we need to configure the ASP.NET Core web API project so that it exports the OTLP traces to Jaeger. Open the `Program.cs` file and update the `WithTracing()` method, as follows:

```
.WithTracing(tracing =>
{
    tracing.AddAspNetCoreInstrumentation()
        .AddHttpClientInstrumentation()
        .AddConsoleExporter()
        .AddOtlpExporter(options =>
        {
            options.Endpoint = new Uri("http://localhost:4317");
```

```
      });
   });
```

We use the `AddOtlpExporter` method to add the exporter for Jaeger. As a best practice, it is recommended to use the configuration system to set the URL, rather than hard-coding it. As an example, you can define it in the `appsettings.json` file.

Restart the three applications and send some `POST` requests to the `/api/Orders` endpoint. Here is a payload example:

```
{
  "id": 0,
  "orderNumber": "string",
  "contactName": "string",
  "description": "string",
  "amount": 0,
  "customerId": 1,
  "orderDate": "2023-10-15T08:57:54.724Z",
  "dueDate": "2023-10-15T08:57:54.724Z",
  "orderItems": [
    {
      "id": 1,
      "orderId": 0,
      "productId": 1,
      "quantity": 0,
      "unitPrice": 0
    }
  ],
  "status": 0
}
```

Navigate to `http://localhost:16686/`; you will see the Jaeger UI. In the **Search** tab, choose **Service**, then **Operation**, and then click the **Find Traces** button. You will see the traces, as shown here:

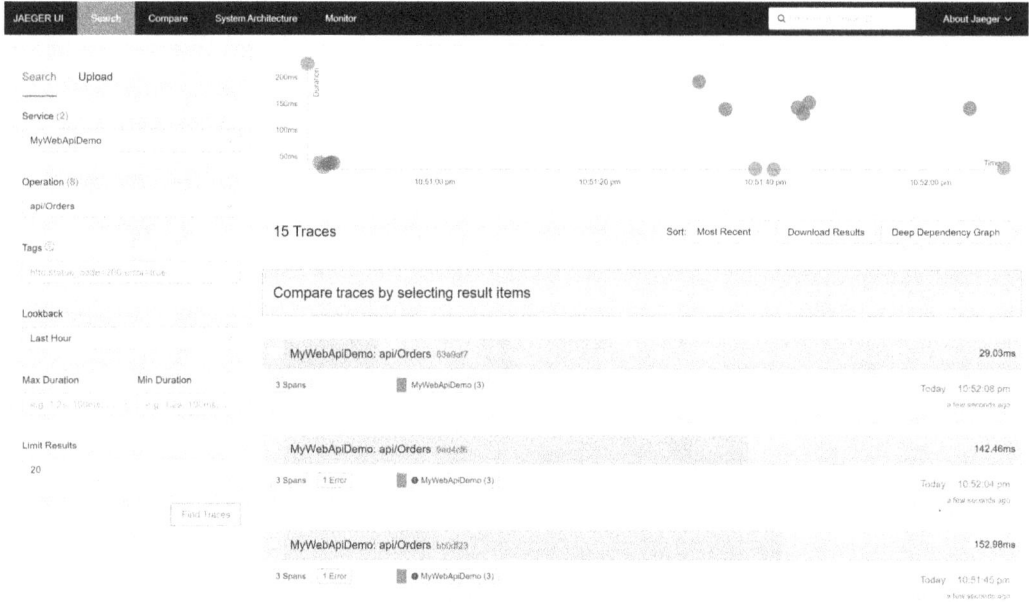

Figure 16.23 – Jaeger traces

Click on a trace to view its details:

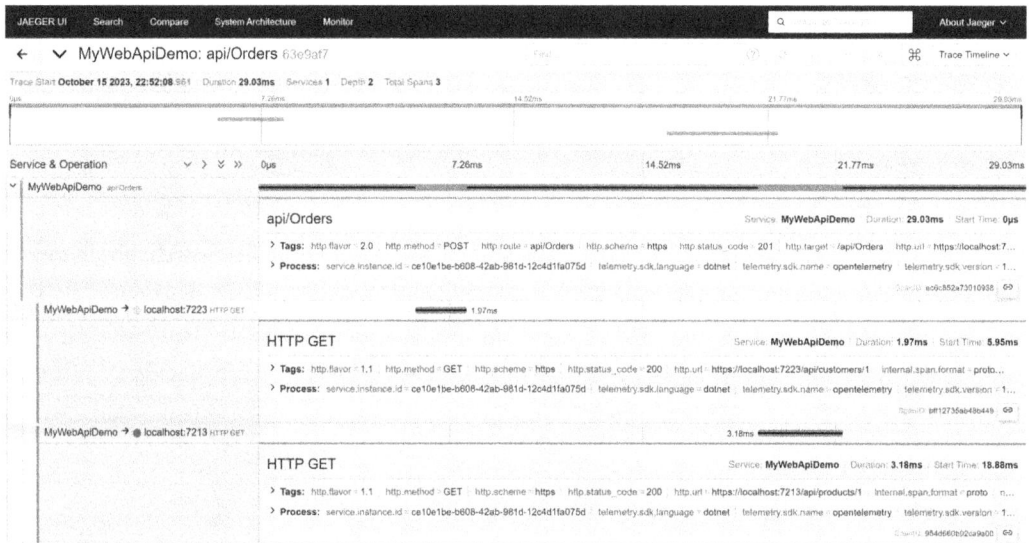

Figure 16.24 – A trace's details

You will see that this request includes three spans. The parent is the inbound request, and it has two outbound requests to other services.

We can enable traces in the dependent services to better understand how these requests are processed. Configure `ProductService` and `CustomerService` following the same methods. These traces should be sent to one Jaeger instance so that Jaeger can correlate the requests across different services.

Check the Jaeger UI now. You will find that one `/api/Orders` call has five spans now:

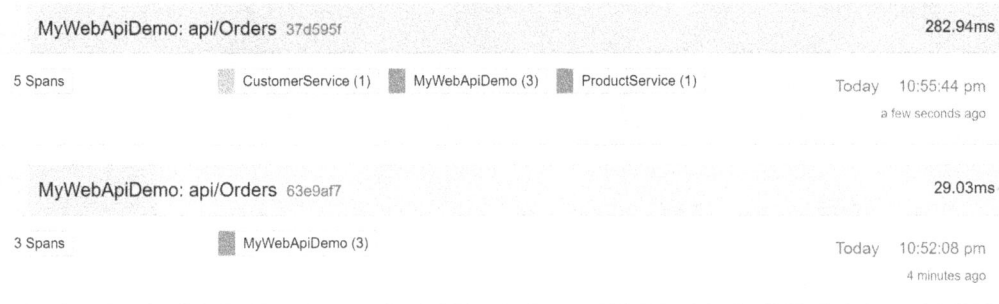

Figure 16.25 – The traces across multiple services

We can also check the latency for each span, as shown in *Figure 16.26*:

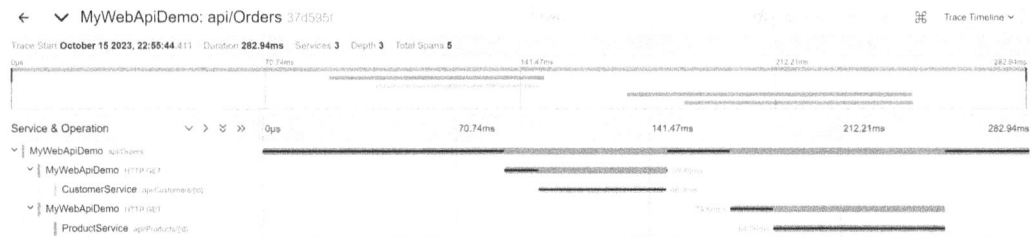

Figure 16.26 – The latency for each span

Using traces can help us understand how requests are handled by the application. We can also use traces to find performance bottlenecks. It is especially useful for microservice architectures. For more information about Jaeger, please refer to the official documentation: `https://github.com/PacktPublishing/Web-API-Development-with-ASP.NET-Core-8/tree/main/samples/chapter16/MyWebApiDemo`.

Next, we will recap logging and discuss how to propagate the trace context in the logs.

Using HTTP logging

In *Chapter 4*, we discussed how to use the ILogger interface to log messages. Sometimes, we want to log the HTTP requests and responses for troubleshooting purposes. In this section, we will discuss HTTP logging.

Follow *Chapter 4* to configure the logging system. You can use Serilog to send logs to Seq. To enable HTTP logging, we need to use the HTTP logging middleware. The middleware will log the inbound requests and outbound responses. The updated code is as follows:

```
var logger = new LoggerConfiguration().WriteTo.Seq("http://
localhost:5341").CreateLogger();
builder.Logging.AddSerilog(logger);
builder.Services.AddHttpLogging(logging =>
{
    logging.LoggingFields = HttpLoggingFields.All;
});
// Omitted for brevity
app.UseHttpLogging();
```

In the preceding code, we specify HttpLoggingFields to log all fields. Be careful when you use this option in production because it may potentially impact the performance and log sensitive information. We should not log **personally identifiable information (PII)** and any sensitive information. We're using it for demonstration purposes only here.

We can also update the appsettings.json file to specify the log levels. Add the following code to the LogLevel section of the appsettings.json file so that we can see information logs:

```
"Microsoft.AspNetCore.HttpLogging.HttpLoggingMiddleware":
"Information"
```

Configure the logging in the CustomerService and ProductService projects using the same methods.

Run the three applications and send some requests to the /api/Orders endpoint. You will see the following logs in the Seq dashboard:

16 Oct 2023 20:22:10.190	Request finished HTTP/2 GET https://localhost:7003/api/Invoices?page=1&pageSize=10 - 200 null application/json; charset=utf-8 2407.0476ms
16 Oct 2023 20:22:10.186	Duration: 2399.1753ms
16 Oct 2023 20:22:10.183	ResponseBody: [{"id":"eb4ad088-456a-4afd-a17d-035ecb0a362a","invoiceNumber":"test","contactName":"test","description":"test","amount":0.00,"invoiceDat...
16 Oct 2023 20:22:10.181	Executed endpoint 'MyWebApiDemo.Controllers.InvoicesController.Get (MyWebApiDemo)'
16 Oct 2023 20:22:10.181	Executed action MyWebApiDemo.Controllers.InvoicesController.Get (MyWebApiDemo) in 2366.3307ms
16 Oct 2023 20:22:10.177	Response: StatusCode: 200 Content-Type: application/json; charset=utf-8
16 Oct 2023 20:22:10.161	Executing OkObjectResult, writing value of type 'System.Collections.Generic.List`1[[MyWebApiDemo.Models.Invoice, MyWebApiDemo, Version=1.0.0.0, C...
16 Oct 2023 20:22:10.156	Executed action method MyWebApiDemo.Controllers.InvoicesController.Get (MyWebApiDemo), returned result Microsoft.AspNetCore.Mvc.OkObjectResu...
16 Oct 2023 20:22:10.084	Executed DbCommand (49ms) [Parameters=[@__p_0='?' (DbType = Int32), @__p_1='?' (DbType = Int32)], CommandType='Text', CommandTimeout='30'] S...
16 Oct 2023 20:22:08.740	The query uses a row limiting operator ('Skip'/'Take') without an 'OrderBy' operator. This may lead to unpredictable results. If the 'Distinct' operator is used ...
16 Oct 2023 20:22:08.735	The query uses a row limiting operator ('Skip'/'Take') without an 'OrderBy' operator. This may lead to unpredictable results. If the 'Distinct' operator is used ...
16 Oct 2023 20:22:07.938	Executing action method MyWebApiDemo.Controllers.InvoicesController.Get (MyWebApiDemo) - Validation state: Valid
16 Oct 2023 20:22:07.811	Route matched with {action = "Get", controller = "Invoices"}. Executing controller action with signature System.Threading.Tasks.Task`1[Microsoft.AspNet...
16 Oct 2023 20:22:07.787	Executing endpoint 'MyWebApiDemo.Controllers.InvoicesController.Get (MyWebApiDemo)'
16 Oct 2023 20:22:07.786	Request: Protocol: HTTP/2 Method: GET Scheme: https PathBase: Path: /api/Invoices Accept: text/plain Host: localhost:7003 User-Agent: Mozilla/5.0 (Windows NT 10.0; Win64; x64) AppleWebKit/537.36 (KHTML, like Gecko) Chrome/117.0.0.0 Safari/537.36 Accept-Encoding: gzip, deflate, br Accept-Language: en-US,en;q=0.9 Referer: [Redacted] sec-ch-ua: [Redacted] sec-ch-ua-mobile: [Redacted] sec-ch-ua-platform: [Redacted] sec-ch-ua-site: [Redacted] sec-fetch-mode: [Redacted] sec-fetch-dest: [Redacted]

Event ✓   Level (Information) ✓   Type (0xF0588704) ✓   Trace (0437...) ✓   Export ✓

✓ ✕	Accept	text/plain
✓ ✕	Accept-Encoding · @Properties[...	gzip, deflate, br
✓ ✕	Accept-Language · @Properties[...	en-US,en;q=0.9
✓ ✕	ConnectionId	0HMUE30KCM3OV
✓ ✕	EventId	{Id: 1, Name: 'RequestLog'}
✓ ✕	Host	localhost:7003
✓ ✕	HttpLog	Request:

Figure 16.27 – HTTP logging

In the logs, you will find details about the HTTP requests and responses. If you want to change the logging fields, you can change the `LoggingFields` property of `HttpLoggingOptions` in the `AddHttpLogging()` method. The `LoggingFields` property is an enum. You can choose `RequestPath`, `RequestQuery`, `RequestMethod`, `RequestStatusCode`, `RequestBody`, `RequestHeaders`, `ResponseHeaders`, `ResponseBody`, `Duration`, and so on. The `HttpLoggingOptions` class has other properties, such as `RequestHeaders`, `ResponseHeaders`, `RequestBodyLogLimit`, `ResponseBodyLogLimit`, and others. You can use these properties to configure the logging system.

Since we enabled HTTP logging for all requests, we can filter the logs by trace ID. Check the Jaeger UI and click on a trace. You will find the trace ID in the URL:

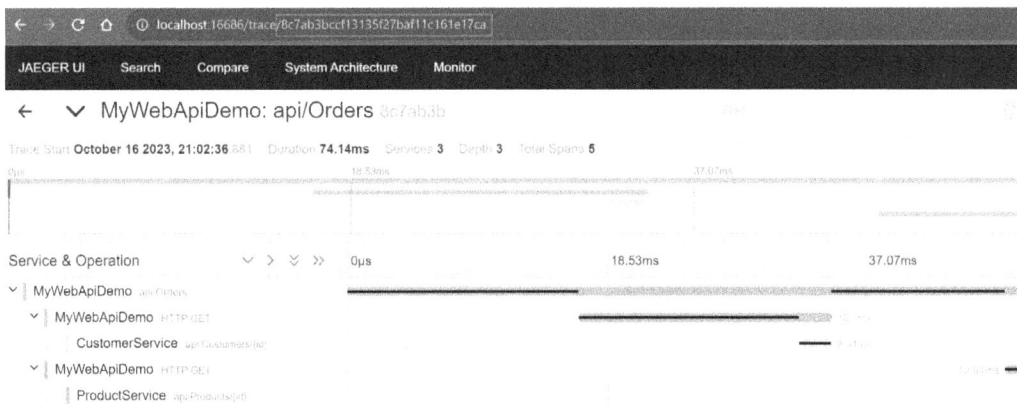

Figure 16.28 – The trace ID in Jaeger

In the preceding screenshot, the trace ID is `8c7ab3bccf13135f27baf11c161e17ca`. Copy this trace ID and use the following query in the Seq dashboard:

```
@TraceId = '8c7ab3bccf13135f27baf11c161e17ca'
```

Click the green **Go** button to filter the logs. You will see the logs for this trace:

Figure 16.29 – Filtering the logs by trace ID

Figure 16.29 provides a comprehensive view of all the logs for the trace, including HTTP requests and responses for three services. This is an invaluable resource for troubleshooting.

Using Azure Application Insights

In the preceding sections, we explored how to use OpenTelemetry to collect metrics and traces. We also discussed how to leverage open-source tools such as Prometheus, Grafana, Jaeger, and Seq to visualize the metrics, traces, and logs. Now, we'll look at how to use Azure Application Insights to create a unified dashboard for monitoring an ASP.NET Core web API application.

Azure Application Insights is an extensible APM service for monitoring applications. It can collect and analyze logs, metrics, and traces from multiple sources. To follow this section, you need to have an Azure subscription. If you do not have one, you can create a free account here: `https://azure.microsoft.com/en-us/free/`.

Go to the Azure portal and create a new Application Insights resource. You can find the Application Insights service in the **Monitoring** category. Choose the **Application Insights** service and click the **Create** button. On the next page, you need to specify the resource group, name, region, and pricing tier, as shown in *Figure 16.30*:

Application Insights   ...
Monitor web app performance and usage

Basics   Tags   Review + create

Create an Application Insights resource to monitor your live web application. With Application Insights, you have full observability into your application across all components and dependencies of your complex distributed architecture. It includes powerful analytics tools to help you diagnose issues and to understand what users actually do with your app. It's designed to help you continuously improve performance and usability. It works for apps on a wide variety of platforms including .NET, Node.js and Java EE, hosted on-premises, hybrid, or any public cloud. Learn More

PROJECT DETAILS

Select a subscription to manage deployed resources and costs. Use resource groups like folders to organize and manage all your resources.

Subscription * ⓘ	
Resource Group * ⓘ	devops-lab
	Create new

INSTANCE DETAILS

Name * ⓘ	ai-mywebapidemo
Region * ⓘ	(US) East US
Resource Mode * ⓘ	Classic Workspace-based

WORKSPACE DETAILS

Subscription * ⓘ	
Log Analytics Workspace * ⓘ	(new) DefaultWorkspace-

Figure 16.30 – Creating an Application Insights resource

Once the Application Insights resource has been created, navigate to the **Overview** page. You will find the **Instrumentation Key** and **Connection String** values. **Instrumentation Key** is used to identify the Application Insights resource, while **Connection String** is used to connect to the Application Insights resource. We will use **Connection String** to configure the ASP.NET Core web API applications.

Open the `appsettings.json` file in the `MyWebApiDemo` project. Add the following setting:

```
"APPLICATIONINSIGHTS_CONNECTION_STRING": "InstrumentationKey=xxxxx"
```

Please replace the connection string with your own value. This is for demonstration purposes only. It is recommended to use different Application Insights resources for different environments. This will ensure that metrics and traces are not mixed.

Next, we need to install the `Azure.Monitor.OpenTelemetry.AspNetCore` package using the following command:

```
dotnet add package Azure.Monitor.OpenTelemetry.AspNetCore --prerelease
```

This package is used to export metrics and traces to Azure Application Insights. At the time of writing, the package was still in preview. If you are reading this book after the package has been released, you can omit the `--prerelease` option.

Then, update the `Program.cs` file, as follows:

```
builder.Services.AddOpenTelemetry()
    // Omitted for brevity
    .UseAzureMonitor()
```

This method will read the `APPLICATIONINSIGHTS_CONNECTION_STRING` setting from the configuration system and export the metrics and traces to Azure Application Insights.

Configure the `CustomerService` and `ProductService` projects using the same methods. Run the three applications and send some `POST` requests to the `/api/Orders` endpoint. Then, navigate to the Application Insights resource in the Azure portal. You will see the following output:

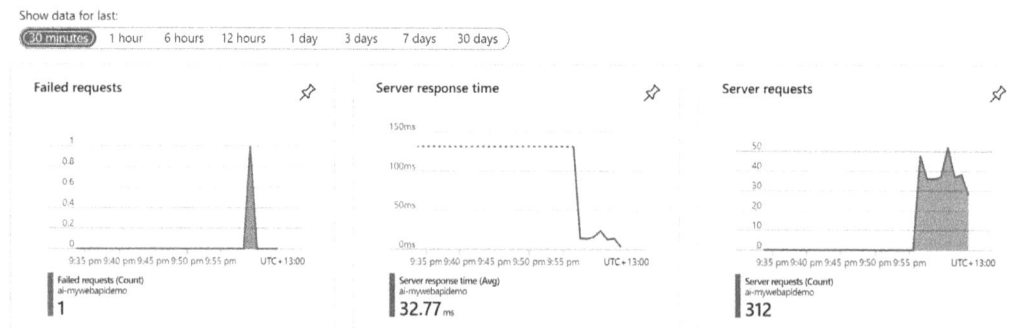

Figure 16.31 – Overview of Azure Application Insights

You can find even more information about logs, metrics, and traces. Click the **Logs** tab; you will see the following page:

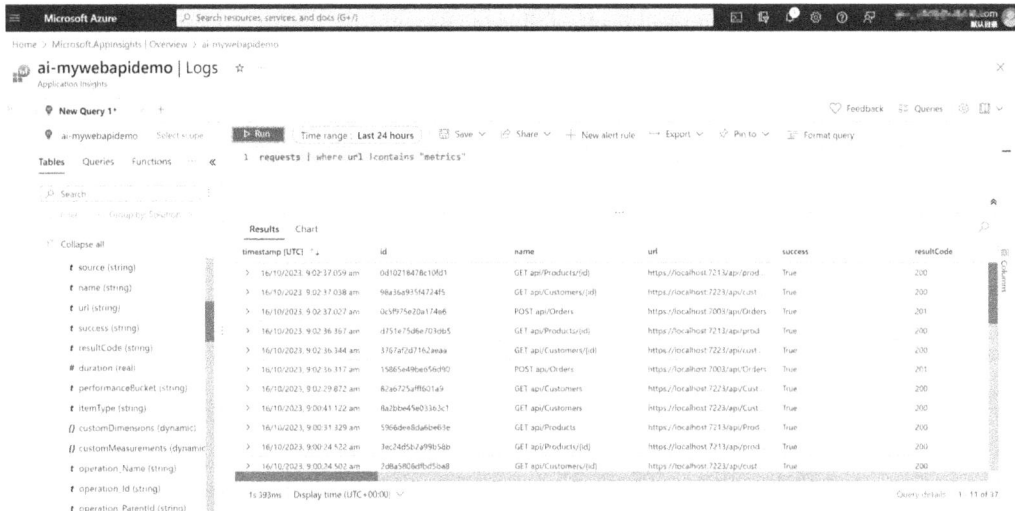

Figure 16.32 – Azure Application Insights logs

In *Figure 16.32*, we use `requests | where url !contains "metrics"` to query the logs. This query will filter the logs that do not contain the `metrics` keyword. You can also use `traces` to query the traces.

The **Metrics** tab shows the available metrics, as follows:

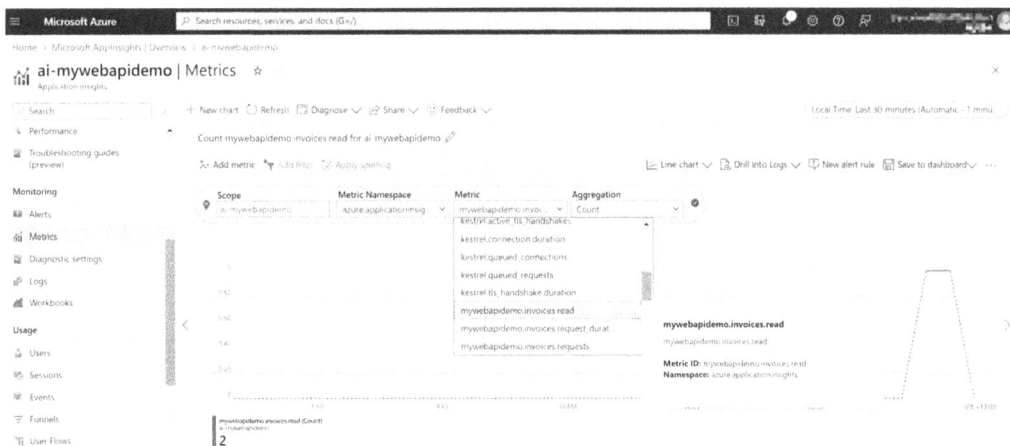

Figure 16.33 – Azure Application Insights metrics

Here, you can find the metrics we defined for the `api/Invoices` endpoint. If you cannot see the metrics, send some requests to the `api/Invoices` endpoint and wait a few minutes.

Click the **Application map** tab; you will see the following page:

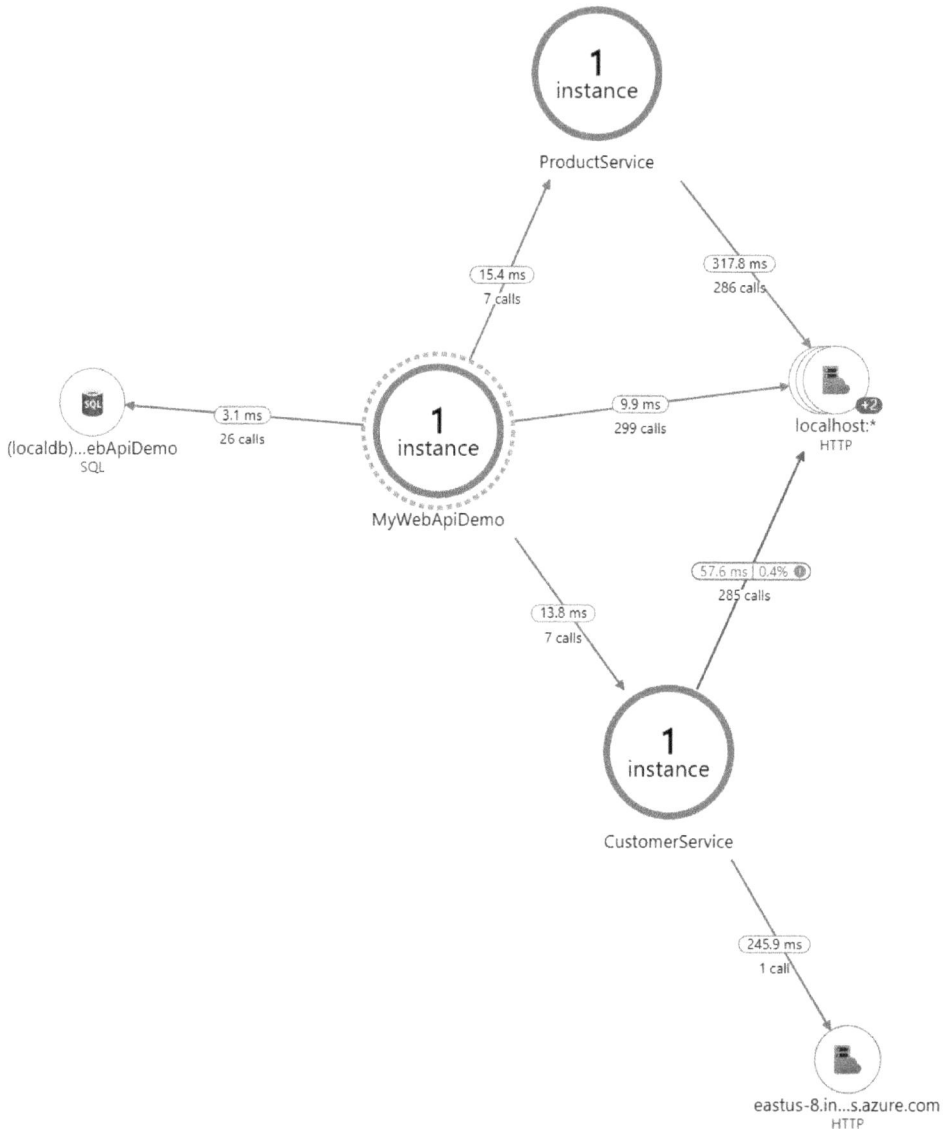

Figure 16.34 – Overview of the Azure Application Insights application map

Figure 16.34 shows the requests flow across multiple services. You can also find the latency for each service.

Upon clicking any request in the diagram, you will see details such as the response time, dependency count, performance histogram, and dependencies, as follows:

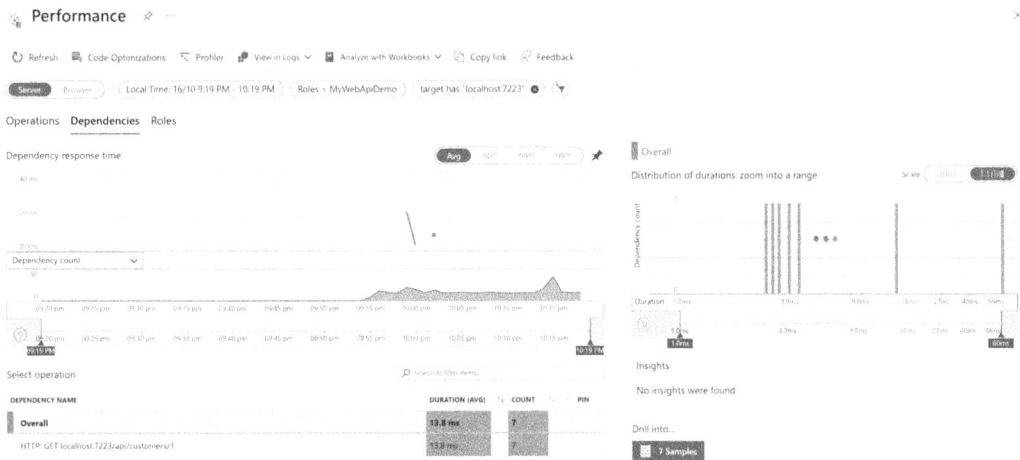

Figure 16.35 – Overview of Azure Application Insights request details

Click the **Performance** tab; you will see the overall performance of the application:

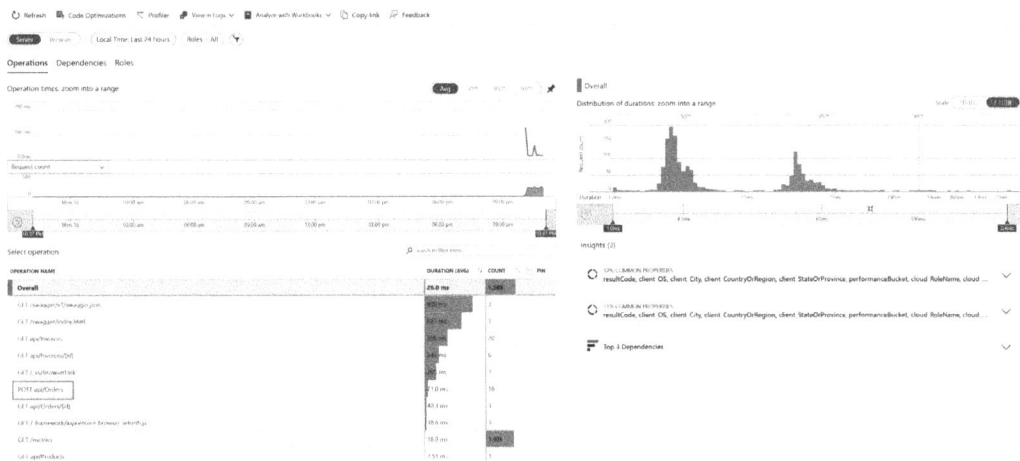

Figure 16.36 – Overview of Azure Application Insights performance

Clicking on one operation, such as POST api/Orders, will allow you to view the performance of that operation. For further details, click the **xx Samples** button located under the **Drill into...** label in the bottom-right corner. You will see a list of all the requests for that operation on the right-hand side of the screen. Clicking on one of these requests will allow you to view the details of that request, including the request and response body, as shown in *Figure 16.37*:

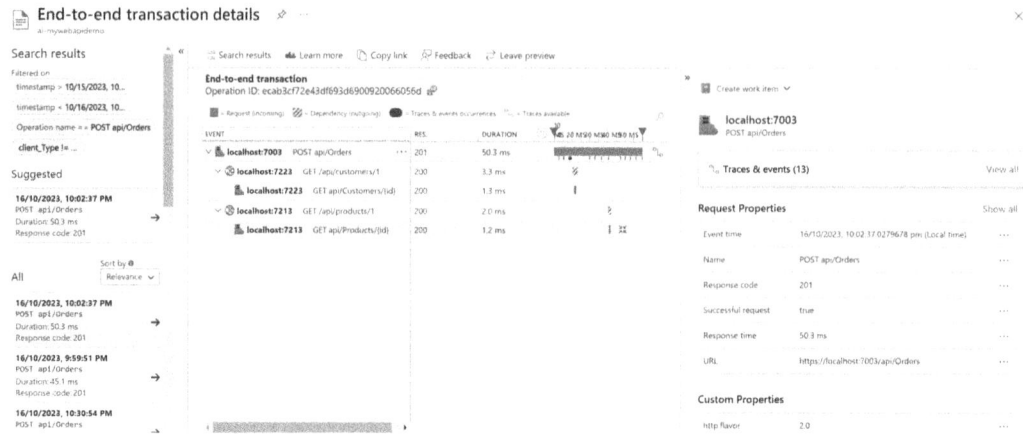

Figure 16.37 – Azure Application Insights end-to-end transaction details

In *Figure 16.37*, you can see how the request is processed by multiple services, similar to what the Jaeger UI does.

Azure Application Insights is a super powerful tool for monitoring applications. The benefits of using Azure Application Insights are as follows:

- It is a managed service. You do not need to maintain the infrastructure.

- It provides a unified dashboard for monitoring applications. You do not need to use multiple tools. Application Insights can provide a centralized view of the metrics, traces, and logs.

- It is easy to integrate with your applications. Configuring one connection string is much easier than configuring multiple tools.

- It provides a powerful query language to query metrics, traces, and logs. You can use the query language to create custom dashboards.

- It offers more features, such as alerting, failure analysis, funnel analysis, user flows, and so on.

Note that Azure Application Insights is not free. It is part of Azure Monitor, a comprehensive monitoring solution for enterprise applications. You can find its pricing here: `https://azure.microsoft.com/en-us/pricing/details/monitor/`.

Summary

In this chapter, we discussed monitoring and observability in ASP.NET Core web API applications. We explored how to handle errors and exceptions and return proper error responses. We also discussed how to implement health checks to determine the status of the application. Then, we learned about the basic concepts of observability, including logs, metrics, and traces, and how to integrate with OpenTelemetry and define custom metrics. We also explored some open-source tools, such as Prometheus, Grafana, Jaeger, and Seq, to collect and visualize metrics, traces, and logs. Finally, we introduced Azure Application Insights, a managed service for monitoring applications in one place.

Monitoring and observability are complex topics that require a deeper understanding of distributed systems and microservice architecture. In this chapter, we only introduced the basic concepts. To gain a more comprehensive understanding of these topics, further study is necessary.

In the next chapter, we will explore advanced topics related to architecture and design patterns. These include **domain-driven design** (**DDD**), clean architecture, and cloud-native patterns such as CQRS, resilience patterns, and more. This will provide you with a comprehensive overview of the various approaches to architecture and design.

17

Cloud-Native Patterns

In the preceding chapters, we have covered a range of fundamental skills for web API development using ASP.NET Core. We discussed different styles of API development, such as REST, gRPC, and GraphQL, and how to implement the data access layer using Entity Framework Core. We also introduced how to secure a web API using the ASP.NET Core Identity framework. Additionally, we learned how to write unit tests and integration tests for web API applications, as well as common practices for API development, such as testing, caching, observability, and more. We also discussed how to deploy the containerized web API application to the cloud by using CI/CD pipelines. These are all essential skills for web API development.

However, this is just the beginning of the journey. As we wrap up our exploration of the fundamental concepts of web API development using ASP.NET Core, it is time to embark on a journey to explore more advanced topics. In this chapter, we will transition from the basics to delve into topics that are important for developers aspiring to master web API development. Now, let us elevate our skills to the next level.

We will delve into the following topics in this chapter:

- Domain-driven design
- Clean architecture
- Microservices
- Web API design patterns

By the end of this chapter, you will have a high-level understanding of these topics and be able to explore them further on your own.

Technical requirements

The code example in this chapter can be found at `https://github.com/PacktPublishing/Web-API-Development-with-ASP.NET-Core-8/tree/main/samples/chapter17.`

Domain-driven design

The term **domain-driven design**, also known as **DDD**, was coined by Eric Evans in his book *Domain-Driven Design: Tackling Complexity in the Heart of Software*, published in 2003. DDD consists of a set of principles and practices that focus on the domain model and domain logic, which help developers manage the complexity and build flexible and maintainable software. DDD is not bound to any particular technology or framework. You can use it in any software project, including web API development.

In Eric Evans' book, he defines three important principles of DDD:

- Focusing on the core domain and domain logic

- Basing complex designs on the domain models

- Collaborating with technical and domain experts to iteratively refine the model that solves domain problems

A domain is a subject area that the software system is built for. The domain model is a conceptual model of the domain, which incorporates both data and behavior. Developers build the domain model based on the domain knowledge from domain experts. The domain model is the core of the software system that can be used to solve domain problems.

In the following subsections, we will introduce the basic concepts of DDD and how to apply them to web API development. Note that domain-driven design is a comprehensive topic that cannot be covered in a single chapter. So, the subsections are not intended to be a complete guide to DDD. Instead, it will provide a high-level overview of DDD and explain some of the key concepts of DDD. If you want to learn more about DDD, you can refer to other resources, such as Eric Evans' DDD book.

Ubiquitous language

One of the core concepts of DDD is that to build a software system for a complex business domain, we need to build ubiquitous language and a domain model that reflects the business domain. Under domain-driven design, the software structure and code, such as class names, class methods, and so on, should match the business domain. The domain terms should be embedded in the code. When developers talk to domain experts, they should use the same terms. For example, if we are building a web API for a banking system, we may have a `Banking` domain. When we discuss the requirements with domain experts, we may hear terms such as `Account`, `Transaction`, `Deposit`, `Withdrawal`, and so on. In a banking system, an `Account` object can have different types, such as `SavingAccount`, `LoanAccount`, `CreditCardAccount`, and so on. A `SavingAccount` may have a `Deposit()` method and a `Withdrawal()` method. In the code of the system, we should use the same terms as the domain experts.

The use of a ubiquitous language is an essential pillar of DDD. This language provides a common understanding between domain experts, developers, and users, allowing them to effectively communicate system requirements, design, and implementation. By consciously using the ubiquitous language in the code, developers can build a domain model that accurately reflects the business domain. Without this, the code may become disconnected from the business domain and become difficult to manage.

Bounded context

In the realm of DDD, the concept of a bounded context is essential. A bounded context is a boundary that defines a domain model and serves as a delineated area of responsibility within a software system. It is like a linguistic territory in which a specific model holds meaning and relevance. By encapsulating a distinct understanding of the domain, bounded contexts promote clarity and precision in communication between domain experts and developers.

Consider a scenario where we are building a web API for a banking system. Without bounded contexts, the term `Account` could be interpreted differently in the `Banking` domain and the `customer relationship management` (CRM) domain. This ambiguity can lead to confusion, misaligned expectations, and ultimately, a fragmented understanding of the entire system. To avoid this, bounded contexts should be used to clearly define the scope of the domain model.

In many cases, a domain consists of several subdomains, each of which may refer to a distinct part of the business domain, thus creating different bounded contexts. These bounded contexts communicate with each other through programmatic interfaces, such as web APIs and message queues.

DDD layers

A DDD solution is often represented as a layered architecture. Each layer has a specific responsibility. The following diagram shows the typical layers of a DDD application:

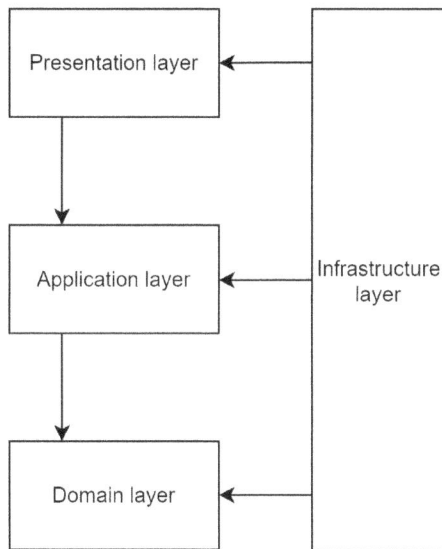

Figure 17.1 – Typical layers of a DDD application

In the preceding diagram, there are four layers:

- **Presentation layer**: This layer is responsible for presenting the data to the user and receiving user input. Normally, this layer is implemented as a user interface, such as a web application, a mobile application, or a desktop application. In this book, we mainly focus on web API applications that do not have a user interface. In this case, the presentation layer can be a client application that consumes the web API.

- **Application layer**: This layer is responsible for coordinating the application's activity. It receives user input from the presentation layer, invokes the domain layer to perform business logic, and returns the results to the presentation layer. In our case, the application layer is the web API application, which receives HTTP requests from the client application, invokes the domain layer to execute business logic, and returns the results to the client application.

- **Domain layer**: This layer is the heart of the application. It contains the domain model and domain logic, which is independent of the application layer and the infrastructure layer. In other words, the domain layer does not depend on any other layers. For example, if we need to implement a `Deposit` method for the `SavingAccount` class, the logic in the domain layer does not have any knowledge of how to save data to the database. Instead, it only focuses on the business logic of the `Deposit` method using abstractions and interfaces. This layer often contains entities, value objects, aggregates, repositories, and domain services.

- **Infrastructure layer**: This layer implements the application infrastructure, such as data access, caching, logging, messaging, and so on. It often integrates with external systems as dependencies, such as database, message queues, and so on. In our case, the infrastructure layer can include the data access layer, which uses EF Core to access the database.

DDD focuses primarily on the domain and application layers. This is because the UI layer and the infrastructure layer are not specific to DDD and can be implemented with any technology or framework. For example, the UI layer can be implemented using **ASP.NET Core MVC**, **Blazor**, **React**, **WPF**, or any other UI framework on various platforms, while the core domain logic remains the same. Similarly, DDD does not dictate data storage, which can be a relational database, a NoSQL database, or any other data storage. The domain layer uses the repository pattern to access the data, which is independent of data storage. Another example is the logging mechanism, which is also not specific to DDD, as the domain layer needs to log the business events but does not care about the logging system used.

DDD building blocks

DDD has a set of building blocks that can be used to build the domain model. These building blocks include entities, value objects, aggregates, repositories, and domain services. In the following subsections, we will introduce these building blocks and how to use them to build the domain model:

Entity

You may have essential knowledge of **object-oriented programming (OOP)** and **object-relational mapping (ORM)** if you have read the previous chapters. In OOP, an object is an instance of a class. An object has a state and behavior. The state is represented by the properties of the object, while the behavior is represented by the methods of the object.

In DDD, the entity is similar to an object in OOP, but it is more than that. An **entity** is an object that has a unique identity and is defined by its identity, not its attributes. Normally, an entity is mapped to a table in the database.

The identity of an entity is normally represented by an ID property. The ID property is immutable, which means that once it is set, it cannot be changed. The ID property can be a primitive type, such as an integer, a string, or a GUID. It can also be composite keys.

If two entities have the same properties but different identities, they are considered different entities.

For example, in a banking system, `Account` is an entity. It has a unique identity that can be represented by an `Id` property. Two accounts cannot have the same `Id` property.

Value object

A **value object** is another type of object in DDD. It is identified by its properties, rather than a unique identity. Normally, a value object is immutable, which means that its properties cannot be changed once it is created. If two value objects have the same properties, they are considered the same value object.

For example, `Address` is a value object. It is identified by its properties, such as `Street`, `City`, `State`, and `ZipCode`. If two addresses have the same `Street`, `City`, `State`, and `ZipCode`, they are considered the same address.

Aggregate

An **aggregate** is a cluster of associated objects, including entities and value objects, which are treated as a unit for data changes. An aggregate has a root entity, which is the only object that can be accessed from outside the aggregate. The root entity is responsible for maintaining the consistency and integrity of the aggregate. It is important to note that if the external objects need to access the objects inside the aggregate or modify the objects inside the aggregate, they must go through the root entity.

For example, in an invoicing system, an `Invoice` entity is an aggregate root. It contains a list of `InvoiceItem` entities, which are the items of the invoice. To add an item to the invoice, the external objects must go through the `Invoice` entity, as shown in the following code:

```
public class Invoice
{
    public int Id { get; private set; } // Aggregate root Id, which
should not be changed once it is set
```

```
    public DateTime Date { get; set; }
    public InvoiceStatus Status { get; private set; }
    public decimal Total { get; private set; } // The total amount of
the invoice, which should be updated when an item is added or removed,
but cannot be changed directly
    // Other properties
    public List<InvoiceItem> Items { get; private set; }

    public void AddItem(InvoiceItem item)
    {
        // Add the item to the invoice
        Items.Add(item);
        // Update the invoice total, etc.
        // ...
    }

    public void RemoveItem(InvoiceItem item)
    {
        // Remove the item from the invoice
        Items.Remove(item);
        // Update the invoice total, etc.
        // ...
    }

    public void Close()
    {
        // Close the invoice
        Status = InvoiceStatus.Closed;
    }
}
```

In the preceding example, if we need to add or remove an item from the invoice, we must get the `Invoice` entity first, and then call the `AddItem()` or `RemoveItem()` method to add or remove the item. We cannot directly add or remove an item from the `Items` property because the `Items` property is private and can only be accessed from inside the `Invoice` entity. In this way, the domain logic is encapsulated inside the `Invoice` entity, and the consistency and integrity of the invoice are maintained. Similarly, we cannot change the `Total` property directly. Instead, the `AddItem` or `RemoveItem` method can update the `Total` property.

Repository

A **repository** is an abstraction layer used to access the data persistence layer. It encapsulates the data access logic and provides a way to query and save data. To ensure the domain layer does not depend on any specific data access technology, a repository is typically implemented as an interface. The infrastructure layer can then use a specific data access technology, such as EF Core or Dapper, to implement the repository interface and access different data sources, such as relational databases or NoSQL databases. This decouples the domain layer from the data access technology and data storage.

An example of a repository interface is shown in the following code:

```
public interface IInvoiceRepository
{
    Task<Invoice> GetByIdAsync(Guid id);
    Task<List<Invoice>> GetByCustomerIdAsync(Guid customerId);
    Task AddAsync(Invoice invoice);
    Task UpdateAsync(Invoice invoice);
    Task DeleteAsync(Invoice invoice);
}
```

We introduced the repository pattern in *Chapter 9*. It is not a specific DDD pattern. However, it is often used in DDD to decouple the domain layer from the data access layer.

Domain service

A **domain service** is a stateless service that contains domain logic that does not belong to any specific entity or value object. It is often used to implement complex domain logic that involves multiple entities or value objects. To access the data persistence layer, a domain service may depend on one or more repositories. Additionally, it may also depend on other external services. These dependencies are injected into the domain service through the dependency injection mechanism.

For example, in a banking system, the TransferService domain service is responsible for the logic of transferring money from one account to another. To do this, it relies on the AccountRepository to access the Account entity. Additionally, it may need to use an external service to send a notification to the account holder after the transfer is complete. If the accounts are in different banks, the TransferService domain service may also need to use an external service to transfer money between them.

The following code shows an example of a domain service:

```
public class TransferService
{
    private readonly IAccountRepository _accountRepository;
    private readonly ITransactionRepository _transactionRepository;
    private readonly INotificationService _notificationService;
    private readonly IBankTransferService _bankTransferService;

    public TransferService(IAccountRepository accountRepository,
    ITransactionRepository transactionRepository, INotificationService
    notificationService, IBankTransferService bankTransferService)
    {
```

```
        _accountRepository = accountRepository;
        _transactionRepository = transactionRepository;
        _notificationService = notificationService;
        _bankTransferService = bankTransferService;
    }


    public async Task TransferAsync(Guid fromAccountId, Guid
toAccountId, decimal amount)
    {
        // Get the account from the repository
        var fromAccount = await _accountRepository.
GetByIdAsync(fromAccountId);
        var toAccount = await _accountRepository.
GetByIdAsync(toAccountId);
        // Transfer money between the accounts
        fromAccount.Withdraw(amount);
        toAccount.Deposit(amount);
        // Save the changes to the repository
        await _accountRepository.UpdateAsync(fromAccount);
        await _accountRepository.UpdateAsync(toAccount);
        // Create transaction records
        await _transactionRepository.AddAsync(new Transaction
        {
            FromAccountId = fromAccountId,
            ToAccountId = toAccountId,
            Amount = amount,
            Date = DateTime.UtcNow
        });
        await _transactionRepository.AddAsync(new Transaction
        {
            FromAccountId = toAccountId,
            ToAccountId = fromAccountId,
            Amount = -amount,
            Date = DateTime.UtcNow
        });
        // Send a notification to the account holder
        await _notificationService.SendAsync(fromAccount.HolderId,
$"You have transferred {amount}to {toAccount.HolderId}");
        await _notificationService.SendAsync(toAccount.HolderId, $"You
have received {amount} from{fromAccount.HolderId}");
        // Transfer money between the banks
        // await _bankTransferService.TransferAsync(fromAccount.
BankId, toAccount.BankId, amount);
    }
}
```

The preceding code shows a `TransferService` domain service. It has four dependencies: `IAccountRepository`, `ITransactionRepository`, `INotificationService`, and `IBankTransferService`. The `TransferAsync` method transfers money from one account to another. It first obtains the accounts from `IAccountRepository`, and then transfers money between the accounts. After that, it saves the changes to `IAccountRepository` and creates transaction records in `ITransactionRepository`. Finally, it sends a notification to the account holders using `INotificationService`.

> **Important note**
>
> The preceding example is simplified for demonstration purposes. The actual implementation to transfer money between two accounts is much more complicated. For example, it may need to check the balance of the accounts, check the daily transfer limit, and so on. It may also need to transfer money between different banks, which involves a lot of complex logic to handle any errors that may occur during the transfer. If any error occurs, it may need to roll back the transaction. This is a typical example of a domain service that implements complex domain logic.

Unit of work

In the preceding example, when transferring money between two accounts, the process involves multiple steps. What if an error occurs during the process? In order to prevent any money from being lost during the process of transferring funds between two accounts, it is necessary to wrap the process in a transaction. This will ensure that in the event of an error occurring, the transaction will be rolled back, and the funds will remain secure. For example, if the `TransferAsync()` method throws an exception after the money has been withdrawn from `fromAccount` but before it is deposited to `toAccount`, the transaction will be rolled back, and the money will not be lost.

The term **transaction** is often used in the context of databases. This kind of transaction is called a **unit of work** in DDD. A unit of work is a sequence of operations that must be performed as a whole. All the steps in a unit of work must succeed or fail together. If any step fails, the entire unit of work must be rolled back. This prevents the data from being left in an inconsistent state.

A unit of work can be implemented in various ways. In many scenarios, a unit of work is implemented as a database transaction. Another example is a message queue. When a message is received, it is processed as a unit of work. If the process is successful, the message is removed from the queue. Otherwise, the message remains in the queue and will be processed again at a later time.

Application service

The application service is responsible for managing the application process. It receives user input from the presentation layer, invokes the domain service to execute business logic, and returns the results to the Presentation Layer. In a web API application, the Application Service can be implemented as a web API controller or a separate service that is invoked by the web API controller.

The application service should be thin and delegate most of the work to the domain service. Typically, the application service uses **data transfer objects (DTOs)** to transfer data between the presentation layer and the domain layer. A DTO is a simple object that contains data and does not have any behavior. The DTOs are often mapped to the entities using mapping tools such as `AutoMapper`. For example, an `InvoiceDto` class may contain the properties of an invoice, such as `Id`, `Date`, `Status`, `Total`, and so on. It does not have any method to add or remove an invoice item or close the invoice. It is purely a data container. If a property of the `Invoice` entity is not needed in the presentation layer, it should not be included in `InvoiceDto`.

The presentation layer can send a DTO to the application service when it needs to create or update an entity. The application service will then map the DTO to the entity and invoke the domain service to execute the necessary business logic. Finally, the application service will map the entity back to the DTO and return it to the presentation layer.

Here is a simple example of an application service:

```
[Route("api/[controller]")]
[ApiController]
public class InvoicesController : ControllerBase
{
    private readonly IInvoiceService _invoiceService;

    public InvoicesController(IInvoiceService invoiceService)
    {
        _invoiceService = invoiceService;
    }

    [HttpPost]
    public async Task<IActionResult> CreateAsync(InvoiceDto
invoiceDto)
    {
        var invoice = await _invoiceService.CreateAsync(invoiceDto);
        return Ok(invoice);
    }

    // Omitted other methods
}

public interface IInvoiceService
{
    Task<InvoiceDto> CreateAsync(InvoiceDto invoiceDto);
    // Omitted other methods
}
```

```csharp
public class InvoiceService : IInvoiceService
{
    private readonly IInvoiceRepository _invoiceRepository;
    private readonly IMapper _mapper;

    public InvoiceService(IInvoiceRepository invoiceRepository,
IMapper mapper)
    {
        _invoiceRepository = invoiceRepository;
        _mapper = mapper;
    }

    public async Task<InvoiceDto> CreateAsync(InvoiceDto invoiceDto)
    {
        var invoice = _mapper.Map<Invoice>(invoiceDto);
        await _invoiceRepository.AddAsync(invoice);
        return _mapper.Map<InvoiceDto>(invoice);
    }

    // Omitted other methods
}
```

In the preceding example, the `IInvoiceService` interface defines the methods of the application service. The `InvoiceService` class implements the `IInvoiceService` interface. It has two dependencies: `IInvoiceRepository` and `IMapper`. `IInvoiceRepository` is used to access the `Invoice` entity, while `IMapper` is used to map `InvoiceDto` to the `Invoice` entity and vice versa. The `CreateAsync()` method receives `InvoiceDto` from the presentation layer via the controller, maps it to the `Invoice` entity, and then invokes the `AddAsync()` method of `IInvoiceRepository` to add the `Invoice` entity to the database. Finally, it maps the `Invoice` entity back to `InvoiceDto` and returns it to the presentation layer.

> **Important note**
>
> In the preceding example, there is no domain service. This is because the logic to create an invoice is simple. In this case, the application service layer can directly invoke the repository to add the invoice to the database. However, if the logic is more complex, involving multiple entities or aggregates, it is better to use a domain service to implement the logic.

DDD focuses on how to build a domain model that reflects the business domain and how to maintain the consistency and integrity of the domain model. It is not used to produce reports or user interfaces. Reports may need complex queries that are not suitable for the domain model. For this case, you may need to use a separate reporting database or reporting service. Similarly, the user interface may need to display data in a different way than the domain model. However, the domain model should remain the same no matter how the data are displayed.

DDD can help you to manage the complexity and build a flexible and maintainable software system. But keep in mind that DDD is not a silver bullet. Typically, DDD is used for complex business domains. Developers must implement a lot of isolation, abstraction, and encapsulation to maintain the model. This may lead to a lot of effort and complexity. If your project is simple, DDD may be a bit overkill. In this case, a simple layered architecture may be a better choice.

Clean architecture

Clean architecture is a software architecture that was proposed by Robert C. Martin (also known as Uncle Bob) in his book *Clean Architecture: A Craftsman's Guide to Software Structure and Design*, published in 2017. It is a layered architecture that focuses on the separation of concerns. Similar to DDD, clean architecture is not a specific technology or framework. It is a set of principles and practices that can be applied to any software project.

Clean architecture is also called onion architecture because the layers are arranged in a circular shape, like an onion. The following diagram shows the typical layers of clean architecture:

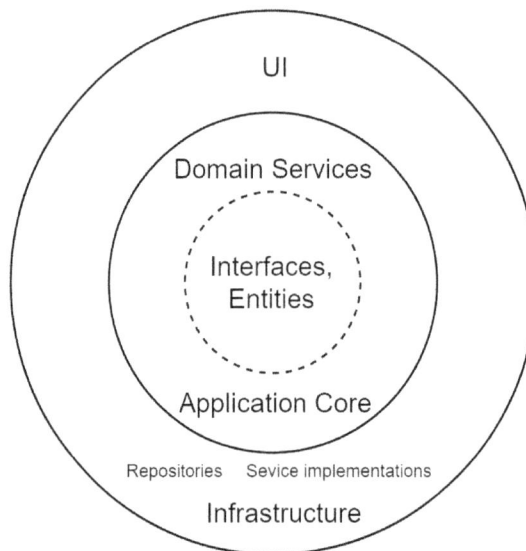

Figure 17.2 – Typical layers of clean architecture

The preceding diagram illustrates the dependencies flowing from the outer layers to the inner layers. At the center of the architecture lies the application core layer, which contains the entities and interfaces for business logic. Additionally, this layer contains domain services that implement the interfaces. It does not depend on any other layers. Surrounding the application core layer is the Infrastructure Layer and UI layer, both of which depend on the application core layer. This architecture ensures that the application core layer is unaware of how the data are stored or presented to the user. In addition, the Infrastructure Layer and UI layer can be replaced without impacting the application core layer.

Clean architecture shares some similarities with DDD. Both of them are layered architectures that focus on the separation of concerns. They both use dependency injection (or inversion of control) to decouple the layers. DDD focuses on the domain layer, while clean architecture prioritizes the importance of isolating the core business logic from the external dependencies. The separation of concerns allows for modifications to external components without affecting the core business logic, making it easier to adapt to evolving requirements.

DDD and clean architecture complement each other and can be used together. While DDD guides how to build a domain model and understand the business domain, clean architecture offers a blueprint for organizing and structuring the codebase. Combining these approaches can lead to a flexible and maintainable software system.

DDD and clean architecture are both layered architectures that focus on a business domain. Next, let us discuss the architecture of the entire software system. In the next section, we will introduce microservices, which is a popular architecture for building scalable and maintainable software systems.

Microservices

Many traditional applications are built as a monolith. A monolithic application is deployed as a single unit on a single server. The monolithic application is easy to develop and deploy. However, as the application grows, it becomes more and more difficult to maintain and scale. A small change in the application may require the entire application to be rebuilt, retested, and redeployed. Moreover, if one part of the application needs to be scaled, the application must be scaled as a whole, which is not cost-effective. In addition, if one part of the application fails, it may affect the entire application.

This is where microservices come in. A microservice is a small, independent service that is responsible for a specific business domain. Each microservice has its own database and dependencies. It can be developed, deployed, and scaled independently. These microservices communicate with each other through programmatic interfaces, such as web APIs or message queues.

Microservices provide several benefits:

- **Single responsibility**: Each microservice is responsible for a specific business domain. It has its own dependencies and database.

- **Resilience and fault tolerance**: Microservices are designed to be resilient and fault tolerant. If one microservice fails, it does not affect other microservices.

- **Scalability**: Microservices can be scaled independently based on demand. If one microservice has a high workload, we can increase the number of instances of that microservice to handle the workload.

- **Technology diversity**: Each microservice can be built using different technologies and frameworks as long as they communicate with each other through standard interfaces, such as HTTP APIs or gRPC.

- **CI/CD**: Microservices facilitate CI/CD by allowing individual microservices to be built, tested, and deployed independently, minimizing disruption to the entire system.

Microservices is not a new concept; it has been around for decades. However, it has become more popular in recent years, especially with the rise in cloud computing. Cloud computing provides a scalable and cost-effective infrastructure for microservices. In addition, the emergence of container technology, such as Docker, makes it easier to build and deploy microservices. By using containers and container orchestration tools, such as **Kubernetes**, developers can easily build and deploy microservices to the cloud. The orchestration tools can automatically scale the microservices based on the workload. This makes it easier to build a scalable and cost-effective software system.

Microservices do not have to be exclusive to other architectures. In fact, they can be used in conjunction with other architectures to create a more robust and efficient system. You can use layers, such as DDD and clean architecture, to build each microservice. By leveraging the benefits of both architectures, organizations can create a powerful and reliable system that meets their needs. This approach can be especially beneficial for organizations that require a high degree of scalability and flexibility.

For example, in an online shopping system, we may have the following microservices:

- **Product service**: This service is responsible for managing the products, such as adding a new product, updating a product, deleting a product, and so on. It has its own database to store the product data.

- **Order service**: This service is responsible for managing the orders, such as creating a new order, updating an order, deleting an order, and so on. It also has its own database to store the order data.

- **Payment service**: This service is responsible for processing payments, such as credit card payments, PayPal payments, and so on. It has its own database to store the payment data. It may also need to integrate with external payment services, such as PayPal, Stripe, online banking services, and so on.

- **Shipping service**: This service is responsible for shipping the products, such as shipping a product to a customer and tracking the shipment. It needs to integrate with external shipping services, such as FedEx, UPS, and so on.

- **Notification service**: This service is responsible for sending notifications to the customers, such as sending an email or a text message notification, and so on. It needs to integrate with external notification services, such as SendGrid, Twilio, and so on.

- **Identity service**: This service is responsible for managing users, such as creating a new user, updating a user, deleting a user, and so on. It may provide third-party authentication, such as from Microsoft, Google, Facebook, and so on.

- **Gateway service**: This service is responsible for routing the requests to the appropriate microservices. It is the entry point of the system. It does not have its own database. Instead, it routes the requests to the appropriate microservices based on the request URL. It can also implement rate-limiting, authentication, authorization, and so on.

- **Client applications**: These are the client applications that consume the microservices. They can be web applications, mobile applications, or desktop applications.

Each service is responsible for a specific business domain and has its own dependencies. Developers can use different technologies and frameworks to build the services because they communicate with each other through standard HTTP APIs or gRPC. If one service needs to be scaled, it can be scaled independently. For example, if the `Order` service has a high workload, we can increase the number of instances of the `Order` service to handle the workload. This is much more cost-effective than scaling the entire application. Moreover, if one service fails, it does not affect other services. For example, if the `Payment` service fails, the `Order` service and `Product` service can still work. It can still receive orders and allow users to view products. When the `Payment` service is back online, it can process the orders that have not been processed.

Microservices have become more and more popular in recent years. However, it increases the complexity of the system. Before adopting microservices, you should carefully consider whether it is suitable for your project. Consider the following challenges to microservices:

- **Distributed system complexity**: Microservices are distributed systems. They are more complex than monolithic applications. For example, if one service needs to call another service, you need to consider how to handle the communication between the services and how to maintain the consistency of the data. In addition, you need to handle network failures, partial failures, cascading failures, and so on.

- **Data management**: Each microservice has its own database. This makes it difficult to maintain data consistency, as transactions that span multiple microservices are not supported. To query data from multiple microservices, a distributed query mechanism must be implemented, which can be a complex process.

- **Service discovery**: In a microservice architecture, each service has its own URL. They need to know the URLs of other services in order to communicate with them. This is called service discovery. There are many ways to implement service discovery, such as using a service registry, using a service mesh, and so on. The container orchestration tools, such as Kubernetes, can also be used to implement service discovery, as they can maintain the internal service URLs of the microservices.

- **Testing**: Testing a microservice architecture is more complex than testing a monolithic application. In addition to unit testing, integration testing, and end-to-end testing, you also need to test the communication between the microservices.

- **Monitoring**: Monitoring a microservice architecture requires a well-designed monitoring system. You need to monitor the health of each microservice, as well as the communication between the microservices. The tracing mechanism can be used to trace the requests between the microservices.

In summary, if your application is simple, do not overcomplicate it by using microservices. As your application grows, you can consider refactoring it into microservice architecture step by step.

Next, let us discuss some common design patterns for web API applications.

Web API design patterns

To build a flexible, scalable, and maintainable web API application, leveraging well-established design patterns is essential. These patterns address common challenges encountered in web API development, providing effective solutions. Microsoft's comprehensive guide offers insights into these design patterns, and you can find more details at the following link: `https://learn.microsoft.com/en-us/azure/architecture/patterns/`.

These design patterns are not exclusive to ASP.NET Core; they can be applied to any web API, regardless of the underlying technology or framework. In the following sub-sections, we will introduce some key design patterns, outlining the problems they solve, their implementation details, and considerations for their usage. These patterns cover solution design and implementation, messaging, reliability, and so on, including the following:

- **Command query responsibility segregation (CQRS)**
- **Publish/subscribe (pub/sub)**
- **Backend for frontend (BFF)**
- Timeout
- Rate limiting
- Retry
- Circuit breaker

CQRS

CQRS is a powerful tool for addressing the challenge of scaling and optimizing read and write operations. By separating the responsibilities for handling commands (writes) and queries (reads), CQRS enables each operation to be optimized independently, resulting in improved scalability and efficiency.

Traditionally, the data model of an application is designed to support both read and write operations. However, the requirements for read and write operations are often different. The read operations may execute different queries, resulting in different DTO models. The write operations may need to update multiple tables in the database. This may lead to a complex data model that is difficult to maintain. In addition, the read operations and write operations may have different performance requirements.

CQRS divides the application's data model into separate models for reading and writing. This enables the use of different storage mechanisms and optimizations tailored to the specific needs of each operation. CQRS uses queries to read data and commands to update data. Queries do not change the state of the system, while commands do.

To better separate the read and write operations, CQRS can also use different data stores for reading and writing. For example, the read store can use multiple read-only replicas of the write store, which can improve the performance of the read operations. The replicas must be kept in sync with the write store, which can be done by using built-in database replication features or an event-driven mechanism.

The following diagram shows a typical CQRS architecture:

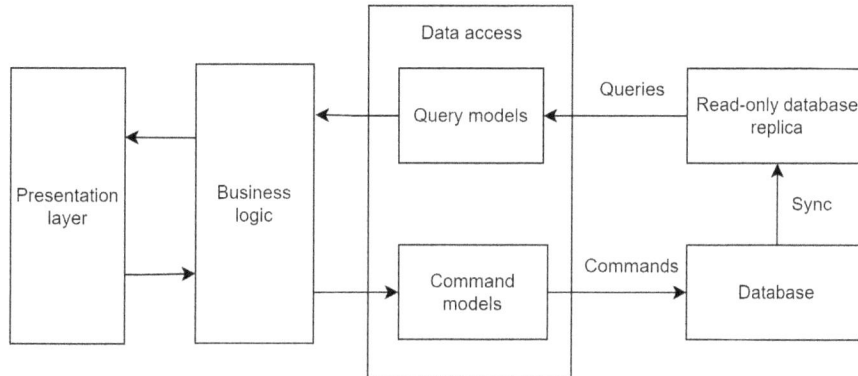

Figure 17.3 – Typical CQRS architecture

To implement CQRS in ASP.NET Core web API applications, you can use the `MediatR` library, which is a simple mediator implementation in .NET. This library is a simple mediator implementation in .NET that enables the use of the mediator pattern. The mediator pattern is a behavioral design pattern that enables objects to interact without having to refer to each other explicitly. Instead, they communicate through the mediator, which decouples the objects and allows for greater flexibility.

The following diagram shows a typical CQRS architecture using the `MediatR` library:

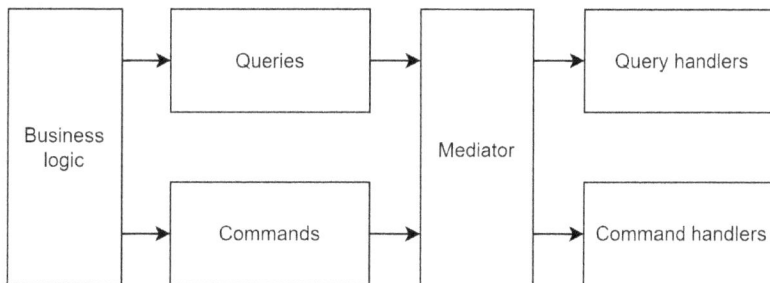

Figure 17.4 – Typical CQRS architecture using the MediatR library

In the preceding diagram, the mediator is responsible for receiving the commands and queries from the business logic layer and then invoking the corresponding handlers to execute the commands and queries. Then, the handlers can use the repositories to access the data persistence layer for reading and writing data. The business logic layer does not need to know how the mediator invokes the handlers. It only needs to send the commands and queries to the mediator. This decouples the business logic layer from the data persistence layer. This pattern also makes it easier to send commands and queries to multiple handlers. For example, if we have a command to send an email notification to the customer and we need to add a text message notification, we can simply add a new handler to handle the command without changing the client code.

You can find a sample application that demonstrates how to implement CQRS in ASP.NET Core web API applications in the `/chapter17/CqrsDemo` folder of the source code.

> **Important note**
>
> The sample project has a separate infrastructure project to implement the data persistence layer by following clean architecture. When you run the `dotnet ef` command to add a migration or update the database, you need to specify the startup project. For example, to add a migration, you need to navigate to the `CqrsDemo.Infrastructure` project and run the following command:
>
> **dotnet ef migrations add InitialCreate --startup-project ../CqrsDemo.WebApi**

To learn more about the `dotnet ef` command, you can refer to the following link: `https://learn.microsoft.com/en-us/ef/core/cli/dotnet#target-project-and-startup-project`.

To follow the next steps, you can use the project in the `/chapter17/CqrsDemo/start` folder of the source code. This project contains a basic ASP.NET Core web API application to manage the invoices. It contains the following projects:

- `CqrsDemo.WebApi`: This is the ASP.NET Core web API project. It contains the controllers and application configurations.

- `CqrsDemo.Core`: This is the core project that contains the domain models, interfaces of repositories, services, and so on.

- `CqrsDemo.Infrastructure`: This project contains the implementation of the repositories.

Implementing the model mapping

In the core project, note that the service layer uses DTOs, as shown in the following:

```
public interface IInvoiceService
{
    Task<InvoiceDto?> GetAsync(Guid id, CancellationToken
cancellationToken = default);
```

```
    Task<List<InvoiceWithoutItemsDto>> GetPagedListAsync(int
pageIndex, int pageSize, CancellationToken cancellationToken =
default);
    Task<InvoiceDto> AddAsync(CreateOrUpdateInvoiceDto invoice,
CancellationToken cancellationToken = default);
    Task<InvoiceDto?> UpdateAsync(Guid id, CreateOrUpdateInvoiceDto
invoice, CancellationToken cancellationToken = default);
    // Omitted
}
```

These methods use different DTO types for reading and writing. To map the entities to DTOs and vice versa, we can use AutoMapper, which is a popular object-to-object mapper library. The following code shows how to configure AutoMapper in the InvoiceProfile.cs file:

```
public InvoiceProfile()
{
    CreateMap<CreateOrUpdateInvoiceItemDto, InvoiceItem>();
    CreateMap<InvoiceItem, InvoiceItemDto>();
    CreateMap<CreateOrUpdateInvoiceDto, Invoice>();
    CreateMap<Invoice, InvoiceWithoutItemsDto>();
    CreateMap<Invoice, InvoiceDto>();
}
```

Then, we can register AutoMapper in the Program.cs file as follows:

```
builder.Services.AddAutoMapper(typeof(InvoiceProfile));
```

To use the mapper, just simply inject the IMapper interface into the service layer, as shown in the following:

```
public class InvoiceService(IInvoiceRepository invoiceRepository,
IMapper mapper) : IInvoiceService
{
    public async Task<InvoiceDto?> GetAsync(Guid id, CancellationToken
cancellationToken = default)
    {
        var invoice = await invoiceRepository.GetAsync(id,
cancellationToken);
        return invoice == null ? null : mapper.
Map<InvoiceDto>(invoice);
    }
    // Omitted
}
```

Using AutoMapper can save us a lot of time for mapping the entities to DTOs and vice versa. Next, we can implement the queries and commands using the MediatR library.

Implementing queries

Next, we will implement the CQRS pattern using the MediatR library. Follow these steps:

1. First, we need to install the MediatR NuGet package. Run the following command in the terminal window to install the MediatR package:

   ```
   dotnet add package MediatR
   ```

   You need to install the MediatR package to the CqrsDemo.Core project and the CqrsDemo.WebApi project.

   MediatR provides the following interfaces:

   - IMediator: This is the main interface of the MediatR library. It can be used to send requests to the handlers. It can also be used to publish events to multiple handlers.

   - ISender: This interface is used to send a request through the mediator pipeline to be handled by a single handler.

   - IPublisher: This interface is used to publish a notification or event through the mediator pipeline to be handled by multiple handlers.

   The IMediator interface can be used to send all requests or events. For a clearer indication of the purpose of the request or event, it is recommended to use the ISender interface for requests handled by a single handler and the IPublisher interface for notifications or events that require multiple handlers.

2. Create a Queries folder in the CqrsDemo.Core project. Then, create a GetInvoiceByIdQuery.cs file in the Queries folder with the following code:

   ```
   public class GetInvoiceByIdQuery(Guid id) :
   IRequest<InvoiceDto?>
   {
       public Guid Id { get; set; } = id;
   }
   ```

   The preceding code defines a GetInvoiceByIdQuery class that implements the IRequest<InvoiceDto?> interface. This interface is used to indicate that this is a query that returns an InvoiceDto object. The Id property is used to specify the ID of the invoice to be retrieved.

3. Similarly, create a GetInvoiceListQuery.cs file in the Queries folder with the following code:

   ```
   public class GetInvoiceListQuery(int pageIndex, int pageSize) :
   IRequest<List<InvoiceWithoutItemsDto>>
   {
       public int PageIndex { get; set; } = pageIndex;
       public int PageSize { get; set; } = pageSize;
   }
   ```

Note that the GetInvoiceListQuery query returns a list of InvoiceWithoutItemsDto objects. This is because we do not need the invoice items when listing the invoices. This is an example to show how to use different DTOs for reading and writing.

4. Next, create a Handlers folder in the Queries folder. Then, create a GetInvoiceByIdQueryHandler.cs file in the Handlers folder with the following code:

```
public class GetInvoiceByIdQueryHandler(IInvoiceService
invoiceService) : IRequestHandler<GetInvoiceByIdQuery,
InvoiceDto?>
{
    public Task<InvoiceDto?> Handle(GetInvoiceByIdQuery request,
CancellationToken cancellationToken)
    {
        return invoiceService.GetAsync(request.Id,
cancellationToken);
    }
}
```

The GetInvoiceByIdQueryHandler class implements the IRequestHandler<GetInvoiceByIdQuery, InvoiceDto?> interface. This interface is used to indicate that this handler handles the GetInvoiceByIdQuery query and returns an InvoiceDto object. The Handle() method receives the GetInvoiceByIdQuery query and invokes the GetAsync() method of IInvoiceService to get the invoice by using the ID.

The IInvoiceService interface can be injected into the handler. Alternatively, you may choose to inject the IInvoiceRepository interface directly into the handler and implement business logic there. Ultimately, it is your decision where to store the logic. It is important to keep in mind that the goal is to separate business logic from the data persistence layer.

5. Similarly, create a GetInvoiceListQueryHandler.cs file in the Handlers folder with the following code:

```
public class GetInvoiceListQueryHandler(IInvoiceService
invoiceService) : IRequestHandler<GetInvoiceListQuery,
List<InvoiceWithoutItemsDto>>
{
    public Task<List<InvoiceWithoutItemsDto>>
Handle(GetInvoiceListQuery request, CancellationToken
cancellationToken)
    {
        return invoiceService.GetPagedListAsync(request.
PageIndex, request.PageSize, cancellationToken);
    }
}
```

Now, we have two handlers to handle the `GetInvoiceByIdQuery` query and the `GetInvoiceListQuery` query. Next, we need to update the controllers to use the `MediatR` library.

6. Update the `InvoicesController.cs` file in the `CqrsDemo.WebApi` project with the following code:

```
[Route("api/[controller]")]
[ApiController]
public class InvoicesController(IInvoiceService invoiceService,
ISender mediatorSender) : ControllerBase
{
    // Omitted
}
```

The preceding code injects the `ISender()` interface into the controller. You can also inject the `IMediator` interface instead. In this example, we will use the `ISender` interface to send the requests to the handlers.

7. Update the `GetInvoice()` method of the `InvoicesController` class with the following code:

```
[HttpGet("{id}")]
public async Task<ActionResult<InvoiceDto>> GetInvoice(Guid id)
{
    var invoice = await mediatorSender.Send(new
GetInvoiceByIdQuery(id));
    return invoice == null ? NotFound() : Ok(invoice);
}
```

The preceding code creates a `GetInvoiceByIdQuery` object that contains the `id` parameter. The `ISender` interface will invoke the `GetInvoiceByIdQueryHandler` handler to handle the query. Then, the handler will invoke the `GetAsync` method of the `IInvoiceService` to get the invoice via the ID. So, the controller is decoupled from the service layer.

8. Similarly, update the `GetInvoices` method of the `InvoicesController` class with the following code:

```
[HttpGet]
[Route("paged")]
public async
Task<ActionResultIEnumerableInvoiceWithoutItemsDto>>>
GetInvoices(int pageIndex, int pageSize)
{
    var invoices = await mediatorSender.Send(new
GetInvoiceListQuery(pageIndex, pageSize));
    return Ok(invoices);
}
```

The preceding code creates a `GetInvoiceListQuery` object that contains the `pageIndex` and `pageSize` parameters. The `ISender` interface will invoke the `GetInvoiceListQueryHandler` handler to handle the query. Then, the handler will invoke the `GetPagedListAsync()` method of the `IInvoiceService` to get the list of invoices.

9.   Next, we need to register the `MediatR` in the `Program.cs` file as follows:

```
builder.Services.AddMediatR(cfg => cfg.
RegisterServicesFromAssembly(typeof(GetInvoiceByIdQueryHandler).
Assembly));
```

The preceding code registers all three `MediatR` interfaces and the handlers in the `CqrsDemo.Core` project.

Now, we use the queries to implement the read operations. You can run the application and test the endpoints, such as `/api/invoices/{id}` and `/api/invoices/paged`. These endpoints should work as before.

Implementing commands

Next, we will implement the write operations using commands. Follow these steps:

1.   Create a `Commands` folder in the `CqrsDemo.Core` project. Then, create a `CreateInvoiceCommand.cs` file in the `Commands` folder with the following code:

```
public class CreateInvoiceCommand(CreateOrUpdateInvoiceDto
invoice) : IRequest<InvoiceDto>
{
    public CreateOrUpdateInvoiceDto Invoice { get; set; } =
invoice;
}
```

The preceding code defines a `CreateInvoiceCommand` class that implements the `IRequest<InvoiceDto>` interface.

2.   Create a `Handlers` folder in the `Commands` folder. Then, create a `CreateInvoiceCommandHandler.cs` file in the `Handlers` folder with the following code:

```
public class CreateInvoiceCommandHandler(IInvoiceService
invoiceService) : IRequestHandler<CreateInvoiceCommand,
InvoiceDto>
{
    public Task<InvoiceDto> Handle(CreateInvoiceCommand request,
CancellationToken cancellationToken)
    {
        return invoiceService.AddAsync(request.Invoice,
cancellationToken);
    }
}
```

3. Update the `InvoicesController` class with the following code:

```
[HttpPost]
public async Task<ActionResult<InvoiceDto>>
CreateInvoice(CreateOrUpdateInvoiceDto invoice)
{
    var result = await mediatorSender.Send(new
CreateInvoiceCommand(invoice));
    return CreatedAtAction(nameof(GetInvoice), new { id =
result.Id }, result);
}
```

Now, run the application and send a POST request to the `/api/invoices` endpoint. You should be able to create a new invoice.

We will not implement all the commands and queries in this example. You can work on the remaining commands and queries as an exercise.

`MediatR` makes it easy to implement the CQRS pattern in ASP.NET Core web API applications. However, it is not the only way to implement CQRS. You can also implement CQRS without using the `MediatR` library.

One benefit of using the `MediatR` library is that it can send requests to multiple handlers. For example, we can create a command to send an email notification and a text message notification to the customer. Then, we can create two handlers to handle the command. Follow these steps to implement this feature:

1. Add the two properties to the invoice models, as shown in the following:

```
public string ContactEmail { get; set; } = string.Empty;
public string ContactPhone { get; set; } = string.Empty;
```

You need to update the `Invoice` class, `CreateOrUpdateInvoiceDto` class, `InvoiceWithoutItemsDto` class, and `InvoiceDto` class. You can also define a `Contact` class for better encapsulation.

2. Add the database migration and update the database. You may also need to update the seed data. Note that you need to specify the startup project when running the `dotnet ef` command. For example, to add a migration, you need to navigate to the `CqrsDemo.Infrastructure` project and run the following command:

```
dotnet ef migrations add AddContactInfo --startup-project ../
CqrsDemo.WebApi
```

3. Then, update the database:

```
dotnet ef database update --startup-project ../CqrsDemo.WebApi
```

4. Create a `Notification` folder in the `CqrsDemo.Core` project. Then, create a `SendInvoiceNotification` class in the `Notification` folder with the following code:

```
public class SendInvoiceNotification(Guid invoiceId) :
INotification
{
    public Guid InvoiceId { get; set; } = invoiceId;
}
```

The preceding code defines a `SendInvoiceNotification` class that implements the `INotification` interface. This interface is used to indicate that this is a notification that does not return any result.

5. Create a `Handlers` folder in the `Notification` folder. Then, create a `SendInvoiceE-mailNotificationHandler` class in the `Handlers` folder with the following code:

```
public class SendInvoiceEmailNotificationHandler(IInvoiceService
invoiceService) : INotificationHandler<SendInvoiceNotification>
{
    public async Task Handle(SendInvoiceNotification
notification, CancellationToken cancellationToken)
    {
        // Send email notification
        var invoice = await invoiceService.
GetAsync(notification.InvoiceId, cancellationToken);
        if (invoice is null || string.
IsNullOrWhiteSpace(invoice.ContactEmail))
        {
            return;
        }
        // Send email notification
        Console.WriteLine($"Sending email notification to
{invoice.ContactEmail} for invoice {invoice.Id}");
    }
}
```

In the preceding code, we use `IInvocieService` to obtain the invoice via the ID. Then, we check if the invoice exists and if the contact email is specified. If so, we send an email notification to the customer. For simplicity, we just print a message to the console.

6. Similarly, create a `SendInvoiceTextMessageNotificationHandler` class in the `Handlers` folder with the following code:

```
public class
SendInvoiceTextMessageNotificationHandler(IInvoiceService
invoiceService) : INotificationHandler<SendInvoiceNotification>
{
```

```
        public async Task Handle(SendInvoiceNotification
notification, CancellationToken cancellationToken)
    {
        // Send text message notification
        var invoice = await invoiceService.
GetAsync(notification.InvoiceId, cancellationToken);
        if (invoice is null || string.
IsNullOrWhiteSpace(invoice.ContactPhone))
        {
            return;
        }
        // Send text message notification
        Console.WriteLine($"Sending text message notification to
{invoice.ContactPhone} for invoice {invoice.Id}");
    }
}
```

The preceding code is similar to the previous handler. It sends a text message notification to the customer.

7. Inject the IPublisher interface into the InvoicesController class, as shown in the following:

```
public class InvoicesController(IInvoiceService invoiceService,
ISender mediatorSender, IPublisher mediatorPublisher) :
ControllerBase
{
    // Omitted
}
```

The IPublisher interface is used to publish a notification or event through the mediator pipeline to be handled by multiple handlers.

8. Update the CreateInvoice method in the InvoicesController class with the following code:

```
[HttpPost]
public async Task<ActionResult<InvoiceDto>>
CreateInvoice(CreateOrUpdateInvoiceDto invoiceDto)
{
    //var invoice = await invoiceService.AddAsync(invoiceDto);
    var invoice = await mediatorSender.Send(new
CreateInvoiceCommand(invoiceDto));
    await mediatorPublisher.Publish(new
SendInvoiceNotification(invoice.Id));
    return CreatedAtAction(nameof(GetInvoice), new { id =
invoice.Id }, invoice);
}
```

In the preceding code, when creating a new invoice, we send a `SendInvoiceNotification` notification to the `IPublisher` interface. The `IPublisher` interface will invoke the `SendInvoiceEmailNotificationHandler` handler and the `SendInvoiceTextMessageNotificationHandler` handler to handle the notification. Then, they will send the email notification and text message notification to the customer. If we need more notifications, we can simply add more handlers to handle the notification without changing the controller code.

Run the application and send a `POST` request to the `/api/invoices` endpoint to create a new invoice. You should be able to see the console messages for the email notification and text message notifications.

This is just a simple example to demonstrate how to use the `MediatR` library to implement the CQRS pattern. CQRS and `MediatR` allow us to separate the read and write concerns and decouple the business logic layer from the data persistence layer. You can also try to use different databases for reading and writing or even for different projects. However, note that using different databases may lead to data consistency issues. You can use the event-sourcing pattern with the CQRS pattern to maintain data consistency and full audit trails. We will not cover the event-sourcing pattern in this book. You can find more details about the event-sourcing pattern at the following link: `https://learn.microsoft.com/en-us/azure/architecture/patterns/event-sourcing`.

Next, we will introduce a popular pattern for asynchronous communication between microservices: the pub/sub pattern.

Pub/sub

In a microservice architecture, the microservices communicate with each other through standard interfaces, such as HTTP APIs or gRPC. Sometimes, a microservice may need to communicate with other services in an asynchronous way. It may also need to broadcast an event to multiple services. The pub/sub pattern can be used to address the need for loosely coupled communication between microservices. It facilitates broadcasting events or messages to multiple subscribers without them being directly aware of each other.

The pub/sub pattern is a communication model that facilitates the exchange of messages between publishers and subscribers without requiring them to be aware of each other. It consists of three components: publishers, subscribers, and a message broker. Publishers are responsible for publishing events or messages to the message broker, which then distributes them to subscribers. Subscribers, in turn, subscribe to the message broker and receive the events or messages that have been published. This pattern allows for asynchronous communication between publishers and subscribers, enabling them to remain independent of each other.

Many message brokers can be used to implement the pub/sub pattern. Some popular message brokers include the following:

- **RabbitMQ**: RabbitMQ is an open-source, cross-platform message broker that is widely used in microservice architectures. It is lightweight and easy to deploy on-premises and in the cloud. For more details, refer to the following link: `https://rabbitmq.com/`.

- **Redis**: Redis is an open-source in-memory data structure store. It is versatile and has high performance. Redis is a popular choice for various use cases, such as key-value databases, caches, and message brokers. We learned how to use Redis as a cache in *Chapter 15*. It can also be used as a message broker to implement the pub/sub pattern. For more details, refer to the following link: `https://redis.io/`.

- **Apache Kafka**: Apache Kafka is an open-source, distributed event-streaming platform. It is a reliable and scalable message broker that can be used to implement the pub/sub pattern. It ensures the durable and reliable storage of event streams in a scalable, fault-tolerant, and secure manner. You can manage it yourself or use a managed service provided by a variety of cloud providers. For more details, refer to the following link: `https://kafka.apache.org/`.

- **Azure Service Bus**: Azure Service Bus is a fully managed enterprise message broker provided by Microsoft Azure. It supports message queues and topics. For more details, refer to the following link: `https://learn.microsoft.com/en-us/azure/service-bus-messaging/`.

The pub/sub pattern decouples the microservices from each other. It also improves the scalability and reliability. All the messages or events are handled in an asynchronous way. This helps the service continue to function even if the workload increases or one of the services fails. However, it also increases the complexity of the system. You need to manage the message ordering, message priority, message duplication, message expiration, dead-letter queues, and so on. To learn more about the pub/sub pattern, you can refer to the following link: `https://learn.microsoft.com/en-us/azure/architecture/patterns/publisher-subscriber`.

Backends for frontends

Backends for frontends (**BFFs**) address the challenge of efficiently serving diverse client interfaces with distinct requirements. This is useful when applications need to serve multiple client types, such as web, mobile, and desktop. Each client type may need a different data format. In this case, a monolithic backend may struggle to cater to the unique needs of each client. Specifically, if the backend includes multiple microservices, each microservice may need to provide multiple endpoints to serve different client types. This can lead to a complex and inefficient system.

BFF architecture is a useful solution for applications that need to serve multiple client types, such as web, mobile, and desktop. Each client type may have distinct requirements for a data format, which can be difficult to manage with a monolithic backend. If the backend includes multiple microservices, each microservice may need to provide multiple endpoints to serve different client types, resulting in a complex and inefficient system. BFFs can help address this challenge by efficiently serving diverse client interfaces with distinct requirements.

BFF introduces dedicated backend services tailored for specific frontend clients. Each frontend client has its corresponding backend, enabling fine-grained control over data retrieval, processing, and presentation. This allows for a more efficient and flexible system that can better meet the needs of each client.

The following diagram shows a typical BFF architecture:

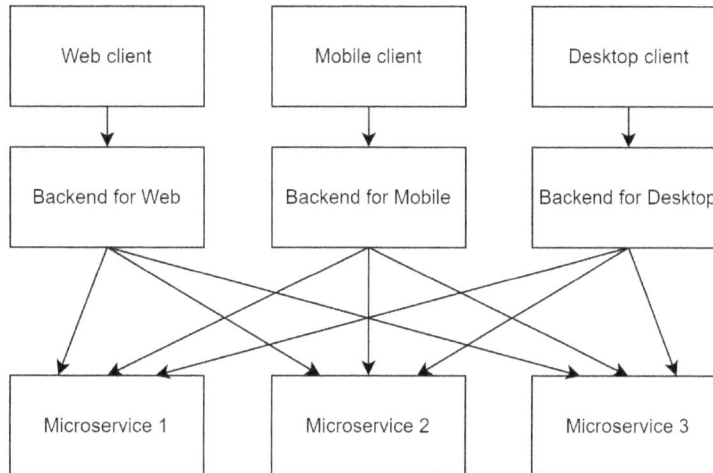

Figure 17.5 – Typical BFF architecture

In *Figure 17.5*, each BFF service is responsible for a specific frontend client. It can retrieve data from multiple microservices and combine the data into a single response. Each BFF service is fine-tuned to meet the specific needs of the frontend client. It also illustrates how each BFF service is responsible for a particular frontend client. Each BFF service is tailored to meet the specific requirements of the frontend client. It can retrieve data from multiple microservices and combine them into a single response.

BFFs should be lightweight. They can contain client-specific logic but should not contain business logic. The main purpose of BFFs is to tailor the data for each frontend client. However, this may lead to code duplication. If the data format is similar for multiple frontend clients, BFFs may not be required.

Resilience patterns

In a microservice architecture, resilience and reliability are essential for a successful system. web APIs are often subject to unpredictable environments, such as network latency, transient failures, service unavailability, high traffic, and so on. To ensure that these APIs are resilient and reliable, several patterns can be implemented. These include retry, rate-limiting, time out, circuit breaker, and so on. In this section, we will discuss how to use the `Polly` library to implement these patterns in ASP.NET Core web API applications.

You can find a sample project in the /chapter17/PollyDemo folder. This project contains two basic ASP.NET Core web API applications:

- PollyServerWebApi, which behaves as a server
- PollyClientWebApi, which is also a web API application but also behaves as a client

We will use these two applications to demonstrate how to use the Polly library to implement the rate-limiting, retry, timeout, and circuit breaker. Polly is a popular .NET resilience and transient-fault-handling library. You can find more details about Polly at the following link: https://www.thepollyproject.org/.

To use Polly in ASP.NET Core web API applications, you need to install the Polly NuGet package. Navigate to the PollyClientWebApi project and run the following command in the terminal window to install the Polly package:

```
dotnet add package Polly.Core
```

Polly provides a resilience pipeline builder to build a resilience pipeline. A resilience pipeline runs a series of resilience policies. Each policy is responsible for handling a specific type of issue. The following code shows how to create a resilience pipeline builder:

```
var pipeline = new ResiliencePipelineBuilder();
```

Next, we will explore several resilience policies provided by Polly.

Timeout

The **timeout** pattern is a common pattern used to handle slow or unresponsive services. When a service is slow or unresponsive, the client may wait for an extended period before receiving a response. To avoid this, a timeout can be set for the service. If the service cannot respond within the given time frame, the client can return an error to the user, thus preventing them from waiting unnecessarily.

In *Chapter 4*, we introduced a RequestTimeout middleware to set the timeout for the ASP.NET Core web API application. The RequestTimeout middleware is applied to the endpoints or actions that need to be timed out. Sometimes, we may need to set the timeout for a specific method call, such as calling a REST API or querying a database. Let us explore other ways to set the timeout.

The HttpClient class in .NET Core provides a timeout feature. You can set the timeout for the HttpClient object by setting the Timeout property. The following code shows how to set the timeout for the HttpClient object:

```
var httpClient = httpClientFactory.CreateClient();
httpClient.Timeout = TimeSpan.FromSeconds(10);
```

The preceding code creates an `HttpClient` object and sets the timeout to 10 seconds. If the service does not respond within 10 seconds, the `HttpClient` object will throw an exception. You can catch the exception and return an error to the user.

Setting the timeout for the `HttpClient` object is useful for simple tasks, such as calling a REST API. However, it is not suitable for more complex tasks that do not use `HttpClient`, such as a database query. For other tasks, such as a database query, you can use the `CancellationToken` to set the timeout. The following code shows how to set the timeout for a database query:

```
var cancellationToken = new CancellationTokenSource(TimeSpan.
FromSeconds(10));
var invoice = await invoiceRepository.GetAsync(id, cancellationToken.
Token);
```

The preceding code creates a `CancellationTokenSource` object and sets the timeout to 10 seconds. If the database query is not complete within 10 seconds, the `GetAsync()` method will throw an exception. This prevents the client from waiting for an extended period before receiving a response.

Sometimes, there may be multiple services that need to be called. Moreover, setting the timeout for each service call may be tedious. To simplify this, we can use the `Polly` library to implement the timeout policy.

`Polly` provides a timeout policy that can be used to set the timeout for a service. Follow these steps to implement the timeout policy:

1. Create an endpoint in the `PollyServerWebApi` application to simulate a slow service. Open the `Program.cs` file and add the following code:

    ```
    app.MapGet("/api/slow-response", async () =>
    {
        var random = new Random();
        var delay = random.Next(1, 20);
        await Task.Delay(delay * 1000);
        return Results.Ok($"Response delayed by {delay} seconds");
    });
    ```

    The preceding code defines a minimal API endpoint that simulates a slow service. It generates a random delay between 1 and 20 seconds. This endpoint will return a response after the delay. This is just an example of simulating a slow service. In a real-world application, the service may be slow due to network latency, high traffic, and so on.

2. Create a controller in the `PollyClientWebApi` application to call the slow service. Add a `PollyController` class in the `Controllers` folder with the following code:

    ```
    namespace PollyClientWebApi.Controllers;
    [Route("api/[controller]")]
    [ApiController]
    ```

```
public class PollyController(ILogger<PollyController> logger,
IHttpClientFactory httpClientFactory) : ControllerBase
{
    [HttpGet("slow-response")]
    public async Task<IActionResult> GetSlowResponse()
    {
        var client = httpClientFactory.
CreateClient("PollyServerWebApi");
        var response = await client.GetAsync("api/slow-
response");
        var content = await response.Content.
ReadAsStringAsync();
        return Ok(content);
    }
}
```

This controller uses the IHttpClientFactory to create an HttpClient object. Then, it calls the slow service and returns the response to the client.

3. Run the two applications and send a request to the /api/polly/slow-response endpoint of the PollyClientWebApi application. You should be able to see the response after a random delay between 1 and 20 seconds.

4. Next, we will implement the timeout policy using Polly. For example, we can set the timeout to 5 seconds, which means if the service does not respond within 5 seconds, the client will return an error to the user instead of waiting for a long time. Update the GetSlowResponse() method of the PollyController class as follows:

```
[HttpGet("slow-response")]
public async Task<IActionResult> GetSlowResponse()
{
    var pipeline = new ResiliencePipelineBuilder().
AddTimeout(TimeSpan.FromSeconds(5)).Build();
    try
    {
        var response = await pipeline.ExecuteAsync(async
cancellationToken =>
            await client.GetAsync("api/slow-response",
cancellationToken));
        var content = await response.Content.
ReadAsStringAsync();
        return Ok(content);
    }
    catch (Exception e)
    {
        logger.LogError(e.Message);
```

```
            return Problem(e.Message);
        }
    }
```

The preceding code uses `Polly` to create a `ResiliencePipelineBuilder` object. Then, it adds a timeout policy with a timeout of 5 seconds. The `ExecuteAsync()` method is used to execute the pipeline. If the service does not respond within 5 seconds, the `ExecuteAsync()` method will throw an exception. The `catch` block is used to catch the exception and return an error to the user.

5. Note that in the `ExecuteAsync()` method, the cancellation token is passed to the `GetAsync()` method of the `HttpClient` object. If it does not, the `HttpClient` will continue to wait even if the timeout occurs. It is important to respect the cancellation token from the `Polly` resilience pipeline.

6. Run the two applications and send a request to the `/api/polly/slow-response` endpoint of the `PollyClientWebApi` application. You should be able to see the error message after 5 seconds.

In the preceding example, we defined the timeout policy in the controller. To reuse the timeout policy, we can define a global timeout policy in the `Program.cs` file and then use dependency injection to inject the policy into the controller. Follow these steps to implement the global timeout policy:

1. Install the `Polly.Extensions` NuGet package. Navigate to the `PollyClientWebApi` project and run the following command in the terminal window to install the `Polly.Extensions` package:

    dotnet add package Polly.Extensions

2. Open the `Program.cs` file of the `PollyClientWebApi` application and add the following code:

    ```
    builder.Services.AddResiliencePipeline("timeout-5s-pipeline",
    configure =>
    {
        configure.AddTimeout(TimeSpan.FromSeconds(5));
    });
    ```

    The preceding code defines a global timeout policy with a timeout of 5 seconds. The policy is named `timeout-5s-pipeline`. You can use any name you like. The `AddResiliencePipeline()` method is used to add the timeout policy to the pipeline.

3. Inject the `ResiliencePipelineProvider<string>` class into the `PollyController` class, as shown in the following:

    ```
    public class PollyController(ILogger<PollyController>
    logger, IHttpClientFactory httpClientFactory,
    ResiliencePipelineProvider<string> resiliencePipelineProvider) :
    ControllerBase
    ```

```
    {
        // Omitted
    }
```

The `ResiliencePipelineProvider<string>` class is used to retrieve the global timeout policy. The `string` type parameter specifies the type of the policy name.

4. Update the `GetSlowResponse()` method of the `PollyController` class as follows:

    ```
    var pipeline = resiliencePipelineProvider.GetPipeline("timeout-
    5s-pipeline");
    // Omitted
    ```

    In this way, we can reuse the global timeout policy by its name.

`Polly` supports many other resilience patterns. Next, let us discuss rate-limiting.

Rate-limiting

The **rate-limiting** pattern is a common pattern used to limit the number of requests that can be made to a service. The rate should be set to a reasonable value to avoid overloading the service. You can run a performance test to determine the optimal rate limit. The performance of the service depends on many factors, such as the hardware, network, and the complexity of business logic. Once you have determined the optimal rate limit, you can apply it to the service to ensure that it can handle the workload.

For example, if a service can handle 100 requests per second when the number of requests exceeds 100, the service may become slow or even unavailable. The client may encounter a timeout error. To avoid this, we can set the rate limit for the service. When the number of requests exceeds the rate limit, the service will reject the requests and return an error to the client. This can prevent the service from being overloaded.

ASP.NET Core provides a rate-limiting middleware that can be used to configure rate-limiting in various policies, such as `fixed window`, `sliding window`, `token bucket`, and `concurrency`. We introduced rate-limiting middleware in *Chapter 4*. You can find more details about rate-limiting middleware at the following link: `https://learn.microsoft.com/en-us/aspnet/core/performance/rate-limit`.

You can open the `PollyDemo` solution in the `/chapter17/PollyDemo/end` folder. In the `Program.cs` file of the `PollyServerWebApi` project, you can find the following code:

```
builder.Services.AddRateLimiter(options =>
{
    options.AddFixedWindowLimiter("FiveRequestsInThreeSeconds",
limiterOptions =>
    {
        limiterOptions.PermitLimit = 5;
        limiterOptions.Window = TimeSpan.FromSeconds(3);
    });
```

```
    options.RejectionStatusCode = StatusCodes.
Status429TooManyRequests;
    options.OnRejected = async (context, _) =>
    {
        await context.HttpContext.Response.WriteAsync("Too many
requests. Please try later.", CancellationToken.None);
    };
});
// Omitted
app.UseRateLimiter();
```

The rate-limiting policy is applied to the WeatherForecastController class:

```
[EnableRateLimiting("FiveRequestsInThreeSeconds")]
[ApiController]
[Route("[controller]")]
public class
WeatherForecastController(ILogger<WeatherForecastController> logger) :
ControllerBase
{
    // Omitted
}
```

The preceding code configures a fixed window rate limiter with a rate limit of five requests per 3 seconds. Of course, this is just an example for demonstration purposes. When the PollyClientWebApi application sends more than five requests per 3 seconds to the PollyServerWebApi application, the PollyServerWebApi application will return a 429 Too Many Requests error to the client. The OnRejected callback is used to handle the rejected requests. In this example, we simply return a message to the client.

Use the dotnet run command to run the PollyServerWebApi application and the PollyClientWebApi application. Then, send more than five requests per 3 seconds to the /weatherforecast endpoint of the PollyClientWebApi application. You should be able to see the 429 Too Many Requests error in the PollyClientWebApi application. In this way, we can limit the number of requests to the PollyServerWebApi service so that it can handle the workload without being overloaded.

We can also use Polly to implement the rate-limiting pattern. Follow these steps to implement the rate-limiting pattern using Polly:

1. Install the Polly.RateLimiting NuGet package for the PollyClientWebApi project by running the following command in the terminal window:

   dotnet add package Polly.RateLimiting

The `Polly.RateLimiting` package is a wrapper for the `System.Threading.RateLimiting` package provided by Microsoft. It also depends on the `Polly.Core` package. So, if you have not installed the `Polly.Core` package, it will be installed automatically.

2. Create a `/api/normal-response` endpoint in the `PollyServerWebApi` application to simulate a normal service. Open the `Program.cs` file and add the following code:

```
app.MapGet("/api/normal-response", async () =>
{
    var random = new Random();
    var delay = random.Next(1, 1000);
    await Task.Delay(delay);
    return Results.Ok($"Response delayed by {delay}
milliseconds");
});
```

This endpoint will return a response after a random delay between 1 and 1000 milliseconds, which means, in the worst case, it may take 1 second to return a response. To limit the number of requests to this endpoint, we can use the rate-limiting policy for the `PollyClientWebApi` application.

3. We will use the dependency injection to inject the rate-limiting policy for convenience. Define a rate-limiting policy in the `Program.cs` as follows:

```
builder.Services.AddResiliencePipeline("rate-limit-5-requests-
in-3-seconds", configure =>
{
    configure.AddRateLimiter(new FixedWindowRateLimiter(new
FixedWindowRateLimiterOptions
    { PermitLimit = 5, Window = TimeSpan.FromSeconds(3) }));
});
```

The preceding code defines a fixed window rate limiter with a rate limit of 5 requests per 3 seconds. The policy is named `rate-limit-5-requests-in-3-seconds`. You can use any name you like.

4. In this example, we create a separate `Polly` pipeline for the rate-limiting policy. You can also combine multiple policies into a single pipeline. For example, you can combine the rate-limiting policy and the timeout policy into a single pipeline using the following code:

```
builder.Services.AddResiliencePipeline("combined-resilience-
policy", configure =>
{
    configure.AddRateLimiter(
        // Omitted
    );
    configure.AddTimeout(
        // Omitted
    );
```

```
        // You can add more policies here
    });
```

5. Inject the ResiliencePipelineProvider<string> class into the PollyController class of the PollyClientWebApi project, as shown in the following:

```
[HttpGet("rate-limit")]
public async Task<IActionResult>
GetNormalResponseWithRateLimiting()
{
    var client = httpClientFactory.
CreateClient("PollyServerWebApi");
    try
    {
        var pipeline = resiliencePipelineProvider.
GetPipeline("rate-limit-5-requests-in-3-seconds");
        var response = await pipeline.ExecuteAsync(async
cancellationToken =>
            await client.GetAsync("api/normal-response",
cancellationToken));
        var content = await response.Content.
ReadAsStringAsync();
        return Ok(content);
    }
    catch (Exception e)
    {
        logger.LogError($"{e.GetType()} {e.Message}");
        return Problem(e.Message);
    }
}
```

You will find that the code is quite similar to the timeout policy.

6. Run the two applications and send more than 5 requests per 3 seconds to the /api/polly/ rate-limit endpoint of the PollyClientWebApi application. Sometimes, you may see an error message in the console window of the PollyClientWebApi application as follows:

```
Polly.RateLimiting.RateLimiterRejectedException The operation
could not be executed because it was rejected by the rate
limiter. It can be retried after '00:00:03'.
```

7. Similarly, you can use Polly to implement other rate-limiting policies, such as sliding window, concurrency, and token bucket. Here is an example of the sliding window rate limiter:

```
configure.AddRateLimiter(new SlidingWindowRateLimiter(new
SlidingWindowRateLimiterOptions
{ PermitLimit = 100, Window = TimeSpan.FromMinutes(1) }));
```

The preceding code defines a sliding window rate limiter with a rate limit of 100 requests per minute.

8. As the Polly `RateLimiter` is a disposable resource, it is a good practice to dispose of it when it is no longer needed. `Polly` provides an `OnPipelineDisposed` callback that can be used to dispose of the `RateLimiter` object. For example, we can dispose of the `RateLimiter` object in the `OnPipelineDisposed` callback as follows:

```
builder.Services.AddResiliencePipeline("rate-limit-5-requests-
in-3-seconds", (configure, context) =>
{
    var rateLimiter = new FixedWindowRateLimiter(new
FixedWindowRateLimiterOptions
    { PermitLimit = 5, Window = TimeSpan.FromSeconds(3) });
    configure.AddRateLimiter(rateLimiter);
    // Dispose the rate limiter when the pipeline is disposed
    context.OnPipelineDisposed(() => rateLimiter.Dispose());
});
```

In this way, we can dispose of the `RateLimiter` object when the pipeline is disposed of so that it does not consume resources unnecessarily.

Retry

Next, let us discuss **retry**. When the client API gets an error, such as a `429 Too Many Requests` error, or a `500 Internal Server Error` error, it can retry the request after a delay because the error may be caused by a temporary issue, such as rate-limiting or a network glitch. The next time the client API sends the request, it may succeed. This is called retry.

The retry pattern is a common approach to addressing transient failures in communication between microservices. This pattern is particularly useful in a microservice architecture, where network glitches or the temporary unavailability of a service can cause communication failures. By implementing retry mechanisms, these transient issues can be managed, and the overall reliability of the system can be improved.

Follow these steps to implement the retry pattern using `Polly`:

1. Update the `Get()` method of the `WeatherForecastController` class as follows:

```
[HttpGet(Name = "GetWeatherForecast")]
public async Task<ActionResult<IEnumerable<WeatherForecast>>>
Get()
{
    var httpClient = httpClientFactory.
CreateClient("PollyServerWebApi");
    var pollyPipeline = new ResiliencePipelineBuilder()
    .AddRetry(new Polly.Retry.RetryStrategyOptions()
    {
```

```
            ShouldHandle = new PredicateBuilder().
    Handle<Exception>(),
            MaxRetryAttempts = 3,
            Delay = TimeSpan.FromMilliseconds(500),
            MaxDelay = TimeSpan.FromSeconds(5),
            OnRetry = args =>
            {
                logger.LogWarning($"Retry {args.AttemptNumber}, due
    to: {args.Outcome.Exception?.Message}.");
                return default;
            }
        })
        .Build();
        HttpResponseMessage? response = null;
        await pollyPipeline.ExecuteAsync(async _ =>
        {
            response = await httpClient.GetAsync("/
    WeatherForecast");
            response.EnsureSuccessStatusCode();
        });
        if (response != null & response!.IsSuccessStatusCode)
        {
            var result = await response.Content.
    ReadFromJsonAsync<IEnumerable<WeatherForecast>>();
            return Ok(result);
        }
        return StatusCode((int)response.StatusCode, response.
    ReasonPhrase);
    }
```

The preceding code creates a `ResiliencePipelineBuilder` object to build a resilience pipeline. Then, it adds a retry policy to the pipeline. The retry policy will retry the request three times if it fails. The delay between retries is 500 milliseconds. The `MaxDelay` property is used to specify the maximum delay duration. The `OnRetry` callback is used to log the retry attempts. Finally, it executes the pipeline to send the request to the `PollyServerWebApi` application.

2. Run the two applications and send more than five requests per 3 seconds to the / weatherforecast endpoint of the `PollyClientWebApi` application. Sometimes, you may see that the request takes a longer time to complete. This is because the request is retried if it fails. You should also be able to see the retry attempts in the console window of the `PollyClientWebApi` application, as shown in the following:

```
warn: PollyClientWebApi.Controllers.WeatherForecastController[0]
      Retry 2, due to: Response status code does not indicate
success: 429 (Too Many Requests)..
```

In this way, we can automatically retry the request if it fails. This can improve the reliability of the system.

3. The retry policy can be configured in various ways. For example, we can configure the retry policy to retry the request only when the response status code is 429 as follows:

```
ShouldHandle = new PredicateBuilder().Handle<Exception>().
Or<HttpRequestException>(ex => ex.StatusCode == HttpStatusCode.
TooManyRequests),
```

We can also delay the retry attempts with an exponential backoff strategy. This is because the service may be overloaded, and the retry attempts may fail again. In this case, we can delay the retry attempts to avoid overloading the service.

4. To use the exponential backoff strategy, we can specify the `BackoffType` property of the `RetryStrategyOptions` object as follows:

```
BackoffType = DelayBackoffType.Exponential,
```

The `BackoffType` property is a `DelayBackoffType` enum that can be set to `Constant`, `Linear`, or `Exponential`. The `Constant` strategy will delay the retry attempts with a constant delay. The `Linear` strategy will delay the retry attempts with a linear delay. The `Exponential` strategy will delay the retry attempts with an exponential delay. The default strategy is `Constant`.

There are a few considerations when using the retry pattern:

- The retry pattern should only be used to handle transient failures. If you want to implement a repeatable operation, you should use some scheduling mechanism, such as a background service, or suitable tools, such as **Quartz.NET** and **Hangfire**. Do not use the `Polly` retry to implement a scheduled repeating operation.

- Consider using different retry policies for different types of errors. For example, an API call may involve HTTP requests, database queries, and JSON deserialization. If the HTTP request fails due to a network glitch, you can retry the request. However, if the JSON deserialization fails, it is unlikely to succeed even if you retry the JSON deserialization method. In this case, you can use `ShouldHandle` to specify the types of errors that should be retried.

Circuit breaker

The **circuit breaker** pattern is a useful tool for preventing service overload and failure. In the event that a service is becoming seriously overloaded, a client should stop sending requests for a period of time in order to allow the service to recover. This is known as the circuit breaker pattern and can help to avoid a service crash or total failure.

We can use `Polly` to implement the circuit breaker pattern. As we have already learned how to use Polly to implement the timeout pattern, the rate-limiting pattern, and the retry pattern, you should be able to understand the following steps:

1. Create a new `/api/random-failure-response` endpoint in the `PollyServerWebApi` application to simulate an overloaded service. Open the `Program.cs` file and add the following code:

```
app.MapGet("/api/random-failure-response", () =>
{
    var random = new Random();
    var delay = random.Next(1, 100);
    return Task.FromResult(delay > 20 ? Results.Ok($"Response
is successful.") : Results.StatusCode(StatusCodes.
Status500InternalServerError));
});
```

This endpoint will return a `500 Internal Server Error` error with an 80% chance (approximately). This is just an example of simulating an overloaded service. In a real-world application, the service may be overloaded due to high traffic, network latency, and so on.

2. Add the following code to the `Program.cs` file of the `PollyClientWebApi` application:

```
builder.Services.AddResiliencePipeline("circuit-breaker-5-
seconds", configure =>
{
    configure.AddCircuitBreaker(new
CircuitBreakerStrategyOptions
    {
        FailureRatio = 0.7,
        SamplingDuration = TimeSpan.FromSeconds(10),
        MinimumThroughput = 10,
        BreakDuration = TimeSpan.FromSeconds(5),
        ShouldHandle = new PredicateBuilder().
Handle<Exception>()
    });
});
```

The preceding code defines a circuit breaker policy named `circuit-breaker-5-seconds` with a failure ratio of 0.7. This means if the failure ratio is greater than 0.7, the circuit breaker will open. The `SamplingDuration` property is used to specify the duration of the sampling over which the failure ratios are calculated. The `MinimumThroughput` property means that at least 10 requests must be made within the sampling duration. The `BreakDuration` property means that the circuit breaker will stay open for 5 seconds if it opens. The `ShouldHandle` property is used to specify the types of errors that should be handled by the circuit breaker.

3. Create a new action in the `PollyController` class of the `PollyClientWebApi` application to call the overloaded service. Add the following code:

```
[HttpGet("circuit-breaker")]
public async Task<IActionResult>
GetRandomFailureResponseWithCircuitBreaker()
{
    var client = httpClientFactory.
CreateClient("PollyServerWebApi");
    try
    {
        var pipeline = resiliencePipelineProvider.
GetPipeline("circuit-breaker-5-seconds");
        var response = await pipeline.ExecuteAsync(async
cancellationToken =>
        {
            var result = await client.GetAsync("api/random-
failure-response", cancellationToken);
            result.EnsureSuccessStatusCode();
            return result;
        });
        var content = await response.Content.
ReadAsStringAsync();
        return Ok(content);
    }
    catch (Exception e)
    {
        logger.LogError($"{e.GetType()} {e.Message}");
        return Problem(e.Message);
    }
}
```

The preceding code uses `result.EnsureSuccessStatusCode()` to throw an exception if the response status code is not successful. As the overloaded service has an 80% chance of returning an error, the circuit breaker will open after a couple of requests. Then, the circuit breaker will stay open for 5 seconds. During this period, the client will not send any requests to the overloaded service. After 5 seconds, the circuit breaker will close, and the client will send requests to the overloaded service again.

4. Run the two applications and send more than 10 requests to the `/api/polly/circuit-breaker` endpoint of the `PollyClientWebApi` application. Sometimes, you will see a `500 Internal Server Error` error as follows:

```
{
    "type": "https://tools.ietf.org/html/rfc9110#section-15.6.1",
    "title": "An error occurred while processing your request.",
```

```
  "status": 500,
  "detail": "Response status code does not indicate success: 500
(Internal Server Error).",
  "traceId": "00-c5982555dbf0e66d5ca79fd83aa3837c-
46cd1cd7f6acb851-00"
}
```

5. Send more requests, and you will see that the circuit breaker opens and returns a different error message as follows:

```
{
  "type": "https://tools.ietf.org/html/rfc9110#section-15.6.1",
  "title": "An error occurred while processing your request.",
  "status": 500,
  "detail": "The circuit is now open and is not allowing
calls.",
  "traceId": "00-1b6dc3f8912f5ebd4e67a39a89dd605a-
495d67559eaf22b7-00"
}
```

You can see that the error message is different from the previous one, which indicates that the circuit breaker is open, so any requests to the overloaded service will be rejected. You need to wait for 5 seconds before sending more requests to the overloaded service. During these 5 seconds, all the requests to the `/api/polly/circuit-breaker` endpoint will not be sent to the overloaded service and will return the same error message instead.

Circuit breakers are not the same as the retry pattern. The retry pattern expects the operation to succeed eventually. However, the circuit breaker pattern prevents the operation from being executed if it is likely to fail, which can save resources and allow the external service to recover. You can use these two patterns together. But note that the retry logic should check the exception type thrown by the circuit breaker. If the circuit breaker indicates that the operation failure is not a transient issue, the retry logic should not retry the operation.

`Polly` is a powerful library that implements many resilience patterns. This section cannot cover all the patterns provided by `Polly`. You can find more examples at the following link: `https://www.pollydocs.org/index.html`.

In addition to the design patterns discussed in this chapter, there are more patterns for microservice architecture. As many of these patterns are beyond the scope of this book, we will not discuss them in detail. You can find more details about these patterns from Microsoft Learn: `https://learn.microsoft.com/en-us/azure/architecture/patterns/`.

Summary

In this chapter, we explored several concepts and patterns for microservice architecture, including domain drive design, clean architecture, CQRS, pub/sub, and BFF, and resilience patterns, such as timeout, rate-limiting, retry, and circuit breaker. These patterns can help us design and implement a maintainable, reliable, and scalable microservice architecture. Although this chapter does not cover all the patterns for microservice architecture, it should provide a basic understanding of what they are and how they can be used. These patterns are essential for developers who wish to progress beyond the basic knowledge of ASP.NET Core web API.

In the next chapter, we will discuss some open-source frameworks that can be used to build ASP.NET Core web API applications. You can check the chapter out at the following link: `https://github.com/PacktPublishing/Web-API-Development-with-ASP.NET-Core-8/tree/main/samples/chapter18`.

Further reading

To learn more about microservice architecture, the following resources from Microsoft Learn are highly recommended:

- Architectural principles: `https://learn.microsoft.com/en-us/dotnet/architecture/modern-web-apps-azure/architectural-principles`.

- Common web application architectures: `https://learn.microsoft.com/en-us/dotnet/architecture/modern-web-apps-azure/common-web-application-architectures`.

- .NET Microservices: Architecture for Containerized .NET Applications: `https://docs.microsoft.com/en-us/dotnet/architecture/microservices/`.

- Architect Modern Web Applications with ASP.NET Core and Azure: `https://learn.microsoft.com/en-us/dotnet/architecture/modern-web-apps-azure/`.

- Architecting Cloud Native .NET Applications for Azure: `https://learn.microsoft.com/en-us/dotnet/architecture/cloud-native/`.

- Serverless apps: Architecture, patterns, and Azure implementation: `https://learn.microsoft.com/en-us/dostnet/architecture/serverless/`.

- Cloud Design Patterns: `https://docs.microsoft.com/en-us/azure/architecture/patterns/`.

Index

‹packt›

packtpub.com

Subscribe to our online digital library for full access to over 7,000 books and videos, as well as industry leading tools to help you plan your personal development and advance your career. For more information, please visit our website.

Why subscribe?

- Spend less time learning and more time coding with practical eBooks and Videos from over 4,000 industry professionals

- Improve your learning with Skill Plans built especially for you

- Get a free eBook or video every month

- Fully searchable for easy access to vital information

- Copy and paste, print, and bookmark content

Did you know that Packt offers eBook versions of every book published, with PDF and ePub files available? You can upgrade to the eBook version at packtpub.com and as a print book customer, you are entitled to a discount on the eBook copy. Get in touch with us at customercare@packtpub.com for more details.

At www.packtpub.com, you can also read a collection of free technical articles, sign up for a range of free newsletters, and receive exclusive discounts and offers on Packt books and eBooks.

Other Books You May Enjoy

If you enjoyed this book, you may be interested in these other books by Packt:

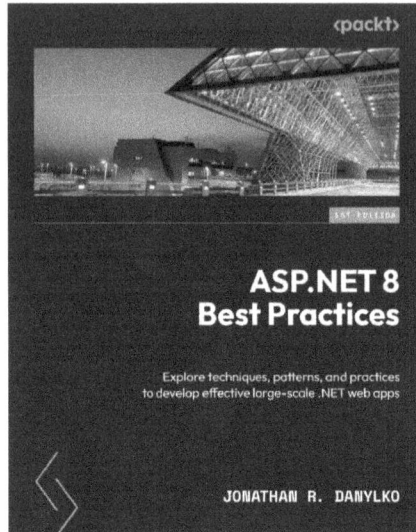

ASP.NET 8 Best Practices

Jonathan R. Danylko

ISBN: 978-1-83763-212-1

- Explore the common IDE tools used in the industry
- Identify the best approach for organizing source control, projects, and middleware
- Uncover and address top web security threats, implementing effective strategies to protect your code
- Optimize Entity Framework for faster query performance using best practices
- Automate software through continuous integration/continuous deployment
- Gain a solid understanding of the .NET Core coding fundamentals for building websites
- Harness HtmlHelpers, TagHelpers, ViewComponents, and Blazor for component-based development

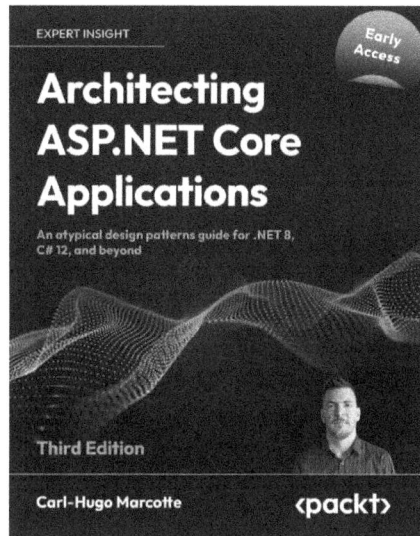

Architecting ASP.NET Core Applications

Carl-Hugo Marcotte

ISBN: 978-1-80512-338-5

- Apply the SOLID principles for building flexible and maintainable software
- Test your apps effectively with automated tests, including black-box testing
- Enter the path of ASP.NET Core dependency injection mastery
- Work with GoF design patterns such as strategy, decorator, facade, and composite
- Design REST APIs using Minimal APIs and MVC
- Discover layering techniques and the tenets of clean architecture
- Use feature-oriented techniques as an alternative to layering
- Explore microservices, CQRS, REPL, vertical slice architecture, and many more patterns

Packt is searching for authors like you

If you're interested in becoming an author for Packt, please visit `authors.packtpub.com` and apply today. We have worked with thousands of developers and tech professionals, just like you, to help them share their insight with the global tech community. You can make a general application, apply for a specific hot topic that we are recruiting an author for, or submit your own idea.

Share Your Thoughts

Now you've finished *Web API Development with ASP.NET Core 8*, we'd love to hear your thoughts! Scan the QR code below to go straight to the Amazon review page for this book and share your feedback or leave a review on the site that you purchased it from.

`https://packt.link/r/1-804-61095-X`

Your review is important to us and the tech community and will help us make sure we're delivering excellent quality content.

Download a free PDF copy of this book

Thanks for purchasing this book!

Do you like to read on the go but are unable to carry your print books everywhere? Is your eBook purchase not compatible with the device of your choice?

Don't worry, now with every Packt book you get a DRM-free PDF version of that book at no cost.

Read anywhere, any place, on any device. Search, copy, and paste code from your favorite technical books directly into your application.

The perks don't stop there, you can get exclusive access to discounts, newsletters, and great free content in your inbox daily

Follow these simple steps to get the benefits:

1. Scan the QR code or visit the link below

https://packt.link/free-ebook/9781804610954

2. Submit your proof of purchase
3. That's it! We'll send your free PDF and other benefits to your email directly